Handbook of Experimental Pharmacology

Volume 228

More information about this series at
http://www.springer.com/series/164

Kathleen M. Kantak • Joseph G. Wettstein
Editors

Cognitive Enhancement

Springer

Editors
Kathleen M. Kantak
Department of Psychological and Brain Sciences
Boston University
Boston
Massachusetts
USA

Joseph G. Wettstein
Roche Pharmaceutical Research and Early Development
Neuroscience, Ophthalmology and Rare Diseases
F. Hoffmann-La Roche Ltd.
Basel
Switzerland

ISSN 0171-2004 ISSN 1865-0325 (electronic)
Handbook of Experimental Pharmacology
ISBN 978-3-319-16521-9 ISBN 978-3-319-16522-6 (eBook)
DOI 10.1007/978-3-319-16522-6

Library of Congress Control Number: 2015938577

Springer Cham Heidelberg New York Dordrecht London

Printed on acid-free paper

Springer International Publishing AG Switzerland is part of Springer Science+Business Media (www.springer.com)

Preface to Cognitive Enhancement

Cognitive enhancement is a treatment goal for a variety of neuropsychiatric and neurological illnesses characterized by deficits in one or more cognitive domains. One difficulty in pursuing drug development is identifying valid methods and measures of cognitive enhancement in both human laboratory and clinical settings as well as in animals. An additional challenge for animal work is the development of models and measures with translation to humans. Using animals in research offers several advantages, including uncovering relevant drug targets, signaling pathways of importance, and brain plasticity changes that underlie cognitive enhancement.

To assemble this volume for the Handbook of Experimental Pharmacology, we invited a group of eminent international scientists with interdisciplinary expertise in cognitive enhancement to write review chapters. The chapters they contributed highlight the behavioral and neurobiological issues relevant for drug development (Part I), review the current status of cognitive enhancing drugs across multiple cognitive domains (Part II), and present perspectives on multiple topics ranging from therapeutic drug use in special populations, non-pharmacological approaches to cognitive enhancement, and emerging technologies (Part III). It is our hope that this volume promotes a mind-set change in the way basic and clinical research in cognitive enhancement is viewed and conducted worldwide. Underlying this international perspective is the notion that cognitive health is a global medical issue, and therefore, the neuroethics of cognitive enhancement will require worldwide consideration.

Boston, MA — Kathleen M. Kantak
Basel, Switzerland — Joseph G. Wettstein

Contents

Part I

Basic Approaches and Perspectives

Introduction

The study of cognition and cognitive enhancement has held the attention of a wide group of scientists for many decades. Cognition is affected by numerous intrinsic and extrinsic factors and is altered across various disease states such as Alzheimer's, schizophrenia, and Down syndrome. Going forward, it is important to continue to improve our clinical methodology for cognition assessment with a focus on the different cognitive domains altered in each disease or illness. Aiding this enhancement will be the study of new experimental drugs in animals as we now have procedures in hand for rodents and nonhuman primates that mirror those used in humans and, indeed, were developed for the specific study of domains such as attention, working memory, visual–spatial learning, recognition memory, and emotional memory. This approach will increase the efficiency of translation of animal data to human studies. To date, it has been difficult to find drugs that improve cognition in disease states. There are now, however, a number of receptor targets and intracellular signaling pathways that have been linked to cognition, and hopefully, further research will identify the cognitive domain most affected by modulation of some of these targets and pathways. One associated phenomenon is the process of hippocampal neurogenesis, a key to brain plasticity. A better understanding of the physiology and biochemistry of neurogenesis could lead to new therapies for cognitive enhancement or biomarkers for cognition change tracking.

Chapter 1 (Harvey and Keefe) outlines both the basic and more intricate methods needed to study cognition and cognitive enhancement in people. With all the methods at hand, one may assume that the study of cognition is relatively straightforward. However, the authors point out the absolute necessity to control clinical studies at their outset and over time. As one begins to assess cognition in people with various neurologic or psychiatric illnesses, having the right paradigms in place and the correct control groups is paramount to the initialization of a quality study.

Critical to this is the choice of which cognitive domain to study and for what duration. For the latter, it is usually for longer periods of time, often up to 1 year or 2, and attention to detail over this time period is vital. Associated with assessment over time is experimental drug administration. This in itself has inherent characteristics that require attention such as drug–drug interactions with the medicines that the affected individuals are taking or any compromised metabolic function that affects the pharmacokinetics of the drug in question. Importantly, the cognition tests themselves can be quite complex for ill patients, so it is important to focus on the appropriate test that best matches the specific cognitive domain altered in the illness of the group of people under study.

Chapter 2 (Wallace, Ballard, and Glavis-Bloom) defines the need to better understand the underlying biology in a given disease and its effect on cognition. With this in mind, Wallace et al. focus their attention on the use of methods in rodents and nonhuman primates that are relevant to the different domains of cognitive function. The highlighted domains of cognition are affected both differently across illnesses and across the chronicity of a single illness. Thus, when studying new drugs for their effect on cognition, it is paramount to use the appropriate preclinical method that can be best translated to a clinical setting. Moreover, it is desirable to work with an animal, whether it be mouse, rat, or monkey, that has perturbed brain function in the direction of the human illness in question. Over time, the generation of animal and human data from similar cognition tests and the integration of both will lead to a better choice of experimental drug testing in humans.

Chapter 3 (Meynard, Gaudreau, and Quirion) discusses various cell signaling mechanisms and pathways that have been associated with cognitive function. Over the past 20 years, a strong case has been made linking glutamatergic tone to cognition. A number of experimental drugs acting on NMDA, AMPA, and mGlu receptors have evolved from this research. More recently, attention has focused on various intracellular signaling pathways, those downstream from receptor activation or blockade such as CaMKII and mTOR. There is another developing body of literature that focuses directly on gene expression and the resulting proteins that affect synaptic plasticity. Some of the most advanced clinical studies on cognition have used drugs that upregulate cholinergic tone. Further understanding of the cholinergic system may lead to the development of more beneficial drugs with fewer side effects. Lastly, there is an interesting, evolving data set that links dynorphins to age-related cognitive decline and specific domains of cognition. Together, it is clear that there are a number of pharmacologic targets that demand further development so that exploratory clinical studies can be pursued.

Chapter 4 (Costa, Lugert, and Jagasia) highlights hippocampal neurogenesis, now a well-known, critical aspect of structural plasticity in both young and aging brains. Cognition and cognitive performance are highly dependent and, in part, regulated by hippocampal function. There are a number of factors that control neurogenesis in the hippocampus, and some of these are possible drug targets. Other factors such as general exercise and enriched environments promote neurogenesis and enhance cognition. One illness, major depressive disorder, is used as an example linking stress, neurogenesis, and cognition. Also, aging and Alzheimer's disease are possibly impacted by lack of or impaired neurogenesis. Therefore, there is a growing interest in finding mechanisms and perhaps small-molecule drugs that are neurogenic. In order to test their efficacy, however, it will likely be critical to find and employ biomarkers of neurogenesis.

Together, Part I shows that accurate cognitive assessment in humans and animals may be best served if proper attention can be placed on specific cognitive domains. Associated with this is the need to develop new drugs that impact signaling pathways or plasticity in a manner that is specific for the various cognitive domains. Research is ongoing in these areas and should prove fruitful in the years to come.

Methods for Delivering and Evaluating the Efficacy of Cognitive Enhancement

Philip D. Harvey and Richard S.E. Keefe

Contents

P.D. Harvey (✉)
University of Miami Miller School of Medicine, 1120 NW 14th Street, Suite 1450, Miami, FL 33136, USA

Research Service, Bruce W. Carter VA Medical Center, Miami, FL, USA
e-mail: pharvey@med.miami.edu

R.S.E. Keefe
Department of Psychiatry and Behavioral Sciences, Duke University Medical Center, Durham, NC, USA

K.M. Kantak, J.G. Wettstein (eds.), *Cognitive Enhancement*, Handbook of Experimental Pharmacology 228, DOI 10.1007/978-3-319-16522-6_1

Abstract

Cognitive deficits are related to impaired everyday functioning in multiple conditions and in healthy individuals. Treatment of cognitive functioning can be facilitated through either pharmacological or remediation strategies. The critical goals of cognitive enhancement are to improve everyday functioning in multiple domains. This chapter describes the strategies that are most desirable for the treatment of cognitive impairments and detection of potential benefits of treatment in cognitive and functional domains. These strategies include the use of performance-based assessments of cognition and functioning and the appropriate use of observational strategies to detect changes. Finally, we define several outcome-related goals and discuss the practicality of their measurement.

Keywords

Cognition • Neuropsychology • Functional capacity • Disability • Schizophrenia • Dementia • Bipolar disorder

1 Introduction

Cognitive functioning has been known to be a major predictor of multiple domains of everyday functioning for over three decades (Heaton and Pendleton 1981). These domains include productive functioning such as work and school, self-care and residential functioning, and social functioning (Green et al. 2000). As a result, treatment of cognitive functioning has multiple potential beneficial effects. These effects can apply to individuals who have a variety of neuropsychiatric conditions, as well as healthy aging, and it is entirely possible that cognitive functions could be improved in people with no illness.

This chapter addresses the practical and scientific issues associated with delivering and measuring cognitive enhancement. There are several issues that require attention across multiple potential populations who could be targeted for these treatments. These include strategies for inclusion of the correct people to be treated and measurement of the effects of cognitive enhancement on cognitive functioning and everyday functioning. Other issues include dosing of the intervention, frequency of delivery, side effects of treatment, and the possible interaction between cognitive enhancement treatments and other concurrent interventions.

1.1 Cognitive Enhancement as a Concept

The central goal of cognitive enhancement is the persistent improvement of cognitive performance through an externally applied intervention (Harvey 2009). There are multiple viable strategies to improve cognitive functioning, including pharmacological and behavioral interventions. Both of these will be reviewed below. Cognitive enhancement to date has largely been restorative in its focus: targeting functioning that is currently poorer than at some point in the past (Keshavan et al. 2014). Cognitive enhancement could also be facilitative: improving functions that are currently performed similarly to lifelong levels of functioning, with a goal of improving skills past their current level, like an exercise intervention aimed at improved aerobic or anaerobic fitness in healthy people (Strassnig et al. in press).

Examples of restorative cognitive enhancement could include treatment of cognitive deficits associated with dementia, neuropsychiatric conditions such as schizophrenia or bipolar illness, or a traumatic brain injury. The intervention would target functions that are viewed as adversely affected by the occurrence of the condition. Thus, memory deficits seen in traumatic brain injury could be the target of a memory-oriented training program, and an intervention for schizophrenia might be aimed at processing speed. An example of a facilitative intervention could be self-administration of a computerized cognitive training program in order to improve memory, concentration, or even social cognition. Even if the processes are performed in an average manner prior to treatment, the facilitative intervention is aimed at improving them.

2 Delivery of Cognitive Enhancement

Cognitive enhancement can be delivered with drugs, with computer programs, in person with exercises delivered by a therapist, or, at least theoretically, through other wellness interventions such as exercise that augment cognitive functioning. While these delivery systems are quite different, the scientific issues regarding treatment delivery and measurement of benefits are likely quite similar. As the literature on all three of these approaches is large, it cannot be reviewed here and is covered elsewhere in the book. We will focus here on the issues of how to maximize the possibility that an intervention will be effective and how to determine whether it is working.

2.1 Dosing

Cognitive remediation is an intervention, and no cognitive enhancement interventions to date have been shown that they can lead to sustained improvement after a single, vaccine-like treatment. Evaluation of the efficacy of cognitive enhancement interventions requires consideration of dosing, including both timing and duration. Remediation strategies and pharmacological interventions both have dosing considerations, although they may be different because of the delivery strategy. Remediation interventions are typically delivered by a schedule. While medications are typically administered daily, remediation interventions are administered in sessions, with each session having a certain, generally limited, duration. Interventions occur with a frequency that can be widely variable, including one or more sessions (Fisher et al. 2010) each day down to a level of once or twice per week (McGurk et al. 2007). Differences in the target population receiving the intervention may have an impact on both the needs for frequent sessions and the likelihood that participants can cooperate with them.

Any plan to deliver a remediation intervention needs to consider the match between the population and the planned dosage. Individuals who are highly motivated to engage in a facilitative intervention may be able to train themselves daily in a variety of settings without prompts or encouragement and may be able to engage in the intervention with a high session frequency. However, very impaired populations such as patients with dementia or severe mental illnesses like schizophrenia may have other interfering factors such as disorganization and low motivation that can limit the frequency of sessions that can be tolerated and the capacity of the individual to complete the session without assistance. These individuals may be unable to self-administer the intervention at home and may need to be trained in person at a treatment center. Noncognitive factors such as the availability of a caregiver or financial considerations regarding transportation have been previously shown to impact the ability to dose cognitive remediation interventions (Keefe et al. 2012).

In terms of pharmacological strategies, dosing also requires consideration. Some of the mechanisms of action of pharmacological strategies may have a nonlinear dose relationship (e.g., stimulant medications for ADHD, cholinesterase inhibitors for dementia). As a result, all pharmacological cognitive enhancers need to be evaluated for dosing as well. Depending on the characteristics of the pharmacological agents, the daily delivery strategies for dosing of the medications may require adjustment and consideration. Drugs with short half-lives may require dosing more than once a day (Freedman et al. 2008) which may require special considerations for some populations when adherence may be an issue.

Some potential cognitive enhancers also require highly specialized delivery. For instance, the dopamine D1 agonist dihydrexidine (George et al. 2007) requires IV administration, so it does not seem to be particularly practical as a widely used treatment. In contrast, cholinesterase inhibitors are now available in a skin-patch delivery system, which allows for highly controlled delivery of a medication that has the potential to be poorly dosed due to nonadherence associated with side effects (Silver et al. 2009). Thus, the route of administration of a medication may also interact with dosing concerns.

Some medications for the treatment of neuropsychiatric and medical conditions have long-acting formulations. For instance, injectable long-acting antipsychotic medications can be dosed as infrequently as every 30 days. This is clearly an important long-term goal for cognitive-enhancing treatments. Long-acting medications bypass issues of adherence, which can impact the actual amount of medication that is delivered to the targeted neurotransmitter systems and key neural circuitry, and may also be quite useful in populations whose cognitive limitations have been shown to be related to their ability to manage medications (Jeste et al. 2003).

2.2 Duration

The duration of treatment is a multifaceted concern and may differ across pharmacological and cognitive remediation strategies. Clearly, interventions need to be continued until there is either evidence of treatment response or futility is acknowledged. Futility is an issue that is more relevant when the intervention has the potential to have side effects, to have a high acquisition cost, or to require substantial effort on the part of the participant receiving treatment. The risk-benefit ratio for computerized and in-person cognitive remediation therapy seems favorable. Risks for drug therapies will be specifically associated with the specific drugs and are difficult to discuss in general. Some drugs may have greater risk with longer duration of therapy or higher doses, and the duration of therapy trials for determination of efficacy for an individual participant may have to be adjusted accordingly.

The duration of a therapeutic trial will also be affected by the desired treatment outcome. If changes in global functioning are the target, then treatment may require a longer duration than if the treatment target is improvement in a single cognitive performance domain. If changes in real-world functioning are desired, a substantial time period may be required before cognitive benefits can lead to real-world functioning improvements (Rosenheck et al. 2006). Assessment of intermediate outcomes (improved cognition) will also be important even if the treatment goal is improved everyday functioning. Improvements in intermediate outcomes (or lack thereof) will provide an important signal regarding whether the intervention (pharmacological or behavioral) is providing "target engagement" (Insel and Gogtay 2014) prior to exerting a beneficial effect on the overall treatment target.

2.3 Maintenance

After successful treatment, the question arises as to how long the treatment should be continued. Antipsychotic, antidepressant, and mood stabilizer treatments are only effective while being administered. Thus, they do not truly provide a restorative effect, at least in individuals with an established course of illness. For these treatments, continued treatment is required for continued benefit, and maintenance therapy is requirement to sustain clinical gains.

The long-term treatment strategy for cognitive-enhancing interventions may be quite different. There is considerable evidence that certain types of cognitive enhancement lead to persistent changes. These have included sustained or even accelerated functional gains following the end of therapy, particularly in individuals with a positive therapeutic response (Bowie et al. 2012; McGurk et al. 2007). These persistent changes may be associated with effects on neural plasticity and may be resulting in changes in brain function. These effects seem even more pronounced in individuals who receive concurrent psychosocial interventions (Wykes et al. 2011), which may mean that there is a cascade of benefits wherein improvements in functioning also lead to improvements in the ability to make further functional gains. As a simple example, an individual who benefits cognitively from an intervention may be able to further their education, thus making them in a better position for occupational gains.

Research on cognitive enhancement will need to determine whether benefits are persistent after the end of treatment and whether renewed therapy, such as booster sessions or rechallenges with medication therapy, is required. It remains to be seen whether this may be a point of divergence between pharmacological and cognitive remediation strategies; more research is needed to determine the differences in persistence following successful treatment. If a pharmacological intervention in the short term leads to functional gains which are self-sustaining, then it may be the case that pharmacological cognitive enhancement can also be delivered in a time-limited manner with treatment responders then manifesting sustained or accelerated functional gains after the cessation of therapy.

2.4 Adherence

Adherence to cognitive remediation interventions is easy to index because of the nature of the way the service is often delivered (i.e., in person) and because remediation software routinely tracks participation. Adherence to pharmacological interventions is critical to index in order to understand whether actually taking the medication is associated with greater initial or persistent gains. Any research involving cognitive enhancement needs to carefully consider both the extent and patterns of adherence.

3 Measurement of Cognitive Change

Indexing the gains associated with cognitive enhancement therapy is a complex and multidimensional task. This measurement requires a careful conceptualization of the goals of treatment and the possible benefits of the treatment. Cognition is complex, and there are multiple assessment strategies that can be employed, with tremendous variation in the complexity of the assessment strategies, and associated challenges of indexing the treatment goals and populations targeted by the treatment (Nuechterlein et al. 2008).

In addition, since cognitive enhancement has the big picture goal of improving everyday functioning, a consideration of the complexities of assessing everyday functioning is important. Everyday functioning is under the control of many factors other than cognition (Harvey et al. 2009a, b), thus successful cognitive enhancement does not necessarily promise real-world functional gains. Separation of the environmental influences on functioning versus those influences derived from cognition and other skills is critical for quantifying the effects of cognitive enhancement therapy.

3.1 Cognition

The scientific construct of cognition includes a broad range of psychological functions: sensation and perception, attention and concentration, learning and memory, reasoning and problem solving, as well as crystallized knowledge and speed of processing (Harvey 2012). These different aspects, or domains, of cognition require very different strategies for treatment and have very different potential levels of benefit. For instance, certain pharmacological compounds have demonstrated benefits in transmitter systems with specific cognitive benefits (acetylcholine and episodic memory; Risacher et al. 2013). Other compounds, such as amphetamine, have wide-ranging cognitive benefits that are not specifically related to the primary pharmacological effects of the compound (see Sostek et al. 1980 for a classic early study).

3.2 Domains

It is beyond the scope of this chapter to fully describe the ranges of functioning that can be considered cognitive in nature. However, we will review the idea that cognition can be conceptualized in terms of separable cognitive domains, which are then amenable to measurement with specialized assessments (Nuechterlein et al. 2004). The history of clinical neuropsychology, which is the origin of many of the assessments that are used to generate outcome measures in clinical cognitive enhancement research, was partially based on the assessment of individuals with focal brain injuries or strokes (Harvey 2012). Thus, tasks were developed to be sensitive to deficits in specific brain regions when performance is preserved in other regions of the brain.

More recent conceptualizations of cognitive performance are focused on networks, which can affect an array of cognitive abilities. For instance, impairments in striatal regions, such as those induced by Huntington's disease and related conditions, impact an array of cognitive functions. These include processing speed, concentration and attention, and learning and memory (Paulsen et al. 1995). Traditionally these tasks were generally assumed to measure the functions of several different regions, but impairments can typically occur in concert.

Traditional domains of cognitive functioning are presented in Table 1. These domains of functioning are differentially impaired in various neuropsychiatric conditions and have been the target of previous cognitive enhancement interventions. However, although these domains can be defined and measured with psychological tests, there are two important points to consider. First, in

Table 1 Typical domains of cognitive functioning

• Sensation
• Perception
• Sustained attention (vigilance)
• Selective attention
• Working memory
• Episodic memory
– Learning
– Delayed recall
– Delayed recognition
• Executive functioning
• Processing speed

many studies of neuropsychiatric conditions, the cognitive tests that are targeted at different domains of functioning are often highly intercorrelated (Dickinson et al. 2007), and the best fitting factor structure is a single, global factor (Keefe et al. 2006a). On the other hand, performances on tests measuring similar cognitive domains, such as elements of intelligence, are often somewhat discrepant from each other in healthy people (Wechsler 2008).

The notion of separable versus highly related cognitive domains may influence strategies for testing cognitive enhancers. First of all, if measures of different cognitive domains are highly intercorrelated, it may be challenging to develop interventions that are selective in their measured cognitive benefits. Certain interventions can have specific effects on a limited set of cognitive domains (Fisher et al. 2009), but global functioning is also likely to improve. Second, research on the correlates of real-world functioning has consistently suggested that specific measures from individual cognitive domains are much less strongly related to real-world outcomes than composite measures that summarize global performance (Green 1996). Thus, the typical distal goal of improving functional outcomes may be better facilitated by interventions that are effective across multiple cognitive domains (Bowie et al. 2012). As a result, treatments with broad cognitive benefits may be the best ways to improve everyday functioning.

3.3 Measurement Strategies

Cognitive functions have traditionally been measured with performance-based assessments. At the same time, many people become candidates for cognitive enhancement because of their subjective experience of cognitive change or difficulty. Thus, self-reported cognitive functions have been explored as an assessment strategy (Keefe et al. 2006b). Finally, observers are often queried about cognitive functioning, particularly in pathological conditions where the observer may have a long history of interacting and observing the cognitively relevant behaviors of the individual in question (Ventura et al. 2013).

These different strategies have different patterns of strengths and weaknesses regarding their validity and practicality. These strengths and weaknesses tend to be reciprocal in many instances, where the more practical strategies may have some of the weaknesses in validity. Thus, the selection of outcomes assessment strategies may be very different depending on the goals of the investigation and the populations to be assessed.

3.4 Performance-Based Cognitive Assessments

Performance-based assessments do not rely on the opinion of an observer and are highly standardized. Further, these assessments have been refined over time, and their psychometric characteristics can be easily quantified and have been investigated in detail (Nuechterlein et al. 2008). Thus, these measures will have a known range of scores in the healthy population. As a result, performance can immediately be interpreted, and improved performance can be quantified with ease. These tests are often administered together as described below, and the administration of these tests can be taught to be people without advanced degrees, if the interpretation of the scores is performed statistically. Interpretation of the pattern of scores and their clinical meaning usually requires an advanced degree in psychology.

An array of performance-based assessments of everyday functioning skills has been developed. These procedures aim at assessment of skills in the domains of residential, social, and vocational functions (Mausbach et al. 2007; Patterson et al. 2001). The majority of these procedures use realistic assessments of functional activities such as shopping, cooking, managing money, and social interactions. These are administered as test procedures, with systematic administration of the tests, systematic scoring procedures, and normative standards. Thus, much like performance-based cognitive tests, they provide a repeatable index of skills competence that can be used as an outcome measure in clinical trials.

Also like cognitive tests, the issues associated with comprehensiveness, duration of the assessment, and practicality need to be considered. Also similar to cognitive assessments, there are computerized versions of functional capacity measures that are available, with the same caveats and concerns as cognitive assessments (Ruse et al. 2014). So, there is considerable similarity across performance-based strategies for assessment of cognition and functional capacity, including their advantages and their limitations.

3.5 Detail and Comprehensiveness

The standard way to measure cognition in clinical practice and clinical trials is performance assessment with intellectual and neuropsychological tests. The tradition in clinical neuropsychological assessment has been to perform a detailed assessment aimed at examining a variety of cognitive strengths and weaknesses

(Heaton 1992). This approach is most appropriate for assessment of the extent and severity of impairments in individuals who have experienced an illness or accident. Similarly, children with academic problems may receive an extensive educationally oriented assessment of abilities and achievements.

However, in many circumstances an extensive assessment is not required. In some conditions performance is highly intercorrelated across tests; a carefully selected, briefer assessment may provide the same amount of information as a much longer assessment (Keefe et al. 2004). In some instances, a condition can be identified through the presence of a single salient deficit. For instance, diagnostic exclusion of possible Alzheimer's disease can be accomplished through a very abbreviated assessment of delayed recall memory, with this deficit leading to substantial separation from the performance of various other diagnostic groups, including even patients with schizophrenia (Davidson et al. 1996).

An advantage of more detailed assessment is that the identification of multiple effects of a cognitive enhancer is only possible with a wide-ranging assessment. Such an assessment would likely be undertaken in the early development phases of a treatment, as regulatory agencies require investigators to declare their primary outcome measure prior to the initiation of a trial (Buchanan et al. 2011). An additional use of a more detailed assessment in early phase studies is that of the detection of any possible adverse effects of a treatment. For example, if a drug or cognitive remediation procedure was to improve problem solving but induce sedation and slow processing speed, this could not be detected unless a wide-ranging assessment of these domains was used. Regulatory agencies may require relatively comprehensive cognitive batteries so that any deleterious effects of a new medication on cognition can be detected (Buchanan et al. 2005, 2011).

3.6 Duration

Some formal neuropsychological assessment batteries can take 6–12 h or more to complete. Duration of the assessment is generally correlated with level of detail, but an assessment of episodic memory can take an hour itself, while an abbreviated but wide-ranging (Harvey et al. 2009b; Keefe et al. 2004) assessment of cognition often used in clinical trials can take as little as 20 min.

Longer assessments pose challenges from two directions. If a treatment trial has multiple assessments other than cognition, then a cognitive assessment with a long duration may increase the length of a study visit to the point that it is not practical. The other point is that some populations are challenged by long assessments. For instance it is not a surprise to see that children who have difficulty sustaining their attention in school have similar problems tolerating long psychological assessments, which can lead to misleading results. In general, however, even very disorganized patient populations such as those with schizophrenia can provide valid data with cognitive batteries requiring 75 min or more of assessment time (Nuechterlein et al. 2008; Keefe et al. 2011; Bowie et al. 2002).

The take-home principles from this discussion are that the shortest possible assessment that assesses important aspects of cognition is the best strategy; however, attention must be paid to important psychometric characteristics of the data collected, such as whether enough information is being collected to allow meaningful conclusions. There are no generic answers for how long and how broad, which will depend on enhancement strategies, targets, and populations.

3.7 Delivery

Performance-based measures are available in paper and pencil or computerized formats. While computerized assessments would seem to ensure greater fidelity and validity, the results are clearly divided. For populations with significant impairments that may lead to problems in cooperation or effort, there have been several studies showing that computerized assessments generate data that are less complete and less reliable than standard administration procedures (Keefe et al. 2006a; Silver et al. 2006; Iverson et al. 2009), and computerized assessment can serve to mask invalid performance. Some recently developed assessment strategies can detect invalid performance (Harvey et al. 2013b), but the message here is clear: testers need to be as active, observant, and involved in the administration of computerized assessments as they are in the administration of paper and pencil assessments.

3.8 Frequency of Assessments and Related Issues

In a cognitive enhancement study, an estimate of treatment-related cognitive change requires assessment before and after treatment. As we have noted before, there are several situations where repeated assessments pose challenges. One is the retest improvement, or "practice," effect, which can be due to exposure to testing, familiarity with the materials, and increased comfort levels. There are several solutions to this problem (Goldberg et al. 2010). One is the use of alternate forms, but alternate forms can be remarkably poorly correlated with each other in certain populations, which can significantly weaken the reliability of assessing cognitive change. Another is the use of a parallel research design, which allows for comparison of changes in performance over time across active and inactive treatments. As long as subjects do not perform at the ceiling of a measure such that improved performance cannot be detected, the difference between active and inactive conditions can index the effect of the treatment minus the effects of repeated testing alone.

Practice effects are challenging because few measures will have information from normative studies that examined practice effects beyond two or three reassessments in the population of interest and even fewer in healthy individuals for normative comparison. While it is generally believed that practice effects habituate after a few assessments, leading to stable performance over time, some other data suggest small but incremental effects across numerous assessment sessions. However, in the absence of ceiling effects, practice effects are preferable over poorly correlated alternate forms, which will prevent the identification of a treatment effect.

3.9 Interview-Based Cognitive Assessments

Interview-based assessments are appealing because they are easy to administer, score, and interpret. Some conditions cannot be diagnosed without a subjective cognitive complaint (i.e., mild cognitive impairment, MCI). At the same time, self-reported cognitive ability can be almost shockingly unrelated to objective performance and the opinions of others, across the multiple conditions where cognitive enhancement would be employed (Durand et al. in press). Even in healthy people, there are substantial response biases which lead to poor self-assessment. In neuropsychiatric conditions, even though objective tests have been found to correlate with the reports of clinician informants and real-world functioning, the correlation between objective test performance and self-reported functioning is essentially zero, meaning that there is no chance of detecting reliable changes from baseline that correspond to objective tests (Keefe et al. 2006a, b; Sabbag et al. 2011). These results are not due to unsystematic self-assessments, because structured interview-based instruments were used in these studies.

3.10 Functioning

The real target of cognitive enhancement is to improve functioning in real-world situations, whether it is the workplace, school, or managing one's life with more efficiency. However, the assessment of real-world outcomes differs substantially in the context of clinical interventions versus treatment studies. An ongoing clinically oriented intervention over an indefinite time period removes many of the challenges of assessment of real-world functioning. A 12-week clinical trial is very different because some functional changes require time and some of these parameters are outside the control of the cognitive enhancement recipient and the clinical treatment team or clinical trial managers. We will address these issues below.

3.11 Real-World Functioning

Assessment of real-world functioning seems to be a trivial task, in that it would be expected that most people would know where they live, what they do for work, and how many friends they have. However, some of the subpopulations targeted for cognitive enhancement may present challenges in these areas. Further, for individuals who have experienced challenges and are functioning suboptimally, there may be a complex array of factors that contribute to real-world functioning.

These include disability compensation, opportunities in the local area, and the complex interaction between care systems, families, and the patients.

3.12 Milestones

In a largely healthy population, achievement of functional milestones like full-time work, living independently, and having friends, family, and social network is expected. Individuals who have achieved these milestones and are seeking cognitive enhancement will be trying to facilitate their functioning. In these individuals, the real-world outcome would be school grades, promotions, and other indicators of greater functional success. In impaired populations, however, the lack of experience with functional demands may prohibit individuals from being able to accurately evaluate their own functioning.

For instance, in a recent study of ours, people with severe mental illness who had never had a job in their lives rated themselves as more socially, vocationally, and residentially capable than other individuals who were employed full time (Harvey et al. 2012). As many populations seeking cognitive enhancement may have a lack of functional success to date, modification of the typical assessment strategy may be required.

3.13 Subthreshold Performance

If one is unemployed at present, there are a variety of functional acts that are preparatory to employment that are positive from the perspective of vocational outcomes. For instance, preparing a resume, applying for jobs, and going on job interviews are positively valenced vocational activities. However, they do not equate to having a full-time job. For a variety of populations where there are long-term aspects of disability and the assessment of job performance is not possible, we are limited to collecting information about the preparation and background activities aimed at real-world functioning. However, these subthreshold milestones have been shown to be positively affected by skills training and cognitive enhancement and are relevant measures related to eventual real-world successes.

Multiple functional outcome rating scales are available. The consistent finding from comparative studies of these scales is that the quality of the informant and their knowledge of the patient are more important than the scales themselves (Sabbag et al. 2011, 2012). Those studies that have conducted ratings with informants who know the patient well, either as a high contact clinician or other caregiver, have found solid correlations with cognitive performance. Studies that have relied on patient self-report have consistently found no correlation between ratings of functional skills and objective information (Bowie et al. 2007). A reasonable conclusion from many of these studies is that the type of clients who are referred for cognitive enhancement is unlikely to be an adequate reporter of either their baseline functioning or their improvements from that baseline.

3.14 Functional Capacity

Because of our knowledge that real-world outcomes are affected by an array of factors other than patient abilities, interest has grown in the area of direct assessment of functional skills. Referred to as "functional capacity," this is the process of assessing the ability to perform critical everyday living skills in simulation settings (Harvey et al. 2007). The ability to perform skilled acts can be contrasted with the actual likelihood of performing those acts (Mausbach et al. 2011; Depp et al. 2010). Like performance-based cognitive tests, these procedures do not rely on self-report, are administered by technicians, and can be evaluated for their psychometric and validity properties. In addition, interviews can also be targeted at functional capacity as well. Like interviews aimed at cognition and real-world functioning, the source of the information may be as important as the specific questions that are asked.

3.15 Cognition or Capacity as the Outcome

The high correlation between assessments of cognitive performance and functional capacity has led to the question (Leifker et al. 2011), partially supported by data, that these are actually different assessments of the same general ability domain (Harvey et al. 2013a). Future research will need to determine the differential suitability of these indices for outcomes assessment in treatment studies. Given that abbreviated assessments of both cognition and functional capacity are available, it would seem prudent to invest the time to assess both of these domains. This issue may change with the future development of computerized functional capacity assessments, as highly validated functional capacity assessments may lead to increased validity and practicality.

3.16 What Defines Improvement?

The definition of improvement in performance following cognitive-enhancing treatments depends on the goal of the assessment. Clinical treatment will have a very different set of standards than a regulatory efficacy trial. Further, improvement can be indexed in several ways. These improvements can be defined, in hierarchical order of rigorousness, as:

- Statistically significant
- Clinically meaningful
- Definitely nonrandom
- Normalization of functioning

Statistically significant is the criterion for demonstrating differences between active treatments and control conditions. This is the lowest bar for empirically defined improvement, because it is largely dependent on the sample size of the study. This criterion does not depend in any way on the level of baseline performance and does not require any predetermined end of study level of performance in order meet this criterion. Although it is the lowest of the bars that we are discussing, it is still important. Any treatment that does not separate from an inactive treatment cannot be seen to provide reliable improvements. This criterion is required for a treatment to receive regulatory approval.

Clinically meaningful is a higher bar than statistical significance. This threshold would imply a certain average degree of improvement for the populations treated. Required as part of this criterion is some notion of what the size of such a change would be, and this requires information obtained from other sources other than statistics. Embedded within this concept is the expectation that a certain amount of improvement in cognition for an individual or group would be associated with a certain amount of improved functioning. For instance, several different studies of functional measures have identified threshold levels of cognitive performance consistent with achievement of functional milestones such as independence in residence. Treatment-related improvements that reach these thresholds would be possible indices of clinically meaningful change.

Definitely Nonrandom When a group of participants receives treatment, even if the benefit is statistically or clinically significant for the group, there is likely to be variation in response among the people treated. The assessment of improvement for individuals differs from that for groups in that to be certain that an individual has improved to a level greater than chance, a host of influences on performance such as practice effects require consideration. The "reliable change index" has been developed in order to quantify whether an improvement in one person exceeds what is expected based upon general influences (Heaton et al. 2001). The reliable change index statistic incorporates the test-retest reliability of the measure and establishes a range of scores that exceeds this level of change. The confidence interval of the reliable change index is typically set at 90 %, meaning that there is only a one in ten chance that the threshold amount of change would have occurred by random factors alone.

With commonly used outcome measures for clinical trials in humans, the typical level of change required to define a definitely nonrandom change is in the range of about 1.0 standard deviations (Leifker et al. 2010). This is a fairly high bar, but in several previous cognitive treatment studies, the group improvements in cognitive outcomes have been as great as 0.8 SD (Bowie et al. 2012; Fisher et al. 2009). This could mean that a number of people treated in those studies manifest individual improvements that are definitely nonrandom.

Normalization of Functioning This is the highest bar and is not necessarily a goal of every treatment. Normalization of functioning would imply two things:

substantial improvement in functioning that is entirely within the normal range and improvement on the part of an individual to at least their pre-illness level of functioning if not better. The normal range of functioning is typically defined as within one standard deviation of the population mean or higher. Further, if an individual's performance was within that range prior to illness, then their posttreatment functioning should be within that range as well. Normalization is a high bar because many individuals whose performance is slightly below the cutoff for normal cognitive functioning (−1.0 SD) are functioning adequately in their lives.

3.17 Practical Concerns

One of the issues that has come to the forefront of treatment with remediation interventions is whether computerized interventions can be self-administered at home (Fisher et al. in press). Like pharmacological interventions, remediation interventions can be delivered outside the clinic setting. As these interventions transition toward wider use, with the anticipated approval of drugs or medical devices for cognitive remediation treatment, assessments may also need to be performed outside of the clinic. This would require the ability to use remotely deliverable cognitive and functional assessment strategies, with the same reliability and validity standards that are conventionally applied to paper and pencil and other in-person testing procedures.

Cognitive tests and functional capacity measures have already been developed for remote administration. The issues associated with computerization and remote delivery of these assessments are the same as in-person assessments. There needs to be considerable evidence supporting the psychometric characteristics of the instruments and a match between the content of the instrument and clinically relevant community outcomes. This is likely to be a major area for future technology development, and these procedures will be more successful if they are flexibly adapted across emerging technology.

3.18 Populations

There are several conditions either defined by cognitive deficits or accompanied by significant enough deficits so that they clearly will be a target population for cognitive enhancement. Within some of these populations, it is acknowledged that all people so classified have deficits; this includes conditions such as schizophrenia, mild cognitive impairment, dementia, and attention deficit disorders. In other conditions, the rates of cognitive impairment are less than 100 %, with these including some proportion of cases known to not manifest cognitive impairments or to only manifest them in a specific clinical state, such as when experiencing a mood episode.

In some situations, therefore, it would seem as if a pretreatment assessment of functioning would be required in order to determine if the person needed treatment. Other conditions would not require specific confirmation of the need but would still require monitoring of treatment outcomes as described above. At the same time, it is possible that even in a population where essentially every person has cognitive deficits, there would be certain variation in treatment response expected. Thus, understanding who would be the better or faster responders might allow for optimized treatment delivery, particularly if treatment was in any way to be rationed or otherwise access controlled.

In the area of cognitive remediation, there are already several leads as far as the detection of better responders. In these studies several factors have been identified.

Age Brain plasticity varies over the life span, and it is known that there are critical, often early, periods for optimal skills acquisition. Some studies have shown that younger patients have a larger or faster clinical response, yet response is still adequate in older people (Bowie et al. 2014; Lindenmayer et al. 2008). This is more of an issue in early-onset lifelong conditions like schizophrenia or bipolar disorder than in mild cognitive impairment or childhood onset ADHD.

Duration of Illness (if not completely overlapping with age) Cases where age and duration of illness do not overlap may include traumatic brain injuries or other neurological events such as stroke. In these conditions, there is often substantial early recovery which can preclude the need for treatment. On the other side, extended time since injury or event, paired with persistent symptoms, may identify cases with poorer prognosis.

Pre-illness Functioning Skills and abilities acquired earlier in life, referred to as "cognitive reserve," are potent predictors of lifelong functioning. In individuals challenged by conditions that adversely impact cognitive functioning, higher levels of cognitive reserve likely identify cases who will have more potential for improvement. It is not clear whether pharmacological or remediation interventions will lead to improvements that surpass pre-illness levels of functioning and adjustment. However, some individuals with intellectual disability may benefit from combined treatments where cognitive remediation is offered as a facilitator of other skills training interventions.

Motivation for Treatment and Functional Improvement Cognitive enhancement with training programs requires attention and effort, as well as occasional self-administration of training (Fisher et al. in press). For all enhancement interventions, real-world functional gains require goals and effort, and patients not looking for increased independence or vocational functioning will not make these gains (Choi and Medalia 2010). If payers are focused on real-world outcomes, then cognitive enhancement studies should be performed on patients who are likely to be motivated to realize external goals.

4 Conclusions

Behavioral and pharmacologic therapies aimed at the enhancement of cognition are broadly applicable, and there are many aspects of cognition and functioning that may experience a potential benefit. Measurement of these potential changes is challenging and potentially complex. There are several strategies for simplifying the measurement of the outcomes of cognitive enhancement, and these strategies will continue to grow and evolve as more successful interventions are developed. It is important that the measurement of the benefits of cognitive enhancement is valid and accurate, such that effective treatments are immediately recognized and ineffective ones not incorrectly endorsed.

References

Bowie CR, Harvey PD, Moriarty PJ et al (2002) Cognitive assessment of geriatric schizophrenic patients with severe impairment. Arch Clin Neuropsychol 17:611–623

Bowie CR, Twamley EW, Anderson H et al (2007) Self-assessment of functional status in schizophrenia. J Psychiatr Res 41:1012–1018

Bowie CR, McGurk SM, Mausbach BT et al (2012) Combined cognitive remediation and functional skills training for schizophrenia: effects on cognition, functional competence, and real-world behavior. Am J Psychiatry 169:710–718

Bowie CR, Grossman M, Gupta M et al (2014) Cognitive remediation in schizophrenia: efficacy and effectiveness in patients with early vs. long-term course of illness. Early Interv Psychiatry 8:32–38

Buchanan RW, Davis M, Goff D, Green MF, Keefe RSE, Leon A, Nuechterlein K, Laughren T, Levin R, Stover E, Fenton W, Marder S (2005) A Summary of the FDA-NIMH-MATRICS workshop on clinical trial designs for neurocognitive drugs for schizophrenia. Schizophr Bull 31:5–19

Buchanan RW, Keefe RS, Umbricht D et al (2011) The FDA-NIMH-MATRICS guidelines for clinical trial design of cognitive-enhancing drugs: what do we know 5 years later? Schizophr Bull 37:1209–1217

Choi J, Medalia A (2010) Intrinsic motivation and learning in a schizophrenia spectrum sample. Schizophr Res 118:12–19

Davidson M, Harvey PD, Welsh K et al (1996) Cognitive impairment in old-age schizophrenia: a comparative study of schizophrenia and Alzheimer's disease. Am J Psychiatry 153:1274–1279

Depp CA, Mausbach BT, Harvey PD et al (2010) Social competence and observer-rated social functioning in bipolar disorder. Bipolar Disord 12:843–850

Dickinson D, Ramsey M, Gold JM (2007) Overlooking the obvious: a meta-analytic comparison of digit symbol coding tasks and other cognitive measures in schizophrenia. Arch Gen Psychiatry 64:532–542

Durand D, Strassnig M, Sabbag S et al (2015) Factors influencing self-assessment of cognition and functioning in schizophrenia: implications for treatment studies. Eur Neuropsychopharmacol 25:185–191

Fisher M, Holland C, Merzenich MM et al (2009) Using neuroplasticity-based auditory training to improve verbal memory in schizophrenia. Am J Psychiatry 166:805–811

Fisher M, Holland C, Subramaniam K et al (2010) Neuroplasticity-based cognitive training in schizophrenia: an interim report on the effects 6 months later. Schizophr Bull 36:869–879

Fisher M, Loewy R, Carter C et al (2015) Neuroplasticity-based auditory training via laptop computer improves cognition in young individuals with recent onset schizophrenia. Schizophr Bull 41:250–258.

Freedman R, Olincy A, Buchanan RW et al (2008) Initial phase 2 trial of a nicotinic agonist in schizophrenia. Am J Psychiatry 165:1040–1047

George MS et al (2007) A single 20 mg dose of dihydrexidine (DAR-0100), a full dopamine D1 agonist, is safe and tolerated in patients with schizophrenia. Schizophr Res 93:42–50

Goldberg TE, Keefe RSE, Goldman R et al (2010) Circumstances under which practice does not make perfect: a review of the practice effect literature in schizophrenia and its relevance to clinical treatment studies. Neuropsychopharmacology 35:1053–1062

Green MF (1996) What are the functional consequences of neurocognitive deficits in schizophrenia? Am J Psychiatry 153:321–330

Green MF, Kern RS, Braff DL et al (2000) Neurocognitive deficits and functional outcome in schizophrenia: are we measuring the 'right stuff'? Schizophr Bull 26:119–136

Harvey PD (2009) Pharmacological cognitive enhancement in schizophrenia. Neuropsychol Rev 19:324–335

Harvey PD (2012) Clinical applications of neuropsychological assessment. Dialogues Clin Neurosci 14:91–99

Harvey PD, Velligan DI, Bellack AS (2007) Performance-based measures of functional skills: usefulness in clinical treatment studies. Schizophr Bull 33:1138–1148

Harvey PD, Helldin L, Bowie CR et al (2009a) Performance-based measurement of functional disability in schizophrenia: a cross-national study in the United States and Sweden. Am J Psychiatry 166:821–827

Harvey PD, Keefe RS, Patterson TL et al (2009b) Abbreviated neuropsychological assessments in schizophrenia: association with different outcomes measures. J Clin Exp Neuropsychol 31:462–471

Harvey PD, Sabbag S, Prestia D et al (2012) Functional milestones and clinician ratings of everyday functioning in people with schizophrenia: overlap between milestones and specificity of ratings. J Psychiatr Res 46:1546–1552

Harvey PD, Raykov T, Twamley EW et al (2013a) The factor structure of neurocognition and functional capacity in schizophrenia: a multidimensional examination of temporal stability. J Int Neuropsychol Soc 2013(19):656–663

Harvey PD, Siu C, Hsu J et al (2013b) Effect of lurasidone on neurocognitive performance in patients with schizophrenia: a short-term placebo- and active-controlled study followed by a 6-month double-blind extension. Eur Neuropsychopharmacol 11:1373–1382

Heaton RK (1992) Comprehensive norms for an expanded Halstead-Reitan battery: demographic corrections, research findings, and clinical applications. Psychological Assessment Resources, Lakeland, FL

Heaton RK, Pendleton MG (1981) Use of neuropsychological tests to predict patients everyday functioning. J Consult Clin Psychol 49:807–821

Heaton RK, Temkin N, Dikmen S et al (2001) Detecting change: a comparison of three neuropsychological methods, using normal and clinical samples. Arch Clin Neuropsychol 16:75–91

Insel TR, Gogtay N (2014) National institute of mental health clinical trials: new opportunities, new expectations. JAMA Psychiatry 71:745–746

Iverson GL, Brooks BL, Ashton VL et al (2009) Does familiarity with computers affect computerized neuropsychological test performance? J Clin Exp Neuropsychol 31(5):594–604

Jeste SD, Patterson TL, Palmer BW et al (2003) Cognitive predictors of medication adherence among middle-aged and older outpatients with schizophrenia. Schizophr Res 63:49–58

Keefe RS, Goldberg TE, Harvey PD et al (2004) The brief assessment of cognition in schizophrenia: reliability, sensitivity, and comparison with a standard neurocognitive battery. Schizophr Res 68:283–297

Keefe RSE, Bilder RM, Harvey PD et al (2006a) Baseline neurocognitive deficits in the CATIE schizophrenia trial. Neuropsychopharmacology 31:2033–2046

Keefe RSE, Poe M, Walker TM et al (2006b) The schizophrenia cognition rating scale: an interview-based assessment and its relationship to cognition, real-world functioning, and functional capacity. Am J Psychiatry 163:426–432

Keefe RSE, Fox KH, Harvey PD, Cucchiaro J, Siu C, Loebel A (2011) Characteristics of the MATRICS Consensus Cognitive Battery in a 29-site Antipsychotic Schizophrenia Clinical Trial. Schizophr Res 125:161–168

Keefe RSE, Vinogradov S, Medalia A et al (2012) Feasibility and pilot efficacy results from the multi-site Cognitive Remediation in the Schizophrenia Trials Network (CRSTN) study. J Clin Psychiatry 73:1016–1022

Keshavan MS, Vinogradov S, Rumsey J et al (2014) Cognitive training in mental disorders: update and future directions. Am J Psychiatry 171:510–522

Leifker FR, Patterson TL, Bowie CR et al (2010) Psychometric properties of performance-based measurements of functional capacity. Schizophr Res 119:246–252

Leifker FR, Patterson TL, Heaton RK et al (2011) Validating measures of real-world outcome: the results of the VALERO Expert Survey and RAND Panel. Schizophr Bull 37:334–343

Lindenmayer JP, McGurk SR, Mueser KT et al (2008) A randomized controlled trial of cognitive remediation among inpatients with persistent mental illness. Psychiatr Serv 59:241–247

Mausbach BT, Harvey PD, Goldman SR et al (2007) Development of a brief scale of everyday functioning in persons with serious mental illness. Schizophr Bull 33:1364–1372

Mausbach BT, Depp CA, Bowie CR et al (2011) Sensitivity and specificity of the UCSD performance-based skills assessment (UPSA-B) for identifying functional milestones in schizophrenia. Schizophr Res 132:165–170

McGurk SR, Mueser KT, Feldman K et al (2007) Cognitive training for supported employment: 2–3 year outcomes of a randomized controlled trial. Am J Psychiatry 164:437–441

Nuechterlein KH, Barch DM, Gold JM et al (2004) Identification of separable cognitive factors in schizophrenia. Schizophr Res 72:29–39

Nuechterlein KH, Green MF, Kern RS et al (2008) The MATRICS consensus cognitive battery, part 1: test selection, reliability, and validity. Am J Psychiatry 165:203–213

Patterson TL, Goldman S, McKibbin CL et al (2001) UCSD performance-based skills assessment: development of a new measure of everyday functioning for severely mentally ill adults. Schizophr Bull 27:235–245

Paulsen JS, Salmon DP, Monsch A et al (1995) Discrimination of cortical from subcortical dementias on the basis of memory and problem-solving tests. J Clin Psychol 51:48–58

Risacher SL, Wang Y, Wishart HA et al (2013) Cholinergic enhancement of brain activation in mild cognitive impairment during episodic memory encoding. Front Psychiatry 4:105. doi:10.3389/fpsyt.2013.00105

Rosenheck R, Leslie D, Keefe R, CATIE Study Investigators Group et al (2006) Barriers to employment for people with schizophrenia. Am J Psychiatry 163:411–417

Ruse SA, Davis VG, Atkins AS et al (2014) Development of a virtual reality assessment of everyday living skills. J Vis Exp 86. doi: 10.3791/51405

Sabbag S, Twamley EW, Vella L et al (2011) Assessing everyday functioning in schizophrenia: not all informants seem equally informative. Schizophr Res 131:250–255

Sabbag S, Twamley E, Vella L et al (2012) Predictors of the accuracy of self-assessment of everyday functioning in people with schizophrenia. Schizophr Res 137:190–195

Silver JM, Koumaras B, Chen M et al (2006) The effects of rivastigmine on cognitive function in patients with traumatic brain injury. Neurology 67:748–755

Silver JM, Koumaras B, Meng X et al (2009) Long-term effects of rivastigmine capsules in patients with traumatic brain injury. Brain Inj 23:123–132

Sostek AJ, Buchsbaum MS, Rapoport JL (1980) Effects of amphetamine on vigilance performance in normal and hyperactive children. J Abnorm Child Psychol 8:491–500

Strassnig M, Signorile J, Gonzalez C et al. (2014) Cognitive training in mental disorders: update and future directions. Schizophr Res Cog 1:112–121

Ventura J, Reise SP, Keefe RS et al (2013) The Cognitive Assessment Interview (CAI): reliability and validity of a brief-based measure of cognition. Schizophr Bull 39(3):583–591

Wechsler D (2008) Wechsler adult intelligence scale, 4th edn. Pearson Assessment, San Antonio, TX

Wykes T, Huddy V, Cellard C et al (2011) A meta-analysis of cognitive remediation for schizophrenia: methodology and effect sizes. Am J Psychiatry 168:472–485

Animal Paradigms to Assess Cognition with Translation to Humans

Tanya L. Wallace, Theresa M. Ballard, and Courtney Glavis-Bloom

Contents

Abstract

Cognition is a complex brain function that represents processes such as learning and memory, attention, working memory, and executive functions amongst others. Impairments in cognition are prevalent in many neuropsychiatric and neurological disorders with few viable treatment options. The development of new therapies is challenging, and poor efficacy in clinical development continues to be one of the most consistent reasons compounds fail to advance, suggesting that traditional animal models are not predictive of human conditions and behavior. An effort to improve the construct validity of neuropsychological

T.L. Wallace (✉) • C. Glavis-Bloom
Center for Neuroscience, SRI International, Menlo Park, CA, USA
e-mail: tanya.wallace@sri.com

T.M. Ballard
Neuroscience, Ophthalmology and Rare Diseases Discovery & Translational Area, Roche Innovation Center, Basel, Switzerland

K.M. Kantak, J.G. Wettstein (eds.), *Cognitive Enhancement*, Handbook of Experimental Pharmacology 228, DOI 10.1007/978-3-319-16522-6_2

testing across species with the intent of facilitating therapeutic development has been strengthening over recent years. With an emphasis on understanding the underlying biology, optimizing the use of appropriate systems (e.g., transgenic animals) to model targeted disease states, and incorporating non-rodent species (e.g., non-human primates) that may enable a closer comparison to humans, an improvement in the translatability of the results will be possible. This chapter focuses on some promising translational cognitive paradigms for use in rodents, non-human primates, and humans.

Keywords
Cognition • Translational models • Mouse • Rat • Monkey • Non-human primate • Human • Neuropsychological tests • Psychiatric disorders • Neurological disorders • Cognitive disorders • Cognitive biomarkers

1 Introduction

Cognition is a complex and multifaceted brain function that represents processes such as attention, sensory gating, perception, learning and memory, working memory, planning, problem solving, and executive functions (Fig. 1). Impairments in cognition are prevalent in many neuropsychiatric and neurological disorders. From autism spectrum disorders beginning early in development through Alzheimer's disease in the aged, cognitive dysfunction is a defining yet heterogeneous feature of

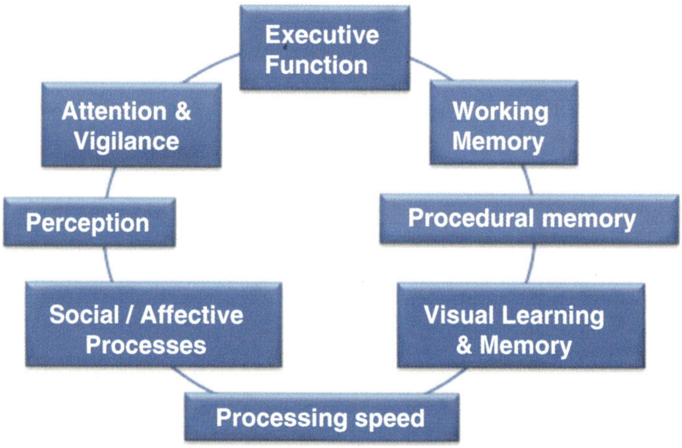

Fig. 1 Cognition is a multifaceted and translatable function that includes several domains that are often interconnected. Verbal learning and memory and semantic memory omitted due to lack of translation

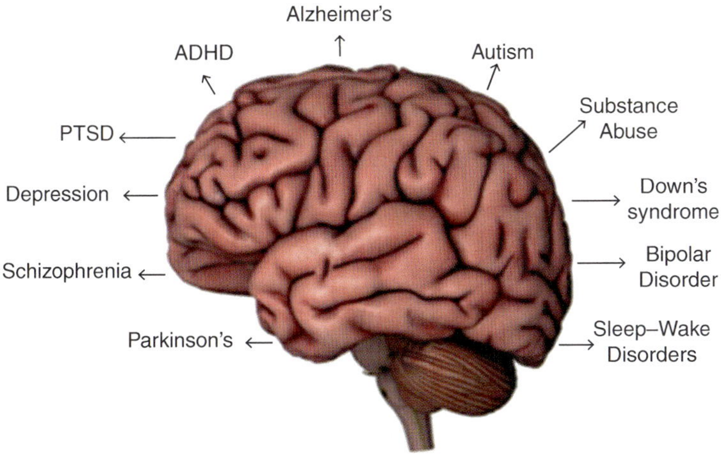

Fig. 2 Some brain conditions associated with cognitive impairment. Cognitive deficits are associated with many neuropsychiatric and neurological disorders, as well as under conditions of stress and general aging

these conditions (Fig. 2). Despite its prevalence, limited treatment options are available for effectively treating cognitive impairments, and these disorders remain a large unmet medical need.

Developing new therapies is challenging, particularly for brain diseases (<10 % success rate once entering clinical phases), and high attrition rates in development threaten the continued investment into research and development for novel therapies (DiMasi et al. 2010; Hay et al. 2014; Kola and Landis 2004; Miller 2010). Many issues can arise along the clinical development pathway to stop advancement of a promising new drug candidate (e.g., safety and tolerability, bioavailability), but a primary reason remains a lack of efficacy in patients. Since the drug candidate likely would not be in development without preclinical data suggestive of a desired effect, it has been repeatedly suggested that animal models are not predictive.

With regard to cognitive-impairing diseases, a growing focus on improving the construct validity of neuropsychological testing across species with the intent of facilitating therapeutic development has strengthened over recent years. In general, a concerted effort toward selecting and developing preclinical paradigms in which a deep understanding of the underlying biology is in place aims to improve translatability of results. In addition, optimizing the use of appropriate systems (e.g., transgenic animals) to model targeted disease states, and incorporating non-rodent species [e.g., non-human primates (NHPs)] that may more closely compare to the cognitive and mental capabilities of humans, is an effective translational strategy (Fig. 3). Ideally, the selection of appropriate animal models can help to inform the clinical study designs for effective doses, identification of sensitive endpoints, and expected side effects. In addition, the clinical data can then help to refine the animal models to improve their sensitivity and validity (i.e., forward and backward translation) (Fig. 4).

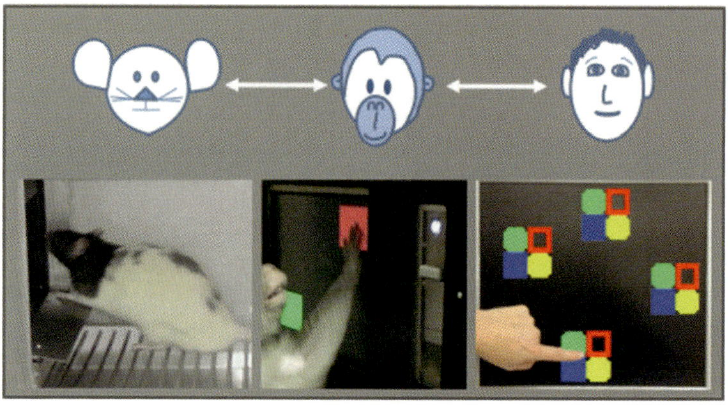

Fig. 3 Effective translation across species. Representative example of rat, non-human primate, and human versions of visual touchscreen-based cognitive testing

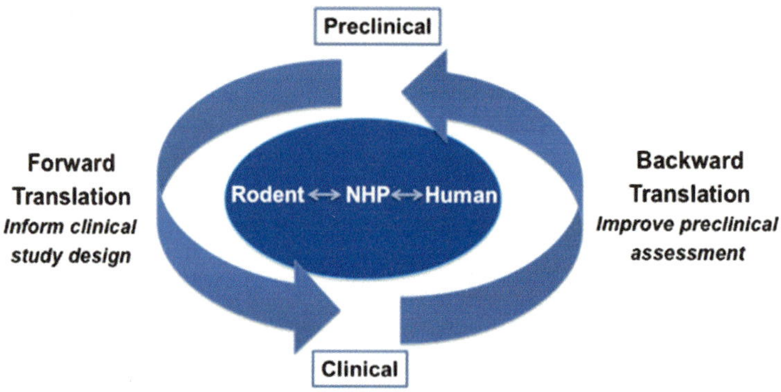

Fig. 4 Translational research across species. Developing an effective translational strategy from animals to humans informs both clinical and preclinical investigations. Preclinical animal investigations can be used to inform clinical study designs for dose selection, endpoint evaluation, and side effect identification (forward translation). Clinical results can be used to improve preclinical assessments, thereby refining the preclinical animal models to improve their sensitivity and validity (backward translation)

In 2002, an initiative known as MATRICS (Measurement and Treatment Research to Improve Cognition in Schizophrenia) was sponsored by the National Institute of Mental Health (NIMH) to assess and define the cognitive deficits in schizophrenic patients. Seven different cognitive domains were selected: attention; working memory; problem solving; processing speed; visual learning and memory; verbal learning and memory; and social cognition (Marder and Fenton 2004; Nuechterlein et al. 2004), and specific neuropsychological tests were suggested to investigate each domain in humans.

Table 1 Psychometric properties for paradigm selection

Relationship to functional outcome
Test-retest reliability
Practicality and tolerability
Pharmacological sensitivity
Utility as a repeated measure

Following the MATRICS, the CNTRICS (Cognitive Neuroscience Treatment Research to Improve Cognition in Schizophrenia) initiative was introduced with a focus of identifying tasks with construct validity across species to improve the translational capability of the test battery from rodents to NHPs and humans. Many psychometric variables important for the selection of paradigms with high translational validity were incorporated (Table 1) to enable sound data collection and to facilitate outcome assessment. Domains that were selected during the CNTRICS initiative were largely modeled from the MATRICS (with the obvious exclusion of verbal learning and memory), and included tests of perception, attention, executive function, working memory, object/relational long-term memory, and social/affective processes (Moore et al. 2013; Young et al. 2009a).

More recently, the NIMH has established the Research Domain Criteria (RDoC) project with a goal to "*Develop, for research purposes, new ways of classifying mental disorders based on dimensions of observable behavior and neurobiological measures.*" At its core, the RDoC is designed to be translational. Its framework uses a matrix design which groups a functional dimension of behavior based on genes, molecules, cells, circuits, physiology, behavior, and self-reports into "*Constructs*" and then into higher-level "*Domains*" that represent system functioning. "Cognitive Systems" is one of five defined Domains in the RDoC matrix and includes the constructs of attention, perception, declarative memory, language, cognitive control, and working memory. Units of analysis are defined to study each level of a construct (e.g., gene, molecules, circuit) for each of the domains. One of the strengths of RDoC is that it provides a framework with which to build and change the structure based on empirical findings over time.

A concerted effort is being made by research scientists and clinicians across academia, government, and the biotechnology-pharmaceutical industry to improve the translation of basic research findings into clinical application to improve human health. One example of this is the European Union Innovative Medicines Initiative (IMI) industrial-academic collaborative project, Novel Methods leading to New Medications in Depression and Schizophrenia (NEWMEDS). One of the major aims of this project was to develop and validate a battery of novel touchscreen-based translational assays of cognition in rodents. By systematically developing and validating paradigms with construct validity across species this goal seems attainable. This chapter will focus on some promising translational cognitive paradigms for use in rodents, NHPs, and humans.

2 Attention

Attention is critical for daily functioning and underlies effective processing of many cognitive domains. When discussing attention, it is important to recognize, it is commonly divided into three subtypes that include: (1) sustained attention, the ability to focus on a specific task for long durations; (2) divided attention, the ability to respond simultaneously to multiple tasks; and (3) selective attention, the ability to focus on certain stimuli while ignoring distractions (Parasuraman and Haxby 1993; Posner and Petersen 1990; Wilkins et al. 1987). Substantial evidence from imaging, lesion, and pharmacology studies suggests that attention is primarily mediated by the prefrontal cortex (PFC). For example, results from positron emission tomography (PET) and functional magnetic resonance imaging (fMRI) studies have demonstrated PFC activation in humans performing attention tasks regardless of which sensory domain the stimuli are presented (e.g., visual, somatosensory, auditory) (Cohen et al. 1988; Coull et al. 1996; Lewin et al. 1996; Pardo et al. 1991). Additionally, several studies have demonstrated that both rodents and patients with frontal lobe lesions are impaired on tasks measuring attention (Maddux and Holland 2011; Passetti et al. 2002; Wilkins et al. 1987).

More specifically, attention is sustained by the basal forebrain cholinergic system (BFCS), and lesions to portions of the BFCS, including the nucleus basalis of Meynert, which provides the major cholinergic innervation to the PFC, produce attention deficits in both rodents and monkeys (McGaughy and Sarter 1998; Mesulam and Geula 1988; Muir et al. 1995; Voytko et al. 1994). Additional evidence for the role of cholinergic control of attention has been demonstrated through pharmacological studies such as reversal of attention deficits in rodents and humans through administration of nicotine and cholinesterase inhibitors such as physostigmine and tacrine (Jones et al. 1992; Muir et al. 1995; Sahakian et al. 1989, 1993). Further, using in vivo microdialysis in rodents, a large increase in acetylcholine release within the PFC following exposure to an attention task has been measured (Dalley et al. 2001; Passetti et al. 2000).

There is also evidence from studies in animal models as well as humans that suggests a role for noradrenergic modulation of attention, with the locus coeruleus (LC) supplying the sole source of norepinephrine (NE) to the cortex, and NE depletion producing deficits in the performance of animals on a variety of attention tasks (Arnsten 1993; Berridge and Waterhouse 2003; Brozoski et al. 1979; Carli et al. 1983; Gamo et al. 2010). Conversely, stimulating the postsynaptic α2A NE receptors in the PFC, either through administration of NE or the selective α2A receptor agonist, guanfacine, increases neuronal PFC firing, and improves attention (Arnsten and Contant 1992; O'Neill et al. 2000; Wang et al. 2007). Consistent with these data, a recent study in aged monkeys demonstrated impaired performance on a sustained attention test that was improved following the acute administration of guanfacine (Decamp et al. 2011).

One of the most widely used tasks to measure sustained attention is the continuous performance test (CPT), which assesses maintenance of attention over long periods of time in response to infrequent, unpredictable, critical events

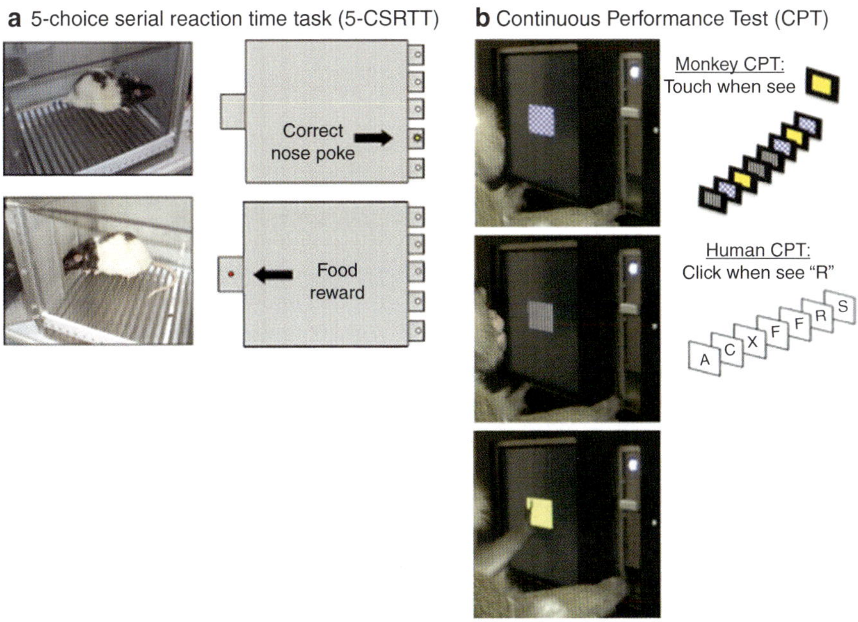

Fig. 5 Sustained attention. (**a**) In the 5-choice serial reaction time task (5-CSRTT), rats are trained to attend to five locations illuminated by a light at each location. In each trial, a different light is illuminated and the rat is given a small food reward for a correct nose poke in the location of the light. Mice also perform the 5-CSRTT. (**b**) In the continuous performance test (CPT), monkeys are presented with a stream of stimuli presented one after another on a touch-sensitive monitor. They are trained to ignore distractors (*top* and *middle panel*) and to respond to targets (*bottom panel*) to earn food rewards. Humans can also perform the CPT, though they are often trained to respond to letters or numbers rather than patterns or shapes

(Beck et al. 1956). First developed for use in humans, the CPT has been effectively translated to non-human primates and rodents (Fig. 5). The most basic version of the CPT requires subjects to continuously attend to a stream of stimuli, and to respond when they detect a previously defined "target" in order to earn a reward, while ignoring all other stimuli ("distractors"). The data are commonly analyzed for two kinds of correct responses, "hits" (correct response to a target) and "correct rejections" (correct withholding of a response to a distractor), and two kinds of incorrect responses, "misses" (incorrect withholding of a response to a target, also called "errors of omission") and "false alarms" (incorrect response to a distractor, also called "errors of commission"). Speed of response can also be measured by assessing reaction time for "hits." Though the basic premise of the CPT remains identical across primate species, colored shapes or symbols are often used when testing non-human primates, whereas letters or numbers are most often used when testing humans. Additionally, there are numerous variants of the CPT in humans, some of which make the task quite difficult, for example, by defining the "target" as a specific letter, but only when it follows a specific series of other letters (i.e.,

CPT-AX). The rapid visual information processing (RVIP) task is also similar to the CPT and has previously been used in both monkeys and humans.

The 5-choice serial reaction time task (5-CSRTT) was adapted from the human version of the CPT for use in rodents (Carli et al. 1983). Like the CPT, the 5-CSRTT also measures speed and accuracy of attentional processes. The 5-CSRTT is mainly used to assess sustained and spatially divided attention. However, selective attention may also be assessed by including irrelevant and distracting (auditory) stimuli during the intertrial interval while the animal has to attend to the five locations. In addition the task can evaluate premature and perseverative responses, which are measures of inhibitory control. Whereas monkeys and humans are tested on the CPT on computer screens, the 5-CSRTT is administered to rodents in a chamber with an array of five spatial locations with a light corresponding to each location. On each trial, one of the five lights is illuminated for a brief period of time (e.g., 100 ms), and the rodent is rewarded for nose-poking in the location of the corresponding light (Fig. 5). There is a wealth of literature on pharmacology and lesion studies in the 5-CSRTT (Lustig et al. 2013; Robbins 2002). The 5-CSRTT has also been successfully implemented on rodent touchscreens by presenting a white square briefly in one of five possible locations (Fitzgerald et al. 2014). Though attention is impaired in a number of neuropsychological and neuropsychiatric diseases, it is most notably affected in patients with attention deficit disorder (ADD) and attention deficit hyperactivity disorder (ADHD). A meta-analysis of children with ADHD determined that CPT performance was impaired relative to performance by age-matched controls, with the ADHD children making twice as many errors of commission and omission than controls (Losier et al. 1996). Administration of methylphenidate significantly decreased the rates of both types of errors on the CPT in adolescents with ADHD (Corbett and Constantine 2006) and ADD (Klorman et al. 1991). Similarly, the performance of rats originally trained to low baseline scores on the 5-CSRTT improved when methylphenidate was administered prior to testing (Navarra et al. 2008; Paine et al. 2007). Overall, the CPT is readily translatable between species, with a large literature of published data in animals and humans alike. There is an extensive amount of normative data (Robbins et al. 1994), and the CPT offers a quick, standardized, objective means of measuring attention processes.

Recently, Young and colleagues have developed the 5-choice continuous performance test (5C-CPT) in mice in which, as the name implies, combines the 5-CSRTT and the CPT (Young et al. 2009b). The 5C-CPT involves a chamber with multiple apertures that are randomly illuminated. The subject is required to respond to one aperture when it is illuminated, and to withhold responding when all five apertures are illuminated. This differs from the 5-CSRTT, since in the 5-CSRTT a rodent does not need to inhibit a response; therefore the 5C-CPT is similar to the human CPT. Thus far, sleep deprivation experiments have demonstrated the use of this task in both mice and humans, and potential future pharmacological and/or lesion experiments may affirm the usefulness of this new task as a translatable paradigm (van Enkhuizen et al. 2014).

In addition to sustained attention, as mentioned earlier, divided attention and selective attention are also assessed across species. Measuring divided attention is accomplished through dual-task performance tests where subjects are required to perform two or more tasks or attend to two or more items, simultaneously. Divided attention abilities are often contrasted to performance on selective attention tasks that reveal greater performance accuracy scores. Both types of attention are assessed primarily in humans using methods requiring the ability to identify letters and numbers. For example, Farran and colleagues used tasks with stimuli consisting of larger letters made up of smaller letters (Farran et al. 2003). In congruent trials, the large letter was drawn with several smaller letters of the same symbol (e.g., the letter "H" was drawn with many smaller "H"s). In incongruent trials, the large letter was drawn with several smaller letters of a different symbol (e.g., the letter "H" was drawn with many smaller "S"s). On each trial of the selective attention task, subjects were asked to identify either the little letters or the big letters. On each trial of the divided attention task, subjects were asked to indicate if they saw a target letter anywhere, regardless of whether it was the small letters or a big letter.

Some work on divided attention has also been done in monkeys and rodents, using clever tasks that do not require language abilities, but that do demand training prior to experimentation. For instance, O'Neill and colleagues trained monkeys to attend simultaneously to two spatial locations on a screen, marked by red circles, and the monkeys were to indicate when small, quickly moving objects entered one or both of the circles. Administration of choline mimetic compounds increased performance on this test of spatial divided attention, particularly in aged monkeys (O'Neill et al. 1999). Divided attention has also been assessed in rodents using cross-modal stimulus presentations (McGaughy et al. 1994). Rats were trained to perform operant conditional discrimination tasks with stimuli presented in the auditory and visual domains. When tested on a block of stimuli presented in one of the sensory domains, performance was better than when stimuli were randomly presented in either domain within a block of trials.

3 Working Memory

Working memory is the basis of abstract thought, and reflects the ability to represent salient information in the absence of sensory stimulation in order to effectively execute a task (Arnsten et al. 2012) (e.g., effectively remembering a telephone number before dialing in humans). The nature of working memory makes it subject to distractibility and interference, and it declines in conditions of stress and in psychiatric and neurological diseases (e.g., schizophrenia, Alzheimer's disease) (Barch and Ceaser 2012; Schroeter et al. 2012). Moreover, working memory underlies other cognitive processes, such as long-term memory and language, amongst others, which also can become impaired with age and disease (Arnsten et al. 2012; Bizon et al. 2012).

The lateral region of the prefrontal cortex (PFC) is critical for effective working memory function in primates (Mishkin 1957). More specifically, it is the persistent firing of a network of interconnected glutamatergic pyramidal neurons (also known as "delay cells") within layer III of the dorsolateral (dl) region of the PFC that subserves visuospatial working memory (Goldman-Rakic 1995, 1996; Wang et al. 2011). Activation of delay cells enables continuous updating of information to maintain and flexibly manipulate information that is no longer present in the environment (i.e., the basis of working memory). Delay cells communicate with response cells, which are activated around a motor response. Whereas delay cells are present in primate dlPFC, the rodent PFC is described to exhibit more response-like cells suggesting that direct comparisons between species should be done cautiously (Caetano et al. 2012).

To assess working memory function in humans and animals, delay tasks are often employed [e.g., delayed alternation (DA), delayed match-to-sample (DMS/DMTS) task]. The basis of these tasks is that information about a previous trial needs to be stored temporarily "online" during an interim period in which the stimulus is absent, and then accessed subsequently during the response period. Longer delays increase the difficulty of the trial, and can lead to decreased performance accuracy.

The DA task has been used successfully in rodents and in monkeys. In the monkey version, animals are presented with two cups separated on a sliding tray in which one of the cups is baited with a food reward. The animals are trained to select one cup to obtain a reward and then are to alternate between the two cups following a delay period ($\geq$5 s) during each trial. In 1957, Mishkin used the DA task to demonstrate the importance of the mid-lateral frontal cortex in delayed-response tasks by lesioning this region in rhesus macaques and subsequently measuring trials to criterion (Mishkin 1957). In this study, half of the animals with mid-lateral lesions did not reach criterion even after completing 1,000 trials, whereas animals with lesions in other areas of the cortex not involved in working memory processes performed the task with 90 % accuracy after less than 300 trials. Subsequent studies have confirmed these results in NHPs following functional depression of the dlPFC (Alexander and Goldman 1978) and in humans with focal lesions in the dlPFC (Brodmann's area 46) (Freedman and Oscar-Berman 1986).

In the rodent the DA task is more commonly a T-maze apparatus, in which one arm of the maze is baited with a food reward and the mouse or rat needs to select the alternate arm from the previous trial to receive a reward. This task exploits the rodents' inherent tendency to alternate arm choices on consecutive trials (Tolman 1945). Although it is not fully understood why rodents alternate on T-mazes (e.g., implicit directional sense, exploration strategy, odor cues), this task has been used extensively to study working memory. The alternation performance is particularly sensitive to hippocampal disruption in rodents and lesioned animals will perform at chance level following relatively short delays (e.g., 15 s) (Dudchenko et al. 2000).

Other tasks used to test working memory in the rodent include the 8-arm radial arm maze (RAM), in which food rewards are placed at the ends of each arm. Rats tend to visit each arm once before returning to a previously visited arm (re-baited in between trials), suggesting the engagement of working memory processes (e.g., remembering which arm they had visited earlier) (Olton and Papas 1979).

Subsequently, introducing delays in between arm entries results in delay-dependent increases in errors in rodents (Bolhuis et al. 1986) similar to results obtained in NHP and human delay tasks. The Morris water maze (MWM) is another commonly used task in rodents to assess working memory, in which the location of a platform submerged in water is switched between sessions on a daily basis. As the animals learn where the platform is using spatial cues, the latency to reach the platform decreases (Morris et al. 1986). Although the MWM and the RAM tasks used in rodents test the working memory construct and enable comparison between species (e.g., NHPs, humans), the tasks themselves traditionally have not been used in NHPs and humans. In addition, it should be noted that rodent working memory tasks, which use long delay intervals, are actually measuring short-term spatial or visual-spatial memory and not the working memory domain as defined for humans.

The DMS task is a test of working memory readily translatable between species. This task occurs in two phases, first a sample phase in which a subject is presented with a stimulus, and then following a variable delay period (e.g., seconds to minutes), the choice phase, in which the subject is reexposed to the original

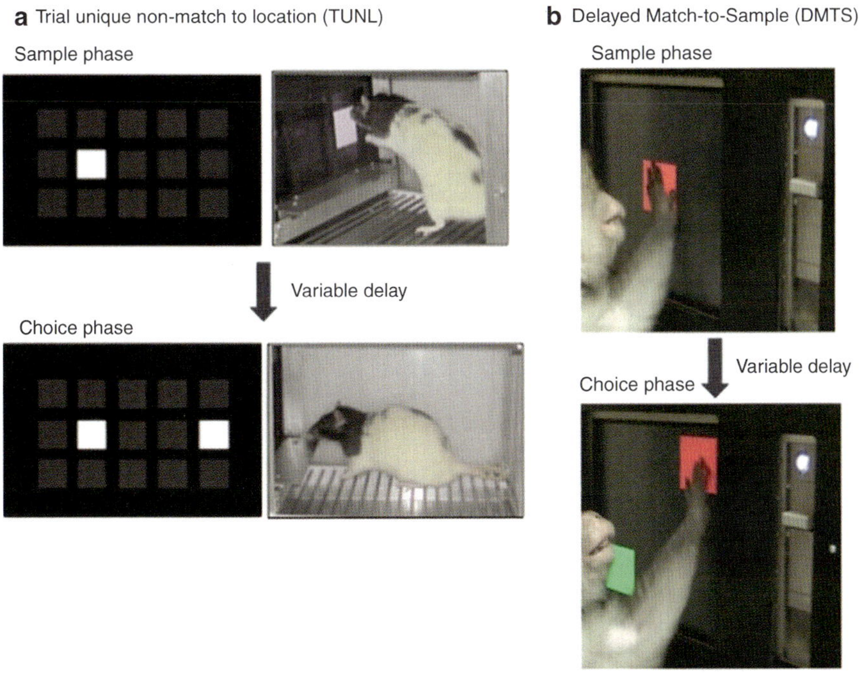

Fig. 6 Working memory. (**a**) In the trial-unique delayed non-matching-to-location (TUNL) task, rats are trained to press a *white square* on the touchscreen during the sample phase and then have to remember this location during the choice phase. The rat is given a small food reward for a response to the correct novel location. Mice can also perform the TUNL task. (**b**) In the delayed match-to-sample (DMTS) task, monkeys are trained to touch a sample stimulus and following a variable delay are rewarded for choosing the same stimulus they were exposed to prior to the delay

stimulus along with the introduction of a distractor stimulus (Fig. 6). The subject is required to match the stimulus that was originally presented in the sample phase to receive positive reinforcement (e.g., food reward for rodents, NHPs). The repeated use of stimuli for all trials ensures the involvement of working memory and not recognition memory processes, the latter of which is engaged when new stimuli are introduced on a per trial basis (Mishkin and Delacour 1975). In a close variant of the DMS task, delayed non-match-to-sample (DNMS) task, the subject is required to select the stimulus that was not presented in the original sample phase. In either version, as the delay increases, so does the task difficulty leading to decreased performance.

The use of visual touchscreen-based working memory procedures such as DMS/DNMS used in humans has been translated successfully to NHPs (Weed et al. 1999). The use of touchscreen tests across species allows for standardization of testing approach, stimuli, and conditions and minimizes experimenter involvement and potential bias. Although some species differences can be present in terms of increased task complexity needed to observe impairments in humans as compared to NHPs (e.g., delay length). In addition, delay tasks to assess working memory may not be suitable for all disease indications. For example, in schizophrenia working memory deficits are prominent; however they do not appear to be delay dependent (Lee and Park 2005). Recent suggestions from the CNTRICS consortium suggested goal maintenance and interference control would be features of working memory that have translational validity and are sensitive for use in schizophrenic patients (Barch and Smith 2008).

More recently, the use of touchscreen working memory tasks has been introduced to rodents with the trial unique non-match-to-location (TUNL) paradigm (Oomen et al. 2013; Talpos et al. 2010). The spatial nature of the TUNL task (Fig. 6) makes it hippocampal dependent; however, the hippocampus likely interacts with the medial prefrontal cortex (mPFC), since lesions of the mPFC have been shown to induce a delay-dependent impairment (McAllister et al. 2013). Traditionally, lever-based operant delayed non-match- or match-to-position paradigms have been plagued with questions of whether rodents are using mediating strategies to bridge the delay period to enable accurate responding instead of employing working memory processes. Many studies have shown that rodents position their bodies to predict the correct location for their response, and even after attempts to minimize this potential (e.g., required nose pokes in between sample and choice phases), biased body position may still be a factor (Chudasama and Muir 1997; Dudchenko and Sarter 1992; Gutnikov et al. 1994). In the TUNL approach, multiple spatial locations are used making it difficult for the rat to predict the correct response, and extensive testing to assess mediating potential has suggested it is unlikely that rats are using this strategy. Although progress has been made, further work is needed for a translational working memory task in rodents, which is mediated by the prefrontal cortex and satisfies the criteria as closely as possible for working memory assessed in humans.

4 Visual-Spatial Learning

Visual-spatial learning is a form of declarative memory that is dependent on the medial temporal lobe. One of the most commonly used paradigms to assess visual-spatial learning across species is the visual-spatial paired-associates learning (vsPAL) task. vsPAL requires subjects to associate a set of objects with particular spatial locations on a trial-by-trial basis and is dependent on the hippocampus in particular (Murray et al. 1993; Owen et al. 1995; Talpos et al. 2009; Wood et al. 2002). Deficits in vsPAL performance have been reported in aging and disease (Rabbitt and Lowe 2000; Robbins et al. 1994). In particular, impaired performance in vsPAL has been demonstrated in patients with mild cognitive impairment who are later diagnosed with Alzheimer's disease (Blackwell et al. 2004; Sahakian et al. 1988; Swainson et al. 2001). Similarly, schizophrenic patients (first episode and following chronic disease) show impaired learning in vsPAL (Barnett et al. 2005; Wood et al. 2002). The sensitivity of the vsPAL task in identifying cognitive impairments associated with neurological and psychiatric disease has enhanced the utility of this paradigm in clinical settings and has driven the need for translation in preclinical species.

Subsequent to its development and use in humans, the vsPAL task has been developed for NHPs, rats and mice using touchscreen-based computerized systems (Bartko et al. 2011; Owen et al. 1995; Spinelli et al. 2005; Taffe et al. 2002; Talpos et al. 2009) (Fig. 7). In humans and NHPs, vsPAL requires subjects to learn to associate a specific stimulus with a particular location on a trial-by-trial basis. The task difficulty increases with the number of stimulus-locations presented, and memory load can vary within a session to allow performance-based assessment at different demands (Taffe et al. 2002). Typically, NHPs have up to 4 stimuli/4 locations in their difficult trials (although this can be higher), whereas humans can generally accommodate more stimulus-location pairs (e.g., up to eight) before the task becomes too difficult and performance decreases (Fowler et al. 1997; Swainson et al. 2001; Taffe et al. 2002).

In the human version of vsPAL, subjects are given verbal/written instructions prior to testing; thus minimal training is needed. The sample phase of vsPAL involves the presentation of a series of identical boxes located around the periphery of a touchscreen monitor (Fig. 7). Sequentially, each box will open to reveal a unique stimulus in a specific location until all the boxes have shown their contents. Following a brief delay, the choice phase begins in which one sample stimulus appears in the center of the screen and the subject is asked to match the stimulus to the location (correct box) in which it was originally presented. Subjects go through each of the sample stimuli until they have responded to them all, and may or may not be rewarded for correct responses.

In the NHP version of vsPAL, animals must acquire the rules through a trial-and-error approach prior to testing; thus training can take many months and modifications to the task may occur as training progresses. During the sample presentation, NHPs learn to touch the sample stimulus as a measure of having attended to it prior to moving to the test phase (Fig. 7). In the vsPAL,

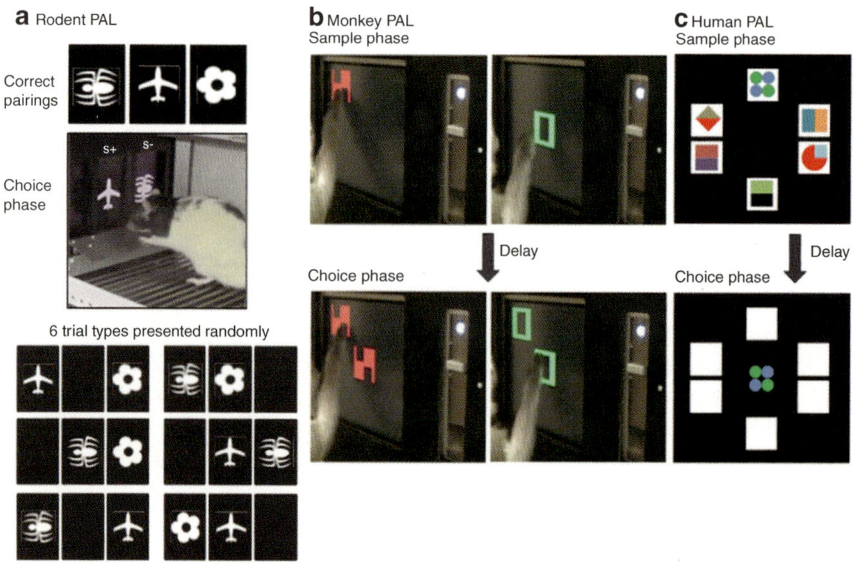

Fig. 7 Visual-spatial memory. (**a**) In the paired associates learning (PAL) task, rats learn which location is associated with each stimulus. Two stimuli are presented at once in three possible locations (i.e., six possible stimulus-location options), with one stimulus in the correct location (S+) and the other stimulus in an incorrect location (S−). The animal is required to select the correct stimulus-location to receive a food reward. Mice can also perform the PAL task. (**b**) In the monkey version of the PAL task, monkeys are presented with a series of stimuli, each in a different location on a touch-sensitive monitor. Following a short delay, each stimulus is presented again, alongside an identical stimulus in a distracter location. The monkey is rewarded for choosing the stimulus in the location that it was originally presented. (**c**) In the human version of the PAL task, subjects are presented with an array of boxes and "open" each box to reveal a unique stimulus in a sample phase. Following a delay, one stimulus is presented in the middle of the array of boxes, and the subjects are required to match the location at which that stimulus was presented in the sample phase

stimulus-locations are randomized (9 locations; 20 stimuli) to avoid memorization, and once animals are trained to a particular criterion, the performance generally remains stable over time as long as the animals are being tested routinely.

In the rodent version of vsPAL, animals learn to associate a particular stimulus with a particular location over an extended training period (Horner et al. 2013). Many of the same approaches in both rats and mice are used, as the mouse version was modeled after the rat. In the rodent vsPAL, animals must learn which location is associated with each stimulus (Fig. 7). Two stimuli are presented at once in three possible locations (i.e., six possible stimulus-location options), with one stimulus being in the correct location (S+) and the other stimulus in an incorrect location (S−). The animal is required to select the correct stimulus-location to receive a food reward (Bartko et al. 2011; Talpos et al. 2009). After performing at a set criterion consistently during the training phase, the animals are given a fixed number of trials within each session; for mice it is generally 36 trials, and for rats it is generally

72 trials. Memory load remains constant throughout the test in the rodent version of vsPAL, but can be manipulated pharmacologically with direct manipulations of the dorsal hippocampus (e.g., lidocaine) (Talpos et al. 2009). Given the length of time and the number of trials across sessions, the rodent version of vsPAL may assess more long-term association formation as compared to the within-session changes of humans and monkeys, which assesses more recent association formation.

5 Recognition Memory

Recognition memory in humans is a subtype of declarative memory and refers to the ability to determine whether or not a stimulus has been previously encountered. A large body of literature, with its beginnings in the 1960s, has identified that recognition memory is critically dependent on medial temporal lobe cortical areas, specifically the perirhinal and entorhinal cortices, and the hippocampus. For instance, neuroimaging studies have demonstrated hippocampal activation during recognition memory tasks in healthy volunteers (Cohen et al. 1999), and lesion studies in animals and patients with specific brain damage have shed more light on this finding. Patients with temporal lobe damage have demonstrated impairments in recognition memory, regardless of whether the damage is extensive (McKee and Squire 1993; Pascalis et al. 2004) or selectively confined to the CA1 region of the hippocampus (Zola-Morgan et al. 1986). Data from several studies in monkeys and rodents have corroborated this finding and have demonstrated that damage specifically to the perirhinal and entorhinal cortices also produces recognition memory impairments (Hadfield et al. 2003; Leonard et al. 1995; Meunier et al. 1993; Suzuki et al. 1993), as can selective damage to the hippocampus (Nemanic et al. 2004; Pascalis and Bachevalier 1999; Zola et al. 2000). Rodent studies have been useful to precisely define which brain regions are required for specific processes. It has been shown that the perirhinal cortex is essential for object recognition memory (Brown et al. 2012), but when tasks require an association of object with contextual or spatial information, then integration of the hippocampus with the perirhinal cortex occurs (Winters et al. 2010b).

Recognition memory tasks in rodents are widely used within many laboratories and are referred to as the object recognition test (ORT) or novel object recognition (NOR) or novel object discrimination (NOD). The "one-trial learning" test which is predominantly used was originally described by Ennaceur and Delacour (Ennaceur and Delacour 1988). The main reason for the popularity of this test is likely due to the minimal equipment required and the ease of running the protocol. Due to the spontaneous nature of the task, no reward or punishment is required. Moreover, most tests require one-trial learning and so lengthy training is not needed, in contrast to operant-based tasks already described in this chapter. In addition, it is possible to determine the effects of compounds on different stages such as encoding, consolidation, and retrieval, which is not always feasible with operant-based tasks. There have been a number of recent reviews that describe variants of the task, including species and gender differences, methodology, apparatus, and

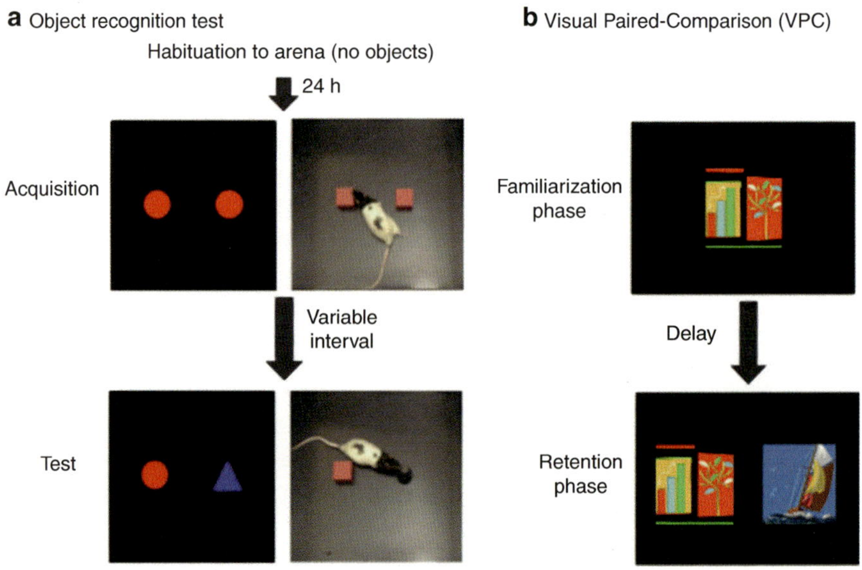

Fig. 8 Recognition memory. (**a**) Rodents are habituated to the test chamber at least once 24 h prior to testing. During acquisition, subjects are presented with two identical objects and the time spent exploring each is recorded. After a variable interval, subjects are returned to the arena and presented with the familiar object and a novel object in the same locations as the acquisition trial. Since there is a natural tendency to explore novel objects, if the rodent remembers the familiar object, then it will spend more time exploring the novel object; (**b**) In the visual paired-comparison (VPC) task, monkeys, or humans are exposed to a stimulus during a familiarization phase. Following a delay, the same stimulus and a novel one are presented side by side. Since both monkeys and humans have a natural preference for novelty, if the subject remembers the familiar stimulus, they will spend more time looking at the novel stimulus

objects (Antunes and Biala 2012; Lyon et al. 2012; van Goethem et al. 2012). The purpose of the recognition memory task in rodents is to assess whether subjects can remember a previously presented object. Usually the test is comprised of a habituation period, acquisition, and test period (Fig. 8). Rodents are habituated to the test chamber at least once 24 h prior to the test. During acquisition, subjects are usually presented with two identical objects and the time spent exploring each is recorded. After an interval, subjects are returned to the arena and presented with a duplicate of the previously explored object and a novel object in the same locations as the acquisition period. Since there is a natural tendency to explore novel objects, if the rodent remembers the familiar object, then it will spend more time exploring the novel object. Object recognition in rodents is sensitive to pharmacological effects modulating dopaminergic, cholinergic, serotonergic, and glutamatergic systems (Lyon et al. 2012). A number of studies have also determined the effects of transgenic manipulations on object recognition. Recognition memory deficits have been found in different transgenic mouse models of Alzheimer's disease; however, the temporal progression differs across the mouse lines and so care must be taken to select the appropriate model and time-point when assessing drug effects (Webster et al. 2014). With regard to recognition memory in rodents, direct translation to NHP and human must be taken with some caution, since in the rodent paradigm it is possible that rodents may not only use visual information to perform the task.

In humans, word and/or picture recognition tasks are primarily employed to test recognition memory. In these tests, subjects are presented with a list of words or a series of pictures, and at a later time, are asked to identify which words or pictures in a new list they had seen previously. While these methods are used extensively in clinical trials, they do not lend themselves well to translating to non-human species because of the critical language component. In primates (i.e., monkey and human), the visual paired-comparison (VPC) paradigm can be used to test recognition memory. This task takes advantage of the subject's innate preference to view novel stimuli, and does not require the subject to learn any rules associated with the task to earn a reward. In the VPC task, recognition memory is measured by the subject's preference to look longer at novel stimuli than to stimuli they have seen a few seconds or minutes earlier. Each trial of a VPC test is divided into a familiarization phase, a delay, and a retention phase (Fig. 8). During the familiarization phase, a sample stimulus is displayed in the center of a screen until the subject looks at it for a predetermined amount of time, after which a delay period ensues. The retention phase follows, during which two images are displayed side by side in a pseudorandom, counterbalanced location; one of the images is the same as the sample stimulus seen in the familiarization phase, and the other image is a novel stimulus. The dependent measure tends to be the total amount of time spent looking at each of the images. Recognition memory is thought to be intact if the subject spends a significantly larger amount of time viewing the novel as opposed to the familiar stimulus. Humans can use both explicit episodic recollection and implicit familiarity, whereas it is thought that rodents predominantly use the latter process.

By manipulating one or many of the task parameters, recognition memory can be enhanced or impaired in humans (Richmond et al. 2004), monkeys (Zeamer et al. 2011), and rodents (Antunes and Biala 2012). The parameters that can be manipulated include the time allowed for familiarization, the length of the delay between the familiarization and retention phases, and the discriminability of the stimuli. Decreasing the length of the familiarization phase, increasing the delay, or increasing the similarity between the stimuli will impair recognition memory, whereas increasing the length of the familiarization phase, decreasing the delay, or decreasing the similarity between the stimuli will enhance recognition memory. There is wide support in the literature demonstrating the applicability of the VPC paradigm to disease state, particularly for Alzheimer's disease. For example, the VPC test has been demonstrated to distinguish those subjects who are aging normally from those with mild cognitive impairment and Alzheimer's disease. Most recently, Zola et al. (2013) reported that scores on the VPC paradigm not only reveal present disease state, but are also predictive of disease progression (Zola et al. 2013). This was demonstrated in subjects who were recruited to participate in the study regardless of whether they were controls, patients with mild cognitive impairment, or patients with Alzheimer's disease. The fact that this paradigm can be used across species (NHPs and humans) and that it is sensitive to disease state suggests it lends itself well to translational research, especially for investigation of novel therapeutics. Rodents have been shown to discriminate visual stimuli on a touchscreen, and acquisition of this task is impaired by perirhinal cortex infusion of $GABA_A$ agonist or muscarinic and NMDA antagonists (Winters et al. 2010a); however, so far it has not been possible to train rodents in a paradigm similar to VPC.

6 Emotional Memory

Acquisition and extinction of fear conditioning require associative learning and have been studied in both animals and humans to understand the neurobiology of emotional memory, as well as to understand the processes underlying disorders such as posttraumatic stress disorder (PTSD). Memory for emotionally arousing stimuli associated with threat or danger is extremely important since it ensures that the individual can learn to predict danger and adapt their behavioral response accordingly. However, repeated exposure to traumatic events may lead to strong emotional memories of the events resulting in PTSD. Patients with PTSD suffer from the occurrence of intrusive memories which are so strong that it is not always possible to dissociate current from past events.

Rodents respond to danger in a species-specific manner by freezing, (i.e., animals will withhold all movement except for respiration), in order to avoid detection. This behavioral response can be utilized to assess associative learning and memory. In fear conditioning experiments during the acquisition period, animals are placed into a chamber and presented with pairings of a conditioned stimulus (CS), such as tone or light cue, with an unconditioned stimulus (US) that is usually a mild foot shock applied through the grid floor (Fig. 9). Animals learn to associate the CS with the US and when animals are placed into a novel chamber and presented with the CS in the absence of the US, this results in the expression of increased freezing [i.e., conditioned response (CR)], which is usually expressed as percent time freezing. If animals are returned to the same chamber where they were previously exposed to the US, then they show increased freezing in this context. The former association is known as cued fear conditioning and the latter is context fear conditioning. Extinction occurs when the CS or context is repeatedly presented in the absence of the US and consequently the CR, (i.e., freezing behavior), diminishes over time. However, studies have shown that the association between the CS and US is not completely abolished because reinstatement can occur later in response to presentation of the CS. Therefore the general consensus is that extinction learning represents new learning during which a new CS-no US memory competes with and inhibits the existing CS-US memory (Bouton 2004). Fear extinction in animals is considered to be a model of exposure therapy in humans, where the fear-provoking stimulus is presented repeatedly in the absence of harm, so that the patient ideally learns to fear it less. Deficits in fear extinction are thought to contribute to PTSD.

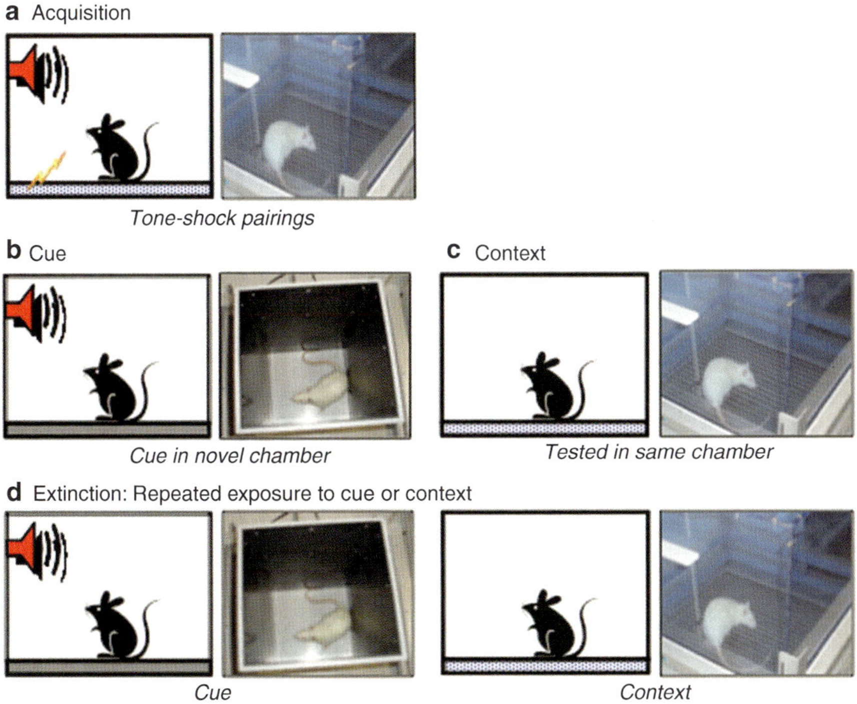

Fig. 9 Fear conditioning and extinction. (**a**) During the acquisition period rodents are placed into a chamber and presented with pairings of a conditioned stimulus (CS), such as tone, with an unconditioned stimulus (US) usually a mild foot shock applied through the grid floor. (**b**) Rodents learn to associate the CS with the US so that when animals are placed into a novel chamber and presented with the CS in the absence of the US, this results in the expression of freezing behavior, i.e., conditioned response (CR). (**c**) If animals are returned to the same chamber where they were previously exposed to the US, they then show freezing behavior in this context. (**d**) Extinction occurs when the CS or context is repeatedly presented in the absence of the US and consequently the CR, (i.e., freezing behavior), diminishes over time

The direct translation of rodent studies has helped to define human paradigms for use in neuroimaging and thus provide further understanding of the underlying circuitry of PTSD (Delgado et al. 2006; Milad et al. 2006; Parsons and Ressler 2013). Note that NHP studies are lacking in the literature, possibly due to the difficulty in setting up a conditioning paradigm without utilizing electric shock as the US. However, there is a recent study describing a novel conditioning paradigm which successfully uses an air puff as the US (Kazama et al. 2013). In human fear conditioning experiments, the CR is usually a psychophysiological response of the sympathetic nervous system in response to fear or perceived threat, which includes heart rate response, electromyographic (EMG) response, or skin conductance response (SCR). SCR is the most frequently used readout and measures perspiration-induced electrical conductance or moisture level of the skin, usually on fingers, palms, or feet. Most studies use a wrist or finger shock as the US and compare response to a conditioned cue (CS+) with a nonconditioned or extinguished cue (CS−). During acquisition of fear conditioning, SCR is higher in response to a CS+ compared to a CS−. This approach can be used in neuroimaging studies and thus aid in defining brain regions required for fear conditioning and extinction in humans (Parsons and Ressler 2013; VanElzakker et al. 2014).

The neural network underlying fear conditioning and extinction has been well characterized in animals and humans, and knowledge has been derived from lesions, pharmacology, genetically modified mice, electrophysiology, and neuroimaging (Fanselow and LeDoux 1999; Fanselow and Poulos 2005; Maren 2001). The main brain region required for acquisition and expression of fear conditioning is the amygdala (lateral and basolateral), which receives sensory input related to CS information for contextual features from the hippocampus and entorhinal and perirhinal cortex and for auditory cues from thalamus (Fanselow and Poulos 2005). The amygdala also has reciprocal projections back to the cortical regions. Output from the amygdala to the brainstem, hypothalamus, periaqueductal gray, and ventral striatum results in reflex modulation, autonomic arousal, stress hormones, freezing, analgesia, and instrumental behavior, respectively. Regarding extinction, there is evidence that the prefrontal cortex is required to modulate this process (Quirk et al. 2006). Patients suffering from PTSD exhibit impaired extinction processes and dysfunctional activation of the fear extinction network, i.e., diminished activity in ventral medial prefrontal cortex, rostral anterior cingulate cortex, hippocampus, and heightened amygdala activation (Shin and Liberzon 2010; VanElzakker et al. 2014). Animal studies have shown that it is possible to modulate extinction by either pharmacological approaches or behavioral methods (Fitzgerald et al. 2014; Kaplan and Moore 2011; Parsons and Ressler 2013). A recent review has stated that over the last 10 years there has been much progress on translational research of fear extinction due to a high degree of coordination between rodent and human researchers (Milad and Quirk 2012).

7 Translatable Biomarkers

In addition to assessing cognition through behavioral measures, investigation of cognitive function through neurophysiological methods and imaging techniques can complement and/or provide additional information to behavioral approaches.

7.1 Electroencephalography

Electroencephalography (EEG) refers to the physiological technique used to examine electrical activity within the brain in response to synchronous activation of pyramidal cells within cortical neural networks. Engagement of these networks occurs through incoming sensory information or cognitive processing or under resting conditions and has been recorded in species such as mice, rats, dogs, and monkeys, and of course, in humans.

Whereas the EEG technique itself lends itself to cross-species utilization, the method in which it is collected may differ. In humans, EEG measurements are generally recorded by multiple electrodes placed on the scalp, and whereas some EEG studies in monkeys have employed scalp electrodes either through specialized skull caps the animals wear (Gil-da-Costa et al. 2013), or by using sharp electrodes placed in the epidermis of the scalp, the vast majority of studies in monkeys and lower species employ electrodes implanted on the skull, the surface of the brain, or within a neural substrate (generally referred to as local field potentials in this case). EEG confers an advantage to many other in vivo tools used to assess neural function (e.g., functional magnetic resonance imaging; fMRI) because of its pronounced temporal resolution (<1 ms) allowing direct assessment of electrical activity in the brain within a defined time frame (Jutzeler et al. 2011). However, depending on where the electrodes are placed (e.g., scalp, skull, brain surface), EEG can lack precise spatial resolution (e.g., >1 mm) due to interference from these substrates. Nevertheless, alterations in neuronal oscillatory activity as measured by EEG underlie many neuropsychiatric and neurological conditions making EEG an important tool for monitoring brain function.

The use of quantitative EEG (qEEG) coupled to spectral frequency analysis has been invaluable for understanding brain function under normal and abnormal conditions, and is used routinely both in preclinical species (mice, rats, monkeys) and in humans. EEG has been traditionally evaluated within specific frequency ranges: delta (0–4 Hz), theta (4–8 Hz), alpha (8–12 Hz), beta (12–30 Hz), and gamma (>30 Hz), and each range is associated with specific brain functions. Oscillations within the delta, theta, alpha, and beta bands are involved in long-range synchronization between cortico-cortical structures, and the oscillations within the gamma band are involved in synchronizations within local circuitry. For instance, it is thought that oscillations in the theta range are involved in perception, learning, and memory (Huerta and Lisman 1993), while oscillations within the alpha frequency band are involved in response inhibition and attention (Uhlhaas and Singer 2010). Beta oscillations have also been implicated in attention, as well as sensory gating, while gamma oscillations have been associated with perception, selective attention, encoding, and retrieval of memories and are being investigated as a translational biomarker for neuropsychiatric diseases such as schizophrenia (Gandal et al. 2012).

In addition to the use of qEEG to investigate cognitive function, the use of event-related potentials (ERPs) assessed by EEG as neurophysiological indices of sensory and cognitive processing has been studied for decades. Human ERPs consist of P50, N100, P200, P300, and mismatch negativity (MMN). Homologous MMN and P300 have been reported between macaques and humans (Gil-da-Costa et al. 2013; Javitt et al. 1992), whereas the timescale for ERPs in mice is approximately 40 % of the human and is represented as the P20, N40, P80, and P120 (Siegel et al. 2003). The decreased latency in rodents may reflect differences in the cortical structure between rodents and humans as well as the smaller relative brain size, although other differences also may contribute. For example, early sensory gating responses in the rat are largely mediated by the hippocampus and are disrupted by certain changes in stimulus parameters (e.g., sound, but not delivery frequency), whereas in humans the hippocampus plays a role in sensory gating at later stages and gating is disrupted by any change to the sensory stimulus (Boutros et al. 1995, 1997; Grunwald et al. 2003).

Overall, alterations in ERPs and selective frequency bands of the EEG have proven to be useful tools to characterize and study disease states, and potential treatment effects in humans. The EEG can be successfully adapted for use in lower species (e.g., rat, NHP) during wake to study cognition with appreciation of some of the aforementioned species differences.

7.2 Eye-Tracking

Eye-tracking is a useful tool for translational research because it can reveal information about the brain that may otherwise remain elusive. The noninvasive manner in which eye-tracking can be employed (e.g., Machado and Nelson 2011), along with the fact that it is independent of language ability, allows the same techniques and equipment to be used across species (e.g., humans, NHPs). These characteristics also allow for translation across developmental stages (i.e., infancy through adulthood) and levels of intelligence.

Some of the many parameters that can be measured with eye-tracking include gaze patterns, saccadic movement, pupil fixation duration and frequency, and pupil dilation. The translatable nature of eye-tracking capitalizes on the fact that both humans and non-human primates rely almost exclusively on vision for nonverbal social communication, demonstrate conserved visual processing strategies revealed as similarities in these measures of eye gaze behavior while observing stimuli (Dahl et al. 2009), and also demonstrate high test–retest reliability (Farzin et al. 2011).

These measures also have demonstrated sensitivity to disease states with known social cognition deficits (e.g., autism) for which abnormal eye gaze patterns are a hallmark feature of the disorder. In fact, when shown photographs of human faces, patients with autism spend less time attending to the faces, and in particular, less time gazing at the eye region of the faces, (Riby and Hancock 2009) than healthy, age-matched controls (Klin et al. 2002; Neumann et al. 2006). Patients with autism also show face-specific recognition deficits (Bradshaw et al. 2011) that are well documented and are speculated to contribute to the social deficits characterized as one of the core deficits of autism in the DSM-IV.

As deficits in social interactions are one of three core deficits that characterize ASD, much consideration has been given to ways to improve social behavior as a potential therapy. To this end, oxytocin, a neurohormone associated with prosocial behavior, has received a great deal of attention (for review Green and Hollander 2010). Of particular interest, intranasal administration of oxytocin to patients with autism increased gaze to the eye region while viewing faces, and was associated functionally with enhanced social behavior (e.g., feelings of trust) (Andari et al. 2010). Similar effects have been identified in healthy volunteers following oxytocin administration, with oxytocin increasing gaze to the eye region while viewing faces (Guastella et al. 2008; Kosfeld et al. 2005). Interestingly, intranasal administration of oxytocin to monkeys also increases gaze to the eye region while viewing faces (Dal Monte et al. 2014), promotes prosocial decisions, and affects visual social orienting (Chang et al. 2011; Ebitz et al. 2013). Taken together, these data suggest pharmacological intervention can improve complex social behaviors and can be monitored through eye gaze patterns as an index of function.

Eye-tracking is a valuable translational approach to assess social cognition for diseases in which social impairments are prevalent. The noninvasive nature of the technology and the ability to measure eye gaze parameters across species, developmental stages, intelligence levels, and language abilities make it a valuable approach to strengthen accurate, earlier diagnosis (Riby and Hancock 2009) and in turn open a larger window of opportunity to treat patients, identify novel therapies, and evaluate the effectiveness of potential therapeutics.

7.3 Functional Imaging: fMRI and In Vivo Oxygen Amperometry

Increases in neuronal activity are accompanied by changes in cerebral blood flow, blood volume, and oxygenation and these changes can be detected using imaging techniques such as blood oxygen level-dependent (BOLD) functional magnetic resonance imaging (fMRI). fMRI can measure the BOLD signal throughout the brain to determine regional brain activity. In humans this method can be used to determine which brain regions are required for specific aspects of cognition during neuropsychological task performance in the scanner. Unfortunately, it is not feasible to determine the relationship between brain region activity and function in rodents since sedation or anesthesia is required to minimize motor movements and so alternative procedures have been developed. An interesting approach is in vivo oxygen amperometry which detects the presence of oxygen in the extracellular fluid in real time in freely moving animals and has been shown to correlate with the fMRI BOLD signal in rats when modulating inspired oxygen levels (Lowry et al. 2010) and can be modified by external factors such as stress and tail pinch (Kealy et al. 2013). Since this method records oxygen levels in specific brain regions longitudinally in freely moving animals, the changes in activity over time can be correlated with ongoing behavior. To determine whether this method would have translational relevance, a study was undertaken to compare nucleus accumbens activation during a reward processing task in rats with data from human fMRI studies using a monetary incentive delay task (Francois et al. 2012). It was demonstrated that the nucleus accumbens was activated during anticipation of reward and this could be modulated by varying the magnitude and/or motivational incentive value of the reward, which was very similar to the human data. The limitation of this technique is that the sensors have to be implanted into a specific brain region and so it is not possible to look at the global effect as in human fMRI studies. However, the method can be extremely useful to probe the function of specific regions during a particular behavior. Moreover, recent work has shown that it is feasible to implant sensors in two brain regions and record simultaneously, for instance, nucleus accumbens and infralimbic cortex during a reward-based learning task (Francois et al. 2014) and dorsal and ventral hippocampus during anxiety and spatial tasks (McHugh et al. 2011).

8 Conclusions

In this chapter, we have provided an overview of several neuropsychological behavioral paradigms with construct validity across cognitive domains and translational relevance across species. In addition, we have highlighted some neurophysiological and imaging techniques that are used routinely to assess cognitive function. For some of the paradigms discussed, there is evidence of strong translational value from rat to NHP to human (e.g., attention tasks). For other paradigms, the translational assessment may be limited to two species (e.g., fear extinction from rodents to humans; or, VPC from NHPs to humans). As mentioned in the Introduction of this chapter, improvement in the area of translational neuroscience is an ongoing process by research scientists and clinicians across academia, government, and the biotechnology/pharmaceutical industry. A continued commitment to the integration of knowledge from animal and human work to strengthen the effectiveness of translational paradigms in the cognitive neurosciences has promise to improve the treatment options for patients with cognitive-impairing disorders.

References

Alexander GE, Goldman PS (1978) Functional development of the dorsolateral prefrontal cortex: an analysis utlizing reversible cryogenic depression. Brain Res 143:233–249

Andari E, Duhamel JR, Zalla T, Herbrecht E, Leboyer M, Sirigu A (2010) Promoting social behavior with oxytocin in high-functioning autism spectrum disorders. Proc Natl Acad Sci USA 107:4389–4394

Antunes M, Biala G (2012) The novel object recognition memory: neurobiology, test procedure, and its modifications. Cogn Process 13:93–110

Arnsten AF (1993) Catecholamine mechanisms in age-related cognitive decline. Neurobiol Aging 14:639–641

Arnsten AF, Contant TA (1992) Alpha-2 adrenergic agonists decrease distractibility in aged monkeys performing the delayed response task. Psychopharmacology (Berl) 108:159–169

Arnsten AF, Wang MJ, Paspalas CD (2012) Neuromodulation of thought: flexibilities and vulnerabilities in prefrontal cortical network synapses. Neuron 76:223–239

Barch DM, Ceaser A (2012) Cognition in schizophrenia: core psychological and neural mechanisms. Trends Cogn Sci 16:27–34

Barch DM, Smith E (2008) The cognitive neuroscience of working memory: relevance to CNTRICS and schizophrenia. Biol Psychiatry 64:11–17

Barnett JH, Sahakian BJ, Werners U, Hill KE, Brazil R, Gallagher O, Bullmore ET, Jones PB (2005) Visuospatial learning and executive function are independently impaired in first-episode psychosis. Psychol Med 35:1031–1041

Bartko SJ, Vendrell I, Saksida LM, Bussey TJ (2011) A computer-automated touchscreen paired-associates learning (PAL) task for mice: impairments following administration of scopolamine or dicyclomine and improvements following donepezil. Psychopharmacology (Berl) 214:537–548

Beck LH, Bransome ED Jr, Mirsky AF, Rosvold HE, Sarason I (1956) A continuous performance test of brain damage. J Consult Psychol 20:343–350
Berridge CW, Waterhouse BD (2003) The locus coeruleus-noradrenergic system: modulation of behavioral state and state-dependent cognitive processes. Brain Res Brain Res Rev 42:33–84
Bizon JL, Foster TC, Alexander GE, Glisky EL (2012) Characterizing cognitive aging of working memory and executive function in animal models. Front Aging Neurosci 4:19
Blackwell AD, Sahakian BJ, Vesey R, Semple JM, Robbins TW, Hodges JR (2004) Detecting dementia: novel neuropsychological markers of preclinical Alzheimer's disease. Dement Geriatr Cogn Disord 17:42–48
Bolhuis JJ, Bijlsma S, Ansmink P (1986) Exponential decay of spatial memory of rats in a radial maze. Behav Neural Biol 46:115–122
Bouton ME (2004) Context and behavioral processes in extinction. Learn Mem 11:485–494
Boutros NN, Torello MW, Barker BA, Tueting PA, Wu SC, Nasrallah HA (1995) The P50 evoked potential component and mismatch detection in normal volunteers: implications for the study of sensory gating. Psychiatry Res 57:83–88
Boutros NN, Bonnet KA, Millana R, Liu J (1997) A parametric study of the N40 auditory evoked response in rats. Biol Psychiatry 42:1051–1059
Bradshaw J, Shic F, Chawarska K (2011) Brief report: face-specific recognition deficits in young children with autism spectrum disorders. J Autism Dev Disord 41:1429–1435
Brown MW, Barker GR, Aggleton JP, Warburton EC (2012) What pharmacological interventions indicate concerning the role of the perirhinal cortex in recognition memory. Neuropsychologia 50:3122–3140
Brozoski TJ, Brown RM, Rosvold HE, Goldman PS (1979) Cognitive deficit caused by regional depletion of dopamine in prefrontal cortex of rhesus monkey. Science 205:929–932
Caetano MS, Horst NK, Harenberg L, Liu B, Arnsten AF, Laubach M (2012) Lost in transition: aging-related changes in executive control by the medial prefrontal cortex. J Neurosci 32:3765–3777
Carli M, Robbins TW, Evenden JL, Everitt BJ (1983) Effects of lesions to ascending noradrenergic neurones on performance of a 5-choice serial reaction task in rats; implications for theories of dorsal noradrenergic bundle function based on selective attention and arousal. Behav Brain Res 9:361–380
Chang SW, Winecoff AA, Platt ML (2011) Vicarious reinforcement in rhesus macaques (macaca mulatta). Front Neurosci 5:27
Chudasama Y, Muir JL (1997) A behavioural analysis of the delayed non-matching to position task: the effects of scopolamine, lesions of the fornix and of the prelimbic region on mediating behaviours by rats. Psychopharmacology (Berl) 134:73–82
Cohen RM, Semple WE, Gross M, Holcomb HJ, Dowling SM, Nordahl T (1988) Functional localization of sustained attention. Neuropsychiatry Neuropsychol Behav Neurol 1:3–20
Cohen NJ, Ryan J, Hunt C, Romine L, Wszalek T, Nash C (1999) Hippocampal system and declarative (relational) memory: summarizing the data from functional neuroimaging studies. Hippocampus 9:83–98
Corbett BA, Constantine LJ (2006) Autism and attention deficit hyperactivity disorder: assessing attention and response control with the integrated visual and auditory continuous performance test. Child Neuropsychol 12:335–348
Coull JT, Frith CD, Frackowiak RS, Grasby PM (1996) A fronto-parietal network for rapid visual information processing: a PET study of sustained attention and working memory. Neuropsychologia 34:1085–1095
Dahl CD, Wallraven C, Bulthoff HH, Logothetis NK (2009) Humans and macaques employ similar face-processing strategies. Curr Biol 19:509–513
Dal Monte O, Noble PL, Costa VD, Averbeck BB (2014) Oxytocin enhances attention to the eye region in rhesus monkeys. Front Neurosci 8:41

Dalley JW, McGaughy J, O'Connell MT, Cardinal RN, Levita L, Robbins TW (2001) Distinct changes in cortical acetylcholine and noradrenaline efflux during contingent and noncontingent performance of a visual attentional task. J Neurosci 21:4908–4914

Decamp E, Clark K, Schneider JS (2011) Effects of the alpha-2 adrenoceptor agonist guanfacine on attention and working memory in aged non-human primates. Eur J Neurosci 34:1018–1022

Delgado MR, Olsson A, Phelps EA (2006) Extending animal models of fear conditioning to humans. Biol Psychol 73:39–48

DiMasi JA, Feldman L, Seckler A, Wilson A (2010) Trends in risks associated with new drug development: success rates for investigational drugs. Clin Pharmacol Ther 87:272–277

Dudchenko P, Sarter M (1992) Behavioral microanalysis of spatial delayed alternation performance: rehearsal through overt behavior, and effects of scopolamine and chlordiazepoxide. Psychopharmacology (Berl) 107:263–270

Dudchenko PA, Wood ER, Eichenbaum H (2000) Neurotoxic hippocampal lesions have no effect on odor span and little effect on odor recognition memory but produce significant impairments on spatial span, recognition, and alternation. J Neurosci 20:2964–2977

Ebitz RB, Watson KK, Platt ML (2013) Oxytocin blunts social vigilance in the rhesus macaque. Proc Natl Acad Sci USA 110:11630–11635

Ennaceur A, Delacour J (1988) A new one-trial test for neurobiological studies of memory in rats. 1: behavioral data. Behav Brain Res 31:47–59

Fanselow MS, LeDoux JE (1999) Why we think plasticity underlying Pavlovian fear conditioning occurs in the basolateral amygdala. Neuron 23:229–232

Fanselow MS, Poulos AM (2005) The neuroscience of mammalian associative learning. Annu Rev Psychol 56:207–234

Farran EK, Jarrold C, Gathercole SE (2003) Divided attention, selective attention and drawing: processing preferences in Williams syndrome are dependent on the task administered. Neuropsychologia 41:676–687

Farzin F, Scaggs F, Hervey C, Berry-Kravis E, Hessl D (2011) Reliability of eye tracking and pupillometry measures in individuals with fragile X syndrome. J Autism Dev Disord 41:1515–1522

Fitzgerald PJ, Seemann JR, Maren S (2014) Can fear extinction be enhanced? A review of pharmacological and behavioral findings. Brain Res Bull 105C:46–60

Fowler KS, Saling MM, Conway EL, Semple JM, Louis WJ (1997) Computerized neuropsychological tests in the early detection of dementia: prospective findings. J Int Neuropsychol Soc 3:139–146

Francois J, Conway MW, Lowry JP, Tricklebank MD, Gilmour G (2012) Changes in reward-related signals in the rat nucleus accumbens measured by in vivo oxygen amperometry are consistent with fMRI BOLD responses in man. Neuroimage 60:2169–2181

Francois J, Huxter J, Conway MW, Lowry JP, Tricklebank MD, Gilmour G (2014) Differential contributions of infralimbic prefrontal cortex and nucleus accumbens during reward-based learning and extinction. J Neurosci 34:596–607

Freedman M, Oscar-Berman M (1986) Bilateral frontal lobe disease and selective delayed response deficits in humans. Behav Neurosci 100:337–342

Gamo NJ, Wang M, Arnsten AF (2010) Methylphenidate and atomoxetine enhance prefrontal function through alpha2-adrenergic and dopamine D1 receptors. J Am Acad Child Adolesc Psychiatry 49:1011–1023

Gandal MJ, Edgar JC, Klook K, Siegel SJ (2012) Gamma synchrony: towards a translational biomarker for the treatment-resistant symptoms of schizophrenia. Neuropharmacology 62:1504–1518

Gil-da-Costa R, Stoner GR, Fung R, Albright TD (2013) Nonhuman primate model of schizophrenia using a noninvasive EEG method. Proc Natl Acad Sci USA 110:15425–15430

Goldman-Rakic PS (1995) Cellular basis of working memory. Neuron 14:477–485

Goldman-Rakic PS (1996) Regional and cellular fractionation of working memory. Proc Natl Acad Sci USA 93:13473–13480

Green JJ, Hollander E (2010) Autism and oxytocin: new developments in translational approaches to therapeutics. Neurotherapeutics 7:250–257

Grunwald T, Boutros NN, Pezer N, von Oertzen J, Fernandez G, Schaller C, Elger CE (2003) Neuronal substrates of sensory gating within the human brain. Biol Psychiatry 53:511–519

Guastella AJ, Mitchell PB, Dadds MR (2008) Oxytocin increases gaze to the eye region of human faces. Biol Psychiatry 63:3–5

Gutnikov SA, Barnes JC, Rawlins JN (1994) Working memory tasks in five-choice operant chambers: use of relative and absolute spatial memories. Behav Neurosci 108:899–910

Hadfield WS, Baxter MG, Murray EA (2003) Effects of combined and separate removals of rostral dorsal superior temporal sulcus cortex and perirhinal cortex on visual recognition memory in rhesus monkeys. J Neurophysiol 90:2419–2427

Hay M, Thomas DW, Craighead JL, Economides C, Rosenthal J (2014) Clinical development success rates for investigational drugs. Nat Biotechnol 32:40–51

Horner AE, Heath CJ, Hvoslef-Eide M, Kent BA, Kim CH, Nilsson SR, Alsio J, Oomen CA, Holmes A, Saksida LM, Bussey TJ (2013) The touchscreen operant platform for testing learning and memory in rats and mice. Nat Protoc 8:1961–1984

Huerta PT, Lisman JE (1993) Heightened synaptic plasticity of hippocampal CA1 neurons during a cholinergically induced rhythmic state. Nature 364:723–725

Javitt DC, Schroeder CE, Steinschneider M, Arezzo JC, Vaughan HG Jr (1992) Demonstration of mismatch negativity in the monkey. Electroencephalogr Clin Neurophysiol 83:87–90

Jones GM, Sahakian BJ, Levy R, Warburton DM, Gray JA (1992) Effects of acute subcutaneous nicotine on attention, information processing and short-term memory in Alzheimer's disease. Psychopharmacology (Berl) 108:485–494

Jutzeler CR, McMullen ME, Featherstone RF, Tatard-Leitman VM, Gandal MJ, Carlson GC, Siegel SJ (2011) Electrophysiological deficits in Schizophrenia: models and mechanisms. In: Uehara T (ed) Psychiatric disorders - trends and developments. InTech, Croatia, pp 19–66

Kaplan GB, Moore KA (2011) The use of cognitive enhancers in animal models of fear extinction. Pharmacol Biochem Behav 99:217–228

Kazama AM, Schauder KB, McKinnon M, Bachevalier J, Davis M (2013) A novel AX+/BX-paradigm to assess fear learning and safety-signal processing with repeated-measure designs. J Neurosci Methods 214:177–183

Kealy J, Bennett R, Lowry JP (2013) Simultaneous recording of hippocampal oxygen and glucose in real time using constant potential amperometry in the freely-moving rat. J Neurosci Methods 215:110–120

Klin A, Jones W, Schultz R, Volkmar F, Cohen D (2002) Visual fixation patterns during viewing of naturalistic social situations as predictors of social competence in individuals with autism. Arch Gen Psychiatry 59:809–816

Klorman R, Brumaghim JT, Fitzpatrick PA, Borgstedt AD (1991) Methylphenidate speeds evaluation processes of attention deficit disorder adolescents during a continuous performance test. J Abnorm Child Psychol 19:263–283

Kola I, Landis J (2004) Can the pharmaceutical industry reduce attrition rates? Nat Rev Drug Discov 3:711–715

Kosfeld M, Heinrichs M, Zak PJ, Fischbacher U, Fehr E (2005) Oxytocin increases trust in humans. Nature 435:673–676

Lee J, Park S (2005) Working memory impairments in schizophrenia: a meta-analysis. J Abnorm Psychol 114:599–611

Leonard BW, Amaral DG, Squire LR, Zola-Morgan S (1995) Transient memory impairment in monkeys with bilateral lesions of the entorhinal cortex. J Neurosci 15:5637–5659

Lewin JS, Friedman L, Wu D, Miller DA, Thompson LA, Klein SK, Wise AL, Hedera P, Buckley P, Meltzer H, Friedland RP, Duerk JL (1996) Cortical localization of human sustained attention: detection with functional MR using a visual vigilance paradigm. J Comput Assist Tomogr 20:695–701

Losier BJ, McGrath PJ, Klein RM (1996) Error patterns on the continuous performance test in non-medicated and medicated samples of children with and without ADHD: a meta-analytic review. J Child Psychol Psychiatry 37:971–987

Lowry JP, Griffin K, McHugh SB, Lowe AS, Tricklebank M, Sibson NR (2010) Real-time electrochemical monitoring of brain tissue oxygen: a surrogate for functional magnetic resonance imaging in rodents. Neuroimage 52:549–555

Lustig C, Kozak R, Sarter M, Young JW, Robbins TW (2013) CNTRICS final animal model task selection: control of attention. Neurosci Biobehav Rev 37:2099–2110

Lyon L, Saksida LM, Bussey TJ (2012) Spontaneous object recognition and its relevance to schizophrenia: a review of findings from pharmacological, genetic, lesion and developmental rodent models. Psychopharmacology (Berl) 220:647–672

Machado CJ, Nelson EE (2011) Eye-tracking with nonhuman primates is now more accessible than ever before. Am J Primatol 73:562–569

Maddux JM, Holland PC (2011) Effects of dorsal or ventral medial prefrontal cortical lesions on five-choice serial reaction time performance in rats. Behav Brain Res 221:63–74

Marder SR, Fenton W (2004) Measurement and treatment research to improve cognition in Schizophrenia: NIMH MATRICS initiative to support the development of agents for improving cognition in schizophrenia. Schizophr Res 72:5–9

Maren S (2001) Neurobiology of Pavlovian fear conditioning. Annu Rev Neurosci 24:897–931

McAllister KA, Saksida LM, Bussey TJ (2013) Dissociation between memory retention across a delay and pattern separation following medial prefrontal cortex lesions in the touchscreen TUNL task. Neurobiol Learn Mem 101:120–126

McGaughy J, Sarter M (1998) Sustained attention performance in rats with intracortical infusions of 192 IgG-saporin-induced cortical cholinergic deafferentation: effects of physostigmine and FG 7142. Behav Neurosci 112:1519–1525

McGaughy J, Turchi J, Sarter M (1994) Crossmodal divided attention in rats: effects of chlordiazepoxide and scopolamine. Psychopharmacology (Berl) 115:213–220

McHugh SB, Fillenz M, Lowry JP, Rawlins JN, Bannerman DM (2011) Brain tissue oxygen amperometry in behaving rats demonstrates functional dissociation of dorsal and ventral hippocampus during spatial processing and anxiety. Eur J Neurosci 33:322–337

McKee RD, Squire LR (1993) On the development of declarative memory. J Exp Psychol Learn Mem Cogn 19:397–404

Mesulam MM, Geula C (1988) Nucleus basalis (Ch4) and cortical cholinergic innervation in the human brain: observations based on the distribution of acetylcholinesterase and choline acetyltransferase. J Comp Neurol 275:216–240

Meunier M, Bachevalier J, Mishkin M, Murray EA (1993) Effects on visual recognition of combined and separate ablations of the entorhinal and perirhinal cortex in rhesus monkeys. J Neurosci 13:5418–5432

Milad MR, Quirk GJ (2012) Fear extinction as a model for translational neuroscience: ten years of progress. Annu Rev Psychol 63:129–151

Milad MR, Rauch SL, Pitman RK, Quirk GJ (2006) Fear extinction in rats: implications for human brain imaging and anxiety disorders. Biol Psychol 73:61–71

Miller G (2010) Is pharma running out of brainy ideas? Science 329:502–504

Mishkin M (1957) Effects of small frontal lesions on delayed alternation in monkeys. J Neurophysiol 20:615–622

Mishkin M, Delacour J (1975) An analysis of short-term visual memory in the monkey. J Exp Psychol Anim Behav Process 1:326–334

Moore H, Geyer MA, Carter CS, Barch DM (2013) Harnessing cognitive neuroscience to develop new treatments for improving cognition in schizophrenia: CNTRICS selected cognitive paradigms for animal models. Neurosci Biobehav Rev 37:2087–2091

Morris RG, Hagan JJ, Rawlins JN (1986) Allocentric spatial learning by hippocampectomised rats: a further test of the "spatial mapping" and "working memory" theories of hippocampal function. Q J Exp Psychol B 38:365–395

Muir JL, Everitt BJ, Robbins TW (1995) Reversal of visual attentional dysfunction following lesions of the cholinergic basal forebrain by physostigmine and nicotine but not by the 5-HT3 receptor antagonist, ondansetron. Psychopharmacology (Berl) 118:82–92

Murray EA, Gaffan D, Mishkin M (1993) Neural substrates of visual stimulus-stimulus association in rhesus monkeys. J Neurosci 13:4549–4561

Navarra R, Graf R, Huang Y, Logue S, Comery T, Hughes Z, Day M (2008) Effects of atomoxetine and methylphenidate on attention and impulsivity in the 5-choice serial reaction time test. Prog Neuropsychopharmacol Biol Psychiatry 32:34–41

Nemanic S, Alvarado MC, Bachevalier J (2004) The hippocampal/parahippocampal regions and recognition memory: insights from visual paired comparison versus object-delayed nonmatching in monkeys. J Neurosci 24:2013–2026

Neumann D, Spezio ML, Piven J, Adolphs R (2006) Looking you in the mouth: abnormal gaze in autism resulting from impaired top-down modulation of visual attention. Soc Cogn Affect Neurosci 1:194–202

Nuechterlein KH, Barch DM, Gold JM, Goldberg TE, Green MF, Heaton RK (2004) Identification of separable cognitive factors in schizophrenia. Schizophr Res 72:29–39

O'Neill J, Fitten LJ, Siembieda DW, Crawford KC, Halgren E, Fisher A, Refai D (1999) Divided attention-enhancing effects of AF102B and THA in aging monkeys. Psychopharmacology (Berl) 143:123–130

O'Neill J, Fitten LJ, Siembieda DW, Ortiz F, Halgren E (2000) Effects of guanfacine on three forms of distraction in the aging macaque. Life Sci 67:877–885

Olton DS, Papas BC (1979) Spatial memory and hippocampal function. Neuropsychologia 17:669–682

Oomen CA, Hvoslef-Eide M, Heath CJ, Mar AC, Horner AE, Bussey TJ, Saksida LM (2013) The touchscreen operant platform for testing working memory and pattern separation in rats and mice. Nat Protoc 8:2006–2021

Owen AM, Sahakian BJ, Semple J, Polkey CE, Robbins TW (1995) Visuo-spatial short-term recognition memory and learning after temporal lobe excisions, frontal lobe excisions or amygdalo-hippocampectomy in man. Neuropsychologia 33:1–24

Paine TA, Tomasiewicz HC, Zhang K, Carlezon WA Jr (2007) Sensitivity of the five-choice serial reaction time task to the effects of various psychotropic drugs in Sprague–Dawley rats. Biol Psychiatry 62:687–693

Parasuraman R, Haxby JV (1993) Attention and brain function in Alzheimer's disease: a review. Neuropsychology 7:242–272

Pardo JV, Fox PT, Raichle ME (1991) Localization of a human system for sustained attention by positron emission tomography. Nature 349:61–64

Parsons RG, Ressler KJ (2013) Implications of memory modulation for post-traumatic stress and fear disorders. Nat Neurosci 16:146–153

Pascalis O, Bachevalier J (1999) Neonatal aspiration lesions of the hippocampal formation impair visual recognition memory when assessed by paired-comparison task but not by delayed nonmatching-to-sample task. Hippocampus 9:609–616

Pascalis O, Hunkin NM, Holdstock JS, Isaac CL, Mayes AR (2004) Visual paired comparison performance is impaired in a patient with selective hippocampal lesions and relatively intact item recognition. Neuropsychologia 42:1293–1300

Passetti F, Dalley JW, O'Connell MT, Everitt BJ, Robbins TW (2000) Increased acetylcholine release in the rat medial prefrontal cortex during performance of a visual attentional task. Eur J Neurosci 12:3051–3058

Passetti F, Chudasama Y, Robbins TW (2002) The frontal cortex of the rat and visual attentional performance: dissociable functions of distinct medial prefrontal subregions. Cereb Cortex 12:1254–1268

Posner MI, Petersen SE (1990) The attention system of the human brain. Annu Rev Neurosci 13:25–42

Quirk GJ, Garcia R, Gonzalez-Lima F (2006) Prefrontal mechanisms in extinction of conditioned fear. Biol Psychiatry 60:337–343

Rabbitt P, Lowe C (2000) Patterns of cognitive ageing. Psychol Res 63:308–316

Riby DM, Hancock PJ (2009) Do faces capture the attention of individuals with Williams syndrome or autism? Evidence from tracking eye movements. J Autism Dev Disord 39:421–431

Richmond J, Sowerby P, Colombo M, Hayne H (2004) The effect of familiarization time, retention interval, and context change on adults' performance in the visual paired-comparison task. Dev Psychobiol 44:146–155

Robbins TW (2002) The 5-choice serial reaction time task: behavioural pharmacology and functional neurochemistry. Psychopharmacology (Berl) 163:362–380

Robbins TW, James M, Owen AM, Sahakian BJ, McInnes L, Rabbitt P (1994) Cambridge Neuropsychological Test Automated Battery (CANTAB): a factor analytic study of a large sample of normal elderly volunteers. Dementia 5:266–281

Sahakian BJ, Morris RG, Evenden JL, Heald A, Levy R, Philpot M, Robbins TW (1988) A comparative study of visuospatial memory and learning in Alzheimer-type dementia and Parkinson's disease. Brain 111(Pt 3):695–718

Sahakian B, Jones G, Levy R, Gray J, Warburton D (1989) The effects of nicotine on attention, information processing, and short-term memory in patients with dementia of the Alzheimer type. Br J Psychiatry 154:797–800

Sahakian BJ, Owen AM, Morant NJ, Eagger SA, Boddington S, Crayton L, Crockford HA, Crooks M, Hill K, Levy R (1993) Further analysis of the cognitive effects of tetrahydroaminoacridine (THA) in Alzheimer's disease: assessment of attentional and mnemonic function using CANTAB. Psychopharmacology (Berl) 110:395–401

Schroeter ML, Vogt B, Frisch S, Becker G, Barthel H, Mueller K, Villringer A, Sabri O (2012) Executive deficits are related to the inferior frontal junction in early dementia. Brain 135:201–215

Shin LM, Liberzon I (2010) The neurocircuitry of fear, stress, and anxiety disorders. Neuropsychopharmacology 35:169–191

Siegel SJ, Connolly P, Liang Y, Lenox RH, Gur RE, Bilker WB, Kanes SJ, Turetsky BI (2003) Effects of strain, novelty, and NMDA blockade on auditory-evoked potentials in mice. Neuropsychopharmacology 28:675–682

Spinelli S, Ballard T, Gatti-McArthur S, Richards GJ, Kapps M, Woltering T, Wichmann J, Stadler H, Feldon J, Pryce CR (2005) Effects of the mGluR2/3 agonist LY354740 on computerized tasks of attention and working memory in marmoset monkeys. Psychopharmacology (Berl) 179:292–302

Suzuki WA, Zola-Morgan S, Squire LR, Amaral DG (1993) Lesions of the perirhinal and parahippocampal cortices in the monkey produce long-lasting memory impairment in the visual and tactual modalities. J Neurosci 13:2430–2451

Swainson R, Hodges JR, Galton CJ, Semple J, Michael A, Dunn BD, Iddon JL, Robbins TW, Sahakian BJ (2001) Early detection and differential diagnosis of Alzheimer's disease and depression with neuropsychological tasks. Dement Geriatr Cogn Disord 12:265–280

Taffe MA, Weed MR, Gutierrez T, Davis SA, Gold LH (2002) Differential muscarinic and NMDA contributions to visuo-spatial paired-associate learning in rhesus monkeys. Psychopharmacology (Berl) 160:253–262

Talpos JC, Winters BD, Dias R, Saksida LM, Bussey TJ (2009) A novel touchscreen-automated paired-associate learning (PAL) task sensitive to pharmacological manipulation of the hippocampus: a translational rodent model of cognitive impairments in neurodegenerative disease. Psychopharmacology (Berl) 205:157–168

Talpos JC, McTighe SM, Dias R, Saksida LM, Bussey TJ (2010) Trial-unique, delayed nonmatching-to-location (TUNL): a novel, highly hippocampus-dependent automated touchscreen test of location memory and pattern separation. Neurobiol Learn Mem 94:341–352

Tolman EC (1945) A stimulus-expectancy need-cathexis psychology. Science 101:160–166

Uhlhaas PJ, Singer W (2010) Abnormal neural oscillations and synchrony in schizophrenia. Nat Rev Neurosci 11:100–113

van Enkhuizen J, Acheson D, Risbrough V, Drummond S, Geyer MA, Young JW (2014) Sleep deprivation impairs performance in the 5-choice continuous performance test: similarities between humans and mice. Behav Brain Res 261:40–48

van Goethem NP, Rutten K, van der Staay FJ, Jans LA, Akkerman S, Steinbusch HW, Blokland A, van't Klooster J, Prickaerts J (2012) Object recognition testing: rodent species, strains, housing conditions, and estrous cycle. Behav Brain Res 232:323–334

VanElzakker MB, Kathryn Dahlgren M, Caroline Davis F, Dubois S, Shin LM (2014) From Pavlov to PTSD: The extinction of conditioned fear in rodents, humans, and anxiety disorders. Neurobiol Learn Mem 113C:3–18

Voytko ML, Olton DS, Richardson RT, Gorman LK, Tobin JR, Price DL (1994) Basal forebrain lesions in monkeys disrupt attention but not learning and memory. J Neurosci 14:167–186

Wang M, Ramos BP, Paspalas CD, Shu Y, Simen A, Duque A, Vijayraghavan S, Brennan A, Dudley A, Nou E, Mazer JA, McCormick DA, Arnsten AF (2007) Alpha2A-adrenoceptors strengthen working memory networks by inhibiting cAMP-HCN channel signaling in prefrontal cortex. Cell 129:397–410

Wang M, Gamo NJ, Yang Y, Jin LE, Wang XJ, Laubach M, Mazer JA, Lee D, Arnsten AF (2011) Neuronal basis of age-related working memory decline. Nature 476:210–213

Webster SJ, Bachstetter AD, Nelson PT, Schmitt FA, Van Eldik LJ (2014) Using mice to model Alzheimer's dementia: an overview of the clinical disease and the preclinical behavioral changes in 10 mouse models. Front Genet 5:88

Weed MR, Taffe MA, Polis I, Roberts AC, Robbins TW, Koob GF, Bloom FE, Gold LH (1999) Performance norms for a rhesus monkey neuropsychological testing battery: acquisition and long-term performance. Brain Res Cogn Brain Res 8:185–201

Wilkins AJ, Shallice T, McCarthy R (1987) Frontal lesions and sustained attention. Neuropsychologia 25:359–365

Winters BD, Bartko SJ, Saksida LM, Bussey TJ (2010a) Muscimol, AP5, or scopolamine infused into perirhinal cortex impairs two-choice visual discrimination learning in rats. Neurobiol Learn Mem 93:221–228

Winters BD, Saksida LM, Bussey TJ (2010b) Implications of animal object memory research for human amnesia. Neuropsychologia 48:2251–2261

Wood SJ, Proffitt T, Mahony K, Smith DJ, Buchanan JA, Brewer W, Stuart GW, Velakoulis D, McGorry PD, Pantelis C (2002) Visuospatial memory and learning in first-episode schizophreniform psychosis and established schizophrenia: a functional correlate of hippocampal pathology? Psychol Med 32:429–438

Young JW, Powell SB, Risbrough V, Marston HM, Geyer MA (2009a) Using the MATRICS to guide development of a preclinical cognitive test battery for research in schizophrenia. Pharmacol Ther 122:150–202

Young JW, Light GA, Marston HM, Sharp R, Geyer MA (2009b) The 5-choice continuous performance test: evidence for a translational test of vigilance for mice. PLoS One 4(1):e4227

Zeamer A, Meunier M, Bachevalier J (2011) Stimulus similarity and encoding time influence incidental recognition memory in adult monkeys with selective hippocampal lesions. Learn Mem 18:170–180

Zola SM, Squire LR, Teng E, Stefanacci L, Buffalo EA, Clark RE (2000) Impaired recognition memory in monkeys after damage limited to the hippocampal region. J Neurosci 20:451–463

Zola SM, Manzanares CM, Clopton P, Lah JJ, Levey AI (2013) A behavioral task predicts conversion to mild cognitive impairment and Alzheimer's disease. Am J Alzheimers Dis Other Demen 28:179–184

Zola-Morgan S, Squire LR, Amaral DG (1986) Human amnesia and the medial temporal region: enduring memory impairment following a bilateral lesion limited to field CA1 of the hippocampus. J Neurosci 6:2950–2967

Signaling Pathways Relevant to Cognition-Enhancing Drug Targets

Caroline Ménard, Pierrette Gaudreau, and Rémi Quirion

Contents

C. Ménard
Douglas Mental Health University Institute, McGill University, Perry Pavilion, 6875 LaSalle Boulevard, Montreal, QC, Canada, H4H 1R3

Department of Psychiatry, McGill University, Montreal, QC, Canada, H3A 1A1

Centre Hospitalier de l'Université de Montréal Research Center, Montreal, QC, Canada, H1W 4A4

Department of Medicine, University of Montreal, Montreal, QC, Canada, H3T 1J4

P. Gaudreau
Centre Hospitalier de l'Université de Montréal Research Center, Montreal, QC, Canada, H1W 4A4

Department of Medicine, University of Montreal, Montreal, QC, Canada, H3T 1J4

R. Quirion (✉)
Douglas Mental Health University Institute, McGill University, Perry Pavilion, 6875 LaSalle Boulevard, Montreal, QC, Canada, H4H 1R3

Department of Psychiatry, McGill University, Montreal, QC, Canada, H3A 1A1
e-mail: remi.quirion@mcgill.ca

K.M. Kantak, J.G. Wettstein (eds.), *Cognitive Enhancement*, Handbook of Experimental Pharmacology 228, DOI 10.1007/978-3-319-16522-6_3

Abstract

Aging is generally associated with a certain cognitive decline. However, individual differences exist. While age-related memory deficits can be observed in humans and rodents in the absence of pathological conditions, some individuals maintain intact cognitive functions up to an advanced age. The mechanisms underlying learning and memory processes involve the recruitment of multiple signaling pathways and gene expression, leading to adaptative neuronal plasticity and long-lasting changes in brain circuitry. This chapter summarizes the current understanding of how these signaling cascades could be modulated by cognition-enhancing agents favoring memory formation and successful aging. It focuses on data obtained in rodents, particularly in the rat as it is the most common animal model studied in this field. First, we will discuss the role of the excitatory neurotransmitter glutamate and its receptors, downstream signaling effectors [e.g., calcium/calmodulin-dependent protein kinase II (CaMKII), protein kinase C (PKC), extracellular signal-regulated kinases (ERK), mammalian target of rapamycin (mTOR), cAMP response element-binding protein (CREB)], associated immediate early gene (e.g., Homer 1a, Arc and Zif268), and growth factors [insulin-like growth factors (IGFs) and brain-derived neurotrophic factor (BDNF)] in synaptic plasticity and memory formation. Second, the impact of the cholinergic system and related modulators on memory will be briefly reviewed. Finally, since dynorphin neuropeptides have recently been associated with memory impairments in aging, it is proposed as an attractive target to develop novel cognition-enhancing agents.

Keywords

Aging • Memory • Synaptic plasticity • Glutamate • Acetylcholine • Dynorphin

1 Introduction

Despite a general lengthening of life span in humans over the last decades, the quality of life still varies substantially among older adults. Some individuals are active and socially engaged, while others have physical or cognitive impairments and/or present depressive symptoms (Rowe and Kahn 1997; Frisardi et al. 2011). A better understanding of the processes leading to individual differences during aging might help identifying new pharmacological targets and develop innovative treatments to favor successful cognitive aging. This chapter summarizes the current knowledge on signaling pathways of particular importance in memory formation (especially spatial memory) in rodents and the changes occurring during normal (non-pathological) aging. Each section covers potential cognition-enhancing drug targets and includes an overview of related published studies, focusing on the data obtained in rats, and compounds already available (Fig. 1).

Normal aging is associated with increasing memory losses that can be detected in middle-aged rats (Deupree et al. 1993) similarly to humans (Davis et al. 2003). However, in aging rats of similar ages, important inter-individual differences in cognitive abilities have been reported (Gallagher et al. 1993, 2003; Aubert et al. 1995; Quirion et al. 1995; Rowe et al. 1998; Wilson et al. 2003; Menard and Quirion 2012b). Sex and strain differences have also been observed (Markowska 1999; Menard et al. 2014b) but will not be specifically addressed in this chapter. Variations of cognitive status in aged rats are not related to neuronal loss, as cell death in the hippocampus and neocortex does not characterize normal aging in rodents (Rapp and Gallagher 1996; Rasmussen et al. 1996; Merrill et al. 2001; Gallagher et al. 2003). Moreover, no regression of dendrites (Turner and Deupree 1991; Flood 1993; Pyapali and Turner 1996) or decrease of spine density (Curcio and Hinds 1983; Markham et al. 2005) has been reported in old rats. Most electrical properties of the neurons remain constant over the life span including resting membrane potential, threshold to reach an action potential, and the width and amplitude of Na^+ action potentials (Segal 1982; Landfield and Pitler 1984; Niesen et al. 1988; Kerr et al. 1989; Barnes et al. 1992; Potier et al. 1992, 1993; Burke and Barnes 2006). These observations suggest that in rodents, age-related memory impairments associated with normal aging might be linked to altered cell signaling and dysregulation of gene expression as reported by us and others (for a review, see Benoit et al. 2011).

2 Types of Memory

The hippocampus-dependent Morris Water Maze (MWM) task (Morris 1984) is one of the most widely used behavioral paradigms in neuroscience. It is particularly efficient to discriminate aged rat subgroups depending on their level of cognitive fitness. However, multiple tests and paradigms have been developed over the years to study different forms of learning and memory. The following sections will

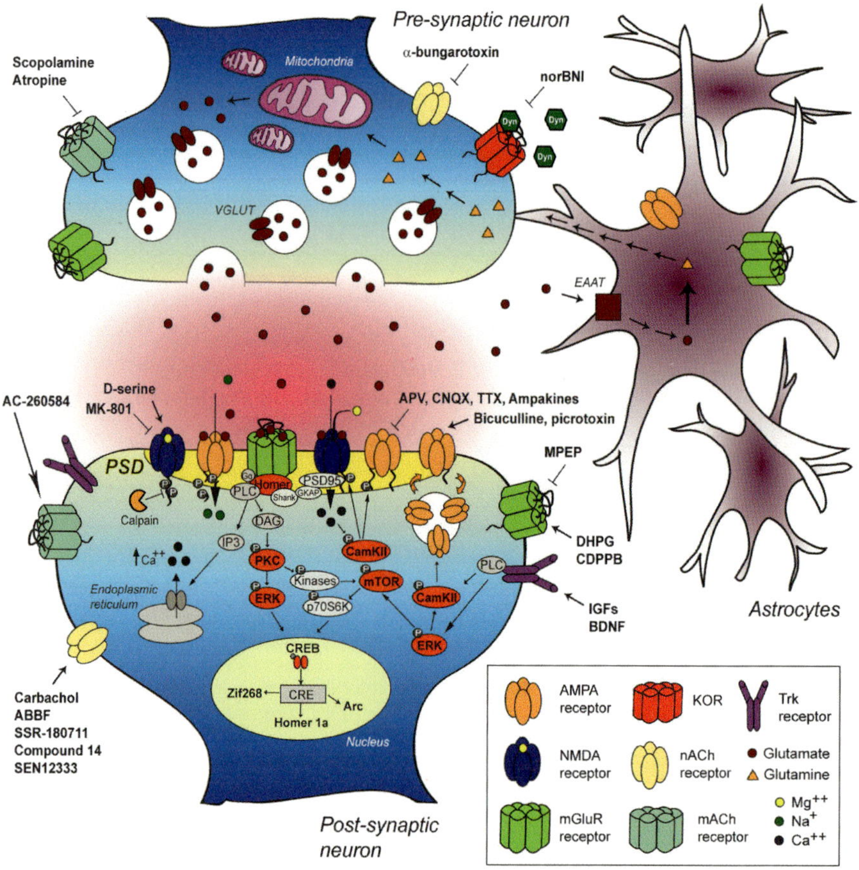

Fig. 1 Summary of the signaling pathways involved in memory formation. Following neurotransmitter glutamate release, postsynaptic ionotropic (AMPA, NMDA) and metabotropic (mGluR) receptors become activated leading to the phosphorylation of downstream signaling effectors, notably PKC, CaMKII, ERK, mTOR and CREB, and immediate early gene expression (Arc, Homer 1a, Zif268). Glutamatergic neurotransmission can be modulated by other neurotransmitters, such as acetylcholine and its nicotinic (nACh) and muscarinic (mACh) receptors, neurotrophins (BDNF) and growth factors (IGF) through the Trk receptors, or neuromodulators such as dynorphins (Dyn) that act on presynaptic κ-opioid receptors (KOR) to block glutamate release. Astrocytes play an active role in the glutamatergic system activity as they recapture glutamate through the excitatory amino-acid transporters (EAATs), where it is amidated and reconverted in glutamate in neuron mitochondria to finally be accumulated in synaptic vesicles through the vesicular glutamate transporters (VGLUTs)

summarize spatial, recognition, social, and fear memory neuronal processes and the behavioral paradigms to which they are associated.

2.1 Spatial Memory

Spatial memory is probably the most studied form of cognition in rodents. It is required to navigate in an environment or to remember where objects have been placed and implies various representations and encoding (Bird and Burgess 2008). Initial information was obtained from epileptic patients showing the devastating effects of bilateral medial temporal lobe or hippocampal damage (Milner and Penfield 1955; Scoville and Milner 1957). In rodents, hippocampal lesions severely impair performances in the MWM task (Moser and Moser 1998) which consist in finding a hidden escape platform in a pool filled with opaque water (Morris 1984). The animals use spatial cues on the walls of the room to orient themselves in the environment and successfully navigate. Following several days of training (multiple trials per day), young rats or mice will reach the platform quickly. Old rat learning curves can then be compared to classify them in aged memory-impaired (AI) and -unimpaired subgroups (AU) (Gallagher et al. 1993, 2003; Aubert et al. 1995; Quirion et al. 1995; Rowe et al. 1998; Wilson et al. 2003; Menard and Quirion 2012b). A probe test for which the platform is removed can subsequently be conducted to confirm cognitive status. In this task, the number of platform crossings and time or distance spent in the target quadrant can be compared to assess memory accuracy. The classical paradigm can be modified to add a second week of training in which the platform is moved to another quadrant of the pool (Menard and Quirion 2012b). Inhibitory and reversal learning, which contributes to the extinction of previously acquired memories and learning of a novel similar task, is strongly altered in AI rats suggesting that adaptative synaptic plasticity is affected and less efficient than in young and AU animals (Menard and Quirion 2012b). This could be related to *place cells* which are an ensemble of cells that fired when an animal is moving in an environment, encoding a cognitive map with a specific spatial-firing pattern (O'Keefe and Dostrovsky 1971). Wrong encoding or recollection of the patterns could lead to cognitive deficits in old rodents (Wilson et al. 2003, 2006).

Other paradigms have been developed to study spatial memory in rodents including the Barnes maze, radial arm maze, and the hole-board task. In the Barnes maze, rodents have to find an escape box using visual cues on a circular surface with up to 20 holes around its circumference (Barnes 1979). The task is based on rodents' aversion of open bright spaces and is considered relatively unstressful and modestly demanding physically. Similarly to the MWM task, various parameters can be measured such as latency to escape, path length, velocity, etc. Some senescent rats exhibit poor performances in this test (Barnes 1979; Harrison et al. 2006; Barrett et al. 2009). Hippocampal lesions induced by traumatic brain injury also lead to memory deficits in this task (Fox et al. 1998). Another interesting behavioral paradigm to evaluate spatial memory is the radial arm maze (Walker and

Olton 1979; Hudon et al. 2002; Webster et al. 2014). In this task, the animals have to find food reward at the end of the baited arms and the design allowed to explore reference and working memory function separately (Roberge et al. 2008; Grayson et al. 2014). Indeed, reference memory errors are associated with visits in non-baited arms, while working memory errors are the results of reentry in a previously visited arm. Nevertheless, the MWM task is generally used to evaluate the impact of normal aging on spatial memory particularly in rats.

2.2 Recognition Memory

While spatial memory is necessary to explore and navigate in old or new environments, other skills such as the ability to discriminate novelty from stimuli that have been previously encountered are also necessary for survival. Interestingly, the ability to recognize a familiar versus novel stimulus (Rowe et al. 1998) or objects (Menard et al. 2013b, 2014b) declines with normal aging in rodents. Recognition memory tests compared time spent exploring familiar versus novel objects, smells, or tastes in distinct spatial locations (Fedulov et al. 2007; Dere et al. 2007; Tse et al. 2007; Menard et al. 2013b). In contrast, spatial memóry which involves circuitry of the hippocampal formation, albeit recognition memory, is linked to the perirhinal cortex (Burke et al. 2012). AI rodents seem to falsely identify novel objects or stimulus as familiar leading to pattern separation deficits (Burke et al. 2010). As for spatial memory and hippocampal lesions, impairments of recognition memory have been reported in rats with lesions of the perirhinal cortex (McTighe et al. 2010). Again no significant loss of neurons has been observed in this brain structure over the life span (Rapp et al. 2002).

2.3 Social Memory

Another form of cognition essential for survivability is social memory. Rodents need to interact with each other, establish social networks, and learn how to respond to stimuli that define hierarchy and mate choice (Berry and Bronson 1992). Basic social interaction between rodents can be studied using video recording to analyze active interaction time between a test animal and a novel unfamiliar mouse or rat. Various behavioral paradigms were developed to study in detail memory formation processes by social recognition and learning (for a review, see van der Kooij and Sandi 2012). In rodents, androgens and estrogens control social information processing by regulating hormones and neuropeptides such as oxytocin and arginine–vasopressin (Winslow et al. 1993; Neumann 2008; Choleris et al. 2009). Interestingly, the memory of mice seems to be far superior to that of rats in social recognition paradigms and it could be related to differences in olfaction (Noack et al. 2010). Aging affects olfactory sensory function in rats, particularly in reversal learning (Schoenbaum et al. 2002; Brushfield et al. 2008), and thus impairs social memory recognition processes (Guan and Dluzen 1994). Social defeat stress can induce behavioral adaptations relevant to depression such as anxiety-like behaviors, social avoidance, and anhedonia (Golden et al. 2011). In humans, depression-related cognitive deficits might be a risk factor for dementia while normal aging could involve memory impairments associated with enhanced anxious behaviors (Bunce et al. 2012).

2.4 Fear Memory

Stressful and emotionally arousing or challenging experiences are generally retained in memory (Schacter 1999; Smith et al. 2004; LaBar and Cabeza 2006; Joels et al. 2011). In fact, fear memory is essential not only to rodents but to all species to avoid dangerous situations and improve coping strategies. Fear learning is fast and efficient: single exposure to a stressful event can lead to the formation of long-lasting fear memories but also lead to detrimental behaviors (Najavits et al. 1998). Fortunately, adaptation and underlying brain plasticity allow for the damping of fear memories. However, these processes are slower than fear learning and often require multiple non-reinforcing expositions to the fear-associated cues of contexts (Myers and Davis 2002, 2007). While the hippocampus is still involved in the formation of fear learning and memory (Radulovic and Tronson 2010), the amygdala is central to these processes (Maren and Quirk 2004; McGaugh 2004; Hermans et al. 2014). Aversive learning can be evaluated with multiple behavioral paradigms including passive avoidance, contextual and cued fear conditioning, eyeblink conditioning, fear-conditioned startle, or taste aversion (for a review, see Crawley 2008). Anxiety-like behaviors and stress responses on the other hand can be measured with methods exploiting the approach-avoidance conflict between rodents' innate desire to explore new environments and fear of open bright space with apparatus such as the light dark box, elevated plus maze, or open field (Bouwknecht and Paylor 2002; Ducottet and Belzung 2005; Crawley 2008; Menard et al. 2013b). Old rodents are generally characterized by exacerbated anxious behaviors and stress responses (Menard et al. 2013b, 2014b) which affects brain synaptic plasticity and memory formation.

3 Synaptic Plasticity Associated with Learning and Memory Formation

Learning and memory processes benefit from brain plasticity and this induces reversible cellular and molecular changes in the central nervous system. These modifications can then be stabilized or consolidated to create long-lasting memories (Dudai 1996; McGaugh 2000; Lamprecht and LeDoux 2004; Frankland and Bontempi 2005; Baudry et al. 2011; Choquet and Triller 2013; Huganir and Nicoll 2013). In this section, the role of the neurotransmitter glutamate, its receptors, and related signaling pathways on cognitive function will be discussed.

3.1 Glutamate Receptors

3.1.1 NMDA Receptors

Glutamate is the main excitatory neurotransmitter in the brain and activation of its *N*-methyl-D-aspartate receptors (NMDAR) play a critical role in synaptic plasticity and memory formation (Morris et al. 1986; Sakimura et al. 1995; Tsien et al. 1996; Kiyama et al. 1998; Nakazawa et al. 2002, 2003; McHugh et al. 2007; Lee and Silva 2009). NMDARs form a heterotetramer composed of two obligatory GluN1 subunits and two modulatory GluN2 (A, B, C and D subtypes) or GluN3 (A or B subtypes) subunits. Receptor subunit composition changes during development (Monyer et al. 1994; Sheng et al. 1994; Bellone and Nicoll 2007) and aging (Kuehl-Kovarik et al. 2000; Zhao et al. 2009; Magnusson et al. 2010), influencing the kinetics of the receptor channel opening. Greater ratios of GluN2B prolonged NMDAR currents enhancing long-term potentiation (LTP) (Foster et al. 2010; Cui et al. 2011; Muller et al. 2013). LTP is a form of synaptic plasticity closely related to learning and memory formation (Bliss and Collingridge 1993) which is altered in the aging brain and could contribute to cognitive decline (Landfield and Lynch 1977; Deupree et al. 1993; Rosenzweig et al. 1997; Shankar et al. 1998; Tombaugh et al. 2002; Barnes 2003; Burke and Barnes 2006). Facilitation, saturation or inhibition of LTP by pharmacological agents or genetic manipulation directly affects behaviors in rodents (Morris et al. 1986; Sakimura et al. 1995; Tsien et al. 1996; Kiyama et al. 1998; Tang et al. 1999). Trafficking of glutamate receptors from the cytoplasm to the membrane and postsynaptic densities (PSD) are crucial to facilitate LTP maintenance and synaptic plasticity (Malinow and Malenka 2002; Rumpel et al. 2005). Accordingly, transgenic mice overexpressing the kinesin-like protein KIF17, a protein involved in GluN2B transport along microtubules, display better spatial learning and working memory performances (Wong et al. 2002). In contrast, degradation of NMDAR by the protease calpain decreases the number of functional receptors in the PSD (Simpkins et al. 2003; Dong et al. 2006; Baudry et al. 2013). Cyclin-dependent kinase 5 (Cdk5) regulates calpain-dependent GluN2B proteolysis (Su and Tsai 2011) and deletion of Cdk5 reduces GluN2B degradation favoring stronger LTP and memory processes (Hawasli et al. 2007). Mice overexpressing GluN2B outperform age-matched controls in hippocampus-dependent memory tasks up to 18 months of age (Cao et al. 2007), suggesting that GluN2B and related downstream signaling pathways could be promising targets for cognition-enhancing drugs (Mony et al. 2009).

Enhancement of NMDAR functioning has been a pharmacological target for cognition for decades (for a review, see Collingridge et al. 2013). Briefly, NMDAR activity can be modulated either directly with agonists or antagonists and regulation of posttranslational modifications such as phosphorylation, palmitoylation, ubiquitination, and proteolysis, or indirectly through its interactions with other receptors and neuromodulators (Collingridge et al. 2013). NMDAR antagonists generally impair NMDA-dependent LTP, learning, and memory (Morris 1989; Manahan-Vaughan et al. 2008; Blot et al. 2013). However, exceptions exist, notably memantine, a fast, voltage-dependent channel blocker (Bresink et al. 1996; Frankiewicz et al. 1996), which is used to treat late-stage Alzheimer's disease as it delays cognitive decline (Danysz and Parsons 2003). NMDAR antagonists may enhance cognition by blocking aberrant activation of the receptors while preserving physiological functions (Frankiewicz and Parsons 1999; Fitzjohn et al. 2008).

Another attractive therapeutic avenue to rescue age-related memory deficits is the potentiation of NMDAR activity via the glycine-binding site (Baxter et al. 1994). Glycine or glycine-like substance such as D-serine acts as a co-agonist of glutamate to open the NMDAR channel (Johnson and Ascher 1987; Kleckner and Dingledine 1988; Mothet et al. 2000) and NMDAR full activation requires agonist binding at two glycine and two glutamate sites of the heterotetramer complex (Benveniste and Mayer 1991; Clements and Westbrook 1991). Age-associated changes in D-serine signaling could contribute to cognitive decline in aging (Billard and Rouaud 2007; Potier et al. 2010). Finally, other glutamatergic receptors such as α-amino-3-hydroxy-5-methyl-4-isoxazole-propionic acid (AMPA) and group 1 metabotropic glutamate receptors (mGluR) interact physically with NMDAR regulating, to some extent, its activity.

3.1.2 AMPA Receptors

Like NMDAR, ionotropic AMPAR consists of four subunits (GluA1–4) that form heteromeric tetrameric complexes (Traynelis et al. 2010; Huganir and Nicoll 2013). GluA1–4 subunits can be phosphorylated on serine, threonine, and tyrosine residues by several protein kinases including Ca^{2+}/calmodulin-dependent protein kinase II (CaMKII) and protein kinase C (PKC) on over 20 different phosphorylation sites (Shepherd and Huganir 2007; Lu and Roche 2012). Phosphorylation of AMPAR subunits regulates its function and intracellular trafficking, raising the hypothesis that posttranslational modifications could mediate synaptic plasticity (Isaac et al. 1995; Liao et al. 1995; Barria et al. 1997; Lee et al. 1998, 2000; Derkach et al. 1999). AMPAR trafficking between the plasma membrane and intracellular compartments is highly dynamic and can be modified by short-term and long-term changes in neuronal activity (for a review, see Bredt and Nicoll 2003; Huganir and Nicoll 2013). Synaptic scaling which is a homeostatic response to long-term changes in a network activity has been associated with AMPAR trafficking regulation by intrinsic activity (for a review, see Turrigiano 2008). Furthermore, AMPAR are mobile within the plasma membrane (Opazo and Choquet 2011), but their mobility decreases when entering the synapse (Borgdorff and Choquet 2002).

AMPAR synaptic levels and responsiveness can be modulated with various pharmacological agents including inhibitors [(2*R*)-amino-5-phosphonopentanoate, APV; 6-cyano-7-nitroquinoxaline-2,3-dione, CNQX; tetrodotoxin, TTX] and activators (bicuculline, picrotoxin) (Lissin et al. 1998; O'Brien et al. 1998; Turrigiano et al. 1998). Auxiliary subunits, known as transmembrane AMPAR regulatory proteins (TARPs), bind to the receptors and ensure proper maturation and delivery at the membrane and synapses (Tomita et al. 2003). TARPs can also affect biophysical and pharmacological properties of AMPAR (Priel et al. 2005; Menuz et al. 2007). For example, in the presence of TARPs, the antagonist CNQX acts as a partial agonist (Menuz et al. 2007). Experiments conducted with ampakines, a class of compounds strongly interacting with AMPAR, suggest that region-specific expression of GluA1–4 and TARPs may explain the variations reported in experimental drug activity (Montgomery et al. 2009). Ampakines potentiate AMPAR-mediated synaptic currents by slowing the receptor deactivation and, consequently, enhance synaptic responses and LTP (Staubli et al. 1994; Arai and Kessler 2007). Early on, ampakines were targeted as cognition-enhancing drugs (Davis et al. 1997; Hampson et al. 1998a, b). Interestingly in pilot experiments, ampakines improved recall memory in aged humans (Lynch et al. 1997).

3.1.3 mGlu Receptors

Our group and others have recently highlighted the importance of group 1 mGluR-related synaptic plasticity in successful cognitive aging (Menard and Quirion 2012b; Menard et al. 2013b, 2014b; Yang et al. 2013a). Eight mGluR have been identified and divided into three groups: group 1 includes postsynaptic mGluR1 and mGluR5, while group 2 (mGluR2, mGluR3) and group 3 (mGluR4, mGluR6, mGluR7, mGluR8) are mainly presynaptic (for a review, see Nicoletti et al. 2011). Activation of presynaptic mGluR2/3 following an excess of glutamate release from neurons or astrocytes inhibits neurotransmitter release, regulating synaptic plasticity and excitatory synaptic transmission (Yokoi et al. 1996; Altinbilek and Manahan-Vaughan 2009). Group 3 mGluRs are localized at the active zone of neurotransmitter release negatively autoregulating glutamate release (Niswender and Conn 2010). Postsynaptic mGluR1s are concentrated in perisynaptic and extrasynaptic areas and coupled to Gq/G11 proteins. Their activation stimulates phospholipase C and intracellular second messenger release, such as inositol-1,4,5-trisphosphate (IP3) and diacylglycerol (DAG) (Nicoletti et al. 2011). Finally, mGluR5s are also coupled to Gq/G11 protein, but their activation stimulates polyphosphoinositide (PI) hydrolysis. These receptors can functionally interact with NMDA receptor GluN2 subunits through a chain of interacting proteins including PSD-95, Shank, and Homer (Tu et al. 1999; Collett and Collingridge 2004).

Group 1 mGluRs are abundant in the hippocampus and cerebral cortex of the adult rat brain (Romano et al. 1996) and involved in hippocampus-dependent spatial learning and LTP (Balschun et al. 1999). Accordingly, mice lacking mGluR5 have reduced LTP and are characterized by cognitive deficits in the MWM task (Lu et al. 1997). Furthermore, spatial memory impairments are exacerbated in a reversal learning paradigm (Xu et al. 2009). This type of memory involves efficient pattern separation and inhibitory learning processes which can be affected by aging (Burke et al. 2010; Menard and Quirion 2012b; Menard et al. 2013b). Stimulation of group 1 mGluRs could act as a molecular switch to facilitate synaptic plasticity (Bortolotto et al. 2005; Manahan-Vaughan and Braunewell 2005; Bikbaev et al. 2008; Neyman and Manahan-Vaughan 2008) particularly in the aging brain (Menard and Quirion 2012a, b; Menard et al. 2013b, 2014b; Yang et al. 2013a).

Long-term depression (LTD), a form of synaptic plasticity involved in learning and memory processes (Ge et al. 2010; Dong et al. 2013; Menard et al. 2013b), can be induced by the group 1 mGluR-specific agonist 3,5-dihydroxyphenylglycine (DHPG) (Palmer et al. 1997). Age-related cognitive deficits have been associated with a reduction of DPHG-induced mGluR-LTD in old mice (Menard et al. 2013b). Over the years, multiple mGluR5 enhancers have been developed (for a review, see Cleva and Olive 2011; Nicoletti et al. 2011). Indeed, positive allosteric modulators can facilitate mGluR-related synaptic plasticity and improve spatial learning (Ayala et al. 2009; Menard et al. 2013b) possibly through NMDAR interaction (Rosenbrock et al. 2010) and/or AMPAR regulation (Uslaner et al. 2009). Conversely, mGluR5 antagonists impair learning and memory in adult (Christoffersen et al. 2008) and aged rodents (Menard et al. 2013b). Nevertheless, negative allosteric modulators are under clinical development because overactive mGluR functioning is thought to play a role in neurological disorders such as Alzheimer's disease and Fragile X syndrome (Luscher and Huber 2010).

3.2 Intracellular Glutamatergic Signaling

Learning and memory processes involve multiple signaling pathways triggered by glutamatergic receptor activation. Following Ca^{2+} entry in the neuron, cascades of kinases become phosphorylated leading to transcription factor activation and gene expression. The next section highlights several proteins essential for long-term synaptic plasticity establishment and maintenance.

3.2.1 CaMKII

The hypothesis that phosphorylation/dephosphorylation of AMPAR subunits regulates receptor function and modulates synaptic transmission was proposed in the early 1990s (Swope et al. 1992; Soderling 1993). Data from a number of studies have shown that protein kinase activity, particularly CaMKII, is required for LTP induction (Malenka et al. 1989; Malinow et al. 1989; Wyllie and Nicoll 1994). CaMKII is considered to be the primary downstream target following Ca^{2+} entry through NMDAR activation and associated with LTP, AMPAR trafficking, and memory formation (Anggono and Huganir 2012; Lisman et al. 2012). In fact, elevation of Ca^{2+} level in the cytoplasm induces recruitment of CaMKII to the PSD where it binds to NMDAR GluN2B subunits (Barria and Malinow 2005; Zhou et al. 2007; Halt et al. 2012) and phosphorylates multiple targets, notably GluN2B, AMPAR GluA1, and PSD-95 (Yoshimura et al. 2000, 2002; Dosemeci and Jaffe 2010). However, activation of CaMKII during LTP lasts only a few minutes (Lee et al. 2009), suggesting that downstream signaling cascades are required for LTP maintenance and memory consolidation.

3.2.2 PKC

Twelve PKC isoforms have been identified in mammals (Sun and Alkon 2010). These serine–threonine kinases are central to many signal transduction pathways and densely expressed in the brain (Saito et al. 1988). PKC isoforms seem to play an essential role in multiple forms of learning and memory processes (Bank et al. 1988; Olds et al. 1989; Coalombo et al. 1997; Colombo and Gallagher 2002; Nelson et al. 2008; Nithianantharajah and Murphy 2009; Zhang et al. 2009; Menard and Quirion 2012b). Inhibition of kinases such as PKC can block LTP induction (Malinow et al. 1989). Phosphorylation of GluA1 by PKC controls synaptic incorporation of GluA1-containing AMPAR into the synapses during LTP (Boehm et al. 2006). Moreover, GluA2 phosphorylation by PKC modifies its binding to scaffolding proteins (Matsuda et al. 1999; Chung et al. 2000) and appears to be essential for LTD (Chung et al. 2000). Activation of both ionotropic and metabotropic glutamate receptors stimulate PKCγ activity (Codazzi et al. 2006). Moreover, mGluR activity can enhance NMDAR currents via a PKC-dependent mechanism (Tyszkiewicz et al. 2004). For example, following training in a spatial memory task PKC gamma (γ) expression increases (Nithianantharajah and Murphy 2009). This kinase was linked to the individual differences observed in the cognitive status of aging rats (Coalombo et al. 1997; Colombo and Gallagher 2002; Menard and Quirion 2012b), and its activation in small groups of hippocampal or cortical neurons improves old rat performances in the MWM task (Zhang et al. 2009). PKCγ activity may promote neuronal interconnections (Menard et al. 2013a) and synaptogenesis (Hongpaisan and Alkon 2007) and protects against neurodegeneration (for a review, see Sun and Alkon 2010). PKC enzymes can be activated by Ca^{2+}, DAG, arachidonic acid, phospholipids, and phorbol esters. The development of cognition-enhancing drugs based on PKC isoform pharmacology was proposed to treat dementias (for a review, see Sun and Alkon 2010).

3.2.3 ERK

The extracellular signal-regulated kinases (ERKs) signaling pathway plays a crucial role in neuronal processes including long-term synaptic plasticity and memory formation (English and Sweatt 1996; Blum et al. 1999; Thomas and Huganir 2004; Davis and Laroche 2006; Ciccarelli and Giustetto 2014). ERKs activities regulate AMPAR transmission, potentiation by CaMKII, and insertion into synapses (Zhu et al. 2002). When activated by phosphorylation, ERKs translocate to the nucleus where they activate downstream transcription factors and immediate early genes (IEG) expression (Thomas and Huganir 2004; Davis and Laroche 2006; Menard and Quirion 2012a; Yang et al. 2013b). Long-term synaptic plasticity can last for weeks and the late phase is dependent on gene transcription activation and synthesis of new proteins (Bliss and Collingridge 1993; Lynch 2004). Following NMDAR or voltage-gated calcium channel activation, Ca^{2+} level increases in the cytoplasm activating ERK through Ras signaling (Rosen et al. 1994). However, Ras GTPases signaling can be induced by other stimuli including activation of tyrosine receptor kinase (Trk receptor) or G-protein-coupled receptors (GPCR) (Ciccarelli and Giustetto 2014). Ca^{2+}-independent co-activation of NMDAR and mGluR5 can also lead to ERK phosphorylation and IEG expression (Yang et al. 2004). ERK signaling is necessary to establish mGluR-LTD in the hippocampus (Gallagher et al. 2004) and seems to be affected by aging (Williams et al. 2006), possibly through age-related changes in Ca^{2+} homeostasis (Burke and Barnes 2010) leading to cognitive deficits (Menard and Quirion 2012b).

3.2.4 mTOR

The mammalian target of rapamycin (mTOR) serine/threonine kinase is another kinase regulating several translation regulatory factors and promoting protein synthesis (Page et al. 2006; Costa-Mattioli et al. 2009). Similarly to ERKs, mTOR inhibition blocks long-term synaptic plasticity and memory formation (Tang et al. 2002; Stoica et al. 2011). mTOR activation via phosphorylation can be triggered by various synaptic signals including glutamatergic agonists and neurotrophic factors such as insulin-like growth factor (IGF) or brain-derived neurotrophic factor (BDNF) (Costa-Mattioli et al. 2009; Costa-Mattioli and Monteggia 2013). mTOR complex 1 (mTORC1) has been associated with translational control, while mTORC2 seems to be involved in the cytoskeleton actin dynamics (for a review, see Costa-Mattioli and Monteggia 2013). Activation of NMDAR and mGluR modulates activity-dependent dendritic synthesis through mTOR activity in hippocampal neurons (Gong et al. 2006). Inhibition of mTOR prevents DHPG-induced mGluR-LTD (Hou and Klann 2004), while maintenance of good performances in the MWM spatial memory task was positively correlated with mTOR phosphorylation in aged rats (Menard and Quirion 2012b). Formation and stability of long-term fear memory is also compromised when mTOR activation is altered (Parsons et al. 2006). Altogether, mTOR function appears to be an attractive target in the cognition-enhancing target space. However, in addition to protein synthesis and actin polymerization, mTOR is involved in autophagy, lipid synthesis, ribosome biogenesis, nutrient support, and other growth-related processes (Costa-Mattioli and Monteggia 2013). Therefore, a better understanding of the various cell mechanisms associated with mTOR activity is necessary if one is to develop highly selective compounds that will improve cognition.

3.2.5 CREB

The transcription factor cAMP response element-binding protein (CREB) has probably been the most intensively studied kinase substrate with regard to cognition (for a review, see Alberini 2009). In fact, CREB-dependent transcription is essential for multiple forms of learning and memory such as fear conditioning and social recognition (Josselyn et al. 2001; Kida et al. 2002; Lonze and Ginty 2002; West et al. 2002; Han et al. 2007; Suzuki et al. 2011). Phosphorylation of CREB at the residue Ser133 regulates gene transcription (Shaywitz and Greenberg 1999) and this posttranslational modification is prevented by ERK inhibition (Wu et al. 2001; Hardingham et al. 2001). CaMKIV can also activate CREB-dependent transcription (Sun et al. 1996). Lower phosphorylation and total protein levels of CREB have been linked to age-related memory impairments in rats (Brightwell et al. 2004; Monti et al. 2005; Menard and Quirion 2012b). Compounds potentiating CREB activation have been identified as potential cognition-enhancing drugs (Tully et al. 2003; Xia et al. 2009). However, like mTOR, CREB is expressed ubiquitously and involved in several critical functions limiting its usefulness (Barco et al. 2003).

An alternative strategy might be to manipulate CREB primary gene targets and therefore, enhance treatment specificity.

3.3 Gene Expression

As mentioned previously, behavioral experience-induced activation of neuronal transmission and subsequent synaptic plasticity require the transcription of essential IEGs for long-term memory formation and consolidation (Marrone et al. 2008). These genes affect cell signaling, cytoskeletal dynamics, protein trafficking and degradation, and posttranslational modifications. In the following sections, the roles of five out of the growing list of genes involved in cognition (Benoit et al. 2011) are discussed.

3.3.1 Arc

The IEG activity-regulated cytoskeleton-associated protein (Arc) (Link et al. 1995; Lyford et al. 1995) is considered a master regulator of synaptic plasticity (Bramham et al. 2008; Shepherd and Bear 2011). In fact, cellular imaging of Arc mRNA and protein induction is currently used to detect the neuronal networks involved in behavioral encoding (Guzowski et al. 2005). Spatial exploration, for example, induces Arc transcription in ~40 % of hippocampal neurons of the hippocampus CA1 region after only 5 min (Guzowski et al. 2005). Several kinases and transcription factors are implicated in Arc expression including CaMKII, ERK, and CREB (Waltereit et al. 2001; Vazdarjanova et al. 2006; Shepherd and Bear 2011). Interestingly, Arc protein can be found in PSD and co-purified with NMDAR (Husi et al. 2000; Steward and Worley 2001). However NMDAR-independent synaptic transmission, notably through group 1 mGluR activity, can also regulate Arc transcription (Park et al. 2008). Our group reported higher Arc expression in memory-unimpaired old mice characterized by intact mGluR-LTD in comparison to aged mice for which mGluR-LTD and cognition were altered (Menard et al. 2013b). Arc mRNA is enriched in the dendrites of active synapses (Steward et al. 1998) possibly to facilitate protein expression, synaptic plasticity, and spine remodeling (Messaoudi et al. 2007). Downregulation of the Arc gene blocks consolidation of spatial memory (Guzowski et al. 2000) and fear conditioning (Ploski et al. 2008) while Arc knockout (KO) mice exhibit impaired long-term memory (Plath et al. 2006). Arc seems also crucial for the late phases and maintenance of LTP (Guzowski et al. 2000). Furthermore, mGluR-LTD requires Arc translation (Waung et al. 2008) which is impaired in Arc KO mice (Park et al. 2008). In fact, Arc affects AMPAR trafficking through interactions with the endocytic machinery (Chowdhury et al. 2006; Waung et al. 2008) and activity-dependent Arc induction is involved in AMPAR-mediated neuronal homeostasis (Shepherd et al. 2006; Beique et al. 2011). Development of cognition-enhancing drugs targeting Arc expression in specific area of the brain may therefore become a promising research avenue.

3.3.2 Homer 1a

Homer 1a is another interesting IEG dynamically regulated in response to synaptic activity and closely related to learning and memory formation (Vazdarjanova et al. 2002; Szumlinski et al. 2004; Celikel et al. 2007; Menard and Quirion 2012b; Menard et al. 2013b, 2014b). As mentioned previously, NMDARs directly interact with mGluRs through PSD-95, Shank, and Homer scaffolding proteins (Tu et al. 1999; Collett and Collingridge 2004). In fact, Homer proteins act as both scaffolding and transduction molecules (Brakeman et al. 1997; Ciruela et al. 2000; Ango et al. 2002; Fagni et al. 2002). Long Homer isoforms are constitutively expressed, enriched in PSD where they form synaptic clusters (Xiao et al. 1998) and facilitate signal transduction (Duncan et al. 2005; Shiraishi-Yamaguchi and Furuichi 2007). In contrast, the Homer 1a short isoform is an IEG produced following neuronal activity (Brakeman et al. 1997; Vazdarjanova et al. 2002) and when bound to mGluRs disrupts the protein clusters by dominant negative competitive binding (Kammermeier and Worley 2007). Homer 1a can also inhibit NMDAR currents by altering Homer–Shank complexes (Bertaso et al. 2010). Overexpression of Homer 1a in the hippocampus impairs LTP maintenance and spatial memory in adult mice (Celikel et al. 2007). Furthermore, elevated Homer 1a protein level has been correlated with cognitive deficits in aged rodents (Menard and Quirion 2012b; Menard et al. 2013b), which may be related to persistent uncoupling of mGluRs with its downstream signaling effectors (Menard and Quirion 2012b). To our knowledge, no drug has been proposed so far to directly modulate Homer protein expression or function.

3.3.3 Zif268

Induction of LTP is associated with a rapid and robust transcription of the IEG Zif268 in the hippocampus (Cole et al. 1989; Wisden et al. 1990; Jones et al. 2001; Alberini 2009). Learning-related increases in Zif268 expression have been reported for spatial (Guzowski et al. 2001) and fear memory (Hall et al. 2001). In mice lacking the Zif268 gene, LTP early phases are intact but late LTP is absent, and long-term memory is impaired in multiple tasks after a 24-h delay (Jones et al. 2001). Thus, expression of Zif268 may be critical for LTP persistence and memory consolidation (Abraham et al. 1993; Jones et al. 2001; Alberini 2009). Interestingly, learning task repetitions seem to reduce Zif268 expression (Guzowski et al. 2001). This observation is in line with similar Zif268 protein levels in aged rats trained for several consecutive weeks in the MWM task despite individual difference in cognitive status (Menard and Quirion 2012b). In a recent study, we reported a negative correlation between NMDAR, mGluR5, Arc, and Zif268 protein levels in old rats, suggesting that persistent transcription of this IEG may be involved in age-related cognitive deficits (Menard et al. 2014b). Such as for Homer 1a, no drug is currently available to modulate Zif268 expression or function.

3.3.4 IGF

As mentioned earlier, mTOR activity can be triggered by the binding of the neurotrophic factor IGF to Trk receptors, initiating intracellular signaling (Costa-

Mattioli et al. 2009; Costa-Mattioli and Monteggia 2013). PKC activity modulates IGF-1-induced activation of the serine–threonine protein kinase Akt (Zheng et al. 2000), a major actor of neuronal survival regulation (Dudek et al. 1997). IGFs play an important role in development, tissue repair, apoptosis, and regeneration (Dore et al. 1997; Werther et al. 1998; Russo et al. 2005) as well as in memory formation, consolidation, enhancement and extinction (Svensson et al. 2006; Agis-Balboa et al. 2011; Chen et al. 2011; Stern et al. 2014). IGF-I and IGF-II are growth-promoting peptides acting on plasma membrane Trk receptors on the cell surface, the type I IGF receptors (IGF-IR) (Russo et al. 2005). IGF binding to the IGF-IR promotes the activation of downstream signaling cascades including ERK (Russo et al. 2005). IGF-II is the most abundantly expressed IGF in the adult brain and is particularly concentrated in the hippocampus (Kar et al. 1993). Interestingly, an IGF-II polymorphism has been associated with cognitive functions in humans (Alfimova et al. 2012) and IGF-II expression declines with aging (Kitraki et al. 1993). Intra-hippocampal injection of recombinant IGF-II enhances memory retention and prevents forgetting via an increase of AMPAR GluA1 subunits and generation of persistent LTP (Chen et al. 2011). Moreover, systemic treatment with IGF-II increases Arc and Zif268 expression in the hippocampus (Stern et al. 2014). These recent studies suggest that IGF-II may represent an attractive target to develop cognition-enhancing drugs.

3.3.5 BDNF

Age-related cognitive deficits might be related to impaired LTP stability (Deupree et al. 1993; Burke and Barnes 2010). Synaptic transmission stimulates the release of BDNF (Balkowiec and Katz 2002; Aicardi et al. 2004), which is associated with rapid modifications of spine actin networks and LTP consolidation (Rex et al. 2007). LTP expression is impaired in BDNF KO mice (Korte et al. 1995), but this deficit can be completely rescued by recombinant BDNF (Patterson et al. 1996). Neurotrophins such as BDNF stimulate process outgrowth during development but also modified the axonal and dendritic cytoskeletons in the mature nervous system directly controlling synaptic plasticity (for reviews, see Huang and Reichardt 2001; Miller and Kaplan 2003). Through Trk receptors activation, neurotrophins regulate CaMKII activity (He et al. 2000) and ERK signaling pathway (Kaplan and Miller 2000). Exogenous BDNF can directly potentiate synaptic transmission (Kang and Schuman 1995) and this effect was proposed to be Arc dependent (Messaoudi et al. 2007). Contextual learning induces a rapid and selective increase of BDNF expression in the hippocampus (Hall et al. 2000) and BDNF-mediated signaling is involved in spatial learning (Mizuno et al. 2003) and fear memory (Andero and Ressler 2012). Physical exercise benefits cognitive processes and neuronal plasticity and this phenomenon seems to be mediated by IGF-1 and signaling cascades triggered by BDNF expression (Ding et al. 2006). Clinical trials have been conducted with BDNF as a therapeutic target for psychiatric diseases with undesirable side effects (Lynch et al. 2008). An alternative strategy would be to increase the production of endogenous BDNF. Ampakines, for example, increase

BDNF production in vitro and in vivo in rodents up to an advanced age (Lauterborn et al. 2000).

4 Cholinergic System and Cognition

The neurotransmitter acetylcholine (ACh) and its receptors play an active role in cognitive processes (Sarter and Parikh 2005). ACh action might be mediated through the regulation of NMDA. Indeed, stimulation of muscarinic ACh receptors (mAChR) potentiates NMDAR responses in the hippocampus (Markram and Segal 1990) and can facilitate NMDAR-LTP induction (Shinoe et al. 2005). In addition, mAChR activation can also promote NMDAR-LTD (Kirkwood et al. 1999; Jo et al. 2010) and induce a NMDAR-independent form of LTD (Dickinson et al. 2009). This last form of plasticity does not appear to involve the same mechanisms as mGluR-LTD (Dickinson et al. 2009). ACh can bind and activate two main classes of receptors: metabotropic mAChRs and nicotinic AChRs (nAChRs) which are ionotropic and permeable to Na^+, K^+, and Ca^{2+} (for a review, see Deiana et al. 2011). An age-related downregulation of the cholinergic system has been proposed to explain the progressive impairments of cognitive abilities associated with normal and pathological aging (Bartus et al. 1982; Bartus 2000; Auld et al. 2002). In fact, increased activation of the cholinergic system generally facilitates learning and memory processes (Scali et al. 1997a, b; Bradley et al. 2010). However, an increased expression of the negative autoreceptor mACh2 was reported in aged rats exhibiting memory deficits (Aubert et al. 1995) and inhibition of these receptors may facilitate spatial memory function (Quirion et al. 1995).

4.1 Impact of AChR Agonists and Antagonists on Memory Function

Early on, studies showed that muscarinic antagonists such as scopolamine or atropine impair cognitive abilities in animals and humans (Deutsch 1971; Drachman 1977). Systemic administration of scopolamine impairs learning acquisition and memory formation in multiple tasks (Aigner and Mishkin 1986; Aigner et al. 1991; Miller and Desimone 1993; Brouillette et al. 2007). Interestingly, IEG Homer 1a expression is enhanced in the hippocampus of amnesic scopolamine-treated rats (Brouillette et al. 2007). Conversely, treatment with a mACh1R allosteric agonist improves cognitive performances (Bradley et al. 2010). Neurotrophins enhance ACh release through TrkA receptor signaling (Auld et al. 2001) and activation of the TrkA receptor with a selective partial agonist can rescue age-related memory deficits in rats through modulation of the cholinergic system (Bruno et al. 2004). IGFs differentially regulate ACh release: IGF-I acts as an inhibitor, while IGF-II potentiated ACh-related currents in rat hippocampal slices (Kar et al. 1997). TTX alters the effect of IGF-I (Kar et al. 1997) suggesting an interaction with AMPAR. Treatment of rat cultured olfactory bulb neuronal cells with carbachol, a cholinergic agonist, increases neuritic outgrowth and this effect is mediated by nAChR since it can be mimicked with nicotinic agonists (Coronas et al. 2000). Furthermore, low concentrations of carbachol can potentiate NMDA responses in the hippocampus (Harvey et al. 1993). These results suggest that nAChR may be actively involved in neuronal plasticity and could represent an attractive target to develop cognition-enhancing drugs.

4.2 α-7 Nicotinic ACh Receptor Agonists and Cognitive Deficits

Multiple nAChR agonists have been examined as possible treatments for memory impairment associated with aging or in psychiatric disorders. In this regard, modulation of the ionotropic α7 nAChR is of particular interest, considering its high density in the hippocampus and cerebral cortex and its implication in cognitive processes (Paterson and Nordberg 2000; Levin and Rezvani 2002; Leiser et al. 2009; Floresco and Jentsch 2011). Treatment with of α7 nAChR agonists such as *N*-[(3*R*)-1-azabicyclo[2.2.2]oct-3-yl]-7-[2-(methoxy)phenyl]-1-benzofuran-2-carboxamide (ABBF), 5-morpholin-4-yl-pentanoic acid (4-pyridin-3-yl-phenyl)-amide (SEN12333), or 4-bromophenyl 1,4diazabicyclo (3.2.2) nonane-4-carboxylate, monohydrochloride (SSR180711) rescue cognitive deficits in spatial (Boess et al. 2007; Pichat et al. 2007), recognition (Wishka et al. 2006; Boess et al. 2007; Pichat et al. 2007; Hashimoto et al. 2008; Roncarati et al. 2009), social (Boess et al. 2007), and fear (Roncarati et al. 2009) memory tasks. Multiple studies have tested the efficacy of α7 nAChR agonists and demonstrated positive cognitive effects following activation of these receptors (for a review, see Leiser et al. 2009). Several clinical trials are currently ongoing notably to treat negative symptoms of schizophrenia (Davis et al. 2014) and Alzheimer's disease (Geerts 2012).

5 Dynorphinergic System and Memory Function

Dynorphins, a class of endogenous opioids peptides expressed in the brain (for a review, see Schwarzer 2009), have been linked to learning and memory processes since the 1990s (McDaniel et al. 1990; Wagner et al. 1993; Sandin et al. 1998). Intra-hippocampal administration of dynorphin in rats impairs spatial learning (McDaniel et al. 1990; Sandin et al. 1998). Encoded by the prodynorphin gene (Pdyn), dynorphin peptides are also involved in emotional control and stress responses (Schwarzer 2009). In humans, Pdyn gene polymorphisms have been associated with episodic memory deficits in the elderly (Kolsch et al. 2009). Furthermore, enhanced dynorphins expression might be related to Alzheimer's disease pathogenesis (Yakovleva et al. 2007). Surprisingly, dynorphin A-(1–13) injection can improve scopolamine-induced cognitive deficits in mice by activating kappa-opioid receptors (KOR) (Itoh et al. 1993) and possibly regulating ACh release (Hiramatsu et al. 1998; Hiramatsu and Watanabe 2006). Pdyn-derived peptides preferentially bind to the postsynaptic GPCR, KOR (Chavkin et al. 1982), modulating PKC (Barg et al. 1993), and ERK signaling pathway activation (Belcheva et al. 1998). Presynaptic KOR can act as an autoreceptor and inhibits the release of dynorphin peptides (Nikolarakis et al. 1989). These peptides can also interact with other opioid receptors (Quirion and Pert 1981; Schwarzer 2009) and NMDAR (Shukla and Lemaire 1994; Schwarzer 2009). Release of endogenous dynorphins inhibits excitatory transmission and blocks LTP induction in the hippocampus (Wagner et al. 1993). Furthermore, dynorphins and activation of presynaptic KORs suppress glutamate release (Drake et al. 1994; Simmons et al. 1994). These findings suggest that dampening of the dynorphinergic system may be a relevant strategy to modulate glutamatergic function and cognition.

5.1 Dynorphins and Age-Related Cognitive Decline

Expression of dynorphins increases with age in the hippocampus of rats (Jiang et al. 1989; Zhang et al. 1991; Kotz et al. 2004) and mice (Menard et al. 2013b) and this upregulation may be associated with cognitive deficits generally observed in old rodents (Jiang et al. 1989; Zhang et al. 1991; Menard et al. 2013b). In line with this idea, knocking down the Pdyn gene improves spatial learning in middle-aged mice (Nguyen et al. 2005). Our group has recently shown that elevated Pdyn expression correlates with age-related body weight gain, memory deficits, and reduced glutamatergic signaling in rats (Menard et al. 2014b). Furthermore, we rescued loss of group 1 mGluR function, related signaling, and cognitive decline in old mice by knocking down the Pdyn gene (Menard et al. 2013b). Whereas aged wild-type (WT) mice developed spatial and recognition memory deficits, aged Pdyn KO mice performances were similar to those of young mice in both tasks (Menard et al. 2013b). Old WT mice performed poorly in an inhibitory learning acquisition task, which has been related to mGluR5 function (Xu et al. 2009). Accordingly, group 1 mGluR protein level was increased and mGluR-LTD unaltered in old KO mice (Menard et al. 2013b). Intact synaptic plasticity and cognition were associated with increased expression of IEG Homer 1a and Arc in aged Pdyn KO mice (Menard et al. 2013b). Pharmacological treatments with 3-cyano-*N*-(1,3-diphenyl-1H-pyrazol-5-yl) benzamide (CDPPB, positive modulator of mGluR5) or norbinaltorphimine (norBNI), a KOR antagonist, rescued memory function in old WT mice. These results are in line with previous studies in which positive modulation of mGluR5 (Uslaner et al. 2009; Reichel et al. 2011; Fowler et al. 2013) as well as norBNI treatment (Bilkei-Gorzo et al. 2014) promoted memory formation. Conversely, mGluR5 antagonism impaired spatial memory of old Pdyn KO mice (Menard et al. 2013b), suggesting that dynorphinergic and glutamatergic systems closely interact to establish memories in the aging brain (Menard et al. 2013b, 2014a, b). Gene expression profiling reveals increased Pdyn expression in the hippocampus of amnesic scopolamine-treated rats (Brouillette et al. 2007), raising the possibility of complex interactions between these systems in cognitive functions.

5.2 Dynorphins and Social Memory

Considering its role in emotional behaviors and stress responses (Schwarzer 2009), the dynorphinergic system might also regulate the strength of social memories. In young mice, genetic deletion of the Pdyn gene enhanced partner recognition ability without affecting recognition memory for objects (Bilkei-Gorzo et al. 2014). Pharmacological blockade of KOR with norBNI enhanced social memory in control animals, whereas KOR activation impaired the abilities of transgenic mice (Bilkei-Gorzo et al. 2014). Emotionally arousing situation such as partner recognition induces higher expression of dynorphins than novel object recognition (Bilkei-Gorzo et al. 2014), raising the possibility that stress-related release of these peptides may affect the formation of social memories.

5.3 Dynorphins, KOR and Stress-Related Memory Deficits

Aging is generally characterized not only by reduced cognitive abilities but also by increased anxiety-related behaviors (Lenze et al. 2001; Lupien et al. 2009; Bedrosian et al. 2011; Menard et al. 2013b, 2014b). Stress exposure over a life span may accelerate cellular aging and promote cognitive dysfunction (Lupien et al. 2009). Furthermore, exacerbated neurobiological sensitivity to threat may even increase the risk of developing age-related diseases (for a review, see O'Donovan et al. 2013). The first association between the dynorphinergic system and anxious behaviors was observed with naloxone, an opioid partial agonist, reversing the effect of benzodiazepines (Billingsley and Kubena 1978). Similar to Pdyn gene deletion, pharmacological treatment with norBNI reduces anxious behaviors and increases exploratory activity in young (Knoll et al. 2007; Wittmann et al. 2009) and aged rodents (Menard et al. 2013b). Conversely, treatment with dynorphin peptides and KOR agonists is anxiogenic (Tsuda et al. 1996; Wittmann et al. 2009; Smith et al. 2012). Endogenous KOR activation has been linked to stress-induced learning and memory deficits (Carey et al. 2009). KOR signaling could also play a role in fear memory extinction (Bilkei-Gorzo et al. 2012). Indeed, mice lacking Pdyn gene are characterized by enhanced cue-dependent fear conditioning, an effect that can be reproduced by blocking KOR before the extinction trials (Bilkei-Gorzo et al. 2012). Interestingly, functional imaging has revealed reduced fear extinction in human volunteers bearing Pdyn polymorphisms (Bilkei-Gorzo et al. 2012), suggesting that dynorphins might be essential to efficient fear memory consolidation. All these results identify the dynorphinergic system as a promising target to develop novel cognition-enhancing drugs that could be efficient in not only in normal but also pathological aging.

6 Conclusions

In summary, despite memory function involving multiple types and processes at synaptic, cellular, and molecular levels, promising targets have been identified that could lead to novel cognition-enhancing drugs. Up to now, glutamatergic and cholinergic receptor modulators have been extensively studied and, in some cases, tested in clinical studies with equivocal results. Here we propose novel targets involved in crucial signaling pathways. Nonetheless, to create efficient tissue-specific and even cell type-specific compounds, modulating these effectors remains a challenge at the chemistry, pharmacokinetic, and formulation levels. However, considering the increase in life span generally observed in various populations, reduction of age-related cognitive deficits represents a biomedical issue deserving a multidisciplinary global approach.

Acknowledgments This research was supported by grants from the Canadian Institutes of Health Research (CIHR) to R.Q. and the Quebec Network for Research on Aging (a network funded by the *Fonds de recherche du Québec—Santé*) to P.G. C.M. was the recipient of a postdoctoral fellowship from CIHR.

References

Abraham WC, Mason SE, Demmer J, Williams JM, Richardson CL, Tate WP, Lawlor PA, Dragunow M (1993) Correlations between immediate early gene induction and the persistence of long-term potentiation. Neuroscience 56:717–727

Agis-Balboa RC, Arcos-Diaz D, Wittnam J, Govindarajan N, Blom K, Burkhardt S, Haladyniak U, Agbemenyah HY, Zovoilis A, Salinas-Riester G, Opitz L, Sananbenesi F, Fischer A (2011) A hippocampal insulin-growth factor 2 pathway regulates the extinction of fear memories. EMBO J 30:4071–4083

Aicardi G, Argilli E, Cappello S, Santi S, Riccio M, Thoenen H, Canossa M (2004) Induction of long-term potentiation and depression is reflected by corresponding changes in secretion of endogenous brain-derived neurotrophic factor. Proc Natl Acad Sci U S A 101:15788–15792

Aigner TG, Mishkin M (1986) The effects of physostigmine and scopolamine on recognition memory in monkeys. Behav Neural Biol 45:81–87

Aigner TG, Mitchell SJ, Aggleton JP, Delong MR, Struble RG, Price DL, Wenk GL, Pettigrew KD, Mishkin M (1991) Transient impairment of recognition memory following ibotenic-acid lesions of the basal forebrain in macaques. Exp Brain Res 86:18–26

Alberini CM (2009) Transcription factors in long-term memory and synaptic plasticity. Physiol Rev 89:121–145

Alfimova MV, Lezheiko TV, Gritsenko IK, Golimbet VE (2012) [Association of the insulin-like growth factor II (IGF2) gene with human cognitive functions]. Genetika 48:993–998

Altinbilek B, Manahan-Vaughan D (2009) A specific role for group II metabotropic glutamate receptors in hippocampal long-term depression and spatial memory. Neuroscience 158:149–158

Andero R, Ressler KJ (2012) Fear extinction and BDNF: translating animal models of PTSD to the clinic. Genes Brain Behav 11:503–512

Anggono V, Huganir RL (2012) Regulation of AMPA receptor trafficking and synaptic plasticity. Curr Opin Neurobiol 22:461–469

Ango F, Robbe D, Tu JC, Xiao B, Worley PF, Pin JP, Bockaert J, Fagni L (2002) Homer-dependent cell surface expression of metabotropic glutamate receptor type 5 in neurons. Mol Cell Neurosci 20:323–329

Arai AC, Kessler M (2007) Pharmacology of ampakine modulators: from AMPA receptors to synapses and behavior. Curr Drug Targets 8:583–602

Aubert I, Rowe W, Meaney MJ, Gauthier S, Quirion R (1995) Cholinergic markers in aged cognitively impaired Long-Evans rats. Neuroscience 67:277–292

Auld DS, Mennicken F, Day JC, Quirion R (2001) Neurotrophins differentially enhance acetylcholine release, acetylcholine content and choline acetyltransferase activity in basal forebrain neurons. J Neurochem 77:253–262

Auld DS, Kornecook TJ, Bastianetto S, Quirion R (2002) Alzheimer's disease and the basal forebrain cholinergic system: relations to beta-amyloid peptides, cognition, and treatment strategies. Prog Neurobiol 68:209–245

Ayala JE, Chen Y, Banko JL, Sheffler DJ, Williams R, Telk AN, Watson NL, Xiang Z, Zhang Y, Jones PJ, Lindsley CW, Olive MF, Conn PJ (2009) mGluR5 positive allosteric modulators facilitate both hippocampal LTP and LTD and enhance spatial learning. Neuropsychopharmacology 34:2057–2071

Balkowiec A, Katz DM (2002) Cellular mechanisms regulating activity-dependent release of native brain-derived neurotrophic factor from hippocampal neurons. J Neurosci 22:10399–10407

Balschun D, Manahan-Vaughan D, Wagner T, Behnisch T, Reymann KG, Wetzel W (1999) A specific role for group I mGluRs in hippocampal LTP and hippocampus-dependent spatial learning. Learn Mem 6:138–152

Bank B, Deweer A, Kuzirian AM, Rasmussen H, Alkon DL (1988) Classical conditioning induces long-term translocation of protein kinase C in rabbit hippocampal CA1 cells. Proc Natl Acad Sci U S A 85:1988–1992

Barco A, Pittenger C, Kandel ER (2003) CREB, memory enhancement and the treatment of memory disorders: promises, pitfalls and prospects. Expert Opin Ther Targets 7:101–114

Barg J, Belcheva MM, Rowinski J, Coscia CJ (1993) kappa-Opioid agonist modulation of [3H] thymidine incorporation into DNA: evidence for the involvement of pertussis toxin-sensitive G protein-coupled phosphoinositide turnover. J Neurochem 60:1505–1511

Barnes CA (1979) Memory deficits associated with senescence: a neurophysiological and behavioral study in the rat. J Comp Physiol Psychol 93:74–104

Barnes CA (2003) Long-term potentiation and the ageing brain. Philos Trans R Soc Lond B Biol Sci 358:765–772

Barnes CA, Rao G, Foster TC, McNaughton BL (1992) Region-specific age effects on AMPA sensitivity: electrophysiological evidence for loss of synaptic contacts in hippocampal field CA1. Hippocampus 2:457–468

Barrett GL, Bennie A, Trieu J, Ping S, Tsafoulis C (2009) The chronology of age-related spatial learning impairment in two rat strains, as tested by the Barnes maze. Behav Neurosci 123:533–538

Barria A, Malinow R (2005) NMDA receptor subunit composition controls synaptic plasticity by regulating binding to CaMKII. Neuron 48:289–301

Barria A, Muller D, Derkach V, Griffith LC, Soderling TR (1997) Regulatory phosphorylation of AMPA-type glutamate receptors by CaM-KII during long-term potentiation. Science 276:2042–2045

Bartus RT (2000) On neurodegenerative diseases, models, and treatment strategies: lessons learned and lessons forgotten a generation following the cholinergic hypothesis. Exp Neurol 163:495–529

Bartus RT, Dean RL 3rd, Beer B, Lippa AS (1982) The cholinergic hypothesis of geriatric memory dysfunction. Science 217:408–414

Baudry M, Bi X, Gall C, Lynch G (2011) The biochemistry of memory: the 26year journey of a 'new and specific hypothesis'. Neurobiol Learn Mem 95:125–133

Baudry M, Chou MM, Bi X (2013) Targeting calpain in synaptic plasticity. Expert Opin Ther Targets 17:579–592

Baxter MG, Lanthorn TH, Frick KM, Golski S, Wan RQ, Olton DS (1994) D-cycloserine, a novel cognitive enhancer, improves spatial memory in aged rats. Neurobiol Aging 15:207–213

Bedrosian TA, Herring KL, Weil ZM, Nelson RJ (2011) Altered temporal patterns of anxiety in aged and amyloid precursor protein (APP) transgenic mice. Proc Natl Acad Sci U S A 108:11686–11691

Beique JC, Na Y, Kuhl D, Worley PF, Huganir RL (2011) Arc-dependent synapse-specific homeostatic plasticity. Proc Natl Acad Sci U S A 108:816–821

Belcheva MM, Vogel Z, Ignatova E, Avidor-Reiss T, Zippel R, Levy R, Young EC, Barg J, Coscia CJ (1998) Opioid modulation of extracellular signal-regulated protein kinase activity is ras-dependent and involves Gbetagamma subunits. J Neurochem 70:635–645

Bellone C, Nicoll RA (2007) Rapid bidirectional switching of synaptic NMDA receptors. Neuron 55:779–785

Benoit CE, Rowe WB, Menard C, Sarret P, Quirion R (2011) Genomic and proteomic strategies to identify novel targets potentially involved in learning and memory. Trends Pharmacol Sci 32:43–52

Benveniste M, Mayer ML (1991) Kinetic analysis of antagonist action at N-methyl-D-aspartic acid receptors. Two binding sites each for glutamate and glycine. Biophys J 59:560–573

Berry RJ, Bronson FH (1992) Life history and bioeconomy of the house mouse. Biol Rev Camb Philos Soc 67:519–550

Bertaso F, Roussignol G, Worley P, Bockaert J, Fagni L, Ango F (2010) Homer1a-dependent crosstalk between NMDA and metabotropic glutamate receptors in mouse neurons. PLoS One 5:e9755

Bikbaev A, Neyman S, Ngomba RT, Conn PJ, Nicoletti F, Manahan-Vaughan D (2008) MGluR5 mediates the interaction between late-LTP, network activity, and learning. PLoS One 3:e2155

Bilkei-Gorzo A, Erk S, Schurmann B, Mauer D, Michel K, Boecker H, Scheef L, Walter H, Zimmer A (2012) Dynorphins regulate fear memory: from mice to men. J Neurosci 32:9335–9343

Bilkei-Gorzo A, Mauer D, Michel K, Zimmer A (2014) Dynorphins regulate the strength of social memory. Neuropharmacology 77:406–413

Billard JM, Rouaud E (2007) Deficit of NMDA receptor activation in CA1 hippocampal area of aged rats is rescued by D-cycloserine. Eur J Neurosci 25:2260–2268

Billingsley ML, Kubena RK (1978) The effects of naloxone and picrotoxin on the sedative and anticonflict effects of benzodiazepines. Life Sci 22:897–906

Bird CM, Burgess N (2008) The hippocampus and memory: insights from spatial processing. Nat Rev Neurosci 9:182–194

Bliss TV, Collingridge GL (1993) A synaptic model of memory: long-term potentiation in the hippocampus. Nature 361:31–39

Blot K, Kimura SI, Bai J, Kemp A, Manahan-Vaughan D, Giros B, Tzavara E, Otani S (2013) Modulation of hippocampus-prefrontal cortex synaptic transmission and disruption of executive cognitive functions by MK-801. Cereb Cortex

Blum S, Moore AN, Adams F, Dash PK (1999) A mitogen-activated protein kinase cascade in the CA1/CA2 subfield of the dorsal hippocampus is essential for long-term spatial memory. J Neurosci 19:3535–3544

Boehm J, Kang MG, Johnson RC, Esteban J, Huganir RL, Malinow R (2006) Synaptic incorporation of AMPA receptors during LTP is controlled by a PKC phosphorylation site on GluR1. Neuron 51:213–225

Boess FG, De Vry J, Erb C, Flessner T, Hendrix M, Luithle J, Methfessel C, Riedl B, Schnizler K, van der Staay FJ, van Kampen M, Wiese WB, Koenig G (2007) The novel alpha7 nicotinic acetylcholine receptor agonist N-[(3R)-1-azabicyclo[2.2.2]oct-3-yl]-7-[2-(methoxy)phenyl]-1-benzofuran-2-carboxa mide improves working and recognition memory in rodents. J Pharmacol Exp Ther 321:716–725

Borgdorff AJ, Choquet D (2002) Regulation of AMPA receptor lateral movements. Nature 417:649–653

Bortolotto ZA, Collett VJ, Conquet F, Jia Z, van der Putten H, Collingridge GL (2005) The regulation of hippocampal LTP by the molecular switch, a form of metaplasticity, requires mGlu5 receptors. Neuropharmacology 49(Suppl 1):13–25

Bouwknecht JA, Paylor R (2002) Behavioral and physiological mouse assays for anxiety: a survey in nine mouse strains. Behav Brain Res 136:489–501

Bradley SR, Lameh J, Ohrmund L, Son T, Bajpai A, Nguyen D, Friberg M, Burstein ES, Spalding TA, Ott TR, Schiffer HH, Tabatabaei A, McFarland K, Davis RE, Bonhaus DW (2010) AC-260584, an orally bioavailable M(1) muscarinic receptor allosteric agonist, improves cognitive performance in an animal model. Neuropharmacology 58:365–373

Brakeman PR, Lanahan AA, O'Brien R, Roche K, Barnes CA, Huganir RL, Worley PF (1997) Homer: a protein that selectively binds metabotropic glutamate receptors. Nature 386:284–288

Bramham CR, Worley PF, Moore MJ, Guzowski JF (2008) The immediate early gene arc/arg3.1: regulation, mechanisms, and function. J Neurosci 28:11760–11767

Bredt DS, Nicoll RA (2003) AMPA receptor trafficking at excitatory synapses. Neuron 40:361–379

Bresink I, Benke TA, Collett VJ, Seal AJ, Parsons CG, Henley JM, Collingridge GL (1996) Effects of memantine on recombinant rat NMDA receptors expressed in HEK 293 cells. Br J Pharmacol 119:195–204

Brightwell JJ, Gallagher M, Colombo PJ (2004) Hippocampal CREB1 but not CREB2 is decreased in aged rats with spatial memory impairments. Neurobiol Learn Mem 81:19–26

Brouillette J, Young D, During MJ, Quirion R (2007) Hippocampal gene expression profiling reveals the possible involvement of Homer1 and GABA(B) receptors in scopolamine-induced amnesia. J Neurochem 102:1978–1989

Bruno MA, Clarke PB, Seltzer A, Quirion R, Burgess K, Cuello AC, Saragovi HU (2004) Long-lasting rescue of age-associated deficits in cognition and the CNS cholinergic phenotype by a partial agonist peptidomimetic ligand of TrkA. J Neurosci 24:8009–8018

Brushfield AM, Luu TT, Callahan BD, Gilbert PE (2008) A comparison of discrimination and reversal learning for olfactory and visual stimuli in aged rats. Behav Neurosci 122:54–62

Bunce D, Batterham PJ, Mackinnon AJ, Christensen H (2012) Depression, anxiety and cognition in community-dwelling adults aged 70 years and over. J Psychiatr Res 46:1662–1666

Burke SN, Barnes CA (2006) Neural plasticity in the ageing brain. Nat Rev Neurosci 7:30–40

Burke SN, Barnes CA (2010) Senescent synapses and hippocampal circuit dynamics. Trends Neurosci 33:153–161

Burke SN, Wallace JL, Nematollahi S, Uprety AR, Barnes CA (2010) Pattern separation deficits may contribute to age-associated recognition impairments. Behav Neurosci 124:559–573

Burke SN, Ryan L, Barnes CA (2012) Characterizing cognitive aging of recognition memory and related processes in animal models and in humans. Front Aging Neurosci 4:15

Cao X, Cui Z, Feng R, Tang YP, Qin Z, Mei B, Tsien JZ (2007) Maintenance of superior learning and memory function in NR2B transgenic mice during ageing. Eur J Neurosci 25:1815–1822

Carey AN, Lyons AM, Shay CF, Dunton O, McLaughlin JP (2009) Endogenous kappa opioid activation mediates stress-induced deficits in learning and memory. J Neurosci 29:4293–4300

Celikel T, Marx V, Freudenberg F, Zivkovic A, Resnik E, Hasan MT, Licznerski P, Osten P, Rozov A, Seeburg PH, Schwarz MK (2007) Select overexpression of homer1a in dorsal hippocampus impairs spatial working memory. Front Neurosci 1:97–110

Chavkin C, James IF, Goldstein A (1982) Dynorphin is a specific endogenous ligand of the kappa opioid receptor. Science 215:413–415

Chen DY, Stern SA, Garcia-Osta A, Saunier-Rebori B, Pollonini G, Bambah-Mukku D, Blitzer RD, Alberini CM (2011) A critical role for IGF-II in memory consolidation and enhancement. Nature 469:491–497

Choleris E, Clipperton-Allen AE, Phan A, Kavaliers M (2009) Neuroendocrinology of social information processing in rats and mice. Front Neuroendocrinol 30:442–459

Choquet D, Triller A (2013) The dynamic synapse. Neuron 80:691–703

Chowdhury S, Shepherd JD, Okuno H, Lyford G, Petralia RS, Plath N, Kuhl D, Huganir RL, Worley PF (2006) Arc/Arg3.1 interacts with the endocytic machinery to regulate AMPA receptor trafficking. Neuron 52:445–459

Christoffersen GR, Simonyi A, Schachtman TR, Clausen B, Clement D, Bjerre VK, Mark LT, Reinholdt M, Schmith-Rasmussen K, Zink LV (2008) MGlu5 antagonism impairs exploration and memory of spatial and non-spatial stimuli in rats. Behav Brain Res 191:235–245

Chung HJ, Xia J, Scannevin RH, Zhang X, Huganir RL (2000) Phosphorylation of the AMPA receptor subunit GluR2 differentially regulates its interaction with PDZ domain-containing proteins. J Neurosci 20:7258–7267

Ciccarelli A, Giustetto M (2014) Role of ERK signaling in activity-dependent modifications of histone proteins. Neuropharmacology 80C:34–44

Ciruela F, Soloviev MM, Chan WY, McIlhinney RA (2000) Homer-1c/Vesl-1L modulates the cell surface targeting of metabotropic glutamate receptor type 1alpha: evidence for an anchoring function. Mol Cell Neurosci 15:36–50

Clements JD, Westbrook GL (1991) Activation kinetics reveal the number of glutamate and glycine binding sites on the N-methyl-D-aspartate receptor. Neuron 7:605–613

Cleva RM, Olive MF (2011) Positive allosteric modulators of type 5 metabotropic glutamate receptors (mGluR5) and their therapeutic potential for the treatment of CNS disorders. Molecules 16:2097–2106

Coalombo PJ, Wetsel WC, Gallagher M (1997) Spatial memory is related to hippocampal subcellular concentrations of calcium-dependent protein kinase C isoforms in young and aged rats. Proc Natl Acad Sci U S A 94:14195–14199

Codazzi F, DI Cesare A, Chiulli N, Albanese A, Meyer T, Zacchetti D, Grohovaz F (2006) Synergistic control of protein kinase Cgamma activity by ionotropic and metabotropic glutamate receptor inputs in hippocampal neurons. J Neurosci 26:3404–3411

Cole AJ, Saffen DW, Baraban JM, Worley PF (1989) Rapid increase of an immediate early gene messenger RNA in hippocampal neurons by synaptic NMDA receptor activation. Nature 340:474–476

Collett VJ, Collingridge GL (2004) Interactions between NMDA receptors and mGlu5 receptors expressed in HEK293 cells. Br J Pharmacol 142:991–1001

Collingridge GL, Volianskis A, Bannister N, France G, Hanna L, Mercier M, Tidball P, Fang G, Irvine MW, Costa BM, Monaghan DT, Bortolotto ZA, Molnar E, Lodge D, Jane DE (2013) The NMDA receptor as a target for cognitive enhancement. Neuropharmacology 64:13–26

Colombo PJ, Gallagher M (2002) Individual differences in spatial memory among aged rats are related to hippocampal PKCgamma immunoreactivity. Hippocampus 12:285–289

Coronas V, Durand M, Chabot JG, Jourdan F, Quirion R (2000) Acetylcholine induces neuritic outgrowth in rat primary olfactory bulb cultures. Neuroscience 98:213–219

Costa-Mattioli M, Monteggia LM (2013) mTOR complexes in neurodevelopmental and neuropsychiatric disorders. Nat Neurosci 16:1537–1543

Costa-Mattioli M, Sossin WS, Klann E, Sonenberg N (2009) Translational control of long-lasting synaptic plasticity and memory. Neuron 61:10–26

Crawley JN (2008) Behavioral phenotyping strategies for mutant mice. Neuron 57:809–818

Cui Y, Jin J, Zhang X, Xu H, Yang L, Du D, Zeng Q, Tsien JZ, Yu H, Cao X (2011) Forebrain NR2B overexpression facilitating the prefrontal cortex long-term potentiation and enhancing working memory function in mice. PLoS One 6:e20312

Curcio CA, Hinds JW (1983) Stability of synaptic density and spine volume in dentate gyrus of aged rats. Neurobiol Aging 4:77–87

Danysz W, Parsons CG (2003) The NMDA receptor antagonist memantine as a symptomatological and neuroprotective treatment for Alzheimer's disease: preclinical evidence. Int J Geriatr Psychiatry 18:S23–S32

Davis S, Laroche S (2006) Mitogen-activated protein kinase/extracellular regulated kinase signalling and memory stabilization: a review. Genes Brain Behav 5(Suppl 2):61–72

Davis CM, Moskovitz B, Nguyen MA, Tran BB, Arai A, Lynch G, Granger R (1997) A profile of the behavioral changes produced by facilitation of AMPA-type glutamate receptors. Psychopharmacology (Berl) 133:161–167

Davis HP, Small SA, Stern Y, Mayeux R, Feldstein SN, Keller FR (2003) Acquisition, recall, and forgetting of verbal information in long-term memory by young, middle-aged, and elderly individuals. Cortex 39:1063–1091

Davis MC, Horan WP, Marder SR (2014) Psychopharmacology of the negative symptoms: current status and prospects for progress. Eur Neuropsychopharmacol 24:788–799

Deiana S, Platt B, Riedel G (2011) The cholinergic system and spatial learning. Behav Brain Res 221:389–411

Dere E, Huston JP, De Souza Silva MA (2007) The pharmacology, neuroanatomy and neurogenetics of one-trial object recognition in rodents. Neurosci Biobehav Rev 31:673–704

Derkach V, Barria A, Soderling TR (1999) Ca2+/calmodulin-kinase II enhances channel conductance of alpha-amino-3-hydroxy-5-methyl-4-isoxazolepropionate type glutamate receptors. Proc Natl Acad Sci U S A 96:3269–3274

Deupree DL, Bradley J, Turner DA (1993) Age-related alterations in potentiation in the CA1 region in F344 rats. Neurobiol Aging 14:249–258

Deutsch JA (1971) The cholinergic synapse and the site of memory. Science 174:788–794

Dickinson BA, Jo J, Seok H, Son GH, Whitcomb DJ, Davies CH, Sheng M, Collingridge GL, Cho K (2009) A novel mechanism of hippocampal LTD involving muscarinic receptor-triggered interactions between AMPARs, GRIP and liprin-alpha. Mol Brain 2:18

Ding Q, Vaynman S, Akhavan M, Ying Z, Gomez-Pinilla F (2006) Insulin-like growth factor I interfaces with brain-derived neurotrophic factor-mediated synaptic plasticity to modulate aspects of exercise-induced cognitive function. Neuroscience 140:823–833

Dong YN, Wu HY, Hsu FC, Coulter DA, Lynch DR (2006) Developmental and cell-selective variations in N-methyl-D-aspartate receptor degradation by calpain. J Neurochem 99:206–217

Dong Z, Bai Y, Wu X, Li H, Gong B, Howland JG, Huang Y, He W, Li T, Wang YT (2013) Hippocampal long-term depression mediates spatial reversal learning in the Morris water maze. Neuropharmacology 64:65–73

Dore S, Kar S, Quirion R (1997) Rediscovering an old friend, IGF-I: potential use in the treatment of neurodegenerative diseases. Trends Neurosci 20:326–331

Dosemeci A, Jaffe H (2010) Regulation of phosphorylation at the postsynaptic density during different activity states of Ca2+/calmodulin-dependent protein kinase II. Biochem Biophys Res Commun 391:78–84

Drachman DA (1977) Memory and cognitive function in man: does the cholinergic system have a specific role? Neurology 27:783–790

Drake CT, Terman GW, Simmons ML, Milner TA, Kunkel DD, Schwartzkroin PA, Chavkin C (1994) Dynorphin opioids present in dentate granule cells may function as retrograde inhibitory neurotransmitters. J Neurosci 14:3736–3750

Ducottet C, Belzung C (2005) Correlations between behaviours in the elevated plus-maze and sensitivity to unpredictable subchronic mild stress: evidence from inbred strains of mice. Behav Brain Res 156:153–162

Dudai Y (1996) Consolidation: fragility on the road to the engram. Neuron 17:367–370

Dudek H, Datta SR, Franke TF, Birnbaum MJ, Yao R, Cooper GM, Segal RA, Kaplan DR, Greenberg ME (1997) Regulation of neuronal survival by the serine-threonine protein kinase Akt. Science 275:661–665

Duncan RS, Hwang SY, Koulen P (2005) Effects of Vesl/Homer proteins on intracellular signaling. Exp Biol Med (Maywood) 230:527–535

English JD, Sweatt JD (1996) Activation of p42 mitogen-activated protein kinase in hippocampal long term potentiation. J Biol Chem 271:24329–24332

Fagni L, Worley PF, Ango F (2002) Homer as both a scaffold and transduction molecule. Sci STKE 2002(137):re8

Fedulov V, Rex CS, Simmons DA, Palmer L, Gall CM, Lynch G (2007) Evidence that long-term potentiation occurs within individual hippocampal synapses during learning. J Neurosci 27:8031–8039

Fitzjohn SM, Doherty AJ, Collingridge GL (2008) The use of the hippocampal slice preparation in the study of Alzheimer's disease. Eur J Pharmacol 585:50–59

Flood DG (1993) Critical issues in the analysis of dendritic extent in aging humans, primates, and rodents. Neurobiol Aging 14:649–654

Floresco SB, Jentsch JD (2011) Pharmacological enhancement of memory and executive functioning in laboratory animals. Neuropsychopharmacology 36:227–250

Foster KA, McLaughlin N, Edbauer D, Phillips M, Bolton A, Constantine-Paton M, Sheng M (2010) Distinct roles of NR2A and NR2B cytoplasmic tails in long-term potentiation. J Neurosci 30:2676–2685

Fowler SW, Walker JM, Klakotskaia D, Will MJ, Serfozo P, Simonyi A, Schachtman TR (2013) Effects of a metabotropic glutamate receptor 5 positive allosteric modulator, CDPPB, on spatial learning task performance in rodents. Neurobiol Learn Mem 99:25–31

Fox GB, Fan L, Levasseur RA, Faden AI (1998) Effect of traumatic brain injury on mouse spatial and nonspatial learning in the Barnes circular maze. J Neurotrauma 15:1037–1046

Frankiewicz T, Parsons CG (1999) Memantine restores long term potentiation impaired by tonic N-methyl-D-aspartate (NMDA) receptor activation following reduction of Mg2+ in hippocampal slices. Neuropharmacology 38:1253–1259

Frankiewicz T, Potier B, Bashir ZI, Collingridge GL, Parsons CG (1996) Effects of memantine and MK-801 on NMDA-induced currents in cultured neurones and on synaptic transmission and LTP in area CA1 of rat hippocampal slices. Br J Pharmacol 117:689–697

Frankland PW, Bontempi B (2005) The organization of recent and remote memories. Nat Rev Neurosci 6:119–130

Frisardi V, Panza F, Farooqui AA (2011) Late-life depression and Alzheimer's disease: the glutamatergic system inside of this mirror relationship. Brain Res Rev 67:344–355

Gallagher M, Burwell R, Burchinal M (1993) Severity of spatial learning impairment in aging: development of a learning index for performance in the Morris water maze. Behav Neurosci 107:618–626

Gallagher M, Bizon JL, Hoyt EC, Helm KA, Lund PK (2003) Effects of aging on the hippocampal formation in a naturally occurring animal model of mild cognitive impairment. Exp Gerontol 38:71–77

Gallagher SM, Daly CA, Bear MF, Huber KM (2004) Extracellular signal-regulated protein kinase activation is required for metabotropic glutamate receptor-dependent long-term depression in hippocampal area CA1. J Neurosci 24:4859–4864

Ge Y, Dong Z, Bagot RC, Howland JG, Phillips AG, Wong TP, Wang YT (2010) Hippocampal long-term depression is required for the consolidation of spatial memory. Proc Natl Acad Sci U S A 107:16697–16702

Geerts H (2012) alpha7 Nicotinic receptor modulators for cognitive deficits in schizophrenia and Alzheimer's disease. Expert Opin Investig Drugs 21:59–65

Golden SA, Covington HE 3rd, Berton O, Russo SJ (2011) A standardized protocol for repeated social defeat stress in mice. Nat Protoc 6:1183–1191

Gong R, Park CS, Abbassi NR, Tang SJ (2006) Roles of glutamate receptors and the mammalian target of rapamycin (mTOR) signaling pathway in activity-dependent dendritic protein synthesis in hippocampal neurons. J Biol Chem 281:18802–18815

Grayson BE, Fitzgerald MF, Hakala-Finch AP, Ferris VM, Begg DP, Tong J, Woods SC, Seeley RJ, Davidson TL, Benoit SC (2014) Improvements in hippocampal-dependent memory and microglial infiltration with calorie restriction and gastric bypass surgery, but not with vertical sleeve gastrectomy. Int J Obes (Lond) 38:349–356

Guan X, Dluzen DE (1994) Age related changes of social memory/recognition in male Fischer 344 rats. Behav Brain Res 61:87–90

Guzowski JF, Lyford GL, Stevenson GD, Houston FP, McGaugh JL, Worley PF, Barnes CA (2000) Inhibition of activity-dependent arc protein expression in the rat hippocampus impairs the maintenance of long-term potentiation and the consolidation of long-term memory. J Neurosci 20:3993–4001

Guzowski JF, Setlow B, Wagner EK, McGaugh JL (2001) Experience-dependent gene expression in the rat hippocampus after spatial learning: a comparison of the immediate-early genes Arc, c-fos, and zif268. J Neurosci 21:5089–5098

Guzowski JF, Timlin JA, Roysam B, McNaughton BL, Worley PF, Barnes CA (2005) Mapping behaviorally relevant neural circuits with immediate-early gene expression. Curr Opin Neurobiol 15:599–606

Hall J, Thomas KL, Everitt BJ (2000) Rapid and selective induction of BDNF expression in the hippocampus during contextual learning. Nat Neurosci 3:533–535

Hall J, Thomas KL, Everitt BJ (2001) Cellular imaging of zif268 expression in the hippocampus and amygdala during contextual and cued fear memory retrieval: selective activation of hippocampal CA1 neurons during the recall of contextual memories. J Neurosci 21:2186–2193

Halt AR, Dallapiazza RF, Zhou Y, Stein IS, Qian H, Juntti S, Wojcik S, Brose N, Silva AJ, Hell JW (2012) CaMKII binding to GluN2B is critical during memory consolidation. EMBO J 31:1203–1216

Hampson RE, Rogers G, Lynch G, Deadwyler SA (1998a) Facilitative effects of the ampakine CX516 on short-term memory in rats: correlations with hippocampal neuronal activity. J Neurosci 18:2748–2763

Hampson RE, Rogers G, Lynch G, Deadwyler SA (1998b) Facilitative effects of the ampakine CX516 on short-term memory in rats: enhancement of delayed-nonmatch-to-sample performance. J Neurosci 18:2740–2747

Han JH, Kushner SA, Yiu AP, Cole CJ, Matynia A, Brown RA, Neve RL, Guzowski JF, Silva AJ, Josselyn SA (2007) Neuronal competition and selection during memory formation. Science 316:457–460

Hardingham GE, Arnold FJ, Bading H (2001) A calcium microdomain near NMDA receptors: on switch for ERK-dependent synapse-to-nucleus communication. Nat Neurosci 4:565–566

Harrison FE, Reiserer RS, Tomarken AJ, McDonald MP (2006) Spatial and nonspatial escape strategies in the Barnes maze. Learn Mem 13:809–819

Harvey J, Balasubramaniam R, Collingridge GL (1993) Carbachol can potentiate N-methyl-D-aspartate responses in the rat hippocampus by a staurosporine and thapsigargin-insensitive mechanism. Neurosci Lett 162:165–168

Hashimoto K, Ishima T, Fujita Y, Matsuo M, Kobashi T, Takahagi M, Tsukada H, Iyo M (2008) Phencyclidine-induced cognitive deficits in mice are improved by subsequent subchronic administration of the novel selective alpha7 nicotinic receptor agonist SSR180711. Biol Psychiatry 63:92–97

Hawasli AH, Benavides DR, Nguyen C, Kansy JW, Hayashi K, Chambon P, Greengard P, Powell CM, Cooper DC, Bibb JA (2007) Cyclin-dependent kinase 5 governs learning and synaptic plasticity via control of NMDAR degradation. Nat Neurosci 10:880–886

He X, Yang F, Xie Z, Lu B (2000) Intracellular Ca(2+) and Ca(2+)/calmodulin-dependent kinase II mediate acute potentiation of neurotransmitter release by neurotrophin-3. J Cell Biol 149:783–792

Hermans EJ, Battaglia FP, Atsak P, de Voogd LD, Fernandez G, Roozendaal B (2014) How the amygdala affects emotional memory by altering brain network properties. Neurobiol Learn Mem 112:2–16

Hiramatsu M, Watanabe E (2006) Dynorphin A (2–13) improves mecamylamine-induced learning impairment accompanied by reversal of reductions in acetylcholine release in rats. Neuropeptides 40:47–56

Hiramatsu M, Murasawa H, Nabeshima T, Kameyama T (1998) Effects of U-50,488H on scopolamine-, mecamylamine- and dizocilpine-induced learning and memory impairment in rats. J Pharmacol Exp Ther 284:858–867

Hongpaisan J, Alkon DL (2007) A structural basis for enhancement of long-term associative memory in single dendritic spines regulated by PKC. Proc Natl Acad Sci U S A 104:19571–19576
Hou L, Klann E (2004) Activation of the phosphoinositide 3-kinase-Akt-mammalian target of rapamycin signaling pathway is required for metabotropic glutamate receptor-dependent long-term depression. J Neurosci 24:6352–6361
Huang EJ, Reichardt LF (2001) Neurotrophins: roles in neuronal development and function. Annu Rev Neurosci 24:677–736
Hudon C, Dore FY, Goulet S (2002) Spatial memory and choice behavior in the radial arm maze after fornix transection. Prog Neuropsychopharmacol Biol Psychiatry 26:1113–1123
Huganir RL, Nicoll RA (2013) AMPARs and synaptic plasticity: the last 25 years. Neuron 80:704–717
Husi H, Ward MA, Choudhary JS, Blackstock WP, Grant SG (2000) Proteomic analysis of NMDA receptor-adhesion protein signaling complexes. Nat Neurosci 3:661–669
Isaac JT, Nicoll RA, Malenka RC (1995) Evidence for silent synapses: implications for the expression of LTP. Neuron 15:427–434
Itoh J, Ukai M, Kameyama T (1993) Dynorphin A-(1–13) markedly improves scopolamine-induced impairment of spontaneous alternation performance in mice. Eur J Pharmacol 236:341–345
Jiang HK, Owyang VV, Hong JS, Gallagher M (1989) Elevated dynorphin in the hippocampal formation of aged rats: relation to cognitive impairment on a spatial learning task. Proc Natl Acad Sci U S A 86:2948–2951
Jo J, Son GH, Winters BL, Kim MJ, Whitcomb DJ, Dickinson BA, Lee YB, Futai K, Amici M, Sheng M, Collingridge GL, Cho K (2010) Muscarinic receptors induce LTD of NMDAR EPSCs via a mechanism involving hippocalcin, AP2 and PSD-95. Nat Neurosci 13:1216–1224
Joels M, Fernandez G, Roozendaal B (2011) Stress and emotional memory: a matter of timing. Trends Cogn Sci 15:280–288
Johnson JW, Ascher P (1987) Glycine potentiates the NMDA response in cultured mouse brain neurons. Nature 325:529–531
Jones MW, Errington ML, French PJ, Fine A, Bliss TV, Garel S, Charnay P, Bozon B, Laroche S, Davis S (2001) A requirement for the immediate early gene Zif268 in the expression of late LTP and long-term memories. Nat Neurosci 4:289–296
Josselyn SA, Shi C, Carlezon WA Jr, Neve RL, Nestler EJ, Davis M (2001) Long-term memory is facilitated by cAMP response element-binding protein overexpression in the amygdala. J Neurosci 21:2404–2412
Kammermeier PJ, Worley PF (2007) Homer 1a uncouples metabotropic glutamate receptor 5 from postsynaptic effectors. Proc Natl Acad Sci U S A 104:6055–6060
Kang H, Schuman EM (1995) Long-lasting neurotrophin-induced enhancement of synaptic transmission in the adult hippocampus. Science 267:1658–1662
Kaplan DR, Miller FD (2000) Neurotrophin signal transduction in the nervous system. Curr Opin Neurobiol 10:381–391
Kar S, Chabot JG, Quirion R (1993) Quantitative autoradiographic localization of [125I]insulin-like growth factor I, [125I]insulin-like growth factor II, and [125I]insulin receptor binding sites in developing and adult rat brain. J Comp Neurol 333:375–397
Kar S, Seto D, Dore S, Hanisch U, Quirion R (1997) Insulin-like growth factors-I and -II differentially regulate endogenous acetylcholine release from the rat hippocampal formation. Proc Natl Acad Sci U S A 94:14054–14059
Kerr DS, Campbell LW, Hao SY, Landfield PW (1989) Corticosteroid modulation of hippocampal potentials: increased effect with aging. Science 245:1505–1509
Kida S, Josselyn SA, Pena de Ortiz S, Kogan JH, Chevere I, Masushige S, Silva AJ (2002) CREB required for the stability of new and reactivated fear memories. Nat Neurosci 5:348–355
Kirkwood A, Rozas C, Kirkwood J, Perez F, Bear MF (1999) Modulation of long-term synaptic depression in visual cortex by acetylcholine and norepinephrine. J Neurosci 19:1599–1609

Kitraki E, Bozas E, Philippidis H, Stylianopoulou F (1993) Aging-related changes in IGF-II and c-fos gene expression in the rat brain. Int J Dev Neurosci 11:1–9

Kiyama Y, Manabe T, Sakimura K, Kawakami F, Mori H, Mishina M (1998) Increased thresholds for long-term potentiation and contextual learning in mice lacking the NMDA-type glutamate receptor epsilon1 subunit. J Neurosci 18:6704–6712

Kleckner NW, Dingledine R (1988) Requirement for glycine in activation of NMDA-receptors expressed in Xenopus oocytes. Science 241:835–837

Knoll AT, Meloni EG, Thomas JB, Carroll FI, Carlezon WA Jr (2007) Anxiolytic-like effects of kappa-opioid receptor antagonists in models of unlearned and learned fear in rats. J Pharmacol Exp Ther 323:838–845

Kolsch H, Wagner M, Bilkei-Gorzo A, Toliat MR, Pentzek M, Fuchs A, Kaduszkiewicz H, van den Bussche H, Riedel-Heller SG, Angermeyer MC, Weyerer S, Werle J, Bickel H, Mosch E, Wiese B, Daerr M, Jessen F, Maier W, Dichgans M (2009) Gene polymorphisms in prodynorphin (PDYN) are associated with episodic memory in the elderly. J Neural Transm 116:897–903

Korte M, Carroll P, Wolf E, Brem G, Thoenen H, Bonhoeffer T (1995) Hippocampal long-term potentiation is impaired in mice lacking brain-derived neurotrophic factor. Proc Natl Acad Sci U S A 92:8856–8860

Kotz CM, Weldon D, Billington CJ, Levine AS (2004) Age-related changes in brain proDynorphin gene expression in the rat. Neurobiol Aging 25:1343–1347

Kuehl-Kovarik MC, Magnusson KR, Premkumar LS, Partin KM (2000) Electrophysiological analysis of NMDA receptor subunit changes in the aging mouse cortex. Mech Ageing Dev 115:39–59

Labar KS, Cabeza R (2006) Cognitive neuroscience of emotional memory. Nat Rev Neurosci 7:54–64

Lamprecht R, Ledoux J (2004) Structural plasticity and memory. Nat Rev Neurosci 5:45–54

Landfield PW, Lynch G (1977) Impaired monosynaptic potentiation in in vitro hippocampal slices from aged, memory-deficient rats. J Gerontol 32:523–533

Landfield PW, Pitler TA (1984) Prolonged Ca2+-dependent afterhyperpolarizations in hippocampal neurons of aged rats. Science 226:1089–1092

Lauterborn JC, Lynch G, Vanderklish P, Arai A, Gall CM (2000) Positive modulation of AMPA receptors increases neurotrophin expression by hippocampal and cortical neurons. J Neurosci 20:8–21

Lee YS, Silva AJ (2009) The molecular and cellular biology of enhanced cognition. Nat Rev Neurosci 10:126–140

Lee HK, Kameyama K, Huganir RL, Bear MF (1998) NMDA induces long-term synaptic depression and dephosphorylation of the GluR1 subunit of AMPA receptors in hippocampus. Neuron 21:1151–1162

Lee HK, Barbarosie M, Kameyama K, Bear MF, Huganir RL (2000) Regulation of distinct AMPA receptor phosphorylation sites during bidirectional synaptic plasticity. Nature 405:955–959

Lee SJ, Escobedo-Lozoya Y, Szatmari EM, Yasuda R (2009) Activation of CaMKII in single dendritic spines during long-term potentiation. Nature 458:299–304

Leiser SC, Bowlby MR, Comery TA, Dunlop J (2009) A cog in cognition: how the alpha 7 nicotinic acetylcholine receptor is geared towards improving cognitive deficits. Pharmacol Ther 122:302–311

Lenze EJ, Rogers JC, Martire LM, Mulsant BH, Rollman BL, Dew MA, Schulz R, Reynolds CF 3rd (2001) The association of late-life depression and anxiety with physical disability: a review of the literature and prospectus for future research. Am J Geriatr Psychiatry 9:113–135

Levin ED, Rezvani AH (2002) Nicotinic treatment for cognitive dysfunction. Curr Drug Targets CNS Neurol Disord 1:423–431

Liao D, Hessler NA, Malinow R (1995) Activation of postsynaptically silent synapses during pairing-induced LTP in CA1 region of hippocampal slice. Nature 375:400–404

Link W, Konietzko U, Kauselmann G, Krug M, Schwanke B, Frey U, Kuhl D (1995) Somatodendritic expression of an immediate early gene is regulated by synaptic activity. Proc Natl Acad Sci U S A 92:5734–5738

Lisman J, Yasuda R, Raghavachari S (2012) Mechanisms of CaMKII action in long-term potentiation. Nat Rev Neurosci 13:169–182

Lissin DV, Gomperts SN, Carroll RC, Christine CW, Kalman D, Kitamura M, Hardy S, Nicoll RA, Malenka RC, Von Zastrow M (1998) Activity differentially regulates the surface expression of synaptic AMPA and NMDA glutamate receptors. Proc Natl Acad Sci U S A 95:7097–7102

Lonze BE, Ginty DD (2002) Function and regulation of CREB family transcription factors in the nervous system. Neuron 35:605–623

Lu W, Roche KW (2012) Posttranslational regulation of AMPA receptor trafficking and function. Curr Opin Neurobiol 22:470–479

Lu YM, Jia Z, Janus C, Henderson JT, Gerlai R, Wojtowicz JM, Roder JC (1997) Mice lacking metabotropic glutamate receptor 5 show impaired learning and reduced CA1 long-term potentiation (LTP) but normal CA3 LTP. J Neurosci 17:5196–5205

Lupien SJ, McEwen BS, Gunnar MR, Heim C (2009) Effects of stress throughout the lifespan on the brain, behaviour and cognition. Nat Rev Neurosci 10:434–445

Luscher C, Huber KM (2010) Group 1 mGluR-dependent synaptic long-term depression: mechanisms and implications for circuitry and disease. Neuron 65:445–459

Lyford GL, Yamagata K, Kaufmann WE, Barnes CA, Sanders LK, Copeland NG, Gilbert DJ, Jenkins NA, Lanahan AA, Worley PF (1995) Arc, a growth factor and activity-regulated gene, encodes a novel cytoskeleton-associated protein that is enriched in neuronal dendrites. Neuron 14:433–445

Lynch MA (2004) Long-term potentiation and memory. Physiol Rev 84:87–136

Lynch G, Granger R, Ambros-Ingerson J, Davis CM, Kessler M, Schehr R (1997) Evidence that a positive modulator of AMPA-type glutamate receptors improves delayed recall in aged humans. Exp Neurol 145:89–92

Lynch G, Rex CS, Chen LY, Gall CM (2008) The substrates of memory: defects, treatments, and enhancement. Eur J Pharmacol 585:2–13

Magnusson KR, Brim BL, Das SR (2010) Selective vulnerabilities of N-methyl-D-aspartate (NMDA) receptors during brain aging. Front Aging Neurosci 2:11

Malenka RC, Kauer JA, Perkel DJ, Mauk MD, Kelly PT, Nicoll RA, Waxham MN (1989) An essential role for postsynaptic calmodulin and protein kinase activity in long-term potentiation. Nature 340:554–557

Malinow R, Malenka RC (2002) AMPA receptor trafficking and synaptic plasticity. Annu Rev Neurosci 25:103–126

Malinow R, Schulman H, Tsien RW (1989) Inhibition of postsynaptic PKC or CaMKII blocks induction but not expression of LTP. Science 245:862–866

Manahan-Vaughan D, Braunewell KH (2005) The metabotropic glutamate receptor, mGluR5, is a key determinant of good and bad spatial learning performance and hippocampal synaptic plasticity. Cereb Cortex 15:1703–1713

Manahan-Vaughan D, von Haebler D, Winter C, Juckel G, Heinemann U (2008) A single application of MK801 causes symptoms of acute psychosis, deficits in spatial memory, and impairment of synaptic plasticity in rats. Hippocampus 18:125–134

Maren S, Quirk GJ (2004) Neuronal signalling of fear memory. Nat Rev Neurosci 5:844–852

Markham JA, McKian KP, Stroup TS, Juraska JM (2005) Sexually dimorphic aging of dendritic morphology in CA1 of hippocampus. Hippocampus 15:97–103

Markowska AL (1999) Sex dimorphisms in the rate of age-related decline in spatial memory: relevance to alterations in the estrous cycle. J Neurosci 19:8122–8133

Markram H, Segal M (1990) Acetylcholine potentiates responses to N-methyl-D-aspartate in the rat hippocampus. Neurosci Lett 113:62–65

Marrone DF, Schaner MJ, McNaughton BL, Worley PF, Barnes CA (2008) Immediate-early gene expression at rest recapitulates recent experience. J Neurosci 28:1030–1033

Matsuda S, Mikawa S, Hirai H (1999) Phosphorylation of serine-880 in GluR2 by protein kinase C prevents its C terminus from binding with glutamate receptor-interacting protein. J Neurochem 73:1765–1768

McDaniel KL, Mundy WR, Tilson HA (1990) Microinjection of dynorphin into the hippocampus impairs spatial learning in rats. Pharmacol Biochem Behav 35:429–435

McGaugh JL (2000) Memory—a century of consolidation. Science 287:248–251

McGaugh JL (2004) The amygdala modulates the consolidation of memories of emotionally arousing experiences. Annu Rev Neurosci 27:1–28

McHugh TJ, Jones MW, Quinn JJ, Balthasar N, Coppari R, Elmquist JK, Lowell BB, Fanselow MS, Wilson MA, Tonegawa S (2007) Dentate gyrus NMDA receptors mediate rapid pattern separation in the hippocampal network. Science 317:94–99

McTighe SM, Cowell RA, Winters BD, Bussey TJ, Saksida LM (2010) Paradoxical false memory for objects after brain damage. Science 330:1408–1410

Menard C, Quirion R (2012a) Group 1 metabotropic glutamate receptor function and its regulation of learning and memory in the aging brain. Front Pharmacol 3:182

Menard C, Quirion R (2012b) Successful cognitive aging in rats: a role for mGluR5 glutamate receptors, homer 1 proteins and downstream signaling pathways. PLoS One 7:e28666

Menard C, Bastianetto S, Quirion R (2013a) Neuroprotective effects of resveratrol and epigallocatechin gallate polyphenols are mediated by the activation of protein kinase C gamma. Front Cell Neurosci 7:281

Menard C, Tse YC, Cavanagh C, Chabot JG, Herzog H, Schwarzer C, Wong TP, Quirion R (2013b) Knockdown of prodynorphin gene prevents cognitive decline, reduces anxiety, and rescues loss of group 1 metabotropic glutamate receptor function in aging. J Neurosci 33:12792–12804

Menard C, Herzog H, Schwarzer C, Quirion R (2014a) Possible role of dynorphins in Alzheimer's disease and age-related cognitive deficits. Neurodegener Dis 13:82–85

Menard C, Quirion R, Bouchard S, Ferland G, Gaudreau P (2014b) Glutamatergic signaling and low prodynorphin expression are associated with intact memory and reduced anxiety in rat models of healthy aging. Front Aging Neurosci 6:81

Menuz K, Stroud RM, Nicoll RA, Hays FA (2007) TARP auxiliary subunits switch AMPA receptor antagonists into partial agonists. Science 318:815–817

Merrill DA, Chiba AA, Tuszynski MH (2001) Conservation of neuronal number and size in the entorhinal cortex of behaviorally characterized aged rats. J Comp Neurol 438:445–456

Messaoudi E, Kanhema T, Soule J, Tiron A, Dagyte G, da Silva B, Bramham CR (2007) Sustained Arc/Arg3.1 synthesis controls long-term potentiation consolidation through regulation of local actin polymerization in the dentate gyrus in vivo. J Neurosci 27:10445–10455

Miller EK, Desimone R (1993) Scopolamine affects short-term memory but not inferior temporal neurons. Neuroreport 4:81–84

Miller FD, Kaplan DR (2003) Signaling mechanisms underlying dendrite formation. Curr Opin Neurobiol 13:391–398

Milner B, Penfield W (1955) The effect of hippocampal lesions on recent memory. Trans Am Neurol Assoc 42–8

Mizuno M, Yamada K, Takei N, Tran MH, He J, Nakajima A, Nawa H, Nabeshima T (2003) Phosphatidylinositol 3-kinase: a molecule mediating BDNF-dependent spatial memory formation. Mol Psychiatry 8:217–224

Montgomery KE, Kessler M, Arai AC (2009) Modulation of agonist binding to AMPA receptors by 1-(1,4-benzodioxan-6-ylcarbonyl)piperidine (CX546): differential effects across brain regions and GluA1-4/transmembrane AMPA receptor regulatory protein combinations. J Pharmacol Exp Ther 331:965–974

Monti B, Berteotti C, Contestabile A (2005) Dysregulation of memory-related proteins in the hippocampus of aged rats and their relation with cognitive impairment. Hippocampus 15:1041–1049

Mony L, Kew JN, Gunthorpe MJ, Paoletti P (2009) Allosteric modulators of NR2B-containing NMDA receptors: molecular mechanisms and therapeutic potential. Br J Pharmacol 157:1301–1317

Monyer H, Burnashev N, Laurie DJ, Sakmann B, Seeburg PH (1994) Developmental and regional expression in the rat brain and functional properties of four NMDA receptors. Neuron 12:529–540

Morris R (1984) Developments of a water-maze procedure for studying spatial learning in the rat. J Neurosci Methods 11:47–60

Morris RG (1989) Synaptic plasticity and learning: selective impairment of learning rats and blockade of long-term potentiation in vivo by the N-methyl-D-aspartate receptor antagonist AP5. J Neurosci 9:3040–3057

Morris RG, Anderson E, Lynch GS, Baudry M (1986) Selective impairment of learning and blockade of long-term potentiation by an N-methyl-D-aspartate receptor antagonist, AP5. Nature 319:774–776

Moser MB, Moser EI (1998) Functional differentiation in the hippocampus. Hippocampus 8:608–619

Mothet JP, Parent AT, Wolosker H, Brady RO Jr, Linden DJ, Ferris CD, Rogawski MA, Snyder SH (2000) D-serine is an endogenous ligand for the glycine site of the N-methyl-D-aspartate receptor. Proc Natl Acad Sci U S A 97:4926–4931

Muller L, Tokay T, Porath K, Kohling R, Kirschstein T (2013) Enhanced NMDA receptor-dependent LTP in the epileptic CA1 area via upregulation of NR2B. Neurobiol Dis 54:183–193

Myers KM, Davis M (2002) Behavioral and neural analysis of extinction. Neuron 36:567–584

Myers KM, Davis M (2007) Mechanisms of fear extinction. Mol Psychiatry 12:120–150

Najavits LM, Gastfriend DR, Barber JP, Reif S, Muenz LR, Blaine J, Frank A, Crits-Christoph P, Thase M, Weiss RD (1998) Cocaine dependence with and without PTSD among subjects in the National Institute on Drug Abuse Collaborative Cocaine Treatment Study. Am J Psychiatry 155:214–219

Nakazawa K, Quirk MC, Chitwood RA, Watanabe M, Yeckel MF, Sun LD, Kato A, Carr CA, Johnston D, Wilson MA, Tonegawa S (2002) Requirement for hippocampal CA3 NMDA receptors in associative memory recall. Science 297:211–218

Nakazawa K, Sun LD, Quirk MC, Rondi-Reig L, Wilson MA, Tonegawa S (2003) Hippocampal CA3 NMDA receptors are crucial for memory acquisition of one-time experience. Neuron 38:305–315

Nelson TJ, Sun MK, Hongpaisan J, Alkon DL (2008) Insulin, PKC signaling pathways and synaptic remodeling during memory storage and neuronal repair. Eur J Pharmacol 585:76–87

Neumann ID (2008) Brain oxytocin: a key regulator of emotional and social behaviours in both females and males. J Neuroendocrinol 20:858–865

Neyman S, Manahan-Vaughan D (2008) Metabotropic glutamate receptor 1 (mGluR1) and 5 (mGluR5) regulate late phases of LTP and LTD in the hippocampal CA1 region in vitro. Eur J Neurosci 27:1345–1352

Nguyen XV, Masse J, Kumar A, Vijitruth R, Kulik C, Liu M, Choi DY, Foster TC, Usynin I, Bakalkin G, Bing G (2005) Prodynorphin knockout mice demonstrate diminished age-associated impairment in spatial water maze performance. Behav Brain Res 161:254–262

Nicoletti F, Bockaert J, Collingridge GL, Conn PJ, Ferraguti F, Schoepp DD, Wroblewski JT, Pin JP (2011) Metabotropic glutamate receptors: from the workbench to the bedside. Neuropharmacology 60:1017–1041

Niesen CE, Baskys A, Carlen PL (1988) Reversed ethanol effects on potassium conductances in aged hippocampal dentate granule neurons. Brain Res 445:137–141

Nikolarakis KE, Almeida OF, Yassouridis A, Herz A (1989) Presynaptic auto- and allelo-receptor regulation of hypothalamic opioid peptide release. Neuroscience 31:269–273

Niswender CM, Conn PJ (2010) Metabotropic glutamate receptors: physiology, pharmacology, and disease. Annu Rev Pharmacol Toxicol 50:295–322

Nithianantharajah J, Murphy M (2009) Experience on the Barnes spatial maze influences PKCgamma levels in the hippocampus. Int J Neurosci 119:1014–1030

Noack J, Richter K, Laube G, Haghgoo HA, Veh RW, Engelmann M (2010) Different importance of the volatile and non-volatile fractions of an olfactory signature for individual social recognition in rats versus mice and short-term versus long-term memory. Neurobiol Learn Mem 94:568–575

O'Brien RJ, Kamboj S, Ehlers MD, Rosen KR, Fischbach GD, Huganir RL (1998) Activity-dependent modulation of synaptic AMPA receptor accumulation. Neuron 21:1067–1078

O'Donovan A, Slavich GM, Epel ES, Neylan TC (2013) Exaggerated neurobiological sensitivity to threat as a mechanism linking anxiety with increased risk for diseases of aging. Neurosci Biobehav Rev 37:96–108

O'Keefe J, Dostrovsky J (1971) The hippocampus as a spatial map. Preliminary evidence from unit activity in the freely-moving rat. Brain Res 34:171–175

Olds JL, Anderson ML, McPhie DL, Staten LD, Alkon DL (1989) Imaging of memory-specific changes in the distribution of protein kinase C in the hippocampus. Science 245:866–869

Opazo P, Choquet D (2011) A three-step model for the synaptic recruitment of AMPA receptors. Mol Cell Neurosci 46:1–8

Page G, Khidir FA, Pain S, Barrier L, Fauconneau B, Guillard O, Piriou A, Hugon J (2006) Group I metabotropic glutamate receptors activate the p70S6 kinase via both mammalian target of rapamycin (mTOR) and extracellular signal-regulated kinase (ERK 1/2) signaling pathways in rat striatal and hippocampal synaptoneurosomes. Neurochem Int 49:413–421

Palmer MJ, Irving AJ, Seabrook GR, Jane DE, Collingridge GL (1997) The group I mGlu receptor agonist DHPG induces a novel form of LTD in the CA1 region of the hippocampus. Neuropharmacology 36:1517–1532

Park S, Park JM, Kim S, Kim JA, Shepherd JD, Smith-Hicks CL, Chowdhury S, Kaufmann W, Kuhl D, Ryazanov AG, Huganir RL, Linden DJ, Worley PF (2008) Elongation factor 2 and fragile X mental retardation protein control the dynamic translation of Arc/Arg3.1 essential for mGluR-LTD. Neuron 59:70–83

Parsons RG, Gafford GM, Helmstetter FJ (2006) Translational control via the mammalian target of rapamycin pathway is critical for the formation and stability of long-term fear memory in amygdala neurons. J Neurosci 26:12977–12983

Paterson D, Nordberg A (2000) Neuronal nicotinic receptors in the human brain. Prog Neurobiol 61:75–111

Patterson SL, Abel T, Deuel TA, Martin KC, Rose JC, Kandel ER (1996) Recombinant BDNF rescues deficits in basal synaptic transmission and hippocampal LTP in BDNF knockout mice. Neuron 16:1137–1145

Pichat P, Bergis OE, Terranova JP, Urani A, Duarte C, Santucci V, Gueudet C, Voltz C, Steinberg R, Stemmelin J, Oury-Donat F, Avenet P, Griebel G, Scatton B (2007) SSR180711, a novel selective alpha7 nicotinic receptor partial agonist: (II) efficacy in experimental models predictive of activity against cognitive symptoms of schizophrenia. Neuropsychopharmacology 32:17–34

Plath N, Ohana O, Dammermann B, Errington ML, Schmitz D, Gross C, Mao X, Engelsberg A, Mahlke C, Welzl H, Kobalz U, Stawrakakis A, Fernandez E, Waltereit R, Bick-Sander A, Therstappen E, Cooke SF, Blanquet V, Wurst W, Salmen B, Bosl MR, Lipp HP, Grant SG, Bliss TV, Wolfer DP, Kuhl D (2006) Arc/Arg3.1 is essential for the consolidation of synaptic plasticity and memories. Neuron 52:437–444

Ploski JE, Pierre VJ, Smucny J, Park K, Monsey MS, Overeem KA, Schafe GE (2008) The activity-regulated cytoskeletal-associated protein (Arc/Arg3.1) is required for memory consolidation of pavlovian fear conditioning in the lateral amygdala. J Neurosci 28:12383–12395

Potier B, Rascol O, Jazat F, Lamour Y, Dutar P (1992) Alterations in the properties of hippocampal pyramidal neurons in the aged rat. Neuroscience 48:793–806

Potier B, Lamour Y, Dutar P (1993) Age-related alterations in the properties of hippocampal pyramidal neurons among rat strains. Neurobiol Aging 14:17–25

Potier B, Turpin FR, Sinet PM, Rouaud E, Mothet JP, Videau C, Epelbaum J, Dutar P, Billard JM (2010) Contribution of the d-serine-dependent pathway to the cellular mechanisms underlying cognitive aging. Front Aging Neurosci 2:1

Priel A, Kolleker A, Ayalon G, Gillor M, Osten P, Stern-Bach Y (2005) Stargazin reduces desensitization and slows deactivation of the AMPA-type glutamate receptors. J Neurosci 25:2682–2686

Pyapali GK, Turner DA (1996) Increased dendritic extent in hippocampal CA1 neurons from aged F344 rats. Neurobiol Aging 17:601–611

Quirion R, Pert CB (1981) Dynorphins: similar relative potencies on mu, delta- and kappa-opiate receptors. Eur J Pharmacol 76:467–468

Quirion R, Wilson A, Rowe W, Aubert I, Richard J, Doods H, Parent A, White N, Meaney MJ (1995) Facilitation of acetylcholine release and cognitive performance by an M(2)-muscarinic receptor antagonist in aged memory-impaired. J Neurosci 15:1455–1462

Radulovic J, Tronson NC (2010) Molecular specificity of multiple hippocampal processes governing fear extinction. Rev Neurosci 21:1–17

Rapp PR, Gallagher M (1996) Preserved neuron number in the hippocampus of aged rats with spatial learning deficits. Proc Natl Acad Sci U S A 93:9926–9930

Rapp PR, Deroche PS, Mao Y, Burwell RD (2002) Neuron number in the parahippocampal region is preserved in aged rats with spatial learning deficits. Cereb Cortex 12:1171–1179

Rasmussen T, Schliemann T, Sorensen JC, Zimmer J, West MJ (1996) Memory impaired aged rats: no loss of principal hippocampal and subicular neurons. Neurobiol Aging 17:143–147

Reichel CM, Schwendt M, McGinty JF, Olive MF, See RE (2011) Loss of object recognition memory produced by extended access to methamphetamine self-administration is reversed by positive allosteric modulation of metabotropic glutamate receptor 5. Neuropsychopharmacology 36:782–792

Rex CS, Lin CY, Kramar EA, Chen LY, Gall CM, Lynch G (2007) Brain-derived neurotrophic factor promotes long-term potentiation-related cytoskeletal changes in adult hippocampus. J Neurosci 27:3017–3029

Roberge MC, Hotte-Bernard J, Messier C, Plamondon H (2008) Food restriction attenuates ischemia-induced spatial learning and memory deficits despite extensive CA1 ischemic injury. Behav Brain Res 187:123–132

Romano C, van den Pol AN, O'Malley KL (1996) Enhanced early developmental expression of the metabotropic glutamate receptor mGluR5 in rat brain: protein, mRNA splice variants, and regional distribution. J Comp Neurol 367:403–412

Roncarati R, Scali C, Comery TA, Grauer SM, Aschmi S, Bothmann H, Jow B, Kowal D, Gianfriddo M, Kelley C, Zanelli U, Ghiron C, Haydar S, Dunlop J, Terstappen GC (2009) Procognitive and neuroprotective activity of a novel alpha7 nicotinic acetylcholine receptor agonist for treatment of neurodegenerative and cognitive disorders. J Pharmacol Exp Ther 329:459–468

Rosen LB, Ginty DD, Weber MJ, Greenberg ME (1994) Membrane depolarization and calcium influx stimulate MEK and MAP kinase via activation of Ras. Neuron 12:1207–1221

Rosenbrock H, Kramer G, Hobson S, Koros E, Grundl M, Grauert M, Reymann KG, Schroder UH (2010) Functional interaction of metabotropic glutamate receptor 5 and NMDA-receptor by a metabotropic glutamate receptor 5 positive allosteric modulator. Eur J Pharmacol 639:40–46

Rosenzweig ES, Rao G, McNaughton BL, Barnes CA (1997) Role of temporal summation in age-related long-term potentiation-induction deficits. Hippocampus 7:549–558

Rowe JW, Kahn RL (1997) Successful aging. Gerontologist 37:433–440

Rowe WB, Spreekmeester E, Meaney MJ, Quirion R, Rochford J (1998) Reactivity to novelty in cognitively-impaired and cognitively-unimpaired aged rats and young rats. Neuroscience 83:669–680

Rumpel S, Ledoux J, Zador A, Malinow R (2005) Postsynaptic receptor trafficking underlying a form of associative learning. Science 308:83–88

Russo VC, Gluckman PD, Feldman EL, Werther GA (2005) The insulin-like growth factor system and its pleiotropic functions in brain. Endocr Rev 26:916–943
Saito N, Kikkawa U, Nishizuka Y, Tanaka C (1988) Distribution of protein kinase C-like immunoreactive neurons in rat brain. J Neurosci 8:369–382
Sakimura K, Kutsuwada T, Ito I, Manabe T, Takayama C, Kushiya E, Yagi T, Aizawa S, Inoue Y, Sugiyama H et al (1995) Reduced hippocampal LTP and spatial learning in mice lacking NMDA receptor epsilon 1 subunit. Nature 373:151–155
Sandin J, Nylander I, Georgieva J, Schott PA, Ogren SO, Terenius L (1998) Hippocampal dynorphin B injections impair spatial learning in rats: a kappa-opioid receptor-mediated effect. Neuroscience 85:375–382
Sarter M, Parikh V (2005) Choline transporters, cholinergic transmission and cognition. Nat Rev Neurosci 6:48–56
Scali C, Giovannini MG, Bartolini L, Prosperi C, Hinz V, Schmidt B, Pepeu G (1997a) Effect of metrifonate on extracellular brain acetylcholine and object recognition in aged rats. Eur J Pharmacol 325:173–180
Scali C, Giovannini MG, Prosperi C, Bartolini L, Pepeu G (1997b) Tacrine administration enhances extracellular acetylcholine in vivo and restores the cognitive impairment in aged rats. Pharmacol Res 36:463–469
Schacter DL (1999) The seven sins of memory. Insights from psychology and cognitive neuroscience. Am Psychol 54:182–203
Schoenbaum G, Nugent S, Saddoris MP, Gallagher M (2002) Teaching old rats new tricks: age-related impairments in olfactory reversal learning. Neurobiol Aging 23:555–564
Schwarzer C (2009) 30 years of dynorphins—new insights on their functions in neuropsychiatric diseases. Pharmacol Ther 123:353–370
Scoville WB, Milner B (1957) Loss of recent memory after bilateral hippocampal lesions. J Neurol Neurosurg Psychiatry 20:11–21
Segal M (1982) Changes in neurotransmitter actions in the aged rat hippocampus. Neurobiol Aging 3:121–124
Shankar S, Teyler TJ, Robbins N (1998) Aging differentially alters forms of long-term potentiation in rat hippocampal area CA1. J Neurophysiol 79:334–341
Shaywitz AJ, Greenberg ME (1999) CREB: a stimulus-induced transcription factor activated by a diverse array of extracellular signals. Annu Rev Biochem 68:821–861
Sheng M, Cummings J, Roldan LA, Jan YN, Jan LY (1994) Changing subunit composition of heteromeric NMDA receptors during development of rat cortex. Nature 368:144–147
Shepherd JD, Bear MF (2011) New views of Arc, a master regulator of synaptic plasticity. Nat Neurosci 14:279–284
Shepherd JD, Huganir RL (2007) The cell biology of synaptic plasticity: AMPA receptor trafficking. Annu Rev Cell Dev Biol 23:613–643
Shepherd JD, Rumbaugh G, Wu J, Chowdhury S, Plath N, Kuhl D, Huganir RL, Worley PF (2006) Arc/Arg3.1 mediates homeostatic synaptic scaling of AMPA receptors. Neuron 52:475–484
Shinoe T, Matsui M, Taketo MM, Manabe T (2005) Modulation of synaptic plasticity by physiological activation of M1 muscarinic acetylcholine receptors in the mouse hippocampus. J Neurosci 25:11194–11200
Shiraishi-Yamaguchi Y, Furuichi T (2007) The homer family proteins. Genome Biol 8:206
Shukla VK, Lemaire S (1994) Non-opioid effects of dynorphins: possible role of the NMDA receptor. Trends Pharmacol Sci 15:420–424
Simmons ML, Terman GW, Drake CT, Chavkin C (1994) Inhibition of glutamate release by presynaptic kappa 1-opioid receptors in the guinea pig dentate gyrus. J Neurophysiol 72:1697–1705
Simpkins KL, Guttmann RP, Dong Y, Chen Z, Sokol S, Neumar RW, Lynch DR (2003) Selective activation induced cleavage of the NR2B subunit by calpain. J Neurosci 23:11322–11331
Smith AP, Henson RN, Dolan RJ, Rugg MD (2004) fMRI correlates of the episodic retrieval of emotional contexts. Neuroimage 22:868–878

Smith JS, Schindler AG, Martinelli E, Gustin RM, Bruchas MR, Chavkin C (2012) Stress-induced activation of the dynorphin/kappa-opioid receptor system in the amygdala potentiates nicotine conditioned place preference. J Neurosci 32:1488–1495

Soderling TR (1993) Calcium/calmodulin-dependent protein kinase II: role in learning and memory. Mol Cell Biochem 127–128:93–101

Staubli U, Rogers G, Lynch G (1994) Facilitation of glutamate receptors enhances memory. Proc Natl Acad Sci U S A 91:777–781

Stern SA, Kohtz AS, Pollonini G, Alberini CM (2014) Enhancement of memories by systemic administration of insulin-like growth factor II. Neuropsychopharmacology 39(9):2179–2190

Steward O, Worley PF (2001) Selective targeting of newly synthesized Arc mRNA to active synapses requires NMDA receptor activation. Neuron 30:227–240

Steward O, Wallace CS, Lyford GL, Worley PF (1998) Synaptic activation causes the mRNA for the IEG Arc to localize selectively near activated postsynaptic sites on dendrites. Neuron 21:741–751

Stoica L, Zhu PJ, Huang W, Zhou H, Kozma SC, Costa-Mattioli M (2011) Selective pharmacogenetic inhibition of mammalian target of Rapamycin complex I (mTORC1) blocks long-term synaptic plasticity and memory storage. Proc Natl Acad Sci U S A 108:3791–3796

Su SC, Tsai LH (2011) Cyclin-dependent kinases in brain development and disease. Annu Rev Cell Dev Biol 27:465–491

Sun MK, Alkon DL (2010) Pharmacology of protein kinase C activators: cognition-enhancing and antidementic therapeutics. Pharmacol Ther 127:66–77

Sun P, Lou L, Maurer RA (1996) Regulation of activating transcription factor-1 and the cAMP response element-binding protein by Ca2+/calmodulin-dependent protein kinases type I, II, and IV. J Biol Chem 271:3066–3073

Suzuki A, Fukushima H, Mukawa T, Toyoda H, Wu LJ, Zhao MG, Xu H, Shang Y, Endoh K, Iwamoto T, Mamiya N, Okano E, Hasegawa S, Mercaldo V, Zhang Y, Maeda R, Ohta M, Josselyn SA, Zhuo M, Kida S (2011) Upregulation of CREB-mediated transcription enhances both short- and long-term memory. J Neurosci 31:8786–8802

Svensson J, Diez M, Engel J, Wass C, Tivesten A, Jansson JO, Isaksson O, Archer T, Hokfelt T, Ohlsson C (2006) Endocrine, liver-derived IGF-I is of importance for spatial learning and memory in old mice. J Endocrinol 189:617–627

Swope SL, Moss SJ, Blackstone CD, Huganir RL (1992) Phosphorylation of ligand-gated ion channels: a possible mode of synaptic plasticity. FASEB J 6:2514–2523

Szumlinski KK, Dehoff MH, Kang SH, Frys KA, Lominac KD, Klugmann M, Rohrer J, Griffin W 3rd, Toda S, Champtiaux NP, Berry T, Tu JC, Shealy SE, During MJ, Middaugh LD, Worley PF, Kalivas PW (2004) Homer proteins regulate sensitivity to cocaine. Neuron 43:401–413

Tang YP, Shimizu E, Dube GR, Rampon C, Kerchner GA, Zhuo M, Liu G, Tsien JZ (1999) Genetic enhancement of learning and memory in mice. Nature 401:63–69

Tang SJ, Reis G, Kang H, Gingras AC, Sonenberg N, Schuman EM (2002) A rapamycin-sensitive signaling pathway contributes to long-term synaptic plasticity in the hippocampus. Proc Natl Acad Sci U S A 99:467–472

Thomas GM, Huganir RL (2004) MAPK cascade signalling and synaptic plasticity. Nat Rev Neurosci 5:173–183

Tombaugh GC, Rowe WB, Chow AR, Michael TH, Rose GM (2002) Theta-frequency synaptic potentiation in CA1 in vitro distinguishes cognitively impaired from unimpaired aged Fischer 344 rats. J Neurosci 22:9932–9940

Tomita S, Chen L, Kawasaki Y, Petralia RS, Wenthold RJ, Nicoll RA, Bredt DS (2003) Functional studies and distribution define a family of transmembrane AMPA receptor regulatory proteins. J Cell Biol 161:805–816

Traynelis SF, Wollmuth LP, McBain CJ, Menniti FS, Vance KM, Ogden KK, Hansen KB, Yuan H, Myers SJ, Dingledine R (2010) Glutamate receptor ion channels: structure, regulation, and function. Pharmacol Rev 62:405–496

Tse D, Langston RF, Kakeyama M, Bethus I, Spooner PA, Wood ER, Witter MP, Morris RG (2007) Schemas and memory consolidation. Science 316:76–82

Tsien JZ, Huerta PT, Tonegawa S (1996) The essential role of hippocampal CA1 NMDA receptor-dependent synaptic plasticity in spatial memory. Cell 87:1327–1338

Tsuda M, Suzuki T, Misawa M, Nagase H (1996) Involvement of the opioid system in the anxiolytic effect of diazepam in mice. Eur J Pharmacol 307:7–14

Tu JC, Xiao B, Naisbitt S, Yuan JP, Petralia RS, Brakeman P, Doan A, Aakalu VK, Lanahan AA, Sheng M, Worley PF (1999) Coupling of mGluR/Homer and PSD-95 complexes by the Shank family of postsynaptic density proteins. Neuron 23:583–592

Tully T, Bourtchouladze R, Scott R, Tallman J (2003) Targeting the CREB pathway for memory enhancers. Nat Rev Drug Discov 2:267–277

Turner DA, Deupree DL (1991) Functional elongation of CA1 hippocampal neurons with aging in Fischer 344 rats. Neurobiol Aging 12:201–210

Turrigiano GG (2008) The self-tuning neuron: synaptic scaling of excitatory synapses. Cell 135:422–435

Turrigiano GG, Leslie KR, Desai NS, Rutherford LC, Nelson SB (1998) Activity-dependent scaling of quantal amplitude in neocortical neurons. Nature 391:892–896

Tyszkiewicz JP, Gu Z, Wang X, Cai X, Yan Z (2004) Group II metabotropic glutamate receptors enhance NMDA receptor currents via a protein kinase C-dependent mechanism in pyramidal neurones of rat prefrontal cortex. J Physiol 554:765–777

Uslaner JM, Parmentier-Batteur S, Flick RB, Surles NO, Lam JS, McNaughton CH, Jacobson MA, Hutson PH (2009) Dose-dependent effect of CDPPB, the mGluR5 positive allosteric modulator, on recognition memory is associated with GluR1 and CREB phosphorylation in the prefrontal cortex and hippocampus. Neuropharmacology 57:531–538

van der Kooij MA, Sandi C (2012) Social memories in rodents: methods, mechanisms and modulation by stress. Neurosci Biobehav Rev 36:1763–1772

Vazdarjanova A, McNaughton BL, Barnes CA, Worley PF, Guzowski JF (2002) Experience-dependent coincident expression of the effector immediate-early genes arc and Homer 1a in hippocampal and neocortical neuronal networks. J Neurosci 22:10067–10071

Vazdarjanova A, Ramirez-Amaya V, Insel N, Plummer TK, Rosi S, Chowdhury S, Mikhael D, Worley PF, Guzowski JF, Barnes CA (2006) Spatial exploration induces ARC, a plasticity-related immediate-early gene, only in calcium/calmodulin-dependent protein kinase II-positive principal excitatory and inhibitory neurons of the rat forebrain. J Comp Neurol 498:317–329

Wagner JJ, Terman GW, Chavkin C (1993) Endogenous dynorphins inhibit excitatory neurotransmission and block LTP induction in the hippocampus. Nature 363:451–454

Walker JA, Olton DS (1979) Spatial memory deficit following fimbria-fornix lesions: independent of time for stimulus processing. Physiol Behav 23:11–15

Waltereit R, Dammermann B, Wulff P, Scafidi J, Staubli U, Kauselmann G, Bundman M, Kuhl D (2001) Arg3.1/Arc mRNA induction by Ca2+ and cAMP requires protein kinase A and mitogen-activated protein kinase/extracellular regulated kinase activation. J Neurosci 21:5484–5493

Waung MW, Pfeiffer BE, Nosyreva ED, Ronesi JA, Huber KM (2008) Rapid translation of Arc/Arg3.1 selectively mediates mGluR-dependent LTD through persistent increases in AMPAR endocytosis rate. Neuron 59:84–97

Webster SJ, Bachstetter AD, Nelson PT, Schmitt FA, Van Eldik LJ (2014) Using mice to model Alzheimer's dementia: an overview of the clinical disease and the preclinical behavioral changes in 10 mouse models. Front Genet 5:88

Werther GA, Russo V, Baker N, Butler G (1998) The role of the insulin-like growth factor system in the developing brain. Horm Res 49(Suppl 1):37–40

West AE, Griffith EC, Greenberg ME (2002) Regulation of transcription factors by neuronal activity. Nat Rev Neurosci 3:921–931

Williams BJ, Bimonte-Nelson HA, Granholm-Bentley AC (2006) ERK-mediated NGF signaling in the rat septo-hippocampal pathway diminishes with age. Psychopharmacology (Berl) 188:605–618

Wilson IA, Ikonen S, McMahan RW, Gallagher M, Eichenbaum H, Tanila H (2003) Place cell rigidity correlates with impaired spatial learning in aged rats. Neurobiol Aging 24:297–305

Wilson IA, Gallagher M, Eichenbaum H, Tanila H (2006) Neurocognitive aging: prior memories hinder new hippocampal encoding. Trends Neurosci 29:662–670

Winslow JT, Hastings N, Carter CS, Harbaugh CR, Insel TR (1993) A role for central vasopressin in pair bonding in monogamous prairie voles. Nature 365:545–548

Wisden W, Errington ML, Williams S, Dunnett SB, Waters C, Hitchcock D, Evan G, Bliss TV, Hunt SP (1990) Differential expression of immediate early genes in the hippocampus and spinal cord. Neuron 4:603–614

Wishka DG, Walker DP, Yates KM, Reitz SC, Jia S, Myers JK, Olson KL, Jacobsen EJ, Wolfe ML, Groppi VE, Hanchar AJ, Thornburgh BA, Cortes-Burgos LA, Wong EH, Staton BA, Raub TJ, Higdon NR, Wall TM, Hurst RS, Walters RR, Hoffmann WE, Hajos M, Franklin S, Carey G, Gold LH, Cook KK, Sands SB, Zhao SX, Soglia JR, Kalgutkar AS, Arneric SP, Rogers BN (2006) Discovery of N-[(3R)-1-azabicyclo[2.2.2]oct-3-yl]furo[2,3-c]pyridine-5-carboxamide, an agonist of the alpha7 nicotinic acetylcholine receptor, for the potential treatment of cognitive deficits in schizophrenia: synthesis and structure—activity relationship. J Med Chem 49:4425–4436

Wittmann W, Schunk E, Rosskothen I, Gaburro S, Singewald N, Herzog H, Schwarzer C (2009) Prodynorphin-derived peptides are critical modulators of anxiety and regulate neurochemistry and corticosterone. Neuropsychopharmacology 34:775–785

Wong RW, Setou M, Teng J, Takei Y, Hirokawa N (2002) Overexpression of motor protein KIF17 enhances spatial and working memory in transgenic mice. Proc Natl Acad Sci U S A 99:14500–14505

Wu GY, Deisseroth K, Tsien RW (2001) Activity-dependent CREB phosphorylation: convergence of a fast, sensitive calmodulin kinase pathway and a slow, less sensitive mitogen-activated protein kinase pathway. Proc Natl Acad Sci U S A 98:2808–2813

Wyllie DJ, Nicoll RA (1994) A role for protein kinases and phosphatases in the Ca(2+)-induced enhancement of hippocampal AMPA receptor-mediated synaptic responses. Neuron 13:635–643

Xia M, Huang R, Guo V, Southall N, Cho MH, Inglese J, Austin CP, Nirenberg M (2009) Identification of compounds that potentiate CREB signaling as possible enhancers of long-term memory. Proc Natl Acad Sci U S A 106:2412–2417

Xiao B, Tu JC, Petralia RS, Yuan JP, Doan A, Breder CD, Ruggiero A, Lanahan AA, Wenthold RJ, Worley PF (1998) Homer regulates the association of group 1 metabotropic glutamate receptors with multivalent complexes of homer-related, synaptic proteins. Neuron 21:707–716

Xu J, Zhu Y, Contractor A, Heinemann SF (2009) mGluR5 has a critical role in inhibitory learning. J Neurosci 29:3676–3684

Yakovleva T, Marinova Z, Kuzmin A, Seidah NG, Haroutunian V, Terenius L, Bakalkin G (2007) Dysregulation of dynorphins in Alzheimer disease. Neurobiol Aging 28:1700–1708

Yang L, Mao L, Tang Q, Samdani S, Liu Z, Wang JQ (2004) A novel Ca2+-independent signaling pathway to extracellular signal-regulated protein kinase by coactivation of NMDA receptors and metabotropic glutamate receptor 5 in neurons. J Neurosci 24:10846–10857

Yang S, Megill A, Ardiles AO, Ransom S, Tran T, Koh MT, Lee HK, Gallagher M, Kirkwood A (2013a) Integrity of mGluR-LTD in the associative/commissural inputs to CA3 correlates with successful aging in rats. J Neurosci 33:12670–12678

Yang SH, Sharrocks AD, Whitmarsh AJ (2013b) MAP kinase signalling cascades and transcriptional regulation. Gene 513:1–13

Yokoi M, Kobayashi K, Manabe T, Takahashi T, Sakaguchi I, Katsuura G, Shigemoto R, Ohishi H, Nomura S, Nakamura K, Nakao K, Katsuki M, Nakanishi S (1996) Impairment of hippocampal mossy fiber LTD in mice lacking mGluR2. Science 273:645–647

Yoshimura Y, Aoi C, Yamauchi T (2000) Investigation of protein substrates of Ca(2+)/calmodulin-dependent protein kinase II translocated to the postsynaptic density. Brain Res Mol Brain Res 81:118–128

Yoshimura Y, Shinkawa T, Taoka M, Kobayashi K, Isobe T, Yamauchi T (2002) Identification of protein substrates of Ca(2+)/calmodulin-dependent protein kinase II in the postsynaptic density by protein sequencing and mass spectrometry. Biochem Biophys Res Commun 290:948–954

Zhang WQ, Mundy WR, Thai L, Hudson PM, Gallagher M, Tilson HA, Hong JS (1991) Decreased glutamate release correlates with elevated dynorphin content in the hippocampus of aged rats with spatial learning deficits. Hippocampus 1:391–397

Zhang GR, Liu M, Cao H, Kong L, Wang X, O'Brien JA, Wu SC, Cook RG, Geller AI (2009) Improved spatial learning in aged rats by genetic activation of protein kinase C in small groups of hippocampal neurons. Hippocampus 19:413–423

Zhao X, Rosenke R, Kronemann D, Brim B, Das SR, Dunah AW, Magnusson KR (2009) The effects of aging on N-methyl-D-aspartate receptor subunits in the synaptic membrane and relationships to long-term spatial memory. Neuroscience 162:933–945

Zheng WH, Kar S, Quirion R (2000) Stimulation of protein kinase C modulates insulin-like growth factor-1-induced akt activation in PC12 cells. J Biol Chem 275:13377–13385

Zhou Y, Takahashi E, Li W, Halt A, Wiltgen B, Ehninger D, Li GD, Hell JW, Kennedy MB, Silva AJ (2007) Interactions between the NR2B receptor and CaMKII modulate synaptic plasticity and spatial learning. J Neurosci 27:13843–13853

Zhu JJ, Qin Y, Zhao M, Van Aelst L, Malinow R (2002) Ras and Rap control AMPA receptor trafficking during synaptic plasticity. Cell 110:443–455

Role of Adult Hippocampal Neurogenesis in Cognition in Physiology and Disease: Pharmacological Targets and Biomarkers

Veronica Costa, Sebastian Lugert, and Ravi Jagasia

Contents

V. Costa • S. Lugert • R. Jagasia (✉)
Roche Pharmaceutical Research and Early Development, Neuroscience Ophthalmology and Rare Diseases (NORD), Roche Innovation Center Basel, 124 Grenzacherstrasse, 4070 Basel, Switzerland
e-mail: ravi.jagasia@roche.com

K.M. Kantak, J.G. Wettstein (eds.), *Cognitive Enhancement*, Handbook of Experimental Pharmacology 228, DOI 10.1007/978-3-319-16522-6_4

Abstract

Adult hippocampal neurogenesis is a remarkable form of brain structural plasticity by which new functional neurons are generated from adult neural stem cells/precursors. Although the precise role of this process remains elusive, adult hippocampal neurogenesis is important for learning and memory and it is affected in disease conditions associated with cognitive impairment, depression, and anxiety. Immature neurons in the adult brain exhibit an enhanced structural and synaptic plasticity during their maturation representing a unique population of neurons to mediate specific hippocampal function. Compelling preclinical evidence suggests that hippocampal neurogenesis is modulated by a broad range of physiological stimuli which are relevant in cognitive and emotional states. Moreover, multiple pharmacological interventions targeting cognition modulate adult hippocampal neurogenesis. In addition, recent genetic approaches have shown that promoting neurogenesis can positively modulate cognition associated with both physiology and disease. Thus the discovery of signaling pathways that enhance adult neurogenesis may lead to therapeutic strategies for improving memory loss due to aging or disease. This chapter endeavors to review the literature in the field, with particular focus on (1) the role of hippocampal neurogenesis in cognition in physiology and disease; (2) extrinsic and intrinsic signals that modulate hippocampal neurogenesis with a focus on pharmacological targets; and (3) efforts toward novel strategies pharmacologically targeting neurogenesis and identification of biomarkers of human neurogenesis.

Keywords

Adult neurogenesis • Neural stem cell • Hippocampus • Cognition

1 Introduction: Adult Hippocampal Neurogenesis

Purification of prospective neural progenitor cells, which are characterized by their potential to proliferate and give rise to differentiated neural progeny in vitro, has been successfully achieved from many regions of the adult mammalian central nervous system (CNS). However, despite the widespread distribution of neural precursors throughout the adult brain, adult neurogenesis is maintained in only two discrete regions of the adult mammalian brain: the subventricular zone of the lateral ventricles (Altman 1969; Lois and Alvarez-Buylla 1994; Alvarez-Buylla and Garcia-Verdugo 2002) and the subgranular zone of the dentate gyrus of the hippocampal formation (Kaplan and Hinds 1977; Cameron et al. 1993).

From a neuroscientific perspective, hippocampal adult neurogenesis is of major interest as (1) the generation of new hippocampal neurons contributes to hippocampal function including learning and memory and mood regulation; (2) hippocampal neurogenesis is involved in the pathophysiology of depression, schizophrenia, age-related memory impairment, and multiple neuronal developmental disorders including autism spectrum disorder; and (3) the process of developing newborn neurons in a mainly restrictive environment as the adult brain provides an in vivo model to elucidate the molecular and cellular basis of neural regeneration which ultimately could be harnessed to develop novel therapies for neurodegenerative diseases.

The process of generating new granule neurons from adult neural stem cells is a highly dynamic process and at the same time tightly regulated at multiple developmental stages. Developmental stages include neuronal precursors cell proliferation, differentiation, survival, migration, and integration into preexisting hippocampal networks (see Fig. 1). Hippocampal neural stem cells, commonly referred to as radial Type I cells, have the capacity to self-renew and to differentiate into neurons and astroglia (Bonaguidi et al. 2011; Encinas et al. 2011; Lugert et al. 2010). They are localized in the subgranular zone of the dentate gyrus (DG) with a characteristic radial process spanning through the molecular layer and express the radial-glia marker GFAP (Seri et al. 2001; Kriegstein and Alvarez-Buylla 2009). Active Type I cells give rise to transient amplifying neural precursor cells (NPC), referred to as Type II cells, which then develop into mature granule neurons unless negatively selected by hippocampus resident microglia (Sierra et al. 2010). Morphological maturation of newborn granule neurons includes the development of dendritic trees into the molecular cell layer of the dentate gyrus (DG) and the projection of axons toward CA3 to become functionally integrated (Hastings and Gould 1999; Toni et al. 2007). During development, newly generated cells are characterized by a temporally ordered expression of stage-specific markers and changes in morphological and functional properties. From a morphological standpoint, differentiating progenitors located in the subgranular zone show bipolar horizontal processes; after cell cycle exit they develop the primary apical dendrite toward the molecular cell layer and axons toward the CA3 region and show progressive increase in the complexity of the dendritic tree during maturation. Morphological changes are paralleled by dynamic expression of specific marker proteins. Early in the neurogenic lineage transient amplifying Type II cells show the expression of glial markers including the transcription factor SRY (sex determining region Y)-box2 (Sox2). At later stages differentiating progenitors express neuronal transcription factors like NeuroD (neurogenic differentiation) (Kronenberg et al. 2003; Seki 2002) and immature neurons express the microtubule-associated protein doublecortin (DCX) (Brown et al. 2003b).

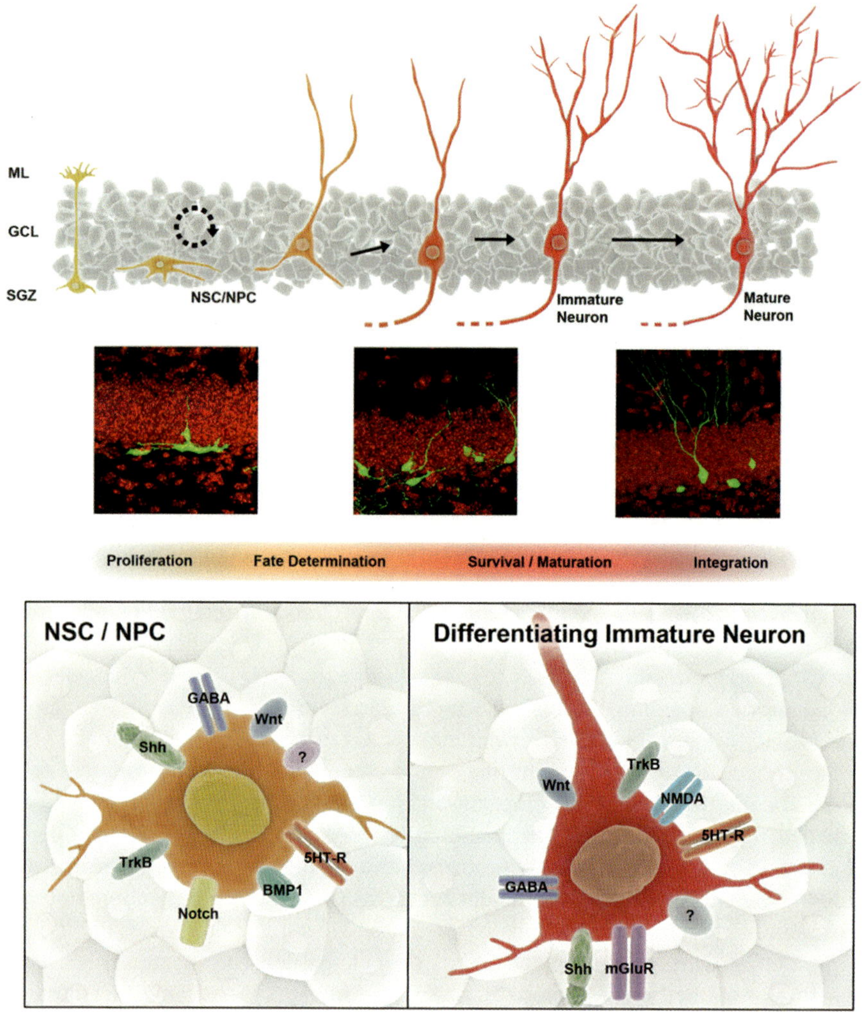

Fig. 1 Hippocampal neurogenesis in the adult rodent brain follows distinct developmental stages. (*Top*) Schematic depiction of developmental stages of hippocampal neurogenesis including proliferation, differentiation, maturation, and integration of newborn neurons. (*Middle*) The morphological development of neurons monitored by GFP expression upon retrovirus-mediated gene transduction in the adult mouse brain. (Proliferation: *left panel*, 3 days postinjection (dpi); differentiation, *middle panel*, 7 dpi, differentiation, mature granule neurons, *right panel*, 28 dpi). Experiments performed by RJ as a fellow in the laboratory of Prof. Dr. Dieter Chichung Lie. (*Bottom*) Depiction of known and potential CNS receptors/targets modulating the process of adult neurogenesis at the proliferation and differentiation stages in vivo

The trajectory of functional maturation includes formation of synaptic connections and the switch from depolarizing to hyperpolarizing action of the neurotransmitter GABA. In particular, while immature adult-born neurons receive synaptic excitatory GABA inputs (Ge et al. 2006, 2007; Karten et al. 2006), at later stages GABAergic inputs become gradually hyperpolarizing (Ge et al. 2006). Concomitantly, immature neurons form spines and receive glutamatergic excitatory inputs and develop mossy fiber boutons around 4–8 weeks after birth (Faulkner et al. 2008).

Adult hippocampal neurogenesis is not exclusive to rodents, but has repetitively been shown to occur in humans (Knoth et al. 2010; Eriksson et al. 1998). In adult humans around 700 new neurons per day are being integrated per dentate gyrus. This corresponds to an annual turnover rate of 1.75 % of the renewing dentate granule cell population (Spalding et al. 2013). This significant number of new hippocampal granule neurons in humans, together with dynamic regulation under physiological conditions, suggests that adult neurogenesis may be integral to brain functions.

2 Factors Affecting Hippocampal Neurogenesis and Plasticity and Correlation to Cognition

Over the past 15 years, substantial converging evidence has indicated that neurogenesis in the adult hippocampus is functionally relevant to hippocampal-dependent cognition. Many factors that modulate cognition, including pharmacological agents and intrinsic pathways involved in brain plasticity, as well as physiological stimuli such as running, learning, and enriched environment, also modulate hippocampal neurogenesis. Multiple studies impairing proliferation and adult neurogenesis in the dentate gyrus using various approaches such as chemically and irradiation-induced blockade of proliferation, genetic ablations of progenitor cells, and optogenetic physiological silencing have demonstrated that neurogenesis contributes to hippocampus-dependent learning and memory. In addition, specifically increasing the amount of newborn neurons in the dentate gyrus improves cognitive processes associated with the hippocampus. We will highlight the role of adult neurogenesis in cognition and discuss intrinsic and extrinsic factors regulating the process of neurogenesis.

2.1 Cognitive Behaviors Dependent on Hippocampal Function

Early evidence of the role of the hippocampus in memory and learning comes from studies in humans showing that bilateral resection or damage of hippocampus and hippocampal gyrus leads to memory impairment (Scoville and Milner 1957). Detailed analysis in animal models using functional imaging approaches, lesioning of specific areas of the brain, and circuit tracing reveals that the hippocampus is required for several forms of memory including declarative memory (the ability for conscious remembering) (Squire 1992), episodic memory (remembering of autobiographical events) (Wixted et al. 2014), contextual association memory (Rudy and Sutherland 1995; Lee et al. 2014), and spatial navigation. The hippocampus combines information about spatial and non-spatial items coming from inputs from the parahippocampal cortex and medial entorhinal cortex and from the perirhinal and lateral entorhinal areas, respectively. In the hippocampus, the dentate gyrus region receives inputs from the entorhinal cortex and DG granule cells project excitatory mossy fibers to the proximal apical dendrites of pyramidal cells in the CA3 area. The dentate gyrus is characterized by structural and functional heterogeneity along the dorso-ventral axis, with the dorsal region being mainly implicated in the regulation of cognition and memory and the ventral region involved in the modulation of mood and stress (Kheirbek et al. 2013; Tannenholz et al. 2014). One major process supported by the dorsal DG is conjunctive encoding which is the processing of multiple unique sensory spatial and non-spatial inputs from the perirhinal cortex and lateral and medial entorhinal cortex to form metric spatial representations (Kesner 2013). Indeed, lesions in the dorsal DG cause impairments in cue-context associations like the ability to associate environmental cues to specific odors (Morris et al. 2013).

Different circuits and regions in the hippocampus are instrumental in the processes of pattern separation and pattern completion, important mechanisms in declarative memory (Yassa and Stark 2011). Pattern separation is the ability to discriminate between similar overlapping representations by differentially encoding small or weak changes from similar inputs such as between two friend's faces (Treves et al. 2008). On the other hand, pattern completion allows accurate reconstruction of incomplete representations based on previously stored representations. The dentate gyrus appears to be critical for the process of pattern separation. This process is facilitated by the distributed pattern of firing activity of the DG cells and the sparse mossy fiber connections onto CA3 pyramidal cells, lowering the probability of two CA3 neurons to receive inputs from the same population of DG neurons (Rolls 1996). On the other hand, local axonal inputs of neurons in the CA3 onto dendrites of cells in the same regions (also called recurrent collaterals) appear to mediate the process of pattern completion. The CA3 area of the hippocampus also receives direct inputs from the entorhinal cortex (perforant path) which are critical for memory retrieval, while inactivation of mossy fiber inputs onto CA3 neurons affects encoding and new learning without altering memory recall (Lassalle et al. 2000; Lee and Kesner 2004; Rolls 2007). In humans, fMRI studies performed during incidental encoding tasks show a correlation between level of activity in the CA3/DG and pattern separation tasks, while activity in CA1, the subiculum, the entorhinal, and parahippocampal cortices correlates with pattern completion (Bakker et al. 2008).

2.2 The Role of Hippocampal Neurogenesis in Cognition

How does adult neurogenesis contribute to memory and hippocampal function and why is this so unique to the dentate gyrus? Adult hippocampal neurogenesis has been implicated in several aspects of contextual and spatial memory. For example, in aged outbred rats there is direct correlation between performance in the water maze test, a hippocampus-dependent spatial learning task, and the amount of neurogenesis in the dentate gyrus. Notably, high performers have a significantly higher number of surviving neurons, based on BrdU (bromodeoxyuridine, a synthetic nucleoside analogue of thymidine incorporated in newly synthesized DNA of replicating cells)-positive cell counting, a readout of proliferation, in the dentate gyrus after learning (Drapeau et al. 2003). Spatial learning tasks such as the Morris water maze modulate hippocampal neurogenesis leading to the question of whether these newborn neurons are integrated in the preexisting memory circuit and reactivated during memory recall.

The evidence that newborn neurons are actively integrated in circuits during specific spatial learning tasks comes from studies analyzing the expression of immediate early genes (like c-fos and arc), a molecular correlate of neuronal firing activity, in newborn cells generated before or after exposure to spatial learning and birthdated with specific thymidine analogues. Interestingly, newborn neurons generated during a specific development window before exposure to the learning task are preferentially activated upon reexposure to the same spatial learning paradigm if compared to mature dentate granule neurons. These studies suggest that the newborn neurons are preferentially recruited in the generation of specific memory circuits compared to mature dentate granule neurons (Kee et al. 2007; Ramirez-Amaya et al. 2006; Tashiro et al. 2007). The enhanced plasticity of newborn young granule cells could potentially facilitate the integration into new memory circuits and upon maturation the increase in threshold for induction of synaptic plasticity could render the connectivity more stable. Thus, sustained hippocampal adult neurogenesis and continuous maturation of pools of immature neurons allow the DG network to achieve both stable analysis of "old" features and adaptation to new environments, supporting precise and distinct representations of new memories throughout life.

To address the causal relationship between neurogenesis and cognition, studies have focused on the analysis of the effect of neurogenesis ablation or enhancement on behavioral performances. X-ray irradiation-mediated ablation of neurogenesis, as well as genetic ablation in the GFAP-TK genetic mouse model (in which a modified herpes simplex virus gene encoding thymidine kinase under the control of the GFAP promoter causes dividing cells to die upon administration of the drug ganciclovir), leads to impairment in context fear conditioning tasks (Saxe et al. 2006; Drew et al. 2010). In another mouse model, where neurogenesis is ablated selectively inducing the expression of Bax, a pro-apoptotic protein, in neural precursors, spatial relational memory is strongly impaired (Dupret et al. 2008). In a mouse model that allowed a transient reduction of the number of adult-born DGCs, it has been shown that reduction of immature neurons confers a deficiency in forming robust, long-term spatial memory and leads to impaired performance in extinction tasks. These results further substantiate that the maturing dentate granule neurons are critical in cognition (Deng et al. 2009). These results were largely confirmed in another study where novel object recognition was impaired by the elimination of 4- to 6-week-old immature neurons (Denny et al. 2012). Recent studies looking at the effect of post-training ablation (retrograde effects) of newborn neurons and silencing of adult-generated neurons on hippocampal memory further highlight the importance of this neuronal population in formation of memory. Ablation of newborn neurons using a diphtheria toxin-based strategy after learning leads to degradation of existing contextual fear and water maze memories, even when the ablation is induced 1 month after learning (Arruda-Carvalho et al. 2011). Along the same line, using an optogenetic approach, it has been shown that silencing newborn neurons affects the retrieval of memory after completion of training. Interestingly, silencing specifically 4-weeks-old but not younger or older neurons leads to memory impairment. This strongly suggests a functional role of newly integrated immature neurons in the hippocampal circuit (Denny et al. 2012) and supports the hypothesis that immature adult-born neurons contribute to proper cognitive processing.

As cells in the dentate gyrus possess low firing rates and are only activated in a sparse manner, it has been hypothesized that the dentate gyrus may possess a supportive function in pattern separation. Indeed, the ablation of neurogenesis through X-irradiation or Bax overexpression impairs the ability to discriminate between two contexts with overlapping features (Clelland et al. 2009). Importantly, by specifically enhancing the survival of newborn neurons through the deletion of Bax, it has been shown that increase in adult hippocampal neurogenesis does not affect the ability to distinguish between two different contexts but significantly improves the ability to discriminate between overlapping contextual representations (Sahay et al. 2011).

Newborn neurons integrate in preexisting hippocampal circuitry competing with already established synaptic connections. Thus beyond modulating formation of novel memory, adult hippocampal neurogenesis may affect memories already stored in these circuits. Indeed, in many species including humans, during infancy, when the degree of neurogenesis is highest, the retrieval of hippocampus-dependent memories is impaired at later time points (Rubin 2000). Recently, a link between neurogenesis and the ability to forget previously acquired memories has been provided. In this study, using a combination of genetic, pharmacologic, and behavioral strategies, the authors show that increase in neurogenesis after learning is responsible for forgetting and leads to the hypothesis that reconfiguration of hippocampal circuits by newborn neurons may reduce the ability to retrieve previously acquired patterns of activity (Akers et al. 2014).

In conclusion, although there is a certain variability in the effect of modulation of neurogenesis on specific behavioral tasks, a number of studies have consistently shown the causal relationship between neurogenesis and hippocampal-dependent cognitive processes. In the next section, we will review physiological, pathological, and pharmacological mechanisms which can modulate neurogenesis and behavior.

2.3 Intrinsic Factors Which Regulate Hippocampal Neurogenesis and Implications in Cognition

Neurogenesis is controlled by interaction of neural progenitor cells and newborn neurons with several components of the dentate gyrus microenvironment, including astrocytes, vasculature, mature granule neurons, and GABAergic interneurons (Song et al. 2002; Palmer et al. 2000; Ma et al. 2009; Ge et al. 2006). Moreover, neurogenesis is tightly regulated by several endogenous signaling molecules including hormones and growth factors. In parallel, the activity of the neuronal network and the release of neurotransmitters from afferent projections onto the dentate gyrus can modulate several aspects of neuronal development. The concerted action of these signaling systems ultimately determines the coordinated functional integration of new neurons in preexisting circuitry (Pathania et al. 2010). Below we will highlight key experimental evidence supporting regulatory roles for some of these factors which are relevant in both physiologically and pharmacologically induced neurogenesis.

2.3.1 Neurotransmitters

Neuronal activity strongly modulates various stages of neurogenesis. Lesions of the entorhinal cortex which is one of the major excitatory afferent on granule cells increases DG cells proliferation (Cameron et al. 1995; Nacher et al. 2001). Furthermore, electrical induction of LTP at the perforant path/granule cells synapses promotes proliferation and survival of 1 and 2 weeks old newborn neurons (Bruel-Jungerman et al. 2006; Chun et al. 2006). Glutamatergic neurotransmission and specifically NMDA receptor activity regulates proliferation and correct functional maturation/integration and survival of newborn neurons (Pathania et al. 2010). In tree shrew DG, pharmacological blockade of NMDA receptors leads to increase in the number of BrdU-positive cells (Gould et al. 1997). Along the same line, activation and blockade of NMDA receptors reduce or promote cell proliferation in adult rat DG, respectively. An important question is to what extent this effect is regulated cell-autonomously rather than indirectly via other signals elicited by neuronal activity in the dentate gyrus. Immature neurons have NMDA receptors and express NR1 and NR2B subunits (Nacher and McEwen 2006; Ambrogini et al. 2004). Deletion of the NMDA subunit NR1 in newborn cells reduces the number of properly integrating/surviving newborn neurons. This effect is due to NMDA-dependent regulation of survival during the third week after neuronal birth and it appears to involve a mechanism of competitive survival between the incoming immature neurons and the preexisting neurons. Indeed, global hippocampal reduction in NMDA signaling can rescue the loss of cells. Importantly, maturing neurons (4–6 weeks after neuronal birth) show increased plasticity and reduced threshold for induction of LTP, a process in part mediated by NR2B (Ge et al. 2007). This suggests that glutamatergic signaling plays multiple roles in modulating neurogenesis and controlling the precise integration of newborn neurons into the hippocampal network.

In the SGZ, GABA neurotransmitter, released by specific populations of interneurons, modulates several aspects of neurogenesis, including precursor cells' proliferation, differentiation, and subsequent neuronal maturation. Like in development, there is a switch between depolarization and hyperpolarization effects of GABA while the newborn cells are maturing and this may alter properties of immature neurons including their synaptic plasticity (Ge et al. 2006). The enhanced synaptic plasticity of immature neurons is likely in part due to a lack of strong GABAergic inhibition (Ge et al. 2008; Markwardt and Overstreet-Wadiche 2008; Pallotto and Deprez 2014). Tonic response of nestin-expressing quiescent radial glia cells to GABA released from parvalbumin interneurons regulates their reactivation and entry into the cell cycle. This is mediated by activation of γ2 containing GABA A receptors since conditional deletion of the subunit induces exit from quiescence and promotes symmetric self-renewal of type I cells (Song et al. 2012). The role of tonic GABA transmission on inhibition of cell proliferation is confirmed in another study upon deletion of the α4 subunit, component of GABA A receptors mediating tonic (extrasynaptic) response (Duveau et al. 2011). Neural progenitors' proliferation is regulated also by GABA B receptors, metabotropic G-protein-coupled receptors located both on pre- and postsynaptic terminals. Both pharmacological blockage and genetic deletion of the B1 subunit of GABA B receptors promote progenitor cells' proliferation (Felice et al. 2012; Giachino et al. 2014). GABA-mediated depolarization, due to high concentration of intracellular Cl^- in immature neurons, induces neuronal differentiation and NeuroD expression in transient amplifying neuronal progenitor (type 2) cells (Tozuka et al. 2005). Deletion of both the α4 and α2 subunits, a component of GABA A receptors mediating synaptic phasic response, causes reduction of dendritic length and complexity in newborn neurons, which is revealed at different stages of differentiation (Duveau et al. 2011). Altering the GABAergic-dependent depolarization/hyperpolarization switch process by genetically modulating the expression levels of the Cl^- importer NKCC1 reveals the key role of this mechanism in the regulation of proper neuronal morphology, differentiation, and synaptic maturation (Jagasia et al. 2009; Ge et al. 2006). The effect of GABA transmission on newborn neurons development is at least in part mediated by activation of downstream signaling events via activity-dependent transcription factors such as CREB (Jagasia et al. 2009).

Loss of cholinergic neurons or blockage of acetylcholine (ACh) receptors in the central nervous system causes learning impairment in experimental and clinical situations in humans (Drachman and Leavitt 1974; Rasmusson and Dudar 1979). Newborn neurons in the dentate gyrus are innervated by forebrain cholinergic fibers (Kaneko et al. 2006) and by septal cholinergic cells as shown using a combination of rabies virus-mediated retrograde tracing and retroviral labeling of new granule cells (Vivar et al. 2012). Neurotoxic and immunotoxic lesion of forebrain cholinergic projections leads to decreased neurogenesis, increased apoptosis and impaired spatial memory (Mohapel et al. 2005; Cooper-Kuhn et al. 2004). Modulation of the cholinergic system using a number of pharmacological approaches further supports the role of the system in regulation of neurogenesis (Veena et al. 2011a).

Table 1 Examples of pharmacological interventions that improve neurogenesis and cognition

Molecule	Effect on neurogenesis	Effect on cognition	Disease model	References
GABA(A) a5 negative allosteric modulator	Correction of hippocampal synaptic plasticity and adult neurogenesis defects	Correction of spatial learning and memory deficits	Ts65Dn mouse model of Down Syndrome	Martinez-Cue et al. (2013)
Isoxazole-9	Enhancement of proliferation and differentiation of neuroblasts, dendritic arborization of immature neurons	Enhancement of memory in Morris Water Maze		Petrik et al. (2012)
P7C3	Enhancement of survival of newborn neurons	Improvement of performance in Morris water maze task in aged rats	Aging	Pieper et al. (2010)
SB216763 (GSK3 beta inhibitor)	Correction of neuronal differentiation and maturation deficits	Improvement in trace conditioning learning test and spatial learning and memory in DMNP radial arm maze	FMR1 ko mouse model of Fragile X syndrome	Guo et al. (2012)
Lithium (GSK3 beta inhibitor)	Enhancement of progenitor cells proliferation and differentiation	Improvement of performance in Morris Water maze task and inhibitory avoidance task	TgCRND8 mouse model of AD	Fiorentini et al. (2010)
Lithium	Correction of hippocampal synaptic plasticity and adult neurogenesis defects	Correction of deficits in performance in fear conditioning, object location, novel object recognition tests	Ts65Dn mouse model of Down Syndrome	Contestabile et al. (2013)
Compound K	Partial correction of neurogenesis impairment	Improvement in passive avoidance and Y maze tests	Cyclophosphamide-treated mice as model of chemotherapy	Hou et al. (2013)

(continued)

Table 1 (continued)

Molecule	Effect on neurogenesis	Effect on cognition	Disease model	References
Fluoxetine	Correction of progenitor cells proliferation in the DG	Improvement in contextual fear conditioning tests	Ts65Dn mouse model of Down Syndrome	Bianchi et al. (2010) (NB: early treatment)
Fluoxetine	Correction of reduced progenitors proliferation in the DG	Improvement in hippocampal-dependent spatial working memory	5-fluorouracil-treated rats as model of chemotherapy	ElBeltagy et al. (2010)
Imipramine	Preservation of proliferation and survival of newborn neurons	Improvement in novel object recognition test	Mouse model of traumatic brain injury	Han et al. (2011)
Amitriptyline	Increase in neurogenesis and BDNF signaling	Improvement in short- and long-term memory retention	3xTgAD mouse model of AD	Chadwick et al. (2011)
Metformin	Increase in neurogenesis	Improvement in spatial reversal learning		Wang et al. (2012)

Stimulation of cholinergic receptors with the cholinergic agonist physostigmine and inhibition of acetylcholinesterase using donepezil induces neurogenesis and promotes proliferation and short-term survival (Mohapel et al. 2005; Kaneko et al. 2006; Kotani et al. 2006). On the other hand, scopolamine, a cholinergic muscarinic receptor blocker, decreases the number of BrdU-positive cells in the DG affecting the survival of newborn neurons (Kotani et al. 2006). Early in their development, adult-born neurons express homomeric α7-containing nicotinic acetylcholine receptors and cell autonomous genetic ablation leads to impairment in dendritic maturation and synaptic integration ultimately resulting in reduced survival (Campbell et al. 2010). In global knockout models of the β2 receptor subunit there is significant reduction in cell proliferation culminating in a net reduction in the size of the dentate granule cell layer (Harrist et al. 2004). In vitro, cholinergic stimulation affects proliferation and survival of rat olfactory bulb and cortical neural precursor cells (Coronas et al. 2000; Ma et al. 2000). Acetylcholine neurotransmission appears to be deregulated with age and in Alzheimer's disease, conditions with reduction in both neurogenesis and cognitive capacity. Notably, pharmacological modulation of the cholinergic activity in aged or stressed animals promotes NSCs' proliferation and corrects cognitive alterations (Itou et al. 2011; Veena et al. 2011b).

The dopaminergic system has been shown to affect proliferation and differentiation of neural progenitor cells during embryonic development and in both adult neurogenic zones. Lesion and pharmacological studies in the SGZ have yielded discrepancy in results (Veena et al. 2011a). Focusing on the pharmacological approaches, depletion of dopamine in rodents reduces proliferation of SGZ neuronal precursor cells and this is reversed by treatment with a D2-like receptor agonist (Hoglinger et al. 2004). Similarly, activation of D2 receptors using quinpirole promotes NSCs' proliferation (Yang et al. 2008). However, administration of haloperidol, a D2-like receptor antagonist, has been reported to induce both positive and negative effects on neurogenesis in the SGZ (Wakade et al. 2002; Wang et al. 2004; Keilhoff et al. 2010; Halim et al. 2004). A recent study demonstrates that dopamine increases adult hippocampal NSCs' proliferation acting on D1-like receptors since the effect is phenocopied by a D1-like receptor agonist but not a D2 agonist (Takamura et al. 2014). On the other hand, stimulating the D3 receptor appears to exert an inhibitory effect on neurogenesis since inhibition of the D3 receptor using the antagonist S33138 increases cell proliferation in the hippocampus and the results are replicated in a D3 KO mouse model (Egeland et al. 2012).

Serotonin and noradrenaline, as well as antidepressant drugs that influence their neurotransmission, play a key role in the regulation of hippocampal neurogenesis and hippocampal-dependent behaviors. This class of molecules will be described more in detail in the section on depression and antidepressant treatments. In Table 1 are listed examples of pharmacological manipulations that have been demonstrated to induce changes in neurogenesis and cognition.

2.3.2 Wnt/Beta-Catenin Pathway

The Wnt signaling pathway is a highly conserved signaling pathway that has been implicated in nervous system development and has multiple functions in the adult brain including a role in hippocampal adult neurogenesis. Disruption of the physiological Wnt signaling pathway has been associated with several CNS pathologies, including schizophrenia, mood disorders, autism, and Alzheimer's disease. A canonical Wnt ligand inhibits glycogen synthase kinase-3β (GSK-3β), which modulates the degradation of β-catenin. In the presence of extracellular Wnt ligand, and subsequent receptor activation, stabilized β-catenin enters the nucleus and associates with TCF/LEF transcription factors, resulting in transcription of Wnt-target genes (Varela-Nallar and Inestrosa 2013).

Based on in vitro and in vivo results, it has been demonstrated that Wnt/β-catenin signaling regulates adult hippocampal NPC proliferation and differentiation (Lie et al. 2005; Kalani et al. 2008). In the hippocampus, lentivirus-mediated expression of Wnt3 or a dominant-negative form of WNT (dnWNT), respectively, increases and almost abolishes adult neurogenesis. Furthermore, expression of dnWNT impairs both long-term retention of spatial memory in the water maze task and performance in a hippocampus-dependent object recognition task (Jessberger et al. 2009). Importantly, the levels of neurogenesis correlate with the performance on specific memory tasks.

Wnt signaling is modulated in diverse physiological conditions characterized by changes in the rate of hippocampal adult neurogenesis. The relationship between aging, neurogenesis, and cognitive impairment will be described in later sections. For example, Wnt signaling shows a reduction during aging when neurogenesis is decreased and, most importantly, modulation of the pathway can counteract age-related neurogenesis and cognitive declines. Aged astrocytes show a reduced expression of multiple canonical Wnt molecules, which in part results in a reduction of adult hippocampal neurogenesis (Miranda et al. 2012). Moreover, the expression of the Wnt antagonist Dickkopf-1 (Dkk1) increases with age and inducible deletion of Dkk1 enhances neurogenesis. Aged mice with a loss of Dkk1 exhibit enhanced spatial working memory and memory consolidation (Seib et al. 2013). Conversely, activity-dependent induction of neurogenesis using electroconvulsive shock leads to reduction in the expression of secreted frizzled-related protein 3 (sFRP3), a naturally secreted Wnt antagonist, in mature dentate granule neurons (Jang et al. 2013a). Deletion of sFRP3 induces the proliferation of precursor cells and promotes newborn neurons maturation, dendritic growth, and dendritic spine formation in the adult mouse hippocampus.

Modulation of Wnt signaling appears to be of therapeutic relevance also in disease conditions. Indeed, sFRP3 deletion alone is sufficient to induce an antidepressant-like behavioral response on the same magnitude of known antidepressants, whose effect, as we will describe later, is at least in part linked to neurogenesis (Jang et al. 2013b). Moreover, in mouse models of Alzheimer's disease characterized by impairment in neurogenesis, treatment with lithium, a pharmacological activator of Wnt/β-catenin signaling acting via GSK3-β inhibition, ameliorates memory loss (Toledo and Inestrosa 2010). In a recent study, in vivo administration of both WASP-1, an activator of Wnt/β-catenin signaling, and FOXY-5, an activator of both Wnt/JNK and Wnt/Ca^{2+} signaling, improves hippocampal-dependent learning and memory processes (Compton et al. 2011). Wnt signaling enhancers would be potentially highly relevant cognitive therapies targeting hippocampal neurogenesis.

2.3.3 Neurotrophic Factors: The Role of BDNF

Neurotrophic factors are extracellular signaling proteins that play critical roles in both the developing nervous system and in adult brain physiology. BDNF and its role in hippocampal neurogenesis have been studied more extensively than any of the other neurotrophins. Chronic infusion of BDNF in the hippocampus of adult rats promotes cell proliferation and neurogenesis (Scharfman et al. 2005). The induction of neurogenesis by BDNF appears to be region specific since it does not affect the process in the SVZ, the other neurogenic niche in the adult rodent brain (Galvao et al. 2008). BDNF affects also later stages of neuronal maturation. Indeed, deletion of TrkB, the receptor of BDNF, in adult neuronal progenitor cells in the hippocampus leads to impairment in dendritic and synaptic growth in newborn neurons and deficits in neurogenesis-dependent LTP (Bergami et al. 2008).

BDNF plays a key role in hippocampus-dependent functions associated with roles in cognition and mood regulation, which will be further discussed in later sections. For example, in pattern separation inhibition of BDNF by infusion of a BDNF-blocking antibody or by antisense oligonucleotide-mediated knockdown impairs the ability to encode and consolidate "pattern separated" memories. On the other hand, acute infusion of recombinant BDNF enhances the separation of representations (Bekinschtein et al. 2013). The effect of BDNF on pattern separation performance is mediated by newborn immature neurons since BDNF infusion has no behavioral effect when neurogenesis is reduced by overexpression of dnWNT (Bekinschtein et al. 2014). As we will describe in more detail later, intact BDNF signaling is critical for learning, exercise, and antidepressants' treatment-induced increase in neurogenesis and effect on behavior (Rossi et al. 2006; Li et al. 2008). BDNF-based therapies would be highly relevant to increase neurogenesis activity and hippocampal function associated with cognition.

2.4 Physical Exercise and Learning: Effect on Hippocampal Neurogenesis, Synaptic Plasticity, and Cognition

The CNS is known to undergo cellular, molecular, and functional changes in response to external social, cognitive and physical stimuli. Voluntary physical exercise has been shown to have beneficial effects on memory and cognition in physiological and pathological conditions in rodents and in humans (Voss et al. 2013). Interestingly, the cognitive amelioration is paralleled by increase in hippocampal neurogenesis and synaptic plasticity. In this section, we will review the effects of exercise on cognitive performance focusing on hippocampal-dependent behaviors and we will describe the effects on adult neurogenesis and synaptic plasticity.

2.4.1 Exercise and Enriched Environment in Animals: From Cognition to Neurogenesis

In adult rodents, physical exercise and exposure to enriched environment (EE), a complex combination of cognitive, physical and social stimulation, improve cognitive functions. Running and EE ameliorate performance in tasks of contextual fear conditioning, novel object recognition, and passive avoidance learning and in tasks assessing hippocampus-dependent memory like spatial memory in the Morris water maze and pattern separation (Kempermann et al. 1997; Fordyce and Farrar 1991; Falls et al. 2010; O'Callaghan et al. 2007; Creer et al. 2010). Hippocampal neurogenesis is one of the most remarkable changes in cellular and synaptic brain plasticity correlating with cognitive improvements upon exercise and EE. The first study analyzing the correlation between exercise and neurogenesis demonstrated that mice housed in an enriched environment with access to a running wheel exhibited better performance in Morris water maze tasks and have a 15 % increase in granule cell neurons in the dentate gyrus (Kempermann et al. 1997). BrdU birthdating experiments in mice show that cell proliferation peaks after 3 days of running, and the effect is still sustained at 10 days (Kronenberg et al. 2006; van der Borght et al. 2006). Running affects cell cycle kinetics of various subpopulations of newborn neurons. It induces both proliferation and cell cycle exit of DCX-positive type 3 precursors, shortens cell cycle in NeuroD1-positive progenitors, and even activates proliferation of radial type 1 stem cells (Brandt et al. 2010; Farioli-Vecchioli et al. 2014; Lugert et al. 2010). Exercise and EE appear to affect specific stages of the neurogenic process. In a study dissecting the role of learning and exercise on neurogenesis, voluntary exercise increases cell proliferation and integration/survival while exposure to an enriched environment, including access to a running wheel, only affects newborn neurons integration/survival in mice (van Praag et al. 1999b). In another study focusing on the effect of different learning paradigms on neurogenesis, the number of adult-generated neurons doubles in the dentate gyrus of rats trained on hippocampus-dependent associative learning tasks like spatial navigation in a Morris water maze and conditioning of the eye blink response using a trace protocol. Learning tasks that do not require the hippocampus fail in eliciting neurogenesis changes. These results suggest that to affect neurogenesis in the hippocampus, animals need to be trained on learning tasks for which the hippocampus is essential (Gould et al. 1999). The increase in newborn neurons upon learning appears to be due to enhanced survival and or integration rather than proliferation (Gould et al. 1999; Kee et al. 2007). Interestingly, a sequential combination of running and EE in mice leads to a 30 % greater increase in neurons than either stimulus alone. This suggests that coupling a stimulus like running which induces precursor cell proliferation to a survival-promoting stimulus like EE can enhance neurogenic pool and then subsequent integration (Fabel et al. 2009). Moreover, the effect of physical exercise and learning on neurogenesis appears to be region specific since generation of new neurons is not observed in the subventricular zone or in the cortex (Brown et al. 2003a; Gould et al. 1999; Ehninger and Kempermann 2003). Studies aimed at understanding whether neurogenesis is necessary for the beneficial effect of exercise and EE on cognition have yielded contradictory results. Reduction of neurogenesis using the antimitotic agent methylazoxymethanol acetate (MAM) in rats prevents the improvement in long-term recognition memory in a novel object recognition task upon EE (Bruel-Jungerman et al. 2005). In mice, gamma irradiation-mediated reduction in neurogenesis has a behavior-specific effect: while running-induced improvements in motor performance (rotarod) and contextual fear conditioning are not affected, spatial memory amelioration is ablated in the absence of neurogenesis (Clark et al. 2008). Interestingly, in very old mice (22 months old) with physiological reduction of neurogenesis which is no longer induced by running, the improvement in spatial pattern separation by voluntary exercise seen in young mice is lost (Creer et al. 2010). However, another study in mice shows that the improvement in spatial learning and the decrease in anxiety-like behavior upon EE are not affected by irradiation-mediated reduction of neurogenesis (Meshi et al. 2006). While several methodological and species differences might contribute to the discrepancies, this work suggests that cognitive improvement might be mediated also by neurogenesis-independent mechanisms such as increase in neurotrophic factors and induction of neuronal and synaptic plasticity.

Upon exercise several growth factors relevant for neuronal function and plasticity like NGF (Neeper et al. 1996), IGF-1 (Carro et al. 2000; Trejo et al. 2001), FGF2 (Gomez-Pinilla et al. 1997), and BDNF are upregulated. The levels of BDNF are increased upon both short- and long-term exercise paradigms (Molteni et al. 2002; Berchtold et al. 2005; Ding et al. 2011) and the increase is sustained up to 2 weeks after exercise has ended in mice (Berchtold et al. 2005). In rats exposed to voluntary running the expression of BDNF in the hippocampus and neocortex positively correlates with the mean distance run per night (Neeper et al. 1996). BDNF is increased in the dentate gyrus also in response to a forced treadmill-running training and this correlates with improved object recognition learning (O'Callaghan et al. 2007). Interestingly, while the performance in this learning task is improved both upon exercise and EE, only exercise can induce an increase in BDNF expression and cell proliferation (Bechara and Kelly 2013). Importantly, genetic ablation of the BDNF receptor TrkB in hippocampal neural progenitor cells ablates neurogenesis in response to exercise (Li et al. 2008). These results indicate that exercise induces BDNF expression, which results in increased neurogenesis in the DG. Interestingly, peripheral neutralization of VEGF abolishes running-induced neurogenesis potentially affecting angiogenesis, a process required in the modulation of the neurogenic niche to sustain greater cell production (discussed below in more detail) (Fabel et al. 2003).

Exercise-induced increase in BDNF often accompanies changes in synaptic plasticity and expression of genes important for neuronal activity and synaptic function (Tong et al. 2001). Voluntary running in rats induces expression of BDNF, NR2B subunit of NMDA receptor, and glutamate receptor 5 and concomitantly alters the induction threshold for synaptic plasticity leading to enhanced short- and long-term potentiation (LTP) in the dentate gyrus (Farmer et al. 2004). In another study, expression analysis in the whole hippocampus after 3 and 7 days of exercise shows an upregulation of NR2A (Molteni et al. 2002), a subunit shown to be necessary for exercise-induced neurogenesis in a genetic mouse model (Kitamura et al. 2003). Similarly, recordings in hippocampal slices from mice exposed to running show an enhancement of LTP specifically in the dentate gyrus (van Praag et al. 1999a). Synaptic transmission properties in the DG and the CA1 area of the hippocampus are modified also in response to learning and EE. For example, electrophysiological recording in freely moving rats shows increase in fEPSPs and in granule cell excitability in the dentate gyrus upon EE exposure (Irvine et al. 2006).

Exercise and learning also affect the morphological maturation of newborn neurons and the structure of already existing neurons in the hippocampus. In newborn DG neurons, running accelerates the formation of mushroom spines and alters spines motility early during differentiation without affecting the total spine density (Zhao et al. 2006). Spatial learning in the Morris water maze increases dendritic arbor complexity in both immature and mature newborn neurons, between 3 weeks and 4 months after birth (Lemaire et al. 2012). Interestingly, spatial and non-spatial environmental cues affect spine morphogenesis in a layer-specific fashion in the DG. Spatial cues induce mushroom spine formation in the middle molecular layer of newborn neurons that receive inputs from the entorhinal cortex (EC) providing spatial information. Conversely, non-spatial components increase mushroom spine formation in the outer molecular layer receiving inputs from the lateral EC (Zhao et al. 2014). Voluntary exercise affects dendritic complexity and spine density not only in the DG but also in afferent populations like pyramidal neurons in the CA1 and layer III pyramidal neurons of the entorhinal cortex (Redila and Christie 2006; Stranahan et al. 2007).

Another brain structural change that correlates with and indirectly supports the increase in neurogenesis is angiogenesis. The brain vasculature is a key component of the neurogenic niche providing extrinsic signals for progenitor cells that are closely associated with blood vessels. Moreover, angiogenic factors can stimulate neurogenesis. Exercise enhances blood flow and blood vessels growth throughout the brain and in the dentate gyrus (Black et al. 1990; van Praag et al. 2005). The growth is at least in part supported by increased expression of angiogenic factors like IGF-1 and VEGF and correlates with increased neurogenesis (Fabel et al. 2003). Intriguingly, experiments using parabiotic animals have identified blood-derived factors that directly regulate neurogenesis in a positive or negative manner, pointing toward systemic factors influencing neurogenesis in adults (Villeda et al. 2011, 2014; Katsimpardi et al. 2014). Interestingly, MRI studies show that increased cerebral blood volume (CBV) in the dentate gyrus can be used as an in vivo correlate of neurogenesis and it is specifically affected by exercise in mice. These findings are confirmed in human where dentate gyrus CBV correlates with cardiorespiratory fitness and cognitive function (Pereira et al. 2007). These data suggest that CBV measurements could represent a correlative biomarker for neurogenesis in humans.

2.4.2 Human Neurogenesis, Cognition and Exercise

Brain imaging studies support the role of the DG/CA3 subfields of the hippocampus in pattern separation in humans. In a study combining functional MRI and ultrahigh-resolution structural MRI, it has been shown that there is a correspondence between CA3 anatomy and functioning and pattern separation, pattern completion and individual differences in episodic memory recall (Chadwick et al. 2014). In non-demented older adults, changes in the activity measured by fMRI in the CA3/DG region correlate with the performance in pattern separation (Yassa et al. 2011). Moreover, DG/CA3 is also involved in pattern separation of emotional information and in patients affected by depression the severity of depressive symptoms negatively correlates with DG/CA3 activity (Leal et al. 2014). As in animals, hippocampal structural changes appear to correlate with training on tasks dependent on the hippocampus. Indeed, the posterior hippocampus stores spatial representations of the environment and people with high dependence on navigational skills like London taxi drivers show increased posterior hippocampal volume (Maguire et al. 2000). These data support the relevance of hippocampal areas and their dynamic regulation in specific cognitive tasks in humans. Although animal studies demonstrate a key role for neurogenesis in pattern separation, the lack of biomarkers for neurogenesis in humans limits conclusive studies. On the other hand, studies in cancer patients show that systemic treatment with chemotherapy agents often results in cognitive impairment and decline in aspects of memory which require hippocampal function. In animal models it has been shown that neural stem cells proliferation in the DG is reduced by chemotherapy, suggesting this as one of the potential mechanisms underlying some of the cognitive deficits related to the treatment (Wigmore 2013).

Several studies in human suggest that aerobic exercise has a positive effect on cognitive performance in healthy individuals and can counteract cognitive impairment during aging or in pathological conditions (Voss et al. 2013). In the healthy population, aerobic exercise can improve executive functions like task switching, selective attention, working memory updating, and inhibitory control in children and young adults (Guiney and Machado 2013). Cardiovascular fitness positively associates with intelligence assessed using tests for logical, verbal, and technical skills (Aberg et al. 2009; Moore et al. 2014), with increased cognitive flexibility and improved action monitoring process (Themanson et al. 2008; Hillman et al. 2008), and with improvement in academic achievements (Chaddock-Heyman et al. 2013; Chaddock et al. 2012).

Exercise has also been shown to specifically improve performance in cognitive tasks known to critically depend on hippocampal function. Young individuals, exposed to a long-term aerobic exercise regime and experiencing a change in fitness, show better performance in visual pattern separation task, visuospatial memory, and positive affect (Dery et al. 2013; Stroth et al. 2009; Herting and Nagel 2012), relational memory (e.g., children show higher ability to remember pairs of faces and houses studied under relational encoding conditions) (Monti et al. 2012; Chaddock et al. 2010, 2011). Interestingly, magnetic resonance imaging shows that performance in relational and visuospatial memory tasks positively correlates with larger hippocampal volumes, suggesting that structural modifications play a role in improved function (Chaddock et al. 2010; Herting and Nagel 2012).

Physical activity has beneficial effects on cognition also in older adults and in conditions associated with cognitive impairment. Older humans exposed to physical exercise improve executive control processes like planning, scheduling, and working memory (Kramer et al. 1999). Similarly to that observed in younger adults, aerobic fitness correlates with increased hippocampal volume and better spatial memory in older individuals (Erickson et al. 2009). Moreover, a longitudinal study showed that in adults over the age of 65 years physical activity correlates with diminished incidence of Alzheimer's disease. In conditions of memory problems or cognitive impairment, exercise still improves cognitive function and positive behavior (Heyn et al. 2004).

Overall these studies support a positive effect of exercise on cognition and on hippocampus-dependent processes like spatial and relational memory and pattern separation. Importantly, cognitive improvements correlate with structural changes in the hippocampus both in animals and in humans. However, a direct link to neurogenesis in humans is lacking, as currently there is no possibility to measure neurogenesis in living humans.

2.5 Depression, Stress, and Antidepressants: Effect on Neurogenesis and Behavior

2.5.1 Cognitive Impairment in Depression

Depression is a widespread disorder and it presents with symptoms which include, among others, low mood, feelings of despair, reduced attention capacity, and suicidal ideation. Patients affected by major depressive disorder (MDD) show high prevalence of cognitive dysfunction, including hippocampal-dependent cognitive processes. Among the most common deficits are memory disturbances, difficulty in making decisions, and reduced cognitive flexibility (Fava et al. 2006; Wagner et al. 2012; Jaeger et al. 2006; McCall and Dunn 2003). Morphometric analysis in patients diagnosed with first episode MDD shows structural alterations in hippocampus, amygdala, and corticolimbic regions (Frodl et al. 2002; Zhu et al. 2011). MDD patients show impaired performance in tests assessing hippocampal-dependent declarative memory and functional imaging analysis shows abnormal activation of hippocampus during verbal memory encoding task (Bremner et al. 2004). As we will describe in detail later, alteration of the hypothalamic–pituitary–adrenal (HPA) axis and in glucocorticoids (GCs) levels are hallmarks of depression pathophysiology. Although not confirmed in all studies, cortisol levels are found to correlate with cognitive deficits in MDD patients and acute administration of GCs in healthy individuals can compromise long-term memory retrieval (Schlosser et al. 2010; Wolf et al. 2009).

Taken together, there is increasing evidence supporting a role for hippocampal dysfunction in cognitive deficits in MDD. Recent anatomical and functional evidence indicates a dissociation of the dorsal and ventral regions of the hippocampus with the dorsal region being responsible for memory and cognition and the ventral part playing a key role in regulation of mood. How alterations in cognitive functions contribute to MDD pathogenesis and to the recovery process is still unclear. In this sections, we will review the effects of stress and antidepressant treatment on neurogenesis and will discuss the role of neurogenesis in the onset and recovery of behavioral phenotypes.

2.5.2 Stress and Neurogenesis

Stressors are considered as experiences and events which challenge the ability of the individual to adapt. Reponses which promote adaptation to stressors include the release of hormones and other cellular factors. However, when the response is deregulated it can induce the activation of pathophysiological processes leading to the onset of neuropsychiatric disorders (McEwen 1998).

Stress has been demonstrated to have deleterious effects on multiple stages of hippocampal neurogenesis, newborn neuron maturation, hippocampal neuron plasticity, and dendritic and synaptic density. Exposure of tree shrews to acute conflict to establish a dominant/subordinate relationship leads to reduction of neurogenesis in the dentate gyrus in the subordinate animal (Gould et al. 1997). Negative regulation of neurogenesis is observed also upon chronic stress paradigms in several animal models. For example, cell proliferation in the dentate gyrus shows a significant reduction in mice subjected to repeated intermittent social defeats (Yap et al. 2006) and in young adult marmosets after forced prolonged social isolation (Cinini et al. 2014). Besides affecting cell proliferation, chronic stress exposure impairs newborn neurons' survival (Tanti et al. 2013; Dagyte et al. 2011) and induces structural abnormalities like dendritic atrophy in CA3 pyramidal neurons (Watanabe et al. 1992; Sousa et al. 2000). At the behavioral level, animal models subjected to chronic stress develop depression-related behaviors. While the effect of stress on neurogenesis has been widely confirmed in several animal models, the role of neurogenesis in the onset of depressive-like behaviors is still a matter of debate. Ablation of neurogenesis in mice via X-ray irradiation of the hippocampus or via pharmacological intervention does not induce behavioral phenotypes relevant to anxiety or depression (Santarelli et al. 2003; David et al. 2009; Bessa et al. 2009). Genetic ablation of neurogenesis via overexpression of the pro-apoptotic protein Bax leads to increase in anxiety-related behaviors while it does not affect depressive behaviors (Revest et al. 2009). Similarly, cyclin D2 knockout mice which lack neurogenesis do not show impaired performance in the forced swimming test (Jedynak et al. 2014). Conversely, in a transgenic model where increase in survival of newborn neurons is achieved by selective deletion of Bax, there are no differences in anxiety and depressive behaviors (Sahay et al. 2011). In contrast to these findings, ablation of radial cell precursors in a GFAP-TK mouse model (expressing under the control of GFAP promoter herpes simplex virus thymidine kinase which renders mitotic cells sensitive to the antiviral drug valganciclovir) leads to depressive behavior in control conditions and increased anxiety in response to stress (Snyder et al. 2011).

2.5.3 Mechanisms Underlying the Effect of Stress on Neurogenesis and Establishment of Depressive Behavior

Several mechanisms have been proposed to mediate the effect of stress on adult neurogenesis and establishment of depressive behavior including alteration in the hypothalamic–pituitary–adrenal (HPA) axis (for a comprehensive review, see Anacker 2014; Bambico and Belzung 2013). Here we will review the evidence supporting a causal link between HPA axis dysfunction and alteration in neurogenesis, synaptic plasticity, and neurotrophic factor signaling in depression.

The hypothalamic–pituitary–adrenal (HPA) axis, part of the neuroendocrine system, plays a crucial role in the response to stress and it is deregulated in depression and chronic stress conditions (Anacker 2014). Upon stress, the hypothalamus releases corticotropin-releasing factor (CRF) which stimulates the pituitary to release adrenocorticotropin (ACTH) which ultimately leads to the synthesis and release of glucocorticoids (GCs) from the adrenal cortex. GCs then bind to their intracellular receptors, namely mineralocorticoid receptors (MRs) and glucocorticoid receptors (GRs). Upon chronic stress, the HPA axis is hyperactive and leads to chronically high levels of glucocorticoids (Anacker and Pariante 2012). Recent anatomical studies have shown that projections from the ventral hippocampus mediate an inhibitory effect on the hypothalamus, whereas the dorsal hippocampus projects to cortical areas where cognitive processes are mediated. Studies in preclinical models support a strong causative link between stress, glucocorticoids, and neurogenesis. Exposure of adult rats to predator odor induces rise in adrenal hormones levels and decrease in neurogenesis which is prevented blocking the increase of corticosterone by adrenalectomy (Tanapat et al. 2001). Interestingly, chronic corticosterone treatment induces depressive behavior in rodents (Gourley and Taylor 2009) and affects cell proliferation reducing the number of BrdU-positive cells in the dentate gyrus in adult mice (David et al. 2009). Besides reducing neurogenesis, it has been proposed that stress and corticosterone treatment promote a cell fate switch of nestin-positive neural stem cells in the DG toward increased oligodendrocytes' differentiation (Chetty et al. 2014). Interestingly, corticosteroid levels and expression of the receptors are also altered during aging, a physiological situation with reduced neurogenesis (Gupta and Morley 2014).

Patients affected by major depressive disorder (MDD) show increased cortisol levels, impaired feedback response to GR activators, and correlation between genetic variation in components of the HPA pathway and clinical manifestations (Young et al. 1991; Belvederi Murri et al. 2014; Schatzberg et al. 2014). Volumetric analysis reveals alterations of several brain regions involved in the control of the HPA axis in MDD patients, like significantly increased volume of adrenal gland and reduced volume of hippocampus (MacMaster et al. 2014; Sheline et al. 1996). At the cellular level, expression of synaptic proteins regulating synapse function and structure as well as glutamate receptor subunits is altered in hippocampi and prefrontal cortex from MDD patients' brains (Duric et al. 2013; Fatemi et al. 2001; Kang et al. 2012). Interestingly, this correlates with alterations in the expression of neurotrophic factors important for the function and plasticity of synapses. Expression levels of BDNF, NGF, and their relative receptors are decreased in hippocampi and PFC from suicide victims (Banerjee et al. 2013; Pandey et al. 2008) and MDD patients (Dunham et al. 2009). Moreover, serum levels of micro-RNAs associated with the regulation of expression of BDNF are altered in depressed patients (Li et al. 2013). Although the mechanisms which link stress and alterations of neurotrophic factors' levels are not completely understood, several studies show that antidepressant treatment corrects neurotrophins levels and that normal BDNF signaling is necessary for antidepressant action (Castren et al. 2007; Adachi et al. 2008). This suggests that these factors play an important role in restoring the physiological function of networks involved in mood disorders (Duman and Duman 2015).

2.5.4 Antidepressants, Neurogenesis, and Synaptic Plasticity

For long time, one of the dominant hypothesis to explain the mechanisms underlying MDD has been the monoamine hypothesis, mainly supported by the observation that antidepressants (ADs) exerting a pharmacological action on the central monoamine systems (monoamine oxidase A inhibitors/MAOI, tricyclic compounds/TCA, serotonin and norepinephrine reuptake inhibitors) were effective in relieving depressive symptoms (Delgado 2000). According to this hypothesis, reduced activity of monoamine neurotransmission is at the core of the pathophysiology of depression and correction of this dysfunction alleviates disease symptoms (Asberg

1976). Through different mechanisms, several ADs classes in clinical use enhance serotonergic (5-HT) and noradrenergic neurotransmission.

Many preclinical and some human postmortem studies demonstrate that antidepressant treatment induces an increase in neurogenesis. Malberg and colleagues showed that chronic but not acute fluoxetine treatment increases the number of dividing cells in the dentate gyrus in rats as measured by BrdU incorporation, while the survival of newborn neurons does not appear to be altered by the treatment (Malberg et al. 2000). In humans, analysis of postmortem brains from MDD patients shows that SSRI and TCA chronic treatment increases the number of neural progenitors in the anterior and mid-dentate gyrus (Boldrini et al. 2009, 2012). However, another study could not replicate these findings (Lucassen et al. 2010). The precise role of specific serotonergic receptors in SSRI-induced neurogenesis is still under investigation. Genetic ablation of 5-HT1A receptor in mouse prevents fluoxetine induction of neurogenesis while a partial reduction is observed with 5-HT4 receptor-specific antagonist (Santarelli et al. 2003; Mendez-David et al. 2014). Moreover, pharmacological studies show that 5-HT1A receptors are involved in regulation of proliferation of neuronal precursors, while 5-HT2 receptors influence proliferation and promote neuronal maturation (Klempin et al. 2010). Induction of neurogenesis has been reported also upon treatment with antidepressants which do not act directly through the serotonin system suggesting some potential convergence in mechanisms of action. For example, the endocannabinoid receptor ligand cannabidiol (CBD) exerts anxiolytic and antidepressant effects and induces hippocampal progenitor proliferation and neurogenesis in mice (Campos et al. 2013). Increase in BrdU-positive cells is observed in the dentate gyrus of mice treated with an antidepressant antagonist of group II metabotropic glutamate receptor (Yoshimizu and Chaki 2004). Moreover, non-pharmacological interventions which elicit an antidepressant response like electroconvulsive shock strongly promote hippocampal neurogenesis (Malberg et al. 2000). Altogether these data support a correlation between ADs and neurogenesis, but is this convergent cellular response necessary for these drugs to have an effect on behavioral symptoms? Ablation of neurogenesis through X-irradiation of hippocampus prevented certain behavioral effects of fluoxetine (SSRI) and imipramine (TCA) in mice (Santarelli et al. 2003). The causal role of neurogenesis in certain specific behavioral effects of antidepressants has been confirmed in several studies (Surget et al. 2008, 2011). Fluoxetine treatment corrects the behavioral and neurogenesis defects in mice exposed to chronic corticosterone treatment and ablation of neurogenesis impairs amelioration in novelty suppressed feeding test, while open field and forced swim test performances are neurogenesis independent (David et al. 2009). Recently, the antidepressant action of cannabidiol has been shown to require neurogenesis since its effects are not observed in GFAP-TK mouse models (Campos et al. 2013). On the other hand, the behavioral effects of a vasopressin V1b antagonist and a CRF1 antagonist (corticotropin-releasing factor 1) which reverse stress-induced suppression of neurogenesis in models of depression appear to be neurogenesis independent (Alonso et al. 2004; Surget et al. 2008).

Chronic antidepressant treatment leads to an increase in the expression levels of neurotrophic factors in the hippocampus and cortical regions, including VEGF, FGF2, and BDNF and its receptor trkB which in turn can positively regulate the hippocampal cellular response (Warner-Schmidt and Duman 2007; Mallei et al. 2002). Taking advantage of genetic mouse models designed to have reduced BDNF signaling (heterozygous BDNF knockout or inducible ablation of trkB receptor in NPC), it has been demonstrated that a BDNF response is necessary for modulation of newborn hippocampal neurons upon chronic imipramine treatment (Li et al. 2008; Sairanen et al. 2005). In addition, behavioral response to antidepressant treatment is impaired in mice with altered BDNF signaling and with conditional ablation of BDNF in forebrain regions (Saarelainen et al. 2003; Monteggia et al. 2004). These data suggest that increasing BDNF activity is a key mechanism mediating antidepressant action.

Interestingly, the BDNF signaling pathway is upregulated in the hippocampus and PFC shortly after treatment with fast-acting antidepressants (Zhou et al. 2013b; Yang et al. 2013; Autry et al. 2011) and it is associated with functional synaptic changes (Tizabi et al. 2012). Ketamine, an antagonist of *N*-methyl-D-aspartate (NMDA) receptors, exerts antidepressant action within few hours after administration with effects lasting up to 2 weeks in depressed patients (Zarate et al. 2006; Price et al. 2009). Administration of NMDA antagonists leads to rapid translation of BDNF in mouse hippocampus and the antidepressant effect is ablated in BDNF conditional knockout mice (Autry et al. 2011). Interestingly, miR-206 regulates the expression of BDNF in response to ketamine (Yang et al. 2014). In light of the role of BDNF in modulation of synaptic plasticity and the increase in BDNF in the hippocampus upon ketamine treatment, it is tempting to speculate that changes in hippocampal synaptic plasticity may contribute, together with other brain regions, to rapid-onset antidepressant effect. However, this hypothesis still needs to be tested.

2.5.5 Antidepressants and Cognitive Impairment

As mentioned above, MDD is associated with alterations in hippocampal-dependent cognitive processes. Since antidepressant treatments have been shown to improve BDNF signaling and neurogenesis, key processes in hippocampus-dependent cognitive functions, antidepressants could potentially ameliorate cognitive symptoms in MDD patients. However, clinical studies have not yielded conclusive results and discrepancies are often attributed to limited study design. Longitudinal studies report persistent cognitive dysfunction in patients after remission from depressive symptomatology (Trivedi and Greer 2014; Hasselbalch et al. 2011; Kuny and Stassen 1995; Weiland-Fiedler et al. 2004; Neu et al. 2005). In one study, cognitive functions have been monitored in patients treated with escitalopram (SSRI) or duloxetine (SNRI); an improvement in working memory, processing speed, and visual episodic memory was best observed with duloxetine (Herrera-Guzman et al. 2009, 2010). To further address the effects of antidepressants on cognitive improvement more studies are needed and understanding this relationship would help design new and more efficacious therapeutic interventions (Goeldner et al. 2013). Along the same line, more research is needed to unravel the role of hippocampal network and neurogenesis in the onset and recovery of cognitive alterations in MDD.

2.6 Cognition and Adult Hippocampal Neurogenic Axis in Physiological and Pathological Aging

Cognitive functions decline with age and are impaired in pathological conditions associated with advanced age like Alzheimer's disease. Impairment of hippocampus-dependent cognitive processes often correlates with structural and functional changes in this brain region and with alterations in neurogenesis. This sections highlights the relevance of neurogenesis in physiological aging and associated cognitive decline in human and animals. Furthermore, the potential pathophysiological role of neurogenesis in cognitive deficits in Alzheimer's disease is highlighted.

2.6.1 Aging and Cognitive Decline in Human: Correlation with Hippocampal Changes

In human, aging affects a broad range of cognitive functions like executive functions (task switching, updating, inhibition) associated with the prefrontal cortex, episodic memory (memory for events which include specific temporal and spatial context) associated with prefrontal cortex and medial temporal lobes, information processing speed, specific aspects of language, and visuospatial functions (for a review, see Alexander et al. 2012). Several studies report specific alterations in hippocampus-dependent processes. Older individuals show impairment in spatial navigation tasks designed to specifically assess hippocampus-dependent ability to develop a cognitive map (a mental representation of the landmarks and paths in the environment) and to use it to reach any target location by any route available (Iaria et al. 2009). These cognitive deficits often correlate with functional and structural alterations in the hippocampus. For example, older individuals tested in a virtual spatial navigation task show differential activation of several brain areas, as measured by voxel-based analysis, if compared to younger controls. Among those differences, old participants show a reduced activation of hippocampus and parahippocampal gyrus (Moffat et al. 2006). Similarly, impairment in hippocampus-dependent spatial and non-spatial functions and recognition memory has been shown to correlate with decreased hippocampal volume and neurochemical properties in older individuals (Driscoll et al. 2003). Along the same line, a specific correlation between hippocampal volume and wayfinding skills (generation and use of a cognitive map) has been described in older individuals (Head and Isom 2010). Moreover, older adults show performance decline compared to younger control individuals in pattern separation task and a bias toward pattern completion. High-resolution (1.5 mm isotropic) blood-oxygenation level-dependent fMRI analysis reveals an altered activation in the CA3/dentate gyrus regions during trials (Yassa et al. 2011). However, a meta-analysis of the correlation between hippocampal size and episodic memory in older adults reveals extremely weak correlation (Van Petten 2004). Moreover, while some studies show a progressive change in the volume of several brain regions including shrinkage of the hippocampus (Raz et al. 2010), others show an age-dependent decline in temporal cortex but not in hippocampal volume (Sullivan et al. 2005).

Moving from the neuroanatomical to the cellular level, early analysis with stereological techniques showed no neuronal loss in several areas of the hippocampus during aging (West et al. 1994). On the other hand, neurogenesis appears to decline with age. The expression of neurogenesis markers and the number of DCX-positive cells are decreased in the dentate gyrus in post-mortem human brains from older donors (Knoth et al. 2010). Moreover, birthdating experiments based on measurements of genomic DNA incorporation of ^{14}C, liberated in the atmosphere during atom bomb testing, show that adult hippocampal neurogenesis in the human brain undergoes a modest decline during aging (Spalding et al. 2013). Whether neurogenesis decline contributes to cognitive deficits in elderly humans remains an open question, although studies in rodents suggest that it might have a significant impact.

2.6.2 Age-Related Cognitive and Neurogenesis Decline in Animals

Decline of cognitive functions has been described in aged rodents and non-human primates. Deficits include impaired performance in spatial learning and memory (Barnes 1979; Gage et al. 1984; Rapp et al. 1997; Lazarov and Marr 2013). Structural MRI analysis shows that hippocampal volume does not change with age (Shamy et al. 2006). However, histologic evaluation shows an imbalance in the volume of different areas of the hippocampus in aged rats with a relative reduction of the volume of the middle portion of the molecular layer (Rapp et al. 1999). Interestingly, early studies demonstrated that in rodents neurogenesis in the subgranular zone declines with age (Seki and Arai 1995; Kuhn et al. 1996; Ben Abdallah et al. 2010) with the major decline taking place during adulthood, before aging (Rao et al. 2005; Demars et al. 2013). Whether the decline in neurogenesis correlates with cognitive impairment is a matter of debate. Studies show that aged rats performing better in Morris water maze and hippocampus-dependent tasks have higher number of proliferating cells and newborn neurons (Driscoll et al. 2006; Drapeau et al. 2003). On the other hand, others show no correlation or negative correlation between the number of proliferating cells and performance in Morris Water Maze in aged rats (Merrill et al. 2003; Bizon et al. 2004). Interestingly, in middle-aged (12 months old) rats, despite the massive reduction of progenitor cell proliferation and newborn cells survival, no deficits in trace fear conditioning are observed (Cuppini et al. 2006).

Several hypotheses have been put forward to explain the mechanisms accounting for neurogenesis decrease in aging and they include both cell-autonomous and non-cell-autonomous processes. Aging is accompanied by changes in the hippocampal stem cell niche vasculature (Hattiangady and Shetty 2008) and reduced expression levels of growth factors important for neurogenesis like VEGF, FGF2, BDNF, and

WNT signaling activity (Shetty et al. 2005; Hattiangady et al. 2005; Bernal and Peterson 2011). Glucocorticoid signaling is altered in aged animals and decreasing corticosterone levels restores the rate of proliferation of neuronal progenitors (Cameron and McKay 1999; Nichols et al. 2001). A heterochronic parabiosis experiment, in which the circulatory system of two animals is connected, shows that blood-borne factors from old mice can induce impairment in synaptic plasticity and cognitive deficits in young animals. Elevated levels of chemokines, including CCL11, are observed in aged animals and increasing peripheral levels of CCL11 in young mice decreases neurogenesis and induces cognitive decline (Villeda et al. 2011). One of the most robust changes observed during aging is a dramatic reduction in progenitor cells' proliferation (Olariu et al. 2007; Walter et al. 2011; McDonald and Wojtowicz 2005; Bondolfi et al. 2004), with the most prominent effect in the ventral hippocampus (Jinno 2011). A meta-analysis comparing the reduction in progenitors' proliferation between different species of rodents, fox, and non-human primates shows that the decline is chronologically equal between species and independent of life span (Amrein et al. 2011). Reduction in neurogenesis in the aged brain has been associated with depletion of the neural stem cells pool as a consequence of their division (Encinas et al. 2011) or to a switch toward a quiescent state (Hattiangady and Shetty 2008; Lugert et al. 2010). Importantly, pro-neurogenic stimuli such as increasing neuronal activity by inducing seizures lead NSCs to reenter the cell cycle and restore proliferation to a level comparable to the one observed in young animals (Lugert et al. 2010). Similarly, exercise and exposure to enriched environment rescue neurogenesis deficits and improve cognitive functions like spatial memory and place recognition memory (van Praag et al. 2005; Kronenberg et al. 2006; Siette et al. 2013). Recently, the chronic administration in aged rats of a blood–brain barrier-permeable peptide derived from the ciliary neurotrophic factor (CNTF), which is known to have neuroprotective properties, restores neurogenesis, synaptic plasticity, and memory (Bolognin et al. 2014). These works suggest that induction of neurogenesis has beneficial effects on cognition during aging. Moreover, targeting the reactivation of neural stem cells in the aged brain could be a promising therapeutic approach to restore cognitive functions. Interestingly, young circulating factors like GDF11, a member of the BMP/TGFβ family, can restore neurogenesis and proper vasculature also in the SVZ, the other neurogenic niche in the adult brain, and improve olfactory discrimination in aged mice (Katsimpardi et al. 2014).

2.6.3 Pathological Aging: Alzheimer's Disease, Cognitive Impairment, and Neurogenesis

Alzheimer's disease (AD) is an age-related neurodegenerative disease characterized by progressive memory loss and cognitive decline (Caselli et al. 2006). AD represents the most common cause of dementia in the elderly. Pathological hallmarks of the disease are extracellular deposition of amyloid-beta (Aβ) plaques, excessive phosphorylation of the cytoskeletal protein tau, formation of intraneuronal fibrillary tangles, and neuronal cell loss in the cerebral cortex and hippocampus (Selkoe 2001; Braak and Braak 1991, 1995). The major environmental risk factor for the onset of the disease is aging, while the greatest genetic factor is apolipoprotein E (ApoE) genotype (Ashford 2004). ApoE exists in three isoforms, with isoform E3 being the wt form and the most common in the population, isoform E2 considered to be protective toward AD, and E4 increasing the risk for the disease and anticipating the age of onset. Hereditary autosomal dominant forms of the disease are caused by mutations in genes encoding for amyloid precursor protein (APP) and presenilin-1 and -2, components of the aspartyl protease γ-secretase complex. Cleavage of APP can follow two pathways: in the non-amyloidogenic pathway cleavage by the protease complex α-secretase releases the soluble fragment of APP (sAPPα) and γ-secretase cuts in the intramembrane domain; in the amyloidogenic pathway cleavage by the aspartyl protease β-site APP cleaving enzyme I (BACE1) and by γ-secretase leads to the formation of Aβ (De Strooper and Woodgett 2003; Selkoe and Wolfe 2007).

The hippocampus and the afferent entorhinal cortex show early neuronal loss that correlates with memory decline (Van Hoesen et al. 1991; Gomez-Isla et al. 1996; West et al. 1994, 2004). Patients affected by amnesic mild cognitive impairment (aMCI) and mild AD show reduced performance in hippocampus-dependent tasks like pattern separation (Ally et al. 2013). Importantly, impairment in performance correlates with CSF levels of Aβ42 and patients carrying the ApoE4 genotype perform worse than others in difficult pattern separation tasks (Wesnes et al. 2014). As discussed previously, performance in pattern separation appears to require hippocampal neurogenesis in the dentate gyrus and growing evidence supports alterations in the neurogenic process in AD. Several proteins associated with AD and mutated in familial forms of the disease are directly involved in the regulation of neural stem cell proliferation and differentiation. Notch1, a master regulator of neural stem cell physiology, is a substrate of PS1/γ-secretase and it is cleaved in response to ligand binding (Alexson et al. 2006; LaVoie and Selkoe 2003; De Strooper et al. 1999). Deletion of *PSEN1*, the gene encoding for PS1 in mice, leads to aberrant neurogenesis during development (Shen et al. 1997). In adult animals, lentiviral-mediated knockdown of PS1 in the dentate gyrus leads to enhanced differentiation of neural stem cells (Gadadhar et al. 2011). Moreover, Notch1 and EGF receptor ligands are substrates of ADAM10, a component of the α-secretase complex (Hartmann et al. 2002; LaVoie and Selkoe 2003). Neural stem cells' proliferation is regulated also by cleavage products of APP, and the intraventricular injection of sAPPα rescues age-dependent decline in neural progenitors' proliferation (Caille et al. 2004; Demars et al. 2013; Lazarov and Demars 2012).

Studies in several mouse models of AD suggest a strong correlation between alterations in neurogenesis, including neural stem cells' proliferation, newborn neurons' survival and maturation, and disease progression. For example, in the transgenic mouse model Tg2576 which overexpresses the human APP isoform with the Swedish double mutation and develops cognitive defects and amyloidosis, aberrant maturation of newborn neurons is observed already in young age (3 months) and this might lead to altered functional integration of new neurons in the hippocampal circuits (Krezymon et al. 2013). Recently, spatial learning and memory impairments have been described in a knock-in mouse model bearing the human pathological domain of ApoE4 gene in young age. Neurogenesis appears to be increased in young animals and decreased with age, while apoptotic markers indicate enhanced cell death. Interestingly, these early cognitive and cellular changes take place in the absence of classical AD pathological hallmarks (Adeosun et al. 2014). In another study, deletion of ApoE in mice results in reduced neurogenesis and knock-in of the human ApoE4 leads to impaired maturation of newborn neurons through a mechanism dependent on GABAergic signaling (Li et al. 2009). Comprehensive reviews of neurogenesis phenotypes in AD models can be found elsewhere (Lazarov and Marr 2010; Verret et al. 2007; Mu and Gage 2011).

Analysis of postmortem brain samples from senile AD patients reveals increased expression of newborn neurons markers including DCX in the subgranular zone, granular layer, and CA1 area. These findings suggest that activation of neurogenesis might represent a mechanism of compensation and replacement of degenerated neurons (Jin et al. 2004). In another study, AD brain samples show altered expression of early progenitor markers in the hippocampus, suggestive of enhanced proliferation, while the number of newborn neurons does not appear to be affected (Perry et al. 2012). Plasma and CSF concentrations of stem cell factor, a growth factor exerting neuroprotective effects and supporting neurogenesis, are decreased in AD patients and the level inversely correlates with the degree of dementia (Laske et al. 2008). On the other hand, in pre-senile AD patients no evidence of altered neurogenesis has been detected, while a proliferative change has been observed in glia and vasculature cells (Boekhoorn et al. 2006).

As described above, the hippocampus and the neurogenic process are strongly altered in AD and this might account for some aspects of cognitive decline in patients. However, growing evidence suggests that molecular and structural prerequisites for activity-dependent hippocampal plasticity are preserved in models of AD. Exposure to enriched environment leads to amyloid plaque load reduction and increased expression of genes involved in protective processes and neurogenesis in a transgenic AD mouse model (Lazarov et al. 2005). Similarly, exposure to EE of mice bearing the Swedish mutation in APP reverses spatial learning and memory deficits and this correlates with restoration of some aspects of the neurogenic process. As opposed to the previous study, amyloid load is not affected by EE in this model (Valero et al. 2011). A study dissecting the role of physical activity and cognitive stimuli on neurogenesis in AD shows that while exposure to EE improves neurogenesis and water maze performance in the APP23 mouse model, exercise alone does not show beneficial effect (Wolf et al. 2006). Besides memory and learning impairment, AD is associated with behavioral symptoms that include anxiety and depression. As described previously, the ventral hippocampus is involved in regulation of the response to stress and depressive behavior and neurogenesis might play a role in this process. In a 3xTgAD mouse model bearing mutations in APP, PS1, and Tau genes, restoration of neurogenesis through chronic overexpression of Wnt3a in the ventral region of the dentate gyrus leads to correction of impairments in danger assessment. The effect is neurogenesis dependent since ablation of neurogenesis via X-irradiation prevents the recovery (Shruster and Offen 2014).

Altogether these works suggest that neurogenesis is pathologically impaired in AD and restoration of the neurogenic process might be a promising therapeutic approach to reverse behavioral and cognitive symptoms in AD patients. Below is highlighted the potential of targeting neurogenesis in disease and drug-discovery approaches for the identification of neurogenic compounds are discussed.

3 Novel Concepts for Targeting Neurogenesis in Cognition and Disease

As described above, compelling preclinical evidence suggests that hippocampal neurogenesis is modulated by a broad range of physiological and pharmacological stimuli, and most excitingly, de novo neurogenesis positively correlates with improved cognitive and emotional states. Three lines of research are emerging that suggest that targeting hippocampal neurogenesis will have translational value in human cognition and warrant future work: (1) recent rodent models demonstrate a cell-autonomous role of neurogenesis in cognition in both physiology and disease; (2) drug screening efforts have led to the identification of novel CNS active compounds with both pro-neurogenic and pro-cognitive effects; and (3) the study of human neurogenesis will potentially lead to the identification of biomarkers to monitor the process in mental illnesses for both diagnostic and therapeutic interventions. Below we highlight discoveries, opportunities, and challenges in these research domains. Research on the role of hippocampal neurogenesis in human physiology and disease will ultimately increase the translational value of this remarkable process of plasticity in the adult human brain.

3.1 Neurogenesis as a Target to Improve Cognition in Physiology and Disease

Recent studies strongly suggest that the discovery of cell-autonomous signaling pathways that enhance adult neurogenesis may lead to therapeutic strategies for improving memory loss due to aging or injury. Here we highlight some of these pathways relevant for the regulation of neurogenesis and implicated in mental disability and we describe the effect of their modulation on cognitive functions. As previously mentioned, the hippocampus plays a key role in declarative memory functioning as a pattern separator. Genetic modulation of the apoptotic pathway in neural stem cells in the DG through ablation of the pro-apoptotic protein Bax leads to increased neurogenesis and improved performance in pattern separation tasks (Sahay et al. 2011). Future studies using this model in disease-relevant backgrounds (e.g., AD models) could potentially elucidate whether the enhancement of neurogenesis changes significantly disease states.

The ERK pathway is important for the modulation of neurogenesis and cognition. ERK5 is a member of the MAP kinase family and it plays a role in prenatal neuronal development and cell fate specification (Cavanaugh et al. 2001; Finegan et al. 2009; Cundiff et al. 2009). ERK5 is expressed in the adult brain in neurogenic regions and inducible conditional ablation of ERK5 in neural progenitors in the adult brain leads to attenuation of hippocampal neurogenesis (Pan et al. 2012a, b). This correlates with alterations in hippocampus-dependent memory tasks like contextual fear conditioning, with defects in learning flexibility and deficits in pattern separation. Conversely, in a genetic mouse model the activation of ERK5 in neuronal progenitors in the DG induces neurogenesis promoting cell survival, neuronal differentiation, and enhanced dendritic structural complexity. The morphological changes are paralleled by improved spatial learning and memory and hippocampus-dependent long-term memory persistence (Wang et al. 2014). The discovery that memory can be prolonged by stimulating adult neurogenesis has important implications for the development of therapeutic strategies to treat memory disorders.

Recent studies show that restoring the function of adult neurogenesis in disease models with clear cognitive dysfunction can reverse the learning deficits. Consequently, treatments directed at this cell population may have a significant impact on disease-relevant cognitive and learning phenotypes. Fragile X syndrome (FXS) is the most common cause of inherited intellectual disability and the leading monogenic cause of autism spectrum disorders. FXS is a trinucleotide repeat disorder caused by a CGG repeat expansion in FMR1, which leads to the loss of its protein product FMRP. Deletion of FMRP in neural stem cells in adult mice affects several aspects of adult hippocampal neurogenesis leading to deficits in proliferation and differentiation of newborn neurons. This is accompanied by reduced performance in hippocampus-dependent learning tasks. Importantly, in FMRP-deficient mice, restoration of FMRP expression specifically in adult neurogenic compartments can rescue these learning deficits (Guo et al. 2011). Although more studies are needed to understand whether postnatal and adult neurogenesis plays a role in the pathogenesis of intellectual disability, these data suggest that FMRP-dependent aberrant neurogenesis could potentially contribute in part to the cognitive deficits observed in Fragile X patients.

Altered neurogenesis appears to contribute to phenotypes observed in DISC1 loss-of-function models. Disrupted-in-schizophrenia 1 (DISC1) is a multifunctional scaffold protein that has been implicated as a susceptibility gene for major mental illness including schizophrenia, bipolar disorders, and autism (Chubb et al. 2008). DISC1 plays an important role in multiple stages of adult neurogenesis in the DG including proliferation, maturation, migration, and synaptic plasticity of newborn neurons. Importantly, the levels of neurogenesis correlate with the performance on specific memory tasks. The cellular phenotypes appear to be in part regulated by disruption in critical signaling pathways modulating adult neurogenesis including the AKT-mTOR, GABA, and GSK3β signaling pathways (Kim et al. 2009, 2012). Indeed, DISC1 has been shown to modulate the proliferation of adult hippocampal precursor cells by inhibiting GSK3β activity (Mao et al. 2009). Pharmacologically targeting these pathways leads to restoration of both neurogenesis and behavioral phenotypes. Notably, in a DISC1 loss-of-function mouse model, treatment with a GSK3β inhibitor, SB-216763, normalizes both the neurogenesis deficits and the schizophrenia- and depression-like behaviors (Mao et al. 2009). In another study, the downregulation of Disc1 specifically during the development of newborn neurons results in defective neuronal maturation, neuronal hyperexcitability, and alteration of neuronal structure. On a molecular level, knockdown of Disc1 results in increased mTOR signaling. Moreover, Disc1 knockdown induces cognitive and affective deficits and behavioral abnormalities are reversed by pharmacological inhibition of the mTOR pathway (Zhou et al. 2013a).

The described roles of adult hippocampal neurogenesis in neurodevelopmental disorders suggest that treatments targeting the adult neural stem cell population may have a significant impact on pathophysiology and endophenotypes and can represent a novel therapeutic approach.

3.2 Potential of Screening for Neurogenic Compounds

Neuro-regenerative approaches aimed at treating pathological cognitive decline could include both the stimulation of endogenous adult neural stem cell populations and the transplantation of exogenous stem cells in the brain. Seminal experiments suggest that both approaches have potential for treating neurological diseases. For example, in a mouse model of hippocampal neurodegeneration that results in significant memory impairment, transplanting neural stem cells improves cognition (Yamasaki et al. 2007). Similarly, transplantation of hippocampal neural stem cells ameliorates complex cognitive deficits in a model of Alzheimer's disease characterized by abundant amyloid/TAU pathology. Interestingly, the behavioral rescue is achieved despite persistent pathological hallmarks and it is mediated by BDNF-dependent increase in hippocampal synaptic density (Blurton-Jones et al. 2009). On the other hand, the ability of neural stem cells to be reactivated in aging and in disease conditions supports the idea that identification of novel pharmacological targets/molecules promoting adult hippocampal neurogenesis represents a promising therapeutic approach.

A search of Thomas Reuter Integrity suggests that there are well over 50 relevant patents claiming neurogenic molecules. A large number of diverse chemical classes are neurogenic and have been patented for diverse CNS therapeutic indications (Rishton 2008). Reported patents claim that besides CNS acting molecules such as anticonvulsants, cognition modifiers, anxiolytics, and mood-stabilizing agents, also antidiabetic, blood pressure-lowering, and cholesterol-lowering agents have surprisingly shown neurogenic properties (Rishton 2008). For example, metformin, which is widely used to treat type II diabetes and other metabolic syndromes, has been shown to have neurogenic properties (Wang et al. 2012). In vitro work on both mouse and human cell cultures shows that metformin promotes neurogenesis and this translates in vivo in increased neurogenesis and enhanced spatial reversal learning in a water maze task.

A promising route for the identification of novel neurogenesis-increasing compounds is the high-throughput screening of chemical libraries using neural stem cell-based assays. Screens have already demonstrated the potential to elucidate known and novel mechanisms regulating adult hippocampal neurogenesis. In an early study, 1,200 compounds with known pharmacological activity were screened on rat neurospheres for proliferative and differentiation capacity leading to the identification of many CNS-relevant targets. Active compounds included modulators of dopamine, serotonin, opioid, glutamate, and vanilloid signaling (Diamandis et al. 2007). Besides targeting known receptors relevant for neurogenesis, molecule screening can help identify novel relevant signal transduction pathways associated with hippocampal plasticity and cognition. In another screen the small molecule isoxazole 9 [Isx-9; *N*-cyclopropyl-5-(thiophen-2-yl) isoxazole-3-carboxamide] was identified. Isx-9 robustly induces neuronal differentiation in an in vitro model of adult neural stem cells. In addition, Isx-9's effects appear to involve myocyte-enhancer factor 2 (Mef2), a family of transcription factors that had never been linked before to adult neurogenesis in vivo (Schneider et al. 2008). In a follow-up study, it has been shown that Isx-9 promotes multiple stages of neurogenesis in vivo and this correlates with enhanced memory in Morris water maze tasks (Petrik et al. 2012). The mechanism of action of Isx-9 on hippocampal neurogenesis and cognition is at least in part cell autonomous since inducible deletion of Mef2 isoforms from neural stem cells and their progeny thwarts the cognitive enhancing effect. The identification of ISX-9 highlights the potential of small molecule screening campaigns to identify novel mechanisms of action relevant to hippocampal neurogenesis and cognition.

A novel in vivo screening campaign has been undertaken identifying a compound called P7C3 with neurogenic properties (Pieper et al. 2010). Further characterization demonstrates that P7C3 can enhance neurogenesis in the dentate gyrus, can modulate mitochondrial physiology and neuronal survival, can increase synaptic activity, and can preserve cognitive capacity in aged rats. The target(s) and pathway(s) of P7C3 remain unknown, which provides both a challenge as well as an interesting opportunity to identify new pathways linked to neurogenesis and pathological cognitive decline. Clearly, the mechanisms mediating the effect of P7C3 on cognition might involve biological processes beyond hippocampal neurogenesis that could regulate broader aspects of brain plasticity. In follow-up studies, modified and more potent analogues of P7C3 demonstrated that the compound class exhibits neuroprotective effect in models of Parkinson's disease, amyotrophic lateral sclerosis, traumatic brain injury, and age-related cognitive decline (MacMillan et al. 2011; Blaya et al. 2014; Walker et al. 2014). Derivatives appended with immobilizing moieties may reveal the protein targets of the P7C3 class of neuroprotective compounds.

Few classes of neurologically active molecules have been discovered in the last 50 years, at least in part because neuroscience drug discovery efforts have been dominated by target-based assays. Unlike target-based approaches, phenotype-based screens can identify compounds hitting on pathways that capture the complexity of the neurobiological tissue. One impressive example of a large-scale HTS screening campaign on primary cells has been recently reported, where one million small-molecule compounds have been screened on primary rat neurospheres, assessing proliferation and differentiation capacity based on the measurement of ATP levels (Liu et al. 2009). Over 5,000 primary hits have been identified to induce proliferation and differentiation. Further characterization of these compounds and a strategy for target identification will potentially provide novel insight into the mechanisms involved in the regulation of neural stem cells and adult neurogenesis.

As highlighted above, several neurodevelopmental disorders show adult neurogenesis alterations that appear to be partly responsible for cognitive deficits. Recent advances in human-induced pluripotent stem cell (hiPSCs) technologies allow the use of human cells with patient-specific genetic alterations for in vitro neuronal disease modeling. In vitro human neuronal differentiation recapitulates several aspects and stages of in vivo neurogenesis and this system can be used to characterize specific disease-related phenotypes observed in vivo. Moreover, hiPSCs represent an unprecedented tool for development of platforms for phenotypic screening using human tissue- and disease-relevant models. Despite the challenges associated with the novelty of the technology and the limited understanding of human neuronal differentiation in vitro, hiPSC disease models hold great promise as tools for innovative drug discovery (Chailangkarn et al. 2012; Bellin et al. 2012).

3.3 Translational Biomarkers of Human Neurogenesis

In CNS translational research, good translational biomarkers are crucial to identify pre-symptomatic early disease states, to improve dose finding, and to allow patient stratification and responder/nonresponders analysis, significantly affecting the success in drug development. As described above, increasing preclinical evidence in animal models of neurodevelopmental disorders, schizophrenia, Alzheimer's disease, stress, and depression demonstrates that alterations in neurogenesis correlate with behavioral and cognitive symptoms. Moreover, multiple pharmacological interventions targeting cognition modulate adult hippocampal neurogenesis. Therefore, the identification of a noninvasive biomarker for human neurogenesis is an area of growing interest and intensive scientific research. The lack of biomarker currently represents a key limitation in the ability to ultimately demonstrate the relevance of the neurogenic process in human health and disease and its role in mediating beneficial effects of therapeutic interventions.

Identification of hippocampal neurogenesis-specific biomarkers in blood or CSF could serve as a proxy for the rates of neurogenesis in the brain. Immature neuronal precursors in the adult brain are characterized by specific transcriptomes, cellular and membrane proteomes, secretomes, metabolomes, and lipidomes (Ramm et al. 2009; Knolhoff et al. 2013). Unbiased approaches to identify neurogenesis-specific transcriptomic profiles have led to the identification of several transcripts expressed during specific stages of adult neurogenesis (Miller et al. 2013; Couillard-Despres et al. 2006; Lim et al. 2006). Moreover, expression levels of miRNA involved in the regulation of neurogenesis appear to be altered in brain samples from AD patients. Importantly, miRNA can be released and detected in the CSF, representing a potential biomarker (Cogswell et al. 2008). Similarly, some neurogenesis-specific proteins are secreted into fluids including the CSF. For example, doublecortin (DCX), which encodes for a microtubule-associated protein specifically expressed in newborn neurons, is detected in CSF of both rodents and humans (Kremer et al. 2013).

Another potential strategy to monitor neurogenesis is brain imaging. A possible approach would be to identify a PET ligand binding molecular markers of adult hippocampal neurogenesis. A PET imaging strategy could be feasible based on anatomical considerations. PET resolution is in the order of 1.5–3 mm, the hippocampus volume in human is ~3 cm^3, and the dentate gyrus is 90 mm^3. It has been reported that around 0.033 % of granular cells are newborn neurons in old primates and 0.07 % in old mice (Aizawa et al. 2009). Considering the sensitivity of PET imaging, shown to reach 30 pM of receptor density for the D2 receptor, the detection of the newborn cells' fraction would be feasible (Kessler et al. 1993; Farde 1996). The challenge is to find a hippocampal neurogenesis-selective protein with a binding pocket that could serve as a PET tracer. 18F-3′-deoxy-3′-fluoro-L-thymidine (FLT) has been described to be a potential PET tracer to monitor neural stem cell division in the rat hippocampus (Rueger et al. 2010). FLT is a thymidine analogue that accumulates in dividing cells and it is a common marker to monitor cell proliferation in tumors. For measurements in the brain, strategies to improve the signal-to-noise ratio and to reduce radioactive doses would significantly implement the approach (Rueger et al. 2010). Monitoring neurogenesis targeting immature neurons could represent another approach considering that in aged macaque monkeys and mice, immature neurons are 4- to 4.5-fold more abundant than dividing NPCs (Aizawa et al. 2009).

MRI/MRS technologies could be a potential way to identify biomarkers of neurogenesis or at least epiphenomena of the process. A conceptually very appealing study, using MRI-based technologies, suggested relative cerebral blood volume (rCBV) increases specific to the dentate gyrus as a putative biomarker for neurogenesis (Pereira et al. 2007). As described previously, neurogenesis and angiogenesis in the DG are tightly linked processes (Palmer et al. 2000; Louissaint et al. 2002). However, the specificity of this method is challenged by the fact that hippocampus hemodynamics observed by fMRI is sensitive to changes due to disease states and drug treatments (Choi et al. 2006; Littlewood et al. 2006; Gozzi et al. 2008).

Using proton MRS, a very exciting study identified a particular resonance at 1.28 ppm with enhanced prevalence in NPCs in vitro (Maletic-Savatic et al. 2008; Manganas et al. 2007). This marker was demonstrated to be a translational biomarker in both rats and human subjects, and the putative NPC-specific peak strongly decreases during brain development from childhood to adulthood (Manganas et al. 2007). If confirmed, this study provides an unparalleled glimpse into the human brain with the promise to provide a truly noninvasive biomarker of adult neurogenesis. Not unexpectedly, this report has aroused a certain amount of controversy on technical and study design aspects (Friedman 2008; Hoch et al. 2008; Jansen et al. 2008; Dong et al. 2009; Ramm et al. 2009). More studies are needed to demonstrate feasibility and practicality of this approach as a tool to investigate the role of neurogenesis in a wide variety of human brain disorders.

4 Conclusions

Adult hippocampal neurogenesis is a complex process that provides an excellent example of the surprising plasticity of the mature brain. Research within the last few years has identified molecular mechanisms that regulate individual steps in neurogenesis and has highlighted its function in various neurological disorders. However, the process of translating basic knowledge into pharmacological interventions is only at the beginning. Key aspects need to be further clarified to support targeting of adult neurogenesis for the treatment of neurological and cognitive disorders in humans: (1) Can results obtained from animal models of neurological disorders be translated to humans?, (2) Can changes in neurogenesis be assessed in humans (i.e., is there a biomarker)?, (3) Is it possible to identify chemical entities or biomolecules to pharmacologically modulate neurogenesis in humans? In this chapter, we have highlighted key studies investigating the role of neurogenesis in cognitive impairment associated with physiological and disease conditions. Moreover, we have described endogenous and pharmacological factors that modulate neurogenesis and have an effect on cognitive functions. Deep understanding of several of these aspects would enable the development of drugs that target and modulate adult neurogenesis in humans to treat patients with neurological disorders.

Acknowledgments VC and SL are supported by the Roche Postdoctoral Fellowship Program. We thank M. Iglesias for artwork.

References

Aberg MA, Pedersen NL, Toren K, Svartengren M, Backstrand B, Johnsson T, Cooper-Kuhn CM, Aberg ND, Nilsson M, Kuhn HG (2009) Cardiovascular fitness is associated with cognition in young adulthood. Proc Natl Acad Sci U S A 106(49):20906–20911. doi:10.1073/pnas.0905307106

Adachi M, Barrot M, Autry AE, Theobald D, Monteggia LM (2008) Selective loss of brain-derived neurotrophic factor in the dentate gyrus attenuates antidepressant efficacy. Biol Psychiatry 63(7):642–649. doi:10.1016/j.biopsych.2007.09.019

Adeosun SO, Hou X, Zheng B, Stockmeier C, Ou X, Paul I, Mosley T, Weisgraber K, Wang JM (2014) Cognitive deficits and disruption of neurogenesis in a mouse model of apolipoprotein E4 domain interaction. J Biol Chem 289(5):2946–2959. doi:10.1074/jbc.M113.497909

Aizawa K, Ageyama N, Yokoyama C, Hisatsune T (2009) Age-dependent alteration in hippocampal neurogenesis correlates with learning performance of macaque monkeys. Exp Anim 58 (4):403–407

Akers KG, Martinez-Canabal A, Restivo L, Yiu AP, De Cristofaro A, Hsiang HL, Wheeler AL, Guskjolen A, Niibori Y, Shoji H, Ohira K, Richards BA, Miyakawa T, Josselyn SA, Frankland

PW (2014) Hippocampal neurogenesis regulates forgetting during adulthood and infancy. Science 344(6184):598–602. doi:10.1126/science.1248903

Alexander GE, Ryan L, Bowers D, Foster TC, Bizon JL, Geldmacher DS, Glisky EL (2012) Characterizing cognitive aging in humans with links to animal models. Front Aging Neurosci 4:21. doi:10.3389/fnagi.2012.00021

Alexson TO, Hitoshi S, Coles BL, Bernstein A, van der Kooy D (2006) Notch signaling is required to maintain all neural stem cell populations—irrespective of spatial or temporal niche. Dev Neurosci 28(1–2):34–48. doi:10.1159/000090751

Ally BA, Hussey EP, Ko PC, Molitor RJ (2013) Pattern separation and pattern completion in Alzheimer's disease: evidence of rapid forgetting in amnestic mild cognitive impairment. Hippocampus 23(12):1246–1258. doi:10.1002/hipo.22162

Alonso R, Griebel G, Pavone G, Stemmelin J, Le Fur G, Soubrie P (2004) Blockade of CRF(1) or V(1b) receptors reverses stress-induced suppression of neurogenesis in a mouse model of depression. Mol Psychiatry 9(3):278–286, 224. doi:10.1038/sj.mp.4001464

Altman J (1969) Autoradiographic and histological studies of postnatal neurogenesis. IV. Cell proliferation and migration in the anterior forebrain, with special reference to persisting neurogenesis in the olfactory bulb. J Comp Neurol 137(4):433–457. doi:10.1002/cne.901370404

Alvarez-Buylla A, Garcia-Verdugo JM (2002) Neurogenesis in adult subventricular zone. J Neurosci 22(3):629–634

Ambrogini P, Lattanzi D, Ciuffoli S, Agostini D, Bertini L, Stocchi V, Santi S, Cuppini R (2004) Morpho-functional characterization of neuronal cells at different stages of maturation in granule cell layer of adult rat dentate gyrus. Brain Res 1017(1–2):21–31. doi:10.1016/j.brainres.2004.05.039

Amrein I, Isler K, Lipp HP (2011) Comparing adult hippocampal neurogenesis in mammalian species and orders: influence of chronological age and life history stage. Eur J Neurosci 34 (6):978–987. doi:10.1111/j.1460-9568.2011.07804.x

Anacker C (2014) Adult hippocampal neurogenesis in depression: behavioral implications and regulation by the stress system. Curr Top Behav Neurosci. doi:10.1007/7854_2014_275

Anacker C, Pariante CM (2012) New models to investigate complex glucocorticoid receptor functions. Front Behav Neurosci 6:90. doi:10.3389/fnbeh.2012.00090

Arruda-Carvalho M, Sakaguchi M, Akers KG, Josselyn SA, Frankland PW (2011) Posttraining ablation of adult-generated neurons degrades previously acquired memories. J Neurosci 31 (42):15113–15127. doi:10.1523/JNEUROSCI.3432-11.2011

Asberg M (1976) Treatment of depression with tricyclic drugs—pharmacokinetic and pharmacodynamic aspects. Pharmakopsychiatr Neuropsychopharmakol 9(1):18–26

Ashford JW (2004) APOE genotype effects on Alzheimer's disease onset and epidemiology. J Mol Neurosci 23(3):157–165. doi:10.1385/JMN:23:3:157

Autry AE, Adachi M, Nosyreva E, Na ES, Los MF, Cheng PF, Kavalali ET, Monteggia LM (2011) NMDA receptor blockade at rest triggers rapid behavioural antidepressant responses. Nature 475(7354):91–95. doi:10.1038/nature10130

Bakker A, Kirwan CB, Miller M, Stark CE (2008) Pattern separation in the human hippocampal CA3 and dentate gyrus. Science 319(5870):1640–1642. doi:10.1126/science.1152882

Bambico FR, Belzung C (2013) Novel insights into depression and antidepressants: a synergy between synaptogenesis and neurogenesis? Curr Top Behav Neurosci 15:243–91. doi:10.1007/7854_2012_234

Banerjee R, Ghosh AK, Ghosh B, Bhattacharyya S, Mondal AC (2013) Decreased mRNA and protein expression of BDNF, NGF, and their receptors in the hippocampus from suicide: an analysis in human postmortem brain. Clin Med Insights Pathol 6:1–11. doi:10.4137/CMPath.S12530

Barnes CA (1979) Memory deficits associated with senescence: a neurophysiological and behavioral study in the rat. J Comp Physiol Psychol 93(1):74–104

Bechara RG, Kelly AM (2013) Exercise improves object recognition memory and induces BDNF expression and cell proliferation in cognitively enriched rats. Behav Brain Res 245:96–100. doi:10.1016/j.bbr.2013.02.018

Bekinschtein P, Kent BA, Oomen CA, Clemenson GD, Gage FH, Saksida LM, Bussey TJ (2013) BDNF in the dentate gyrus is required for consolidation of "pattern-separated" memories. Cell Rep 5(3):759–768. doi:10.1016/j.celrep.2013.09.027

Bekinschtein P, Kent BA, Oomen CA, Clemenson GD, Gage FH, Saksida LM, Bussey TJ (2014) Brain-derived neurotrophic factor interacts with adult-born immature cells in the dentate gyrus during consolidation of overlapping memories. Hippocampus. doi:10.1002/hipo.22304

Bellin M, Marchetto MC, Gage FH, Mummery CL (2012) Induced pluripotent stem cells: the new patient? Nat Rev Mol Cell Biol 13(11):713–726. doi:10.1038/nrm3448

Belvederi Murri M, Pariante C, Mondelli V, Masotti M, Atti AR, Mellacqua Z, Antonioli M, Ghio L, Menchetti M, Zanetidou S, Innamorati M, Amore M (2014) HPA axis and aging in depression: systematic review and meta-analysis. Psychoneuroendocrinology 41:46–62. doi:10.1016/j.psyneuen.2013.12.004

Ben Abdallah NM, Slomianka L, Vyssotski AL, Lipp HP (2010) Early age-related changes in adult hippocampal neurogenesis in C57 mice. Neurobiol Aging 31(1):151–161. doi:10.1016/j.neurobiolaging.2008.03.002

Berchtold NC, Chinn G, Chou M, Kesslak JP, Cotman CW (2005) Exercise primes a molecular memory for brain-derived neurotrophic factor protein induction in the rat hippocampus. Neuroscience 133(3):853–861. doi:10.1016/j.neuroscience.2005.03.026

Bergami M, Rimondini R, Santi S, Blum R, Gotz M, Canossa M (2008) Deletion of TrkB in adult progenitors alters newborn neuron integration into hippocampal circuits and increases anxiety-like behavior. Proc Natl Acad Sci U S A 105(40):15570–15575. doi:10.1073/pnas.0803702105

Bernal GM, Peterson DA (2011) Phenotypic and gene expression modification with normal brain aging in GFAP-positive astrocytes and neural stem cells. Aging Cell 10(3):466–482. doi:10.1111/j.1474-9726.2011.00694.x

Bessa JM, Ferreira D, Melo I, Marques F, Cerqueira JJ, Palha JA, Almeida OF, Sousa N (2009) The mood-improving actions of antidepressants do not depend on neurogenesis but are associated with neuronal remodeling. Mol Psychiatry 14(8):764–773, 739. doi:10.1038/mp.2008.119

Bianchi P, Ciani E, Guidi S, Trazzi S, Felice D, Grossi G, Fernandez M, Giuliani A, Calzà L, Bartesaghi R (2010) Early pharmacotherapy restores neurogenesis and cognitive performance in the Ts65Dn mouse model for Down syndrome. J Neurosci 30(26):8769–8779. doi:10.1523/JNEUROSCI. 0534-10.2010

Bizon JL, Lee HJ, Gallagher M (2004) Neurogenesis in a rat model of age-related cognitive decline. Aging Cell 3(4):227–234. doi:10.1111/j.1474-9728.2004.00099.x

Black JE, Isaacs KR, Anderson BJ, Alcantara AA, Greenough WT (1990) Learning causes synaptogenesis, whereas motor activity causes angiogenesis, in cerebellar cortex of adult rats. Proc Natl Acad Sci U S A 87(14):5568–5572

Blaya MO, Bramlett HM, Naidoo J, Pieper AA, Dietrich WD (2014) Neuroprotective efficacy of a proneurogenic compound after traumatic brain injury. J Neurotrauma 31(5):476–486. doi:10.1089/neu.2013.3135

Blurton-Jones M, Kitazawa M, Martinez-Coria H, Castello NA, Muller FJ, Loring JF, Yamasaki TR, Poon WW, Green KN, LaFerla FM (2009) Neural stem cells improve cognition via BDNF in a transgenic model of Alzheimer disease. Proc Natl Acad Sci U S A 106(32):13594–13599. doi:10.1073/pnas.0901402106

Boekhoorn K, Joels M, Lucassen PJ (2006) Increased proliferation reflects glial and vascular-associated changes, but not neurogenesis in the presenile Alzheimer hippocampus. Neurobiol Dis 24(1):1–14. doi:10.1016/j.nbd.2006.04.017

Boldrini M, Underwood MD, Hen R, Rosoklija GB, Dwork AJ, John Mann J, Arango V (2009) Antidepressants increase neural progenitor cells in the human hippocampus. Neuropsychopharmacology 34(11):2376–2389. doi:10.1038/npp.2009.75

Boldrini M, Hen R, Underwood MD, Rosoklija GB, Dwork AJ, Mann JJ, Arango V (2012) Hippocampal angiogenesis and progenitor cell proliferation are increased with antidepressant use in major depression. Biol Psychiatry 72(7):562–571. doi:10.1016/j.biopsych.2012.04.024

Bolognin S, Buffelli M, Puolivali J, Iqbal K (2014) Rescue of cognitive-aging by administration of a neurogenic and/or neurotrophic compound. Neurobiol Aging 35(9):2134–2146. doi:10.1016/j.neurobiolaging.2014.02.017

Bonaguidi MA, Wheeler MA, Shapiro JS, Stadel RP, Sun GJ, Ming GL, Song H (2011) In vivo clonal analysis reveals self-renewing and multipotent adult neural stem cell characteristics. Cell 145(7):1142–1155. doi:10.1016/j.cell.2011.05.024

Bondolfi L, Ermini F, Long JM, Ingram DK, Jucker M (2004) Impact of age and caloric restriction on neurogenesis in the dentate gyrus of C57BL/6 mice. Neurobiol Aging 25(3):333–340. doi:10.1016/S0197-4580(03)00083-6

Braak H, Braak E (1991) Neuropathological stageing of Alzheimer-related changes. Acta Neuropathol 82(4):239–259

Braak H, Braak E (1995) Staging of Alzheimer's disease-related neurofibrillary changes. Neurobiol Aging 16(3):271–278, discussion 278–284

Brandt MD, Maass A, Kempermann G, Storch A (2010) Physical exercise increases Notch activity, proliferation and cell cycle exit of type-3 progenitor cells in adult hippocampal neurogenesis. Eur J Neurosci 32(8):1256–1264. doi:10.1111/j.1460-9568.2010.07410.x

Bremner JD, Vythilingam M, Vermetten E, Vaccarino V, Charney DS (2004) Deficits in hippocampal and anterior cingulate functioning during verbal declarative memory encoding in midlife major depression. Am J Psychiatry 161(4):637–645

Brown J, Cooper-Kuhn CM, Kempermann G, Van Praag H, Winkler J, Gage FH, Kuhn HG (2003a) Enriched environment and physical activity stimulate hippocampal but not olfactory bulb neurogenesis. Eur J Neurosci 17(10):2042–2046

Brown JP, Couillard-Despres S, Cooper-Kuhn CM, Winkler J, Aigner L, Kuhn HG (2003b) Transient expression of doublecortin during adult neurogenesis. J Comp Neurol 467(1):1–10. doi:10.1002/cne.10874

Bruel-Jungerman E, Laroche S, Rampon C (2005) New neurons in the dentate gyrus are involved in the expression of enhanced long-term memory following environmental enrichment. Eur J Neurosci 21(2):513–521. doi:10.1111/j.1460-9568.2005.03875.x

Bruel-Jungerman E, Davis S, Rampon C, Laroche S (2006) Long-term potentiation enhances neurogenesis in the adult dentate gyrus. J Neurosci 26(22):5888–5893. doi:10.1523/JNEUROSCI.0782-06.2006

Caille I, Allinquant B, Dupont E, Bouillot C, Langer A, Muller U, Prochiantz A (2004) Soluble form of amyloid precursor protein regulates proliferation of progenitors in the adult subventricular zone. Development 131(9):2173–2181. doi:10.1242/dev.01103

Cameron HA, McKay RD (1999) Restoring production of hippocampal neurons in old age. Nat Neurosci 2(10):894–897. doi:10.1038/13197

Cameron HA, Woolley CS, McEwen BS, Gould E (1993) Differentiation of newly born neurons and glia in the dentate gyrus of the adult rat. Neuroscience 56(2):337–344

Cameron HA, McEwen BS, Gould E (1995) Regulation of adult neurogenesis by excitatory input and NMDA receptor activation in the dentate gyrus. J Neurosci 15(6):4687–4692

Campbell NR, Fernandes CC, Halff AW, Berg DK (2010) Endogenous signaling through alpha7-containing nicotinic receptors promotes maturation and integration of adult-born neurons in the hippocampus. J Neurosci 30(26):8734–8744. doi:10.1523/JNEUROSCI.0931-10.2010

Campos AC, Ortega Z, Palazuelos J, Fogaca MV, Aguiar DC, Diaz-Alonso J, Ortega-Gutierrez S, Vazquez-Villa H, Moreira FA, Guzman M, Galve-Roperh I, Guimaraes FS (2013) The anxiolytic effect of cannabidiol on chronically stressed mice depends on hippocampal neurogenesis: involvement of the endocannabinoid system. Int J Neuropsychopharmacol 16(6):1407–1419. doi:10.1017/S1461145712001502

Carro E, Nunez A, Busiguina S, Torres-Aleman I (2000) Circulating insulin-like growth factor I mediates effects of exercise on the brain. J Neurosci 20(8):2926–2933

Caselli RJ, Beach TG, Yaari R, Reiman EM (2006) Alzheimer's disease a century later. J Clin Psychiatry 67(11):1784–1800

Castren E, Voikar V, Rantamaki T (2007) Role of neurotrophic factors in depression. Curr Opin Pharmacol 7(1):18–21. doi:10.1016/j.coph.2006.08.009

Cavanaugh JE, Ham J, Hetman M, Poser S, Yan C, Xia Z (2001) Differential regulation of mitogen-activated protein kinases ERK1/2 and ERK5 by neurotrophins, neuronal activity, and cAMP in neurons. J Neurosci 21(2):434–443

Chaddock L, Erickson KI, Prakash RS, Kim JS, Voss MW, Vanpatter M, Pontifex MB, Raine LB, Konkel A, Hillman CH, Cohen NJ, Kramer AF (2010) A neuroimaging investigation of the association between aerobic fitness, hippocampal volume, and memory performance in preadolescent children. Brain Res 1358:172–183. doi:10.1016/j.brainres.2010.08.049

Chaddock L, Hillman CH, Buck SM, Cohen NJ (2011) Aerobic fitness and executive control of relational memory in preadolescent children. Med Sci Sports Exerc 43(2):344–349. doi:10.1249/MSS.0b013e3181e9af48

Chaddock L, Hillman CH, Pontifex MB, Johnson CR, Raine LB, Kramer AF (2012) Childhood aerobic fitness predicts cognitive performance one year later. J Sports Sci 30(5):421–430. doi:10.1080/02640414.2011.647706

Chaddock-Heyman L, Erickson KI, Voss MW, Knecht AM, Pontifex MB, Castelli DM, Hillman CH, Kramer AF (2013) The effects of physical activity on functional MRI activation associated with cognitive control in children: a randomized controlled intervention. Front Hum Neurosci 7:72. doi:10.3389/fnhum.2013.00072

Chadwick W, Mitchell N, Caroll J, Zhou Y, Park S-S et al (2011) Amitriptyline-mediated cognitive enhancement in aged 3×Tg Alzheimer's disease mice is associated with neurogenesis and neurotrophic activity. PLoS One 6(6):e21660. doi:10.1371/journal.pone.0021660

Chadwick MJ, Bonnici HM, Maguire EA (2014) CA3 size predicts the precision of memory recall. Proc Natl Acad Sci U S A. doi:10.1073/pnas.1319641111

Chailangkarn T, Acab A, Muotri AR (2012) Modeling neurodevelopmental disorders using human neurons. Curr Opin Neurobiol 22(5):785–790. doi:10.1016/j.conb.2012.04.004

Chetty S, Friedman AR, Taravosh-Lahn K, Kirby ED, Mirescu C, Guo F, Krupik D, Nicholas A, Geraghty AC, Krishnamurthy A, Tsai MK, Covarrubias D, Wong AT, Francis DD, Sapolsky RM, Palmer TD, Pleasure D, Kaufer D (2014) Stress and glucocorticoids promote oligodendrogenesis in the adult hippocampus. Mol Psychiatry 19(12):1275–1283. doi:10.1038/mp.2013.190

Choi JK, Mandeville JB, Chen YI, Kim YR, Jenkins BG (2006) High resolution spatial mapping of nicotine action using pharmacologic magnetic resonance imaging. Synapse 60(2):152–157. doi:10.1002/syn.20284

Chubb JE, Bradshaw NJ, Soares DC, Porteous DJ, Millar JK (2008) The DISC locus in psychiatric illness. Mol Psychiatry 13(1):36–64. doi:10.1038/sj.mp.4002106

Chun SK, Sun W, Park JJ, Jung MW (2006) Enhanced proliferation of progenitor cells following long-term potentiation induction in the rat dentate gyrus. Neurobiol Learn Mem 86 (3):322–329. doi:10.1016/j.nlm.2006.05.005

Cinini SM, Barnabe GF, Galvao-Coelho N, de Medeiros MA, Perez-Mendes P, Sousa MB, Covolan L, Mello LE (2014) Social isolation disrupts hippocampal neurogenesis in young non-human primates. Front Neurosci 8:45. doi:10.3389/fnins.2014.00045

Clark PJ, Brzezinska WJ, Thomas MW, Ryzhenko NA, Toshkov SA, Rhodes JS (2008) Intact neurogenesis is required for benefits of exercise on spatial memory but not motor performance or contextual fear conditioning in C57BL/6 J mice. Neuroscience 155(4):1048–1058. doi:10.1016/j.neuroscience.2008.06.051

Clelland CD, Choi M, Romberg C, Clemenson GD Jr, Fragniere A, Tyers P, Jessberger S, Saksida LM, Barker RA, Gage FH, Bussey TJ (2009) A functional role for adult hippocampal neurogenesis in spatial pattern separation. Science 325(5937):210–213. doi:10.1126/science.1173215

Cogswell JP, Ward J, Taylor IA, Waters M, Shi Y, Cannon B, Kelnar K, Kemppainen J, Brown D, Chen C, Prinjha RK, Richardson JC, Saunders AM, Roses AD, Richards CA (2008) Identification of miRNA changes in Alzheimer's disease brain and CSF yields putative biomarkers and insights into disease pathways. J Alzheimers Dis 14(1):27–41

Compton DM, Dietrich KL, Selinger MC, Testa EK (2011) 5-methoxy-N, N-di(iso) propyltryptamine hydrochloride (Foxy)-induced cognitive deficits in rat after exposure in adolescence. Physiol Behav 103(2):203–209. doi:10.1016/j.physbeh.2011.01.021

Contestabile A, Greco B, Ghezzi D, Tucci V, Benfenati F, Gasparini L (2013) Lithium rescues synaptic plasticity and memory in Down syndrome mice. J Clin Invest 123(1):348–361

Cooper-Kuhn CM, Winkler J, Kuhn HG (2004) Decreased neurogenesis after cholinergic forebrain lesion in the adult rat. J Neurosci Res 77(2):155–165. doi:10.1002/jnr.20116

Coronas V, Durand M, Chabot JG, Jourdan F, Quirion R (2000) Acetylcholine induces neuritic outgrowth in rat primary olfactory bulb cultures. Neuroscience 98(2):213–219

Couillard-Despres S, Winner B, Karl C, Lindemann G, Schmid P, Aigner R, Laemke J, Bogdahn U, Winkler J, Bischofberger J, Aigner L (2006) Targeted transgene expression in neuronal precursors: watching young neurons in the old brain. Eur J Neurosci 24 (6):1535–1545. doi:10.1111/j.1460-9568.2006.05039.x

Creer DJ, Romberg C, Saksida LM, van Praag H, Bussey TJ (2010) Running enhances spatial pattern separation in mice. Proc Natl Acad Sci U S A 107(5):2367–2372. doi:10.1073/pnas.0911725107

Cundiff P, Liu L, Wang Y, Zou J, Pan YW, Abel G, Duan X, Ming GL, Englund C, Hevner R, Xia Z (2009) ERK5 MAP kinase regulates neurogenin1 during cortical neurogenesis. PLoS One 4 (4):e5204. doi:10.1371/journal.pone.0005204

Cuppini R, Bucherelli C, Ambrogini P, Ciuffoli S, Orsini L, Ferri P, Baldi E (2006) Age-related naturally occurring depression of hippocampal neurogenesis does not affect trace fear conditioning. Hippocampus 16(2):141–148. doi:10.1002/hipo.20140

Dagyte G, Crescente I, Postema F, Seguin L, Gabriel C, Mocaer E, Boer JA, Koolhaas JM (2011) Agomelatine reverses the decrease in hippocampal cell survival induced by chronic mild stress. Behav Brain Res 218(1):121–128. doi:10.1016/j.bbr.2010.11.045

David DJ, Samuels BA, Rainer Q, Wang JW, Marsteller D, Mendez I, Drew M, Craig DA, Guiard BP, Guilloux JP, Artymyshyn RP, Gardier AM, Gerald C, Antonijevic IA, Leonardo ED, Hen R (2009) Neurogenesis-dependent and -independent effects of fluoxetine in an animal model of anxiety/depression. Neuron 62(4):479–493. doi:10.1016/j.neuron.2009.04.017

De Strooper B, Woodgett J (2003) Alzheimer's disease: mental plaque removal. Nature 423 (6938):392–393. doi:10.1038/423392a

De Strooper B, Annaert W, Cupers P, Saftig P, Craessaerts K, Mumm JS, Schroeter EH, Schrijvers V, Wolfe MS, Ray WJ, Goate A, Kopan R (1999) A presenilin-1-dependent gamma-secretase-like protease mediates release of Notch intracellular domain. Nature 398 (6727):518–522. doi:10.1038/19083

Delgado PL (2000) Depression: the case for a monoamine deficiency. J Clin Psychiatry 61(Suppl 6):7–11

Demars MP, Hollands C, Zhao Kda T, Lazarov O (2013) Soluble amyloid precursor protein-alpha rescues age-linked decline in neural progenitor cell proliferation. Neurobiol Aging 34 (10):2431–2440. doi:10.1016/j.neurobiolaging.2013.04.016

Deng W, Saxe MD, Gallina IS, Gage FH (2009) Adult-born hippocampal dentate granule cells undergoing maturation modulate learning and memory in the brain. J Neurosci 29 (43):13532–13542. doi:10.1523/JNEUROSCI.3362-09.2009

Denny CA, Burghardt NS, Schachter DM, Hen R, Drew MR (2012) 4- to 6-week-old adult-born hippocampal neurons influence novelty-evoked exploration and contextual fear conditioning. Hippocampus 22(5):1188–1201. doi:10.1002/hipo.20964

Dery N, Pilgrim M, Gibala M, Gillen J, Wojtowicz JM, Macqueen G, Becker S (2013) Adult hippocampal neurogenesis reduces memory interference in humans: opposing effects of aerobic exercise and depression. Front Neurosci 7:66. doi:10.3389/fnins.2013.00066

Diamandis P, Wildenhain J, Clarke ID, Sacher AG, Graham J, Bellows DS, Ling EK, Ward RJ, Jamieson LG, Tyers M, Dirks PB (2007) Chemical genetics reveals a complex functional ground state of neural stem cells. Nat Chem Biol 3(5):268–273. doi:10.1038/nchembio873

Ding Q, Ying Z, Gomez-Pinilla F (2011) Exercise influences hippocampal plasticity by modulating brain-derived neurotrophic factor processing. Neuroscience 192:773–780. doi:10.1016/j.neuroscience.2011.06.032

Dong Z, Dreher W, Leibfritz D, Peterson BS (2009) Challenges of using MR spectroscopy to detect neural progenitor cells in vivo. AJNR Am J Neuroradiol 30(6):1096–1101. doi:10.3174/ajnr.A1557

Drachman DA, Leavitt J (1974) Human memory and the cholinergic system. A relationship to aging? Arch Neurol 30(2):113–121

Drapeau E, Mayo W, Aurousseau C, Le Moal M, Piazza PV, Abrous DN (2003) Spatial memory performances of aged rats in the water maze predict levels of hippocampal neurogenesis. Proc Natl Acad Sci U S A 100(24):14385–14390. doi:10.1073/pnas.2334169100

Drew MR, Denny CA, Hen R (2010) Arrest of adult hippocampal neurogenesis in mice impairs single- but not multiple-trial contextual fear conditioning. Behav Neurosci 124(4):446–454. doi:10.1037/a0020081

Driscoll I, Hamilton DA, Petropoulos H, Yeo RA, Brooks WM, Baumgartner RN, Sutherland RJ (2003) The aging hippocampus: cognitive, biochemical and structural findings. Cereb Cortex 13(12):1344–1351

Driscoll I, Howard SR, Stone JC, Monfils MH, Tomanek B, Brooks WM, Sutherland RJ (2006) The aging hippocampus: a multi-level analysis in the rat. Neuroscience 139(4):1173–1185. doi:10.1016/j.neuroscience.2006.01.040

Duman CH, Duman RS (2015) Spine synapse remodeling in the pathophysiology and treatment of depression. Neurosci Lett. doi:10.1016/j.neulet.2015.01.022

Dunham JS, Deakin JF, Miyajima F, Payton A, Toro CT (2009) Expression of hippocampal brain-derived neurotrophic factor and its receptors in Stanley consortium brains. J Psychiatr Res 43 (14):1175–1184. doi:10.1016/j.jpsychires.2009.03.008

Dupret D, Revest JM, Koehl M, Ichas F, De Giorgi F, Costet P, Abrous DN, Piazza PV (2008) Spatial relational memory requires hippocampal adult neurogenesis. PLoS One 3(4):e1959. doi:10.1371/journal.pone.0001959

Duric V, Banasr M, Stockmeier CA, Simen AA, Newton SS, Overholser JC, Jurjus GJ, Dieter L, Duman RS (2013) Altered expression of synapse and glutamate related genes in post-mortem hippocampus of depressed subjects. Int J Neuropsychopharmacol 16(1):69–82. doi:10.1017/S1461145712000016

Duveau V, Laustela S, Barth L, Gianolini F, Vogt KE, Keist R, Chandra D, Homanics GE, Rudolph U, Fritschy JM (2011) Spatiotemporal specificity of GABAA receptor-mediated regulation of adult hippocampal neurogenesis. Eur J Neurosci 34(3):362–373. doi:10.1111/j.1460-9568.2011.07782.x

Egeland M, Zhang X, Millan MJ, Mocaer E, Svenningsson P (2012) Pharmacological or genetic blockade of the dopamine D3 receptor increases cell proliferation in the hippocampus of adult mice. J Neurochem 123(5):811–823. doi:10.1111/jnc.12011

Ehninger D, Kempermann G (2003) Regional effects of wheel running and environmental enrichment on cell genesis and microglia proliferation in the adult murine neocortex. Cereb Cortex 13(8):845–851

ElBeltagy M et al (2010) Fluoxetine improves the memory deficits caused by the chemotherapy agent 5-fluorouracil. Behav Brain Res 208(1):112–117

Encinas JM, Michurina TV, Peunova N, Park JH, Tordo J, Peterson DA, Fishell G, Koulakov A, Enikolopov G (2011) Division-coupled astrocytic differentiation and age-related depletion of neural stem cells in the adult hippocampus. Cell Stem Cell 8(5):566–579. doi:10.1016/j.stem.2011.03.010

Erickson KI, Prakash RS, Voss MW, Chaddock L, Hu L, Morris KS, White SM, Wojcicki TR, McAuley E, Kramer AF (2009) Aerobic fitness is associated with hippocampal volume in elderly humans. Hippocampus 19(10):1030–1039. doi:10.1002/hipo.20547

Eriksson PS, Perfilieva E, Bjork-Eriksson T, Alborn AM, Nordborg C, Peterson DA, Gage FH (1998) Neurogenesis in the adult human hippocampus. Nat Med 4(11):1313–1317. doi:10.1038/3305

Fabel K, Fabel K, Tam B, Kaufer D, Baiker A, Simmons N, Kuo CJ, Palmer TD (2003) VEGF is necessary for exercise-induced adult hippocampal neurogenesis. Eur J Neurosci 18 (10):2803–2812

Fabel K, Wolf SA, Ehninger D, Babu H, Leal-Galicia P, Kempermann G (2009) Additive effects of physical exercise and environmental enrichment on adult hippocampal neurogenesis in mice. Front Neurosci 3:50. doi:10.3389/neuro.22.002.2009

Falls WA, Fox JH, MacAulay CM (2010) Voluntary exercise improves both learning and consolidation of cued conditioned fear in C57 mice. Behav Brain Res 207(2):321–331. doi:10.1016/j.bbr.2009.10.016

Farde L (1996) The advantage of using positron emission tomography in drug research. Trends Neurosci 19(6):211–214

Farioli-Vecchioli S, Mattera A, Micheli L, Ceccarelli M, Leonardi L, Saraulli D, Costanzi M, Cestari V, Rouault JP, Tirone F (2014) Running rescues defective adult neurogenesis by shortening the length of the cell cycle of neural stem and progenitor cells. Stem Cells 32 (7):1968–1982. doi:10.1002/stem.1679

Farmer J, Zhao X, van Praag H, Wodtke K, Gage FH, Christie BR (2004) Effects of voluntary exercise on synaptic plasticity and gene expression in the dentate gyrus of adult male Sprague-Dawley rats in vivo. Neuroscience 124(1):71–79. doi:10.1016/j.neuroscience.2003.09.029

Fatemi SH, Earle JA, Stary JM, Lee S, Sedgewick J (2001) Altered levels of the synaptosomal associated protein SNAP-25 in hippocampus of subjects with mood disorders and schizophrenia. Neuroreport 12(15):3257–3262

Faulkner RL, Jang MH, Liu XB, Duan X, Sailor KA, Kim JY, Ge S, Jones EG, Ming GL, Song H, Cheng HJ (2008) Development of hippocampal mossy fiber synaptic outputs by new neurons in the adult brain. Proc Natl Acad Sci U S A 105(37):14157–14162. doi:10.1073/pnas.0806658105

Fava M, Graves LM, Benazzi F, Scalia MJ, Iosifescu DV, Alpert JE, Papakostas GI (2006) A cross-sectional study of the prevalence of cognitive and physical symptoms during long-term antidepressant treatment. J Clin Psychiatry 67(11):1754–1759

Felice D, O'Leary OF, Pizzo RC, Cryan JF (2012) Blockade of the GABA(B) receptor increases neurogenesis in the ventral but not dorsal adult hippocampus: relevance to antidepressant action. Neuropharmacology 63(8):1380–1388. doi:10.1016/j.neuropharm.2012.06.066

Finegan KG, Wang X, Lee EJ, Robinson AC, Tournier C (2009) Regulation of neuronal survival by the extracellular signal-regulated protein kinase 5. Cell Death Differ 16(5):674–683. doi:10.1038/cdd.2008.193

Fiorentini A, Rosi MC, Grossi C, Luccarini I, Casamenti F (2010) Lithium improves hippocampal neurogenesis, neuropathology and cognitive functions in APP mutant mice. PLoS One 5(12): e14382. doi:10.1371/journal.pone.0014382

Fordyce DE, Farrar RP (1991) Enhancement of spatial learning in F344 rats by physical activity and related learning-associated alterations in hippocampal and cortical cholinergic functioning. Behav Brain Res 46(2):123–133

Friedman SD (2008) Comment on "Magnetic resonance spectroscopy identifies neural progenitor cells in the live human brain". Science 321(5889):640. doi:10.1126/science.1153484, author reply 640

Frodl T, Meisenzahl EM, Zetzsche T, Born C, Groll C, Jager M, Leinsinger G, Bottlender R, Hahn K, Moller HJ (2002) Hippocampal changes in patients with a first episode of major depression. Am J Psychiatry 159(7):1112–1118

Gadadhar A, Marr R, Lazarov O (2011) Presenilin-1 regulates neural progenitor cell differentiation in the adult brain. J Neurosci 31(7):2615–2623. doi:10.1523/JNEUROSCI.4767-10.2011

Gage FH, Dunnett SB, Bjorklund A (1984) Spatial learning and motor deficits in aged rats. Neurobiol Aging 5(1):43–48

Galvao RP, Garcia-Verdugo JM, Alvarez-Buylla A (2008) Brain-derived neurotrophic factor signaling does not stimulate subventricular zone neurogenesis in adult mice and rats. J Neurosci 28(50):13368–13383. doi:10.1523/JNEUROSCI.2918-08.2008

Ge S, Goh EL, Sailor KA, Kitabatake Y, Ming GL, Song H (2006) GABA regulates synaptic integration of newly generated neurons in the adult brain. Nature 439(7076):589–593. doi:10.1038/nature04404

Ge S, Yang CH, Hsu KS, Ming GL, Song H (2007) A critical period for enhanced synaptic plasticity in newly generated neurons of the adult brain. Neuron 54(4):559–566. doi:10.1016/j.neuron.2007.05.002

Ge S, Sailor KA, Ming GL, Song H (2008) Synaptic integration and plasticity of new neurons in the adult hippocampus. J Physiol 586(16):3759–3765. doi:10.1113/jphysiol.2008.155655

Giachino C, Barz M, Tchorz JS, Tome M, Gassmann M, Bischofberger J, Bettler B, Taylor V (2014) GABA suppresses neurogenesis in the adult hippocampus through GABAB receptors. Development 141(1):83–90. doi:10.1242/dev.102608

Goeldner C, Ballard TM, Knoflach F, Wichmann J, Gatti S, Umbricht D (2013) Cognitive impairment in major depression and the mGlu2 receptor as a therapeutic target. Neuropharmacology 64:337–346. doi:10.1016/j.neuropharm.2012.08.001

Gomez-Isla T, Price JL, McKeel DW Jr, Morris JC, Growdon JH, Hyman BT (1996) Profound loss of layer II entorhinal cortex neurons occurs in very mild Alzheimer's disease. J Neurosci 16 (14):4491–4500

Gomez-Pinilla F, Dao L, So V (1997) Physical exercise induces FGF-2 and its mRNA in the hippocampus. Brain Res 764(1–2):1–8

Gould E, McEwen BS, Tanapat P, Galea LA, Fuchs E (1997) Neurogenesis in the dentate gyrus of the adult tree shrew is regulated by psychosocial stress and NMDA receptor activation. J Neurosci 17(7):2492–2498

Gould E, Beylin A, Tanapat P, Reeves A, Shors TJ (1999) Learning enhances adult neurogenesis in the hippocampal formation. Nat Neurosci 2(3):260–265. doi:10.1038/6365

Gourley SL, Taylor JR (2009) Recapitulation and reversal of a persistent depression-like syndrome in rodents. Curr Protoc Neurosci Chapter 9:Unit 9.32. doi:10.1002/0471142301.ns0932s49. Editorial board, Jacqueline N Crawley (et al.)

Gozzi A, Herdon H, Schwarz A, Bertani S, Crestan V, Turrini G, Bifone A (2008) Pharmacological stimulation of NMDA receptors via co-agonist site suppresses fMRI response to phencyclidine in the rat. Psychopharmacology (Berl) 201(2):273–284. doi:10.1007/s00213-008-1271-z

Guiney H, Machado L (2013) Benefits of regular aerobic exercise for executive functioning in healthy populations. Psychon Bull Rev 20(1):73–86. doi:10.3758/s13423-012-0345-4

Guo W, Allan AM, Zong R, Zhang L, Johnson EB, Schaller EG, Murthy AC, Goggin SL, Eisch AJ, Oostra BA, Nelson DL, Jin P, Zhao X (2011) Ablation of Fmrp in adult neural stem cells disrupts hippocampus-dependent learning. Nat Med 17(5):559–565. doi:10.1038/nm.2336

Guo W, Murthy AC, Zhang L et al (2012) Inhibition of GSK3β improves hippocampus-dependent learning and rescues neurogenesis in a mouse model of fragile X syndrome. Hum Mol Genet 21 (3):681–691

Gupta D, Morley JE (2014) Hypothalamic-pituitary-adrenal (HPA) axis and aging. Compr Physiol 4(4):1495–1510. doi:10.1002/cphy.c130049

Halim ND, Weickert CS, McClintock BW, Weinberger DR, Lipska BK (2004) Effects of chronic haloperidol and clozapine treatment on neurogenesis in the adult rat hippocampus. Neuropsychopharmacology 29(6):1063–1069. doi:10.1038/sj.npp.1300422

Han X, Tong J, Zhang J, Farahvar A, Wang E, Yang J, Samadani U, Smith DH, Huang JH (2011) Imipramine treatment improves cognitive outcome associated with enhanced hippocampal neurogenesis after traumatic brain injury in mice. J Neurotrauma 28(6):995–1007. doi:10.1089/neu.2010.1563, PMCID: PMC3113418

Harrist A, Beech RD, King SL, Zanardi A, Cleary MA, Caldarone BJ, Eisch A, Zoli M, Picciotto MR (2004) Alteration of hippocampal cell proliferation in mice lacking the beta 2 subunit of the neuronal nicotinic acetylcholine receptor. Synapse 54(4):200–206. doi:10.1002/syn.20081

Hartmann D, de Strooper B, Serneels L, Craessaerts K, Herreman A, Annaert W, Umans L, Lubke T, Lena Illert A, von Figura K, Saftig P (2002) The disintegrin/metalloprotease ADAM 10 is essential for Notch signalling but not for alpha-secretase activity in fibroblasts. Hum Mol Genet 11(21):2615–2624

Hasselbalch BJ, Knorr U, Kessing LV (2011) Cognitive impairment in the remitted state of unipolar depressive disorder: a systematic review. J Affect Disord 134(1–3):20–31. doi:10.1016/j.jad.2010.11.011

Hastings NB, Gould E (1999) Rapid extension of axons into the CA3 region by adult-generated granule cells. J Comp Neurol 413(1):146–154

Hattiangady B, Shetty AK (2008) Aging does not alter the number or phenotype of putative stem/progenitor cells in the neurogenic region of the hippocampus. Neurobiol Aging 29(1):129–147. doi:10.1016/j.neurobiolaging.2006.09.015

Hattiangady B, Rao MS, Shetty GA, Shetty AK (2005) Brain-derived neurotrophic factor, phosphorylated cyclic AMP response element binding protein and neuropeptide Y decline as early as middle age in the dentate gyrus and CA1 and CA3 subfields of the hippocampus. Exp Neurol 195(2):353–371. doi:10.1016/j.expneurol.2005.05.014

Head D, Isom M (2010) Age effects on wayfinding and route learning skills. Behav Brain Res 209 (1):49–58. doi:10.1016/j.bbr.2010.01.012

Herrera-Guzman I, Gudayol-Ferre E, Herrera-Guzman D, Guardia-Olmos J, Hinojosa-Calvo E, Herrera-Abarca JE (2009) Effects of selective serotonin reuptake and dual serotonergic-noradrenergic reuptake treatments on memory and mental processing speed in patients with major depressive disorder. J Psychiatr Res 43(9):855–863. doi:10.1016/j.jpsychires.2008.10.015

Herrera-Guzman I, Gudayol-Ferre E, Herrera-Abarca JE, Herrera-Guzman D, Montelongo-Pedraza P, Padros Blazquez F, Pero-Cebollero M, Guardia-Olmos J (2010) Major Depressive Disorder in recovery and neuropsychological functioning: effects of selective serotonin reuptake inhibitor and dual inhibitor depression treatments on residual cognitive deficits in patients with Major Depressive Disorder in recovery. J Affect Disord 123(1–3):341–350. doi:10.1016/j.jad.2009.10.009

Herting MM, Nagel BJ (2012) Aerobic fitness relates to learning on a virtual Morris Water Task and hippocampal volume in adolescents. Behav Brain Res 233(2):517–525. doi:10.1016/j.bbr.2012.05.012

Heyn P, Abreu BC, Ottenbacher KJ (2004) The effects of exercise training on elderly persons with cognitive impairment and dementia: a meta-analysis. Arch Phys Med Rehabil 85 (10):1694–1704

Hillman CH, Erickson KI, Kramer AF (2008) Be smart, exercise your heart: exercise effects on brain and cognition. Nat Rev Neurosci 9(1):58–65. doi:10.1038/nrn2298

Hoch JC, Maciejewski MW, Gryk MR (2008) Comment on "Magnetic resonance spectroscopy identifies neural progenitor cells in the live human brain". Science 321(5889):640. doi:10.1126/science.1153058, author reply 640

Hoglinger GU, Rizk P, Muriel MP, Duyckaerts C, Oertel WH, Caille I, Hirsch EC (2004) Dopamine depletion impairs precursor cell proliferation in Parkinson disease. Nat Neurosci 7(7):726–735. doi:10.1038/nn1265

Hou JG, Xue JJ, Lee MR, Sun MQ, Zhao XH, Zheng YN, Sung CK (2013) Compound K is able to ameliorate the impaired cognitive function and hippocampal neurogenesis following chemotherapy treatment. Biochem Biophys Res Commun 436(1):104–109. doi:10.1016/j.bbrc.2013.05.087

Iaria G, Palermo L, Committeri G, Barton JJ (2009) Age differences in the formation and use of cognitive maps. Behav Brain Res 196(2):187–191. doi:10.1016/j.bbr.2008.08.040

Irvine GI, Logan B, Eckert M, Abraham WC (2006) Enriched environment exposure regulates excitability, synaptic transmission, and LTP in the dentate gyrus of freely moving rats. Hippocampus 16(2):149–160. doi:10.1002/hipo.20142

Itou Y, Nochi R, Kuribayashi H, Saito Y, Hisatsune T (2011) Cholinergic activation of hippocampal neural stem cells in aged dentate gyrus. Hippocampus 21(4):446–459. doi:10.1002/hipo.20761

Jaeger J, Berns S, Uzelac S, Davis-Conway S (2006) Neurocognitive deficits and disability in major depressive disorder. Psychiatry Res 145(1):39–48. doi:10.1016/j.psychres.2005.11.011

Jagasia R, Steib K, Englberger E, Herold S, Faus-Kessler T, Saxe M, Gage FH, Song H, Lie DC (2009) GABA-cAMP response element-binding protein signaling regulates maturation and survival of newly generated neurons in the adult hippocampus. J Neurosci 29(25):7966–7977. doi:10.1523/JNEUROSCI.1054-09.2009

Jang MH, Bonaguidi MA, Kitabatake Y, Sun J, Song J, Kang E, Jun H, Zhong C, Su Y, Guo JU, Wang MX, Sailor KA, Kim JY, Gao Y, Christian KM, Ming GL, Song H (2013a) Secreted frizzled-related protein 3 regulates activity-dependent adult hippocampal neurogenesis. Cell Stem Cell 12(2):215–223. doi:10.1016/j.stem.2012.11.021

Jang MH, Kitabatake Y, Kang E, Jun H, Pletnikov MV, Christian KM, Hen R, Lucae S, Binder EB, Song H, Ming GI (2013b) Secreted frizzled-related protein 3 (sFRP3) regulates antidepressant responses in mice and humans. Mol Psychiatry 18(9):957–958. doi:10.1038/mp.2012.158

Jansen JF, Gearhart JD, Bulte JW (2008) Comment on "Magnetic resonance spectroscopy identifies neural progenitor cells in the live human brain". Science 321(5889):640. doi:10.1126/science.1153997, author reply 640

Jedynak P, Kos T, Sandi C, Kaczmarek L, Filipkowski RK (2014) Mice with ablated adult brain neurogenesis are not impaired in antidepressant response to chronic fluoxetine. J Psychiatr Res 56:106–111. doi:10.1016/j.jpsychires.2014.05.009

Jessberger S, Clark RE, Broadbent NJ, Clemenson GD Jr, Consiglio A, Lie DC, Squire LR, Gage FH (2009) Dentate gyrus-specific knockdown of adult neurogenesis impairs spatial and object recognition memory in adult rats. Learn Mem 16(2):147–154. doi:10.1101/lm.1172609

Jin K, Peel AL, Mao XO, Xie L, Cottrell BA, Henshall DC, Greenberg DA (2004) Increased hippocampal neurogenesis in Alzheimer's disease. Proc Natl Acad Sci U S A 101(1):343–347. doi:10.1073/pnas.2634794100

Jinno S (2011) Decline in adult neurogenesis during aging follows a topographic pattern in the mouse hippocampus. J Comp Neurol 519(3):451–466. doi:10.1002/cne.22527

Kalani MY, Cheshier SH, Cord BJ, Bababeygy SR, Vogel H, Weissman IL, Palmer TD, Nusse R (2008) Wnt-mediated self-renewal of neural stem/progenitor cells. Proc Natl Acad Sci U S A 105(44):16970–16975. doi:10.1073/pnas.0808616105

Kaneko N, Okano H, Sawamoto K (2006) Role of the cholinergic system in regulating survival of newborn neurons in the adult mouse dentate gyrus and olfactory bulb. Genes Cells 11 (10):1145–1159. doi:10.1111/j.1365-2443.2006.01010.x

Kang HJ, Voleti B, Hajszan T, Rajkowska G, Stockmeier CA, Licznerski P, Lepack A, Majik MS, Jeong LS, Banasr M, Son H, Duman RS (2012) Decreased expression of synapse-related genes and loss of synapses in major depressive disorder. Nat Med 18(9):1413–1417. doi:10.1038/nm.2886

Kaplan MS, Hinds JW (1977) Neurogenesis in the adult rat: electron microscopic analysis of light radioautographs. Science 197(4308):1092–1094

Karten YJ, Jones MA, Jeurling SI, Cameron HA (2006) GABAergic signaling in young granule cells in the adult rat and mouse dentate gyrus. Hippocampus 16(3):312–320. doi:10.1002/hipo.20165

Katsimpardi L, Litterman NK, Schein PA, Miller CM, Loffredo FS, Wojtkiewicz GR, Chen JW, Lee RT, Wagers AJ, Rubin LL (2014) Vascular and neurogenic rejuvenation of the aging mouse brain by young systemic factors. Science 344(6184):630–634. doi:10.1126/science.1251141

Kee N, Teixeira CM, Wang AH, Frankland PW (2007) Preferential incorporation of adult-generated granule cells into spatial memory networks in the dentate gyrus. Nat Neurosci 10 (3):355–362. doi:10.1038/nn1847

Keilhoff G, Grecksch G, Bernstein HG, Roskoden T, Becker A (2010) Risperidone and haloperidol promote survival of stem cells in the rat hippocampus. Eur Arch Psychiatry Clin Neurosci 260(2):151–162. doi:10.1007/s00406-009-0033-1

Kempermann G, Kuhn HG, Gage FH (1997) More hippocampal neurons in adult mice living in an enriched environment. Nature 386(6624):493–495. doi:10.1038/386493a0

Kesner RP (2013) An analysis of the dentate gyrus function. Behav Brain Res 254:1–7. doi:10.1016/j.bbr.2013.01.012

Kessler RM, Votaw JR, de Paulis T, Bingham DR, Ansari MS, Mason NS, Holburn G, Schmidt DE, Votaw DB, Manning RG et al (1993) Evaluation of 5-[18 F]fluoropropylepidepride as a potential PET radioligand for imaging dopamine D2 receptors. Synapse 15(3):169–176. doi:10.1002/syn.890150302

Kheirbek MA, Drew LJ, Burghardt NS, Costantini DO, Tannenholz L, Ahmari SE, Zeng H, Fenton AA, Hen R (2013) Differential control of learning and anxiety along the dorsoventral axis of the dentate gyrus. Neuron 77(5):955–968. doi:10.1016/j.neuron.2012.12.038

Kim JY, Duan X, Liu CY, Jang MH, Guo JU, Pow-anpongkul N, Kang E, Song H, Ming GL (2009) DISC1 regulates new neuron development in the adult brain via modulation of AKT-mTOR signaling through KIAA1212. Neuron 63(6):761–773. doi:10.1016/j.neuron.2009.08.008

Kim JY, Liu CY, Zhang F, Duan X, Wen Z, Song J, Feighery E, Lu B, Rujescu D, St Clair D, Christian K, Callicott JH, Weinberger DR, Song H, Ming GL (2012) Interplay between DISC1 and GABA signaling regulates neurogenesis in mice and risk for schizophrenia. Cell 148 (5):1051–1064. doi:10.1016/j.cell.2011.12.037

Kitamura T, Mishina M, Sugiyama H (2003) Enhancement of neurogenesis by running wheel exercises is suppressed in mice lacking NMDA receptor epsilon 1 subunit. Neurosci Res 47 (1):55–63

Klempin F, Babu H, De Pietri Tonelli D, Alarcon E, Fabel K, Kempermann G (2010) Oppositional effects of serotonin receptors 5-HT1a, 2, and 2c in the regulation of adult hippocampal neurogenesis. Front Mol Neurosci 3:14. doi:10.3389/fnmol.2010.00014

Knolhoff AM, Nautiyal KM, Nemes P, Kalachikov S, Morozova I, Silver R, Sweedler JV (2013) Combining small-volume metabolomic and transcriptomic approaches for assessing brain chemistry. Anal Chem 85(6):3136–3143. doi:10.1021/ac3032959

Knoth R, Singec I, Ditter M, Pantazis G, Capetian P, Meyer RP, Horvat V, Volk B, Kempermann G (2010) Murine features of neurogenesis in the human hippocampus across the lifespan from 0 to 100 years. PLoS One 5(1):e8809. doi:10.1371/journal.pone.0008809

Kotani S, Yamauchi T, Teramoto T, Ogura H (2006) Pharmacological evidence of cholinergic involvement in adult hippocampal neurogenesis in rats. Neuroscience 142(2):505–514. doi:10.1016/j.neuroscience.2006.06.035

Kramer AF, Hahn S, Cohen NJ, Banich MT, McAuley E, Harrison CR, Chason J, Vakil E, Bardell L, Boileau RA, Colcombe A (1999) Ageing, fitness and neurocognitive function. Nature 400(6743):418–419. doi:10.1038/22682

Kremer T, Jagasia R, Herrmann A, Matile H, Borroni E, Francis F, Kuhn HG, Czech C (2013) Analysis of adult neurogenesis: evidence for a prominent "non-neurogenic" DCX-protein pool in rodent brain. PLoS One 8(5):e59269. doi:10.1371/journal.pone.0059269

Krezymon A, Richetin K, Halley H, Roybon L, Lassalle JM, Frances B, Verret L, Rampon C (2013) Modifications of hippocampal circuits and early disruption of adult neurogenesis in the tg2576 mouse model of Alzheimer's disease. PLoS One 8(9):e76497. doi:10.1371/journal.pone.0076497

Kriegstein A, Alvarez-Buylla A (2009) The glial nature of embryonic and adult neural stem cells. Annu Rev Neurosci 32:149–184. doi:10.1146/annurev.neuro.051508.135600

Kronenberg G, Reuter K, Steiner B, Brandt MD, Jessberger S, Yamaguchi M, Kempermann G (2003) Subpopulations of proliferating cells of the adult hippocampus respond differently to physiologic neurogenic stimuli. J Comp Neurol 467(4):455–463. doi:10.1002/cne.10945

Kronenberg G, Bick-Sander A, Bunk E, Wolf C, Ehninger D, Kempermann G (2006) Physical exercise prevents age-related decline in precursor cell activity in the mouse dentate gyrus. Neurobiol Aging 27(10):1505–1513. doi:10.1016/j.neurobiolaging.2005.09.016

Kuhn HG, Dickinson-Anson H, Gage FH (1996) Neurogenesis in the dentate gyrus of the adult rat: age-related decrease of neuronal progenitor proliferation. J Neurosci 16(6):2027–2033

Kuny S, Stassen HH (1995) Cognitive performance in patients recovering from depression. Psychopathology 28(4):190–207

Laske C, Stellos K, Stransky E, Seizer P, Akcay O, Eschweiler GW, Leyhe T, Gawaz M (2008) Decreased plasma and cerebrospinal fluid levels of stem cell factor in patients with early Alzheimer's disease. J Alzheimers Dis 15(3):451–460

Lassalle JM, Bataille T, Halley H (2000) Reversible inactivation of the hippocampal mossy fiber synapses in mice impairs spatial learning, but neither consolidation nor memory retrieval, in the Morris navigation task. Neurobiol Learn Mem 73(3):243–257. doi:10.1006/nlme.1999.3931

LaVoie MJ, Selkoe DJ (2003) The Notch ligands, Jagged and Delta, are sequentially processed by alpha-secretase and presenilin/gamma-secretase and release signaling fragments. J Biol Chem 278(36):34427–34437. doi:10.1074/jbc.M302659200

Lazarov O, Demars MP (2012) All in the family: how the APPs regulate neurogenesis. Front Neurosci 6:81. doi:10.3389/fnins.2012.00081

Lazarov O, Marr RA (2010) Neurogenesis and Alzheimer's disease: at the crossroads. Exp Neurol 223(2):267–281. doi:10.1016/j.expneurol.2009.08.009

Lazarov O, Marr RA (2013) Of mice and men: neurogenesis, cognition and Alzheimer's disease. Front Aging Neurosci 5:43. doi:10.3389/fnagi.2013.00043

Lazarov O, Robinson J, Tang YP, Hairston IS, Korade-Mirnics Z, Lee VM, Hersh LB, Sapolsky RM, Mirnics K, Sisodia SS (2005) Environmental enrichment reduces Abeta levels and amyloid deposition in transgenic mice. Cell 120(5):701–713. doi:10.1016/j.cell.2005.01.015

Leal SL, Tighe SK, Jones CK, Yassa MA (2014) Pattern separation of emotional information in hippocampal dentate and CA3. Hippocampus. doi:10.1002/hipo.22298

Lee I, Kesner RP (2004) Encoding versus retrieval of spatial memory: double dissociation between the dentate gyrus and the perforant path inputs into CA3 in the dorsal hippocampus. Hippocampus 14(1):66–76. doi:10.1002/hipo.10167

Lee KJ, Park SB, Lee I (2014) Elemental or contextual? It depends: individual difference in the hippocampal dependence of associative learning for a simple sensory stimulus. Front Behav Neurosci 8:217. doi:10.3389/fnbeh.2014.00217

Lemaire V, Tronel S, Montaron MF, Fabre A, Dugast E, Abrous DN (2012) Long-lasting plasticity of hippocampal adult-born neurons. J Neurosci 32(9):3101–3108. doi:10.1523/JNEUROSCI.4731-11.2012

Li Y, Luikart BW, Birnbaum S, Chen J, Kwon CH, Kernie SG, Bassel-Duby R, Parada LF (2008) TrkB regulates hippocampal neurogenesis and governs sensitivity to antidepressive treatment. Neuron 59(3):399–412. doi:10.1016/j.neuron.2008.06.023

Li G, Bien-Ly N, Andrews-Zwilling Y, Xu Q, Bernardo A, Ring K, Halabisky B, Deng C, Mahley RW, Huang Y (2009) GABAergic interneuron dysfunction impairs hippocampal neurogenesis in adult apolipoprotein E4 knockin mice. Cell Stem Cell 5(6):634–645. doi:10.1016/j.stem.2009.10.015

Li YJ, Xu M, Gao ZH, Wang YQ, Yue Z, Zhang YX, Li XX, Zhang C, Xie SY, Wang PY (2013) Alterations of serum levels of BDNF-related miRNAs in patients with depression. PLoS One 8 (5):e63648. doi:10.1371/journal.pone.0063648

Lie DC, Colamarino SA, Song HJ, Desire L, Mira H, Consiglio A, Lein ES, Jessberger S, Lansford H, Dearie AR, Gage FH (2005) Wnt signalling regulates adult hippocampal neurogenesis. Nature 437(7063):1370–1375. doi:10.1038/nature04108

Lim DA, Suarez-Farinas M, Naef F, Hacker CR, Menn B, Takebayashi H, Magnasco M, Patil N, Alvarez-Buylla A (2006) In vivo transcriptional profile analysis reveals RNA splicing and chromatin remodeling as prominent processes for adult neurogenesis. Mol Cell Neurosci 31 (1):131–148. doi:10.1016/j.mcn.2005.10.005

Littlewood CL, Jones N, O'Neill MJ, Mitchell SN, Tricklebank M, Williams SC (2006) Mapping the central effects of ketamine in the rat using pharmacological MRI. Psychopharmacology (Berl) 186(1):64–81. doi:10.1007/s00213-006-0344-0

Liu Y, Lacson R, Cassaday J, Ross DA, Kreamer A, Hudak E, Peltier R, McLaren D, Munoz-Sanjuan I, Santini F, Strulovici B, Ferrer M (2009) Identification of small-molecule modulators of mouse SVZ progenitor cell proliferation and differentiation through high-throughput screening. J Biomol Screen 14(4):319–329. doi:10.1177/1087057109332596

Lois C, Alvarez-Buylla A (1994) Long-distance neuronal migration in the adult mammalian brain. Science 264(5162):1145–1148

Louissaint A Jr, Rao S, Leventhal C, Goldman SA (2002) Coordinated interaction of neurogenesis and angiogenesis in the adult songbird brain. Neuron 34(6):945–960

Lucassen PJ, Stumpel MW, Wang Q, Aronica E (2010) Decreased numbers of progenitor cells but no response to antidepressant drugs in the hippocampus of elderly depressed patients. Neuropharmacology 58(6):940–949. doi:10.1016/j.neuropharm.2010.01.012

Lugert S, Basak O, Knuckles P, Haussler U, Fabel K, Gotz M, Haas CA, Kempermann G, Taylor V, Giachino C (2010) Quiescent and active hippocampal neural stem cells with distinct morphologies respond selectively to physiological and pathological stimuli and aging. Cell Stem Cell 6(5):445–456. doi:10.1016/j.stem.2010.03.017

Ma W, Maric D, Li BS, Hu Q, Andreadis JD, Grant GM, Liu QY, Shaffer KM, Chang YH, Zhang L, Pancrazio JJ, Pant HC, Stenger DA, Barker JL (2000) Acetylcholine stimulates cortical precursor cell proliferation in vitro via muscarinic receptor activation and MAP kinase phosphorylation. Eur J Neurosci 12(4):1227–1240

Ma DK, Kim WR, Ming GL, Song H (2009) Activity-dependent extrinsic regulation of adult olfactory bulb and hippocampal neurogenesis. Ann N Y Acad Sci 1170:664–673. doi:10.1111/j.1749-6632.2009.04373.x

MacMaster FP, Carrey N, Langevin LM, Jaworska N, Crawford S (2014) Disorder-specific volumetric brain difference in adolescent major depressive disorder and bipolar depression. Brain Imaging Behav 8(1):119–127. doi:10.1007/s11682-013-9264-x

MacMillan KS, Naidoo J, Liang J, Melito L, Williams NS, Morlock L, Huntington PJ, Estill SJ, Longgood J, Becker GL, McKnight SL, Pieper AA, De Brabander JK, Ready JM (2011) Development of proneurogenic, neuroprotective small molecules. J Am Chem Soc 133(5):1428–1437. doi:10.1021/ja108211m

Maguire EA, Gadian DG, Johnsrude IS, Good CD, Ashburner J, Frackowiak RS, Frith CD (2000) Navigation-related structural change in the hippocampi of taxi drivers. Proc Natl Acad Sci U S A 97(8):4398–4403. doi:10.1073/pnas.070039597

Malberg JE, Eisch AJ, Nestler EJ, Duman RS (2000) Chronic antidepressant treatment increases neurogenesis in adult rat hippocampus. J Neurosci 20(24):9104–9110

Maletic-Savatic M, Vingara LK, Manganas LN, Li Y, Zhang S, Sierra A, Hazel R, Smith D, Wagshul ME, Henn F, Krupp L, Enikolopov G, Benveniste H, Djuric PM, Pelczer I (2008) Metabolomics of neural progenitor cells: a novel approach to biomarker discovery. Cold Spring Harb Symp Quant Biol 73:389–401. doi:10.1101/sqb.2008.73.021

Mallei A, Shi B, Mocchetti I (2002) Antidepressant treatments induce the expression of basic fibroblast growth factor in cortical and hippocampal neurons. Mol Pharmacol 61(5):1017–1024

Manganas LN, Zhang X, Li Y, Hazel RD, Smith SD, Wagshul ME, Henn F, Benveniste H, Djuric PM, Enikolopov G, Maletic-Savatic M (2007) Magnetic resonance spectroscopy identifies neural progenitor cells in the live human brain. Science 318(5852):980–985. doi:10.1126/science.1147851

Mao Y, Ge X, Frank CL, Madison JM, Koehler AN, Doud MK, Tassa C, Berry EM, Soda T, Singh KK, Biechele T, Petryshen TL, Moon RT, Haggarty SJ, Tsai LH (2009) Disrupted in schizophrenia 1 regulates neuronal progenitor proliferation via modulation of GSK3beta/beta-catenin signaling. Cell 136(6):1017–1031. doi:10.1016/j.cell.2008.12.044

Markwardt S, Overstreet-Wadiche L (2008) GABAergic signalling to adult-generated neurons. J Physiol 586(16):3745–3749. doi:10.1113/jphysiol.2008.155713

Martínez-Cué C, Martínez P, Rueda N, Vidal R, García S, Vidal V, Corrales A, Montero JA, Pazos Á, Flórez J, Gasser R, Thomas AW, Honer M, Knoflach F, Trejo JL, Wettstein JG, Hernández MC (2013) Reducing GABAA α5 receptor-mediated inhibition rescues functional and neuromorphological deficits in a mouse model of down syndrome. J Neurosci 33(9):3953–3966

McCall WV, Dunn AG (2003) Cognitive deficits are associated with functional impairment in severely depressed patients. Psychiatry Res 121(2):179–184

McDonald HY, Wojtowicz JM (2005) Dynamics of neurogenesis in the dentate gyrus of adult rats. Neurosci Lett 385(1):70–75. doi:10.1016/j.neulet.2005.05.022

McEwen BS (1998) Stress, adaptation, and disease. Allostasis and allostatic load. Ann N Y Acad Sci 840:33–44

Mendez-David I, David DJ, Darcet F, Wu MV, Kerdine-Romer S, Gardier AM, Hen R (2014) Rapid anxiolytic effects of a 5-HT(4) receptor agonist are mediated by a neurogenesis-independent mechanism. Neuropsychopharmacology 39(6):1366–1378. doi:10.1038/npp.2013.332

Merrill DA, Karim R, Darraq M, Chiba AA, Tuszynski MH (2003) Hippocampal cell genesis does not correlate with spatial learning ability in aged rats. J Comp Neurol 459(2):201–207. doi:10.1002/cne.10616

Meshi D, Drew MR, Saxe M, Ansorge MS, David D, Santarelli L, Malapani C, Moore H, Hen R (2006) Hippocampal neurogenesis is not required for behavioral effects of environmental enrichment. Nat Neurosci 9(6):729–731. doi:10.1038/nn1696

Miller JA, Nathanson J, Franjic D, Shim S, Dalley RA, Shapouri S, Smith KA, Sunkin SM, Bernard A, Bennett JL, Lee CK, Hawrylycz MJ, Jones AR, Amaral DG, Sestan N, Gage FH, Lein ES (2013) Conserved molecular signatures of neurogenesis in the hippocampal subgranular zone of rodents and primates. Development 140(22):4633–4644. doi:10.1242/dev.097212

Miranda CJ, Braun L, Jiang Y, Hester ME, Zhang L, Riolo M, Wang H, Rao M, Altura RA, Kaspar BK (2012) Aging brain microenvironment decreases hippocampal neurogenesis through Wnt-mediated survivin signaling. Aging Cell 11(3):542–552. doi:10.1111/j.1474-9726.2012.00816.x

Moffat SD, Elkins W, Resnick SM (2006) Age differences in the neural systems supporting human allocentric spatial navigation. Neurobiol Aging 27(7):965–972. doi:10.1016/j.neurobiolaging.2005.05.011

Mohapel P, Leanza G, Kokaia M, Lindvall O (2005) Forebrain acetylcholine regulates adult hippocampal neurogenesis and learning. Neurobiol Aging 26(6):939–946. doi:10.1016/j.neurobiolaging.2004.07.015

Molteni R, Ying Z, Gomez-Pinilla F (2002) Differential effects of acute and chronic exercise on plasticity-related genes in the rat hippocampus revealed by microarray. Eur J Neurosci 16 (6):1107–1116

Monteggia LM, Barrot M, Powell CM, Berton O, Galanis V, Gemelli T, Meuth S, Nagy A, Greene RW, Nestler EJ (2004) Essential role of brain-derived neurotrophic factor in adult hippocampal function. Proc Natl Acad Sci U S A 101(29):10827–10832. doi:10.1073/pnas.0402141101

Monti JM, Hillman CH, Cohen NJ (2012) Aerobic fitness enhances relational memory in preadolescent children: the FITKids randomized control trial. Hippocampus 22(9):1876–1882. doi:10.1002/hipo.22023

Moore RD, Drollette ES, Scudder MR, Bharij A, Hillman CH (2014) The influence of cardiorespiratory fitness on strategic, behavioral, and electrophysiological indices of arithmetic cognition in preadolescent children. Front Hum Neurosci 8:258. doi:10.3389/fnhum.2014.00258

Morris AM, Weeden CS, Churchwell JC, Kesner RP (2013) The role of the dentate gyrus in the formation of contextual representations. Hippocampus 23(2):162–168. doi:10.1002/hipo.22078

Mu Y, Gage FH (2011) Adult hippocampal neurogenesis and its role in Alzheimer's disease. Mol Neurodegener 6:85. doi:10.1186/1750-1326-6-85

Nacher J, McEwen BS (2006) The role of N-methyl-D-asparate receptors in neurogenesis. Hippocampus 16(3):267–270. doi:10.1002/hipo.20160

Nacher J, Rosell DR, Alonso-Llosa G, McEwen BS (2001) NMDA receptor antagonist treatment induces a long-lasting increase in the number of proliferating cells, PSA-NCAM-immunoreactive granule neurons and radial glia in the adult rat dentate gyrus. Eur J Neurosci 13 (3):512–520

Neeper SA, Gomez-Pinilla F, Choi J, Cotman CW (1996) Physical activity increases mRNA for brain-derived neurotrophic factor and nerve growth factor in rat brain. Brain Res 726 (1–2):49–56

Neu P, Bajbouj M, Schilling A, Godemann F, Berman RM, Schlattmann P (2005) Cognitive function over the treatment course of depression in middle-aged patients: correlation with brain

MRI signal hyperintensities. J Psychiatr Res 39(2):129–135. doi:10.1016/j.jpsychires.2004.06.004

Nichols NR, Zieba M, Bye N (2001) Do glucocorticoids contribute to brain aging? Brain Res Brain Res Rev 37(1–3):273–286

O'Callaghan RM, Ohle R, Kelly AM (2007) The effects of forced exercise on hippocampal plasticity in the rat: a comparison of LTP, spatial- and non-spatial learning. Behav Brain Res 176(2):362–366. doi:10.1016/j.bbr.2006.10.018

Olariu A, Cleaver KM, Cameron HA (2007) Decreased neurogenesis in aged rats results from loss of granule cell precursors without lengthening of the cell cycle. J Comp Neurol 501 (4):659–667. doi:10.1002/cne.21268

Pallotto M, Deprez F (2014) Regulation of adult neurogenesis by GABAergic transmission: signaling beyond GABAA-receptors. Front Cell Neurosci 8:166. doi:10.3389/fncel.2014.00166

Palmer TD, Willhoite AR, Gage FH (2000) Vascular niche for adult hippocampal neurogenesis. J Comp Neurol 425(4):479–494

Pan YW, Chan GC, Kuo CT, Storm DR, Xia Z (2012a) Inhibition of adult neurogenesis by inducible and targeted deletion of ERK5 mitogen-activated protein kinase specifically in adult neurogenic regions impairs contextual fear extinction and remote fear memory. J Neurosci 32(19):6444–6455. doi:10.1523/JNEUROSCI.6076-11.2012

Pan YW, Zou J, Wang W, Sakagami H, Garelick MG, Abel G, Kuo CT, Storm DR, Xia Z (2012b) Inducible and conditional deletion of extracellular signal-regulated kinase 5 disrupts adult hippocampal neurogenesis. J Biol Chem 287(28):23306–23317. doi:10.1074/jbc.M112.344762

Pandey GN, Ren X, Rizavi HS, Conley RR, Roberts RC, Dwivedi Y (2008) Brain-derived neurotrophic factor and tyrosine kinase B receptor signalling in post-mortem brain of teenage suicide victims. Int J Neuropsychopharmacol 11(8):1047–1061. doi:10.1017/S1461145708009000

Pathania M, Yan LD, Bordey A (2010) A symphony of signals conducts early and late stages of adult neurogenesis. Neuropharmacology 58(6):865–876. doi:10.1016/j.neuropharm.2010.01.010

Pereira AC, Huddleston DE, Brickman AM, Sosunov AA, Hen R, McKhann GM, Sloan R, Gage FH, Brown TR, Small SA (2007) An in vivo correlate of exercise-induced neurogenesis in the adult dentate gyrus. Proc Natl Acad Sci U S A 104(13):5638–5643. doi:10.1073/pnas.0611721104

Perry EK, Johnson M, Ekonomou A, Perry RH, Ballard C, Attems J (2012) Neurogenic abnormalities in Alzheimer's disease differ between stages of neurogenesis and are partly related to cholinergic pathology. Neurobiol Dis 47(2):155–162. doi:10.1016/j.nbd.2012.03.033

Petrik D, Jiang Y, Birnbaum SG, Powell CM, Kim MS, Hsieh J, Eisch AJ (2012) Functional and mechanistic exploration of an adult neurogenesis-promoting small molecule. FASEB J 26 (8):3148–3162. doi:10.1096/fj.11-201426

Pieper AA, Xie S, Capota E, Estill SJ, Zhong J, Long JM, Becker GL, Huntington P, Goldman SE, Shen CH, Capota M, Britt JK, Kotti T, Ure K, Brat DJ, Williams NS, MacMillan KS, Naidoo J, Melito L, Hsieh J, De Brabander J, Ready JM, McKnight SL (2010) Discovery of a proneurogenic, neuroprotective chemical. Cell 142(1):39–51. doi:10.1016/j.cell.2010.06.018

Price RB, Nock MK, Charney DS, Mathew SJ (2009) Effects of intravenous ketamine on explicit and implicit measures of suicidality in treatment-resistant depression. Biol Psychiatry 66 (5):522–526. doi:10.1016/j.biopsych.2009.04.029

Ramirez-Amaya V, Marrone DF, Gage FH, Worley PF, Barnes CA (2006) Integration of new neurons into functional neural networks. J Neurosci 26(47):12237–12241. doi:10.1523/JNEUROSCI.2195-06.2006

Ramm P, Couillard-Despres S, Plotz S, Rivera FJ, Krampert M, Lehner B, Kremer W, Bogdahn U, Kalbitzer HR, Aigner L (2009) A nuclear magnetic resonance biomarker for neural progenitor cells: is it all neurogenesis? Stem Cells 27(2):420–423. doi:10.1634/stemcells. 2008-0816

Rao MS, Hattiangady B, Abdel-Rahman A, Stanley DP, Shetty AK (2005) Newly born cells in the ageing dentate gyrus display normal migration, survival and neuronal fate choice but endure retarded early maturation. Eur J Neurosci 21(2):464–476. doi:10.1111/j.1460-9568.2005.03853.x

Rapp PR, Kansky MT, Roberts JA (1997) Impaired spatial information processing in aged monkeys with preserved recognition memory. Neuroreport 8(8):1923–1928

Rapp PR, Stack EC, Gallagher M (1999) Morphometric studies of the aged hippocampus: I. Volumetric analysis in behaviorally characterized rats. J Comp Neurol 403(4):459–470

Rasmusson DD, Dudar JD (1979) Effect of scopolamine on maze learning performance in humans. Experientia 35(8):1069–1070

Raz N, Ghisletta P, Rodrigue KM, Kennedy KM, Lindenberger U (2010) Trajectories of brain aging in middle-aged and older adults: regional and individual differences. Neuroimage 51 (2):501–511. doi:10.1016/j.neuroimage.2010.03.020

Redila VA, Christie BR (2006) Exercise-induced changes in dendritic structure and complexity in the adult hippocampal dentate gyrus. Neuroscience 137(4):1299–1307. doi:10.1016/j.neuroscience.2005.10.050

Revest JM, Dupret D, Koehl M, Funk-Reiter C, Grosjean N, Piazza PV, Abrous DN (2009) Adult hippocampal neurogenesis is involved in anxiety-related behaviors. Mol Psychiatry 14 (10):959–967. doi:10.1038/mp.2009.15

Rishton GM (2008) Small molecules that promote neurogenesis in vitro. Recent Pat CNS Drug Discov 3(3):200–208

Rolls ET (1996) A theory of hippocampal function in memory. Hippocampus 6(6):601–620. doi:10.1002/(SICI)1098-1063(1996)6:6<601::AID-HIPO5>3.0.CO;2-J

Rolls ET (2007) An attractor network in the hippocampus: theory and neurophysiology. Learn Mem 14(11):714–731. doi:10.1101/lm.631207

Rossi C, Angelucci A, Costantin L, Braschi C, Mazzantini M, Babbini F, Fabbri ME, Tessarollo L, Maffei L, Berardi N, Caleo M (2006) Brain-derived neurotrophic factor (BDNF) is required for the enhancement of hippocampal neurogenesis following environmental enrichment. Eur J Neurosci 24(7):1850–1856. doi:10.1111/j.1460-9568.2006.05059.x

Rubin DC (2000) The distribution of early childhood memories. Memory 8(4):265–269. doi:10.1080/096582100406810

Rudy JW, Sutherland RJ (1995) Configural association theory and the hippocampal formation: an appraisal and reconfiguration. Hippocampus 5(5):375–389. doi:10.1002/hipo.450050502

Rueger MA, Backes H, Walberer M, Neumaier B, Ullrich R, Simard ML, Emig B, Fink GR, Hoehn M, Graf R, Schroeter M (2010) Noninvasive imaging of endogenous neural stem cell mobilization in vivo using positron emission tomography. J Neurosci 30(18):6454–6460. doi:10.1523/JNEUROSCI.6092-09.2010

Saarelainen T, Hendolin P, Lucas G, Koponen E, Sairanen M, MacDonald E, Agerman K, Haapasalo A, Nawa H, Aloyz R, Ernfors P, Castren E (2003) Activation of the TrkB neurotrophin receptor is induced by antidepressant drugs and is required for antidepressant-induced behavioral effects. J Neurosci 23(1):349–357

Sahay A, Scobie KN, Hill AS, O'Carroll CM, Kheirbek MA, Burghardt NS, Fenton AA, Dranovsky A, Hen R (2011) Increasing adult hippocampal neurogenesis is sufficient to improve pattern separation. Nature 472(7344):466–470. doi:10.1038/nature09817

Sairanen M, Lucas G, Ernfors P, Castren M, Castren E (2005) Brain-derived neurotrophic factor and antidepressant drugs have different but coordinated effects on neuronal turnover, proliferation, and survival in the adult dentate gyrus. J Neurosci 25(5):1089–1094. doi:10.1523/JNEUROSCI.3741-04.2005

Santarelli L, Saxe M, Gross C, Surget A, Battaglia F, Dulawa S, Weisstaub N, Lee J, Duman R, Arancio O, Belzung C, Hen R (2003) Requirement of hippocampal neurogenesis for the behavioral effects of antidepressants. Science 301(5634):805–809. doi:10.1126/science.1083328

Saxe MD, Battaglia F, Wang JW, Malleret G, David DJ, Monckton JE, Garcia AD, Sofroniew MV, Kandel ER, Santarelli L, Hen R, Drew MR (2006) Ablation of hippocampal neurogenesis impairs contextual fear conditioning and synaptic plasticity in the dentate gyrus. Proc Natl Acad Sci U S A 103(46):17501–17506. doi:10.1073/pnas.0607207103

Scharfman H, Goodman J, Macleod A, Phani S, Antonelli C, Croll S (2005) Increased neurogenesis and the ectopic granule cells after intrahippocampal BDNF infusion in adult rats. Exp Neurol 192(2):348–356. doi:10.1016/j.expneurol.2004.11.016

Schatzberg AF, Keller J, Tennakoon L, Lembke A, Williams G, Kraemer FB, Sarginson JE, Lazzeroni LC, Murphy GM (2014) HPA axis genetic variation, cortisol and psychosis in major depression. Mol Psychiatry 19(2):220–227. doi:10.1038/mp.2013.129

Schlosser N, Wolf OT, Fernando SC, Riedesel K, Otte C, Muhtz C, Beblo T, Driessen M, Lowe B, Wingenfeld K (2010) Effects of acute cortisol administration on autobiographical memory in patients with major depression and healthy controls. Psychoneuroendocrinology 35 (2):316–320. doi:10.1016/j.psyneuen.2009.06.015

Schneider JW, Gao Z, Li S, Farooqi M, Tang TS, Bezprozvanny I, Frantz DE, Hsieh J (2008) Small-molecule activation of neuronal cell fate. Nat Chem Biol 4(7):408–410. doi:10.1038/nchembio.95

Scoville WB, Milner B (1957) Loss of recent memory after bilateral hippocampal lesions. J Neurol Neurosurg Psychiatry 20(1):11–21

Seib DR, Corsini NS, Ellwanger K, Plaas C, Mateos A, Pitzer C, Niehrs C, Celikel T, Martin-Villalba A (2013) Loss of Dickkopf-1 restores neurogenesis in old age and counteracts cognitive decline. Cell Stem Cell 12(2):204–214. doi:10.1016/j.stem.2012.11.010

Seki T (2002) Hippocampal adult neurogenesis occurs in a microenvironment provided by PSA-NCAM-expressing immature neurons. J Neurosci Res 69(6):772–783. doi:10.1002/jnr.10366

Seki T, Arai Y (1995) Age-related production of new granule cells in the adult dentate gyrus. Neuroreport 6(18):2479–2482

Selkoe DJ (2001) Alzheimer's disease: genes, proteins, and therapy. Physiol Rev 81(2):741–766

Selkoe DJ, Wolfe MS (2007) Presenilin: running with scissors in the membrane. Cell 131 (2):215–221. doi:10.1016/j.cell.2007.10.012

Seri B, Garcia-Verdugo JM, McEwen BS, Alvarez-Buylla A (2001) Astrocytes give rise to new neurons in the adult mammalian hippocampus. J Neurosci 21(18):7153–7160

Shamy JL, Buonocore MH, Makaron LM, Amaral DG, Barnes CA, Rapp PR (2006) Hippocampal volume is preserved and fails to predict recognition memory impairment in aged rhesus monkeys (Macaca mulatta). Neurobiol Aging 27(10):1405–1415. doi:10.1016/j.neurobiolaging.2005.07.019

Sheline YI, Wang PW, Gado MH, Csernansky JG, Vannier MW (1996) Hippocampal atrophy in recurrent major depression. Proc Natl Acad Sci U S A 93(9):3908–3913

Shen J, Bronson RT, Chen DF, Xia W, Selkoe DJ, Tonegawa S (1997) Skeletal and CNS defects in Presenilin-1-deficient mice. Cell 89(4):629–639

Shetty AK, Hattiangady B, Shetty GA (2005) Stem/progenitor cell proliferation factors FGF-2, IGF-1, and VEGF exhibit early decline during the course of aging in the hippocampus: role of astrocytes. Glia 51(3):173–186. doi:10.1002/glia.20187

Shruster A, Offen D (2014) Targeting neurogenesis ameliorates danger assessment in a mouse model of Alzheimer's disease. Behav Brain Res 261:193–201. doi:10.1016/j.bbr.2013.12.028

Sierra A, Encinas JM, Deudero JJ, Chancey JH, Enikolopov G, Overstreet-Wadiche LS, Tsirka SE, Maletic-Savatic M (2010) Microglia shape adult hippocampal neurogenesis through apoptosis-coupled phagocytosis. Cell Stem Cell 7(4):483–495. doi:10.1016/j.stem.2010.08.014

Siette J, Westbrook RF, Cotman C, Sidhu K, Zhu W, Sachdev P, Valenzuela MJ (2013) Age-specific effects of voluntary exercise on memory and the older brain. Biol Psychiatry 73 (5):435–442. doi:10.1016/j.biopsych.2012.05.034

Snyder JS, Soumier A, Brewer M, Pickel J, Cameron HA (2011) Adult hippocampal neurogenesis buffers stress responses and depressive behaviour. Nature 476(7361):458–461. doi:10.1038/nature10287

Song HJ, Stevens CF, Gage FH (2002) Neural stem cells from adult hippocampus develop essential properties of functional CNS neurons. Nat Neurosci 5(5):438–445. doi:10.1038/nn844

Song J, Zhong C, Bonaguidi MA, Sun GJ, Hsu D, Gu Y, Meletis K, Huang ZJ, Ge S, Enikolopov G, Deisseroth K, Luscher B, Christian KM, Ming GL, Song H (2012) Neuronal circuitry mechanism regulating adult quiescent neural stem-cell fate decision. Nature 489 (7414):150–154. doi:10.1038/nature11306

Sousa N, Lukoyanov NV, Madeira MD, Almeida OF, Paula-Barbosa MM (2000) Reorganization of the morphology of hippocampal neurites and synapses after stress-induced damage correlates with behavioral improvement. Neuroscience 97(2):253–266

Spalding KL, Bergmann O, Alkass K, Bernard S, Salehpour M, Huttner HB, Bostrom E, Westerlund I, Vial C, Buchholz BA, Possnert G, Mash DC, Druid H, Frisen J (2013) Dynamics of hippocampal neurogenesis in adult humans. Cell 153(6):1219–1227. doi:10.1016/j.cell.2013.05.002

Squire LR (1992) Declarative and nondeclarative memory: multiple brain systems supporting learning and memory. J Cogn Neurosci 4(3):232–243. doi:10.1162/jocn.1992.4.3.232

Stranahan AM, Khalil D, Gould E (2007) Running induces widespread structural alterations in the hippocampus and entorhinal cortex. Hippocampus 17(11):1017–1022. doi:10.1002/hipo.20348

Stroth S, Hille K, Spitzer M, Reinhardt R (2009) Aerobic endurance exercise benefits memory and affect in young adults. Neuropsychol Rehabil 19(2):223–243. doi:10.1080/09602010802091183

Sullivan EV, Marsh L, Pfefferbaum A (2005) Preservation of hippocampal volume throughout adulthood in healthy men and women. Neurobiol Aging 26(7):1093–1098. doi:10.1016/j.neurobiolaging.2004.09.015

Surget A, Saxe M, Leman S, Ibarguen-Vargas Y, Chalon S, Griebel G, Hen R, Belzung C (2008) Drug-dependent requirement of hippocampal neurogenesis in a model of depression and of antidepressant reversal. Biol Psychiatry 64(4):293–301. doi:10.1016/j.biopsych.2008.02.022

Surget A, Tanti A, Leonardo ED, Laugeray A, Rainer Q, Touma C, Palme R, Griebel G, Ibarguen-Vargas Y, Hen R, Belzung C (2011) Antidepressants recruit new neurons to improve stress response regulation. Mol Psychiatry 16(12):1177–1188. doi:10.1038/mp.2011.48

Takamura N, Nakagawa S, Masuda T, Boku S, Kato A, Song N, An Y, Kitaichi Y, Inoue T, Koyama T, Kusumi I (2014) The effect of dopamine on adult hippocampal neurogenesis. Prog Neuropsychopharmacol Biol Psychiatry 50:116–124. doi:10.1016/j.pnpbp.2013.12.011

Tanapat P, Hastings NB, Rydel TA, Galea LA, Gould E (2001) Exposure to fox odor inhibits cell proliferation in the hippocampus of adult rats via an adrenal hormone-dependent mechanism. J Comp Neurol 437(4):496–504

Tannenholz L, Jimenez JC, Kheirbek MA (2014) Local and regional heterogeneity underlying hippocampal modulation of cognition and mood. Front Behav Neurosci 8:147. doi:10.3389/fnbeh.2014.00147

Tanti A, Westphal WP, Girault V, Brizard B, Devers S, Leguisquet AM, Surget A, Belzung C (2013) Region-dependent and stage-specific effects of stress, environmental enrichment, and antidepressant treatment on hippocampal neurogenesis. Hippocampus 23(9):797–811. doi:10.1002/hipo.22134

Tashiro A, Makino H, Gage FH (2007) Experience-specific functional modification of the dentate gyrus through adult neurogenesis: a critical period during an immature stage. J Neurosci 27 (12):3252–3259. doi:10.1523/JNEUROSCI.4941-06.2007

Themanson JR, Pontifex MB, Hillman CH (2008) Fitness and action monitoring: evidence for improved cognitive flexibility in young adults. Neuroscience 157(2):319–328. doi:10.1016/j.neuroscience.2008.09.014

Tizabi Y, Bhatti BH, Manaye KF, Das JR, Akinfiresoye L (2012) Antidepressant-like effects of low ketamine dose is associated with increased hippocampal AMPA/NMDA receptor density ratio in female Wistar-Kyoto rats. Neuroscience 213:72–80. doi:10.1016/j.neuroscience.2012.03.052

Toledo EM, Inestrosa NC (2010) Activation of Wnt signaling by lithium and rosiglitazone reduced spatial memory impairment and neurodegeneration in brains of an APPswe/PSEN1DeltaE9 mouse model of Alzheimer's disease. Mol Psychiatry 15(3):272–285, 228. doi:10.1038/mp.2009.72

Tong L, Shen H, Perreau VM, Balazs R, Cotman CW (2001) Effects of exercise on gene-expression profile in the rat hippocampus. Neurobiol Dis 8(6):1046–1056. doi:10.1006/nbdi.2001.0427

Toni N, Teng EM, Bushong EA, Aimone JB, Zhao C, Consiglio A, van Praag H, Martone ME, Ellisman MH, Gage FH (2007) Synapse formation on neurons born in the adult hippocampus. Nat Neurosci 10(6):727–734. doi:10.1038/nn1908

Tozuka Y, Fukuda S, Namba T, Seki T, Hisatsune T (2005) GABAergic excitation promotes neuronal differentiation in adult hippocampal progenitor cells. Neuron 47(6):803–815. doi:10.1016/j.neuron.2005.08.023

Trejo JL, Carro E, Torres-Aleman I (2001) Circulating insulin-like growth factor I mediates exercise-induced increases in the number of new neurons in the adult hippocampus. J Neurosci 21(5):1628–1634

Treves A, Tashiro A, Witter MP, Moser EI (2008) What is the mammalian dentate gyrus good for? Neuroscience 154(4):1155–1172. doi:10.1016/j.neuroscience.2008.04.073

Trivedi MH, Greer TL (2014) Cognitive dysfunction in unipolar depression: implications for treatment. J Affect Disord 152–154:19–27. doi:10.1016/j.jad.2013.09.012

Valero J, Espana J, Parra-Damas A, Martin E, Rodriguez-Alvarez J, Saura CA (2011) Short-term environmental enrichment rescues adult neurogenesis and memory deficits in APP(Sw, Ind) transgenic mice. PLoS One 6(2):e16832. doi:10.1371/journal.pone.0016832

van der Borght K, Ferrari F, Klauke K, Roman V, Havekes R, Sgoifo A, van der Zee EA, Meerlo P (2006) Hippocampal cell proliferation across the day: increase by running wheel activity, but no effect of sleep and wakefulness. Behav Brain Res 167(1):36–41. doi:10.1016/j.bbr.2005.08.012

Van Hoesen GW, Hyman BT, Damasio AR (1991) Entorhinal cortex pathology in Alzheimer's disease. Hippocampus 1(1):1–8. doi:10.1002/hipo.450010102

Van Petten C (2004) Relationship between hippocampal volume and memory ability in healthy individuals across the lifespan: review and meta-analysis. Neuropsychologia 42 (10):1394–1413. doi:10.1016/j.neuropsychologia.2004.04.006

van Praag H, Christie BR, Sejnowski TJ, Gage FH (1999a) Running enhances neurogenesis, learning, and long-term potentiation in mice. Proc Natl Acad Sci U S A 96(23):13427–13431

van Praag H, Kempermann G, Gage FH (1999b) Running increases cell proliferation and neurogenesis in the adult mouse dentate gyrus. Nat Neurosci 2(3):266–270. doi:10.1038/6368

van Praag H, Shubert T, Zhao C, Gage FH (2005) Exercise enhances learning and hippocampal neurogenesis in aged mice. J Neurosci 25(38):8680–8685. doi:10.1523/JNEUROSCI.1731-05.2005

Varela-Nallar L, Inestrosa NC (2013) Wnt signaling in the regulation of adult hippocampal neurogenesis. Front Cell Neurosci 7:100. doi:10.3389/fncel.2013.00100

Veena J, Rao BS, Srikumar BN (2011a) Regulation of adult neurogenesis in the hippocampus by stress, acetylcholine and dopamine. J Nat Sci Biol Med 2(1):26–37. doi:10.4103/0976-9668.82312

Veena J, Srikumar BN, Mahati K, Raju TR, Shankaranarayana Rao BS (2011b) Oxotremorine treatment restores hippocampal neurogenesis and ameliorates depression-like behaviour in chronically stressed rats. Psychopharmacology (Berl) 217(2):239–253. doi:10.1007/s00213-011-2279-3

Verret L, Jankowsky JL, Xu GM, Borchelt DR, Rampon C (2007) Alzheimer's-type amyloidosis in transgenic mice impairs survival of newborn neurons derived from adult hippocampal neurogenesis. J Neurosci 27(25):6771–6780. doi:10.1523/JNEUROSCI.5564-06.2007

Villeda SA, Luo J, Mosher KI, Zou B, Britschgi M, Bieri G, Stan TM, Fainberg N, Ding Z, Eggel A, Lucin KM, Czirr E, Park JS, Couillard-Despres S, Aigner L, Li G, Peskind ER, Kaye JA, Quinn JF, Galasko DR, Xie XS, Rando TA, Wyss-Coray T (2011) The ageing systemic milieu negatively regulates neurogenesis and cognitive function. Nature 477(7362):90–94. doi:10.1038/nature10357

Villeda SA, Plambeck KE, Middeldorp J, Castellano JM, Mosher KI, Luo J, Smith LK, Bieri G, Lin K, Berdnik D, Wabl R, Udeochu J, Wheatley EG, Zou B, Simmons DA, Xie XS, Longo FM, Wyss-Coray T (2014) Young blood reverses age-related impairments in cognitive function and synaptic plasticity in mice. Nat Med 20(6):659–663. doi:10.1038/nm.3569

Vivar C, Potter MC, Choi J, Lee JY, Stringer TP, Callaway EM, Gage FH, Suh H, van Praag H (2012) Monosynaptic inputs to new neurons in the dentate gyrus. Nat Commun 3:1107. doi:10.1038/ncomms2101

Voss MW, Vivar C, Kramer AF, van Praag H (2013) Bridging animal and human models of exercise-induced brain plasticity. Trends Cogn Sci 17(10):525–544. doi:10.1016/j.tics.2013.08.001

Wagner S, Doering B, Helmreich I, Lieb K, Tadic A (2012) A meta-analysis of executive dysfunctions in unipolar major depressive disorder without psychotic symptoms and their changes during antidepressant treatment. Acta Psychiatr Scand 125(4):281–292. doi:10.1111/j.1600-0447.2011.01762.x

Wakade CG, Mahadik SP, Waller JL, Chiu FC (2002) Atypical neuroleptics stimulate neurogenesis in adult rat brain. J Neurosci Res 69(1):72–79. doi:10.1002/jnr.10281

Walker AK, Rivera PD, Wang Q, Chuang JC, Tran S, Osborne-Lawrence S, Estill SJ, Starwalt R, Huntington P, Morlock L, Naidoo J, Williams NS, Ready JM, Eisch AJ, Pieper AA, Zigman JM (2014) The P7C3 class of neuroprotective compounds exerts antidepressant efficacy in mice by increasing hippocampal neurogenesis. Mol Psychiatry. doi:10.1038/mp.2014.34

Walter J, Keiner S, Witte OW, Redecker C (2011) Age-related effects on hippocampal precursor cell subpopulations and neurogenesis. Neurobiol Aging 32(10):1906–1914. doi:10.1016/j.neurobiolaging.2009.11.011

Wang HD, Dunnavant FD, Jarman T, Deutch AY (2004) Effects of antipsychotic drugs on neurogenesis in the forebrain of the adult rat. Neuropsychopharmacology 29(7):1230–1238. doi:10.1038/sj.npp.1300449

Wang J, Gallagher D, DeVito LM, Cancino GI, Tsui D, He L, Keller GM, Frankland PW, Kaplan DR, Miller FD (2012) Metformin activates an atypical PKC-CBP pathway to promote neurogenesis and enhance spatial memory formation. Cell Stem Cell 11(1):23–35. doi:10.1016/j.stem.2012.03.016

Wang W, Pan YW, Zou J, Li T, Abel GM, Palmiter RD, Storm DR, Xia Z (2014) Genetic activation of ERK5 MAP kinase enhances adult neurogenesis and extends hippocampus-dependent long-term memory. J Neurosci 34(6):2130–2147. doi:10.1523/JNEUROSCI.3324-13.2014

Warner-Schmidt JL, Duman RS (2007) VEGF is an essential mediator of the neurogenic and behavioral actions of antidepressants. Proc Natl Acad Sci U S A 104(11):4647–4652. doi:10.1073/pnas.0610282104

Watanabe Y, Gould E, McEwen BS (1992) Stress induces atrophy of apical dendrites of hippocampal CA3 pyramidal neurons. Brain Res 588(2):341–345

Weiland-Fiedler P, Erickson K, Waldeck T, Luckenbaugh DA, Pike D, Bonne O, Charney DS, Neumeister A (2004) Evidence for continuing neuropsychological impairments in depression. J Affect Disord 82(2):253–258. doi:10.1016/j.jad.2003.10.009

Wesnes KA, Annas P, Basun H, Edgar C, Blennow K (2014) Performance on a pattern separation task by Alzheimer's patients shows possible links between disrupted dentate gyrus activity and apolipoprotein [element of]4 status and cerebrospinal fluid amyloid-03B242 levels. Alzheimers Res Ther 6(2):20. doi:10.1186/alzrt250

West MJ, Coleman PD, Flood DG, Troncoso JC (1994) Differences in the pattern of hippocampal neuronal loss in normal ageing and Alzheimer's disease. Lancet 344(8925):769–772

West MJ, Kawas CH, Stewart WF, Rudow GL, Troncoso JC (2004) Hippocampal neurons in pre-clinical Alzheimer's disease. Neurobiol Aging 25(9):1205–1212. doi:10.1016/j.neurobiolaging.2003.12.005

Wigmore P (2013) The effect of systemic chemotherapy on neurogenesis, plasticity and memory. Curr Top Behav Neurosci 15:211–240. doi:10.1007/7854_2012_235

Wixted JT, Squire LR, Jang Y, Papesh MH, Goldinger SD, Kuhn JR, Smith KA, Treiman DM, Steinmetz PN (2014) Sparse and distributed coding of episodic memory in neurons of the human hippocampus. Proc Natl Acad Sci U S A 111(26):9621–9626. doi:10.1073/pnas.1408365111

Wolf SA, Kronenberg G, Lehmann K, Blankenship A, Overall R, Staufenbiel M, Kempermann G (2006) Cognitive and physical activity differently modulate disease progression in the amyloid precursor protein (APP)-23 model of Alzheimer's disease. Biol Psychiatry 60(12):1314–1323. doi:10.1016/j.biopsych.2006.04.004

Wolf OT, Minnebusch D, Daum I (2009) Stress impairs acquisition of delay eyeblink conditioning in men and women. Neurobiol Learn Mem 91(4):431–436. doi:10.1016/j.nlm.2008.11.002

Yamasaki TR, Blurton-Jones M, Morrissette DA, Kitazawa M, Oddo S, LaFerla FM (2007) Neural stem cells improve memory in an inducible mouse model of neuronal loss. J Neurosci 27(44):11925–11933. doi:10.1523/JNEUROSCI.1627-07.2007

Yang P, Arnold SA, Habas A, Hetman M, Hagg T (2008) Ciliary neurotrophic factor mediates dopamine D2 receptor-induced CNS neurogenesis in adult mice. J Neurosci 28(9):2231–2241. doi:10.1523/JNEUROSCI.3574-07.2008

Yang C, Hu YM, Zhou ZQ, Zhang GF, Yang JJ (2013) Acute administration of ketamine in rats increases hippocampal BDNF and mTOR levels during forced swimming test. Ups J Med Sci 118(1):3–8. doi:10.3109/03009734.2012.724118

Yang X, Yang Q, Wang X, Luo C, Wan Y, Li J, Liu K, Zhou M, Zhang C (2014) MicroRNA expression profile and functional analysis reveal that miR-206 is a critical novel gene for the expression of BDNF induced by ketamine. Neuromolecular Med 16(3):594–605. doi:10.1007/s12017-014-8312-z

Yap JJ, Takase LF, Kochman LJ, Fornal CA, Miczek KA, Jacobs BL (2006) Repeated brief social defeat episodes in mice: effects on cell proliferation in the dentate gyrus. Behav Brain Res 172(2):344–350. doi:10.1016/j.bbr.2006.05.027

Yassa MA, Stark CE (2011) Pattern separation in the hippocampus. Trends Neurosci 34(10):515–525. doi:10.1016/j.tins.2011.06.006

Yassa MA, Lacy JW, Stark SM, Albert MS, Gallagher M, Stark CE (2011) Pattern separation deficits associated with increased hippocampal CA3 and dentate gyrus activity in nondemented older adults. Hippocampus 21(9):968–979. doi:10.1002/hipo.20808

Yoshimizu T, Chaki S (2004) Increased cell proliferation in the adult mouse hippocampus following chronic administration of group II metabotropic glutamate receptor antagonist, MGS0039. Biochem Biophys Res Commun 315(2):493–496. doi:10.1016/j.bbrc.2004.01.073

Young EA, Haskett RF, Murphy-Weinberg V, Watson SJ, Akil H (1991) Loss of glucocorticoid fast feedback in depression. Arch Gen Psychiatry 48(8):693–699

Zarate CA Jr, Singh JB, Carlson PJ, Brutsche NE, Ameli R, Luckenbaugh DA, Charney DS, Manji HK (2006) A randomized trial of an N-methyl-D-aspartate antagonist in treatment-resistant major depression. Arch Gen Psychiatry 63(8):856–864. doi:10.1001/archpsyc.63.8.856

Zhao C, Teng EM, Summers RG Jr, Ming GL, Gage FH (2006) Distinct morphological stages of dentate granule neuron maturation in the adult mouse hippocampus. J Neurosci 26(1):3–11. doi:10.1523/JNEUROSCI.3648-05.2006

Zhao C, Jou J, Wolff LJ, Sun H, Gage FH (2014) Spine morphogenesis in newborn granule cells is differentially regulated in the outer and middle molecular layers. J Comp Neurol 522(12):2756–2766. doi:10.1002/cne.23581

Zhou M, Li W, Huang S, Song J, Kim JY, Tian X, Kang E, Sano Y, Liu C, Balaji J, Wu S, Zhou Y, Zhou Y, Parivash SN, Ehninger D, He L, Song H, Ming GL, Silva AJ (2013a) mTOR Inhibition ameliorates cognitive and affective deficits caused by Disc1 knockdown in adult-born dentate granule neurons. Neuron 77(4):647–654. doi:10.1016/j.neuron.2012.12.033

Zhou W, Wang N, Yang C, Li XM, Zhou ZQ, Yang JJ (2013b) Ketamine-induced antidepressant effects are associated with AMPA receptors-mediated upregulation of mTOR and BDNF in rat hippocampus and prefrontal cortex. Eur Psychiatry. doi:10.1016/j.eurpsy.2013.10.005

Zhu X, Wang X, Xiao J, Zhong M, Liao J, Yao S (2011) Altered white matter integrity in first-episode, treatment-naive young adults with major depressive disorder: a tract-based spatial statistics study. Brain Res 1369:223–229. doi:10.1016/j.brainres.2010.10.104

Part II

Cognitive Domains for Pharmacological Intervention: Implications for Neuropsychiatric and Neurological Illnesses

Introduction

Cognitive enhancement is a treatment goal for a variety of neuropsychiatric and neurological illnesses that are characterized by deficits in one or more cognitive domains. Domains include attention, executive function, declarative memory, emotional memory, and social cognition. A basic premise is that pharmacological interventions for cognitive enhancement are likely to vary by cognitive domain. A corollary is that the discovery of a safe and effective cognitive-enhancing therapeutic drug for a particular cognitive domain should offer improvement across neuropsychiatric and neurological illnesses that present with deficits within that same cognitive domain. Part II provides an overview of clinical and preclinical research devoted to domain-specific cognitive enhancement. The current status of that research for selected cognitive domains is presented.

Chapter 5 (Callahan and Terry) provides evidence for the enhancement of attention, focusing on drugs acting on the cholinergic, dopaminergic, noradrenergic, and serotonergic systems. While some promising lead compounds have emerged from preclinical testing (e.g., nicotine, the $\alpha4\beta2$ nicotinic receptor agonist ABT-418, methylphenidate, the dopamine D_1 receptor agonist SKF 38393, the $\alpha1$ adrenergic receptor agonist St-587, and atomoxetine), it is the case that improvements in attention sometimes come at the cost of having a negative impact on either response inhibition or response vigor or being observed only when the demands of the task are high. In many cases, the compounds that have a proven benefit in clinical populations are also effective in the preclinical models, supporting the argument that these models of attention may also have predictive validity.

Chapter 6 (Talpos and Shoaib) explores executive function through the use of animal models. Drugs that enhance attentional set shifting and reversal learning are highlighted. Promising lead compounds that improve extradimensional set-shifting performance include nicotine, a variety of atypical antipsychotic drugs (e.g., risperidone, clozapine), and atomoxetine. Notably, the pro-cognitive effects of these drugs in the attentional set-shifting procedure typically are observed in lesioned animals or in those first pretreated with psychomimetic drugs (e.g., phencyclidine and ketamine). Thus, the opportunity for enhancement in normal subjects may be limited, or alternatively, the attentional set-shift task in animals may not be an ideal model for assessing this purpose. Drugs that enhance reversal learning (mainly by reducing perseverative errors) include methylphenidate, atomoxetine, the serotonin 2c receptor antagonist SB 242084, and the α2 adrenergic receptor agonist atipamezole. Interestingly, an improvement in reversal learning can come at the cost of interference with other aspects of executive functioning such as working memory.

Chapter 7 (Riedel and Blokland) provides an overview of the enhancement of declarative memory in tests conducted in humans and animals, with an emphasis on episodic memory. Drugs influencing dopamine (modafinil, D-amphetamine, methylphenidate) and acetylcholine (cholinesterase inhibitors) neurotransmission as well as glucose consumption have the most consistent effects. Findings with serotonergic and some miscellaneous agents (e.g., polyunsaturated fatty acids) are mixed. Importantly, when considering human and animal data, there is a great discrepancy in findings between species. Some of this discrepancy may be related to type of task used (object recognition in animals vs. verbal learning tests in humans). At issue for enhancing declarative memory is whether shifting to a prevention paradigm for treatment in clinical populations (e.g., the prodromal stage of dementia) or focusing on the memory consolidation stage (e.g., drugs administered after, not before, learning) might yield better outcomes.

Chapter 8 (Sumiyoshi) focuses on verbal memory, an important aspect of declarative memory. Much of the research on the enhancement of verbal memory arises from the study of schizophrenia. Verbal memory in patients with schizophrenia is improved to a greater extent following treatment with atypical than typical antipsychotic drugs. Though studies are few, putative pro-cognitive agents given to patients treated with antipsychotic drugs improve verbal memory as well. These include the serotonin 1a receptor partial agonist tandospirone and the acetylcholinesterase inhibitor galantamine. More research is needed, as improvement in verbal memory has a strong link to an improved functional outcome in schizophrenia.

Chapter 9 (Nader) explores emotional memory from the perspective of memory reactivation (reconsolidation) and its enhancement or disruption by various pharmacological agents. Most research on emotional memory investigates fear memory and the ability of protein synthesis inhibitors, NMDA antagonists, beta-adrenergic antagonists, and other amnestic treatments to block fear memory reconsolidation. This line of research has implications for treatment of a number of clinical conditions involving overly intrusive memories and thoughts, such as post-traumatic stress disorder, obsessive-compulsive disorder, or delusions/hallucinations. Enhancement of emotional memory in animal models is achieved by epinephrine and glucocorticoids, and further drug development may offer viable treatment options for neuropsychiatric and neurological disorders involving underarousal or flat affect.

Chapter 10 (Patin and Hurlemann) covers social cognition, a process impacted in several neuropsychiatric conditions. The pharmacology for the enhancement of social cognition is diverse and includes treatment with oxytocin, 3,4-methylenedioxymethamphetamine (MDMA), modafinil, methylphenidate, and D-cycloserine. With some of these agents, there is a disorder-specific usage for improving social cognition aspects of the illness (e.g., MDMA in post-traumatic stress disorder), while for other agents, the findings for improvement in social cognition are mixed and several contraindications prevail (e.g., modafinil in a range of neuropsychiatric illnesses). An innovative area of investigation is with oxytocin, which has documented prosocial effects in individuals ranging from healthy to those with seriously impaired social functioning, such as patients with autism spectrum disorder and schizophrenia. More standardization of dose, route, and pretreatment time for oxytocin and of the paradigms used to measure basic emotions is needed for accurately determining the full therapeutic potential of oxytocin for the enhancement of social cognition.

From the above overview, it is clear that medication development for cognitive enhancement across a variety of cognitive domains has focused mainly on monoamine and cholinergic systems, with only a few exceptions. More research is needed and likely will require investigation of additional targets and signaling pathways before clinically useful agents are available. For many cognitive domains, animal research may help speed this process, but translatable paradigms in animals have not yet been developed for all cognitive domains.

Attention

Patrick M. Callahan and Alvin V. Terry Jr.

Contents

Abstract

The ability to focus one's attention on important environmental stimuli while ignoring irrelevant stimuli is fundamental to human cognition and intellectual function. Attention is inextricably linked to perception, learning and memory, and executive function; however, it is often impaired in a variety of neuropsychiatric disorders, including Alzheimer's disease, schizophrenia, depression, and attention deficit hyperactivity disorder (ADHD). Accordingly, attention is considered as an important therapeutic target in these disorders. The purpose of

P.M. Callahan • A.V. Terry Jr. (✉)
Department of Pharmacology and Toxicology, CB-3545, Georgia Regents University, 1120 Fifteenth Street, Augusta, GA 30912-2450, USA

Small Animal Behavior Core, Georgia Regents University, Augusta, GA 30912, USA
e-mail: aterry@gru.edu

K.M. Kantak, J.G. Wettstein (eds.), *Cognitive Enhancement*, Handbook of Experimental Pharmacology 228, DOI 10.1007/978-3-319-16522-6_5

this chapter is to provide an overview of the most common behavioral paradigms of attention that have been used in animals (particularly rodents) and to review the literature where these tasks have been employed to elucidate neurobiological substrates of attention as well as to evaluate novel pharmacological agents for their potential as treatments for disorders of attention. These paradigms include two tasks of sustained attention that were developed as rodent analogues of the human Continuous Performance Task (CPT), the Five-Choice Serial Reaction Time Task (5-CSRTT) and the more recently introduced Five-Choice Continuous Performance Task (5C-CPT), and the Signal Detection Task (SDT) which was designed to emphasize temporal components of attention.

Keywords
Sustained attention • Signal detection • Distractibility • Preclinical • Drug development • Animal model

1 Introduction

As is often the case in the fields of cognitive psychology and neuroscience, a precise definition for the concept "attention" is often the subject of debate and controversy. However, in simple terms, attention could be thought of as the allocation or concentration of mental resources on specific (environmentally relevant) stimuli while ignoring other (nonrelevant or less relevant) stimuli. While "attention" is often used as an umbrella or generic term, most theories in cognitive psychology describe at least three or four separate (but interrelated) subcategories of attention, including sustained attention or vigilance (attending to one stimulus over a significant period of time), selective attention (focus directed at one stimulus in lieu of competing, irrelevant stimuli), orienting attention (directional or spatial orientation toward a particular stimulus), and divided attention (simultaneously attending to two or more different stimuli or performing multiple tasks) (Posner and Petersen 1990; Robertson et al. 1996; Parasuraman et al. 1998). It is clear that attention is inextricably linked to intellectual function and the major components of human cognition including perception, learning and memory, and executive function. Moreover, attention is often impaired in a variety of neuropsychiatric disorders, including Alzheimer's disease (Lawrence and Sahakian 1995; Parasuraman et al. 1998), schizophrenia (Laurent et al. 1999), depression (Brown et al. 1994), and attention deficit hyperactivity disorder (ADHD) (Biederman 2005). Accordingly, attention is considered as an important therapeutic target in these disorders.

In selecting preclinical behavioral models of attention, priority should be given to behavioral paradigms that possess construct validity (i.e., the task accurately and specifically measures attentional processes and its performance relies on similar underlying neurophysiological circuitry as in humans), reliability, and task standardization across laboratories. Several preclinical behavioral tasks meet these criteria and have been used extensively to characterize the neural systems associated with attention as well as to assess pharmacological agents that may have therapeutic relevance in alleviating attentional impairments observed in neuropsychiatric disorders. For this review, we have chosen to discuss three commonly used paradigms: (1) the 5-choice serial reaction time task (5-CSRTT; Carli et al. 1983; Robbins 2002), (2) the 5-choice continuous performance task (5C-CPT; Young et al. 2009), and (3) the signal detection task (SDT; Bushnell 1995; McGaughy and Sarter 1995a, b; Rezvani et al. 2002). All of these behavioral tasks have features that are similar to the continuous performance task (CPT; Rosvold et al. 1956) that has been used successfully to detect attention deficits in clinical populations such as ADHD (Riccio et al. 2002; Loo et al. 2004), Alzheimer's Disease (Levinoff et al. 2005), and schizophrenia (Nieuwenstein et al. 2001; Lee and Park 2006). In the CPT, subjects are required to respond to a specific visual stimulus (e.g., the letter X) within a list of letters. Since the letter X occurs less often, subjects must remain attentive during the session. When the letter X is presented the subject is required to press a button or click a computer mouse. This simple response requirement affords the investigator considerable information in addition to attention (correct response) such as false alarms (errors made when no X is presented), processing speed (latency to respond), and impulsivity (responding in the absence of the X stimulus). The behavioral tasks described below incorporate many of the CPT test attributes for measuring attention, information processing, and impulsivity.

2 Five-Choice Serial Reaction Time Task

2.1 Task Description

The 5-CSRTT (see Fig. 1) was developed as a means to assess attention based on Leonard's five-choice sustained attention task in humans (Leonard 1959) and is the preclinical analogue of the CPT task, though there are task dissimilarities (Young et al. 2009). A large literature base exists for the 5-CSRTT with evidence demonstrating construct validity as a model of attention. As a test component within the CANTAB battery, the task has been used in both healthy volunteers and in subjects suffering from neuropsychiatric disorders (Barnett et al. 2010; Cambridge Cognition, camcog.com). The behavioral paradigm was originally developed for rats (Carli et al. 1983), but recently, versions have been developed for mice (Humby et al. 1999; Sanchez-Roige et al. 2012) and nonhuman primates (Weed et al. 1999; Spinelli et al. 2004). The 5-CSRTT assesses the subject's ability to spatially divide its attention across multiple signal locations (usually five locations, but fewer can be used) in order to select the correct target stimulus (light in a single aperture hole location) that, in turn, produces a food reward (see Higgins and Breysse 2008 for paradigm and training description). This behavioral task measures attentiveness to multiple locations over time and, thereby, utilizes both sustained and selective attention (Levin et al. 2011). *Selective attention* occurs

Fig. 1 Diagram of the five-choice serial reaction time task (5-CSRTT). In the 5-CSRTT, subjects are required to scan five apertures in an operant chamber for the appearance of a brief light stimulus (presented pseudorandomly) and to make a nose-poke response in the correct spatial location (i.e., the aperture where the light stimulus was presented) in order to receive a food reward

when the subject faces multiple stimuli and must make a choice among them; the choice is defined by the subject's behavior. In this situation, novelty plays an important factor in determining the subsequent behavior of the animal and, thus, its attentional selection. *Sustained attention* occurs when the subject's behavior is controlled by a single stimulus that occurs unpredictably in time and space. While selective and sustained attention are not independent functions within the 5-CSRTT, these events can be manipulated to preferentially place more "demand" on one function versus the other. For example, a greater demand on sustained attention can be achieved by increasing the temporal unpredictability of the stimulus presentation, whereas increasing the number of potential stimulus locations places greater demand on selective attention. In addition to measuring attention (choice accuracy), the 5-CSRTT can assess a number of other cognitive domains such as impulsivity (premature responses), cognitive flexibility/compulsivity (perseverative and timeout responses), and processing speed (response latency). Task difficulty can be modified by changing the brightness, duration, and temporal predictability of the target stimulus and a distractor stimulus (e.g., auditory tone or white noise) can be interpolated into the protocol to increase task difficulty and place greater attentional demand on the subject.

2.2 Neural Substrates

Considerable scientific work has been devoted to delineating the neural substrates involved in modulating 5-CSRTT performance and its response measures (for reviews, see Robbins 2002; Chudasama and Robbins 2004). Excitotoxic lesions of different subregions of the rat prefrontal cortex differentially affect the behavioral measures associated with the task. Gross lesions of the medial prefrontal cortex (mPFC) that include the dorsal pre-genual anterior cingulated cortex (area Cg1), medial prelimbic cortex (PrL), and to some extent the ventral infralimbic cortex (IL) show profound impairments in choice accuracy, increased perseverative responding, and slower response latencies (Muir et al. 1996). More selective lesions within the rat mPFC have identified precise anatomical loci responsible for controlling specific 5-CSRTT response measures. For example, specific lesions of the dorsal Cg1 area produce deficits in choice accuracy, whereas PrL and orbitofrontal cortex (OFC) lesions result in selective increases in perseverative responding (Passetti et al. 2002; Chudasama et al. 2003). In contrast, increases in premature responding occur following ventral IL cortical lesions (Chudasama et al. 2003). There are also data on the effects of specific anatomical loci of the striatum, subthalamic nucleus, pedunculopontine nucleus, and hippocampus (Chudasama et al. 2012) providing evidence of a systematic "top-down" anatomical connectivity in mediating attention, impulsivity and executive function. Complementing the data from anatomical-based lesion experiments, lesions and neurochemical evaluations/manipulations focused on neurotransmitter pathways (e.g., acetylcholine, dopamine, glutamate, noradrenaline, and serotonin) have provided further insights into the neurobiological substrates of 5-CSRTT performance. The outcome from this seminal research highlighted the dissociable roles that specific neurotransmitter systems have on attention, reaction time, and response control (see Robbins 2002). Probably the most extensively studied neurotransmitter system is the forebrain cholinergic system. Selective lesions of the nucleus basalis magnocellularis (NbM) with the cholinergic immunotoxin 192 - lgG-saporin resulted in decreased PFC acetylcholine levels and choline acetyltransferase activity accompanied by poor choice accuracy, increases in trial omissions, and disruption in response control performance (McGaughy et al. 2002; Lehmann et al. 2003). Additional evidence (e.g., intra-NbM infusion of the GABA agonist muscimol, or intra-PFC infusion of scopolamine) further supported the importance of the basal forebrain cholinergic system and its innervation of the PFC to 5-CSRTT performance (Muir et al. 1992; Robbins 2002).

Studies focused on the catecholamine system have indicated that 6-OHDA depletion of ventral striatal dopamine affects response vigor (i.e., omissions and response latency) with little to no effect on choice accuracy, whereas dorsal striatal dopamine lesions affect only response-related processes. In contrast, 6-OHDA lesions of the mPFC produce minimal effects unless task demands are increased (Robbins 2002). Likewise, lesions of the ascending dorsal noradrenergic (NA) bundle impair attention, but only when greater task demands (e.g., use of distracting stimuli or when the temporal presentation of the stimulus target is unpredictable) are placed on the subject, thereby requiring heightened awareness (Carli et al. 1983; Cole and Robbins 1992). Further support for the engagement of the NA system during challenging situations stems from the observation that increased mPFC noradrenaline efflux occurs only when task contingencies are manipulated, but remain unaltered during baseline conditions (Dalley et al. 2001). Interestingly, the opposite appears to occur for mPFC acetylcholine (i.e., under baseline conditions acetylcholine levels increase but remain unchanged during high task demand) demonstrating a dissociation between the two neurotransmitter systems in controlling specific aspects of attention (Dalley et al. 2001).

Studies assessing the impact of serotonin (5-HT) depletion have also yielded dissociable profiles on 5-CSRTT performance (i.e., impulsivity and choice accuracy) that are related to the particular neurotoxin (PCPA or 5,7-DHT) and brain region (forebrain, dorsal or medial raphe nuclei) investigated (Harrison et al. 1997a, 1997b; Puumala and Sirvio 1998; Dalley et al. 2002). While there is clearly a definitive relationship between 5-HT function and inhibitory control, it appears that different forms of impulsivity (i.e., impulsive choice vs. impulsive action) are mediated by specific 5-HT receptor subtypes within distinct brain regions (Winstanley et al. 2004a, 2004b). Impulsive action (i.e., behavioral inhibition) is typically measured in the 5-CSRTT, and collectively, evidence has demonstrated that reductions in 5-HT activity can either induce or inhibit impulsive behavior which is purely dependent on the 5-HT receptor subtype (i.e., 5-HT_{2A} vs. 5-HT_{2C}) activated (see below).

2.3 Pharmacology

The 5-CSRTT has also been used extensively to evaluate pharmacological agents in rodent models for effects on attention and inhibitory control (see Table 1). These studies have helped to further elucidate the important roles of various neurotransmitter systems on attentional processes and they have facilitated preclinical drug discovery efforts for neuropsychiatric disorders (e.g., ADHD, Alzheimer's disease and schizophrenia). Here we provide an overview of some of the major pharmacological studies conducted to date (see also comprehensive reviews, Robbins 2002; Higgins and Breysse 2008; Barak and Weiner 2011; Sanchez-Roige et al. 2012). As discussed above, considerable evidence from lesion-based experiments supports the argument that the CNS cholinergic system plays a major role in attention. This evidence appears to be supported by pharmacologic experimentation as well. For example, systemic administration of the muscarinic receptor antagonist scopolamine impaired several response measures (accuracy and omissions) in rats (Jones and Higgins 1995; Mirza and Stolerman 2000) and mice (Humby et al. 1999; de Bruin et al. 2006; Pattij et al. 2007). However, task performance appeared to be unaltered following the acetylcholinesterase inhibitors (AChE) physostigmine and donepezil (Mirza and Stolerman 2000; Romberg et al. 2011), the muscarinic M_1 receptor agonist oxotremorine (Mirza and Stolerman 2000), and the mixed AChE-muscarinic M_2 receptor antagonist JWS-USC-75-IX (Terry et al. 2011). Interestingly, JWS-USC-75-IX did attenuate the impairments of choice accuracy and increases in premature responding associated with the NMDA antagonist MK-801 (Terry et al. 2011). Nicotine has been shown to improve attentional processing in humans (Sahakian et al. 1989; White and Levin 1999; Min et al. 2001) and, thus, it has been extensively characterized in the 5-CSRTT in rats (Mirza and Stolerman 1998; Blondel et al. 2000; Grottick and Higgins 2000; Stolerman et al. 2000; Grottick et al. 2003; Bizarro et al. 2004; van Gaalen

Table 1 Effects of pharmacological agents in behavioral models of attention

Drug	Target/mechanism	Task[a]	Task condition	Species	Effect[b]	References
Cholinergic						
Donepezil	AChE inhibitor	SDT	MK-801 reversal	Rats	+	Rezvani et al. (2012)
		5-CSRTT	Standard	Mice	0	Romberg et al. (2011)
Physostigmine	AChE inhibitor	5-CSRTT	Standard	Rats	0	Mirza and Stolerman (2000)
		SDT	Standard	Rats	0	McGaughy and Sarter (1998)
Oxotremorine	M_1 agonist	5-CSRTT	Standard	Rats	0	Mirza and Stolerman (2000)
JWS-USC-75-IX	AChE inhibitor/	5-CSRTT	Standard	Rats	0	Terry et al. (2011)
	M_2 antagonist		MK-801 reversal	Rats	+	Terry et al. (2011)
Nicotine	Nonselective nicotinic agonist	5-CSRTT	Standard	Rats	+	Mirza and Stolerman (1998)
						Stolerman et al. (2000)
						Bizarro et al. (2004)
						Blondel et al. (2000)
						Grottick, and Higgins (2000)
						Grottick et al. (2003)
		5-CSRTT	Standard	Mice	+	de Bruin et al. (2006)
						Pattij et al. (2007)
						Young et al. (2004)
		5C-CPT	Standard/scopolamine reversal	Mice	+	Young et al. (2013)
		SDT	Standard	Rats	0	Turchi et al. (1995)
		SDT	Standard/MK-801 reversal	Rats	+	Rezvani et al. (2002, 2008)
		SDT	Standard/distractor	Rats	0/+	Howe et al. (2010)
		SDT	Standard	Rats	+	Hillhouse and Prus (2013)
Cotinine (Metabolite of nicotine)	Unknown	5-CSRTT	MK-801 reversal	Rats	+	Terry et al. (2012)
A-82696	α4β2 agonist	SDT	Standard	Rats	0	Turchi et al. (1995)

(continued)

Table 1 (continued)

Drug	Target/mechanism	Task[a]	Task condition	Species	Effect[b]	References
ABT-418	α4β2 agonist	5-CSRTT	Standard	Rats	+	Hahn et al. (2003)
		5C-CPT	Standard/scopolamine reversal	Mice	+	Young et al. (2013)
		SDT	Standard	Rats	0	Turchi et al. (1995)
		SDT	Standard	Rats	+	McGaughy et al. (1999)
ABT-594	α4β2 agonist	5-CSRTT	Standard	Rats	+	Mohler et al. (2010)
ADZ 3480	α4β2 agonist	SDT	MK-801 reversal	Rats	+	Rezvani et al. (2012)
Epibatidine	α4β2 agonist	5-CSRTT	Standard	Rats	+	Hahn et al. (2003)
S-38232	α4β2 agonist	SDT	Standard/distractor	Rats	+	Howe et al. (2010)
Sazetidine-A	α4β2 agonist	SDT	Standard/scopolamine, MK-801 reversal	Rats	+	Rezvani et al. (2011, 2012)
SIB-1765F	α4β2 agonist	5-CSRTT	Standard	Rats	+	Grottick and Higgins (2000)
SIB-1553A	α2–4β4 agonist	5-CSRTT	MK-801 reversal	Rats	+	Terry et al. (2002)
ARR-17779	α7 agonist	5-CSRTT	Standard	Rats	0	Grottick and Higgins (2000)
						Grottick et al. (2003)
						Hahn et al. (2003)
PNU 282987	α7 agonist	5C-CPT	Standard/scopolamine reversal	Mice	0	Young et al. (2013)
RG 3487	α7 agonist	SDT	Standard	Rats	+	Rezvani et al. (2009b)
Dopaminergic						
Amphetamine	DA releaser	5-CSRTT	Standard	Rats	+	Grottick and Higgins (2002)
						Bizarro et al. (2004)
		5-CSRTT	Standard	Mice	–	Loos et al. (2010)
						Yan et al. (2011)
		SDT	Standard	Rats	0	McGaughy and Sarter (1995a)

Methylphenidate	DA reuptake inhibitor	5-CSRTT	Standard	Rats	+	Paine et al. (2007)
						Puumala et al. (1996)
						Bizarro et al. (2004)
						Navarra et al. (2008b)
		SDT	Standard/MK-801, scopolamine reversal	Rats	+	Rezvani et al. (2009b)
Modafinil	DA reuptake inhibitor	3-CSRTT	Standard	Rats	+	Morgan et al. (2007)
		5-CSRTT	Standard	Rats	–	Waters et al. (2005)
SKF 38393	DA D_1 agonist	5-CSRTT	Standard	Rats	+	Grannon et al. (2000)
		5C-CPT	Standard	Rats	+	Barnes et al. (2012)
Quinpirole	DA $D_{2/3}$ agonist	5-CSRTT	Standard	Rats	–	Fernando et al. (2012)
Sumanirole	DA D_2 agonist	5-CSRTT	Standard	Rats	–	Fernando et al. (2012)
Noradrenergic						
Atomoxetine	NE reuptake inhibitor	5-CSRTT	Standard	Rats	+	Navarra et al. (2008b)
						Fernando et al. (2012)
						Robinson (2012)
Desipramine	NE reuptake inhibitor	5-CSRTT	Standard	Rats	+	Paine et al. (2007)
						Pattij et al. (2012)
Reboxetine	NE reuptake inhibitor	5-CSRTT	Standard	Rats	+	Robinson (2012)
St-587	α1 agonist	5-CSRTT	Standard	Rats	+	Puumala et al. (1997)
Clonidine	α2 agonist	5-CSRTT	Standard	Rats	+	Pattij et al. (2012)
Guanfacine	α2 agonist	5-CSRTT	Standard	Rats	+	Fernando et al. (2012)
Atipamezole	α2 antagonist	5-CSRTT	Standard	Rats	+	Sirvio et al. (1993)
Clenbuterol	β2 agonist	5-CSRTT	Standard	Rats	+	Pattij et al. (2012)
Serotonergic						
Sibutramine	5-HT reuptake inhibitor	5-CSRTT	Standard	Rats	–	Humpston et al. (2013)
Fluoxetine	5-HT reuptake inhibitor	5-CSRTT	Standard	Rats	–	Humpston et al. (2013)

(continued)

Table 1 (continued)

Drug	Target/mechanism	Task[a]	Task condition	Species	Effect[b]	References
Paroxetine	5-HT reuptake inhibitor	5-CSRTT	Standard	Rats	–	Humpston et al. (2013)
8-OH-DPAT	5-HT_{1A} agonist	5-CSRTT	Standard	Rats	–	Carli and Samanin (2000)
					+	Winstanley et al. (2003)
DOI	5-HT_{2A} agonist	5-CSRTT	Standard	Rats	–	Koskinen and Sirvio (2001)
						Fletcher et al. (2007)
M100907	5-HT_{2A} antagonist	5-CSRTT	Standard	Rats	+	Fletcher et al. (2007)
		5-CSRTT	MK-801, amphetamine, cocaine reversal	Rats	+	Fletcher et al. (2011)
Ro60-0175	5-HT_{2C} agonist	5-CSRTT	Standard	Rats	+	Fletcher et al. (2007)
		5-CSRTT	MK-801, amphetamine, cocaine reversal	Rats	+	Fletcher et al. (2011)
		5-CSRTT	Standard	Mice	+	Fletcher et al. (2013)
WAY-163909	5-HT_{2C} agonist	5-CSRTT	Standard	Rats	+	Navarra et al. (2008b)
Ondansetron	5-HT_3 antagonist	5-CSRTT	Standard	Rats	–	Kirkby et al. (1996)
CMP 42	5-HT_6 antagonist	5-CSRTT	Standard	Rats	–	de Bruin et al. (2013)

[a]Task abbreviations indicate 5-choice serial reaction time task (5-CSRTT), 5-choice continuous performance task (5C-CPT) or signal detection task (SDT)
[b]Symbols indicate performance improvement (+), performance impairment (–) or no effect (0)

et al. 2006; Amitai and Markou 2009) and mice (Young et al. 2004; de Bruin et al. 2006; Pattij et al. 2007). Collectively, these studies suggest that nicotine improves choice accuracy and decreases trial omissions and response latencies, but that it also increases premature responding. Nicotine, therefore, may improve certain aspects of cognitive performance (processing speed, attention) while negatively affecting other cognitive domains (e.g., response inhibition/impulsivity; see Amitai and Markou 2009).

A variety of additional experiments have been conducted to further explore the basis for the effects of nicotine on 5-CSRTT performance. These studies have included evaluations of the major nicotine metabolite, cotinine, as well as nicotinic (subtype-selective) ligands and transgenic receptor knockout mice (i.e., to investigate the role of the specific nicotinic receptor subtypes on 5-CSRTT performance). Cotinine had previously been shown to improve the performance of a standard and distractor version of a delayed-match-to-sample task (DMTS), a working/short-term memory task in nonhuman primates (Terry et al. 2005). Moreover, cotinine reversed the DMTS performance deficits induced by the NMDA antagonist ketamine (Buccafusco and Terry 2009). The positive effects of cotinine (particularly in the distractor version of DMTS) led to further evaluations of cotinine for its effects on attention in the 5-CSRTT. In these studies, cotinine administered alone did not alter 5-CSRTT performance in the rat; however it was effective in attenuating the negative effects of MK-801 on choice accuracy and premature and timeout responses, suggesting that under particular circumstances cotinine might be therapeutically beneficial (Terry et al. 2012).

Studies designed to assess the role of specific nicotinic acetylcholine receptor (nAChR) subtypes on attention in the 5-CSRTT have sometimes been difficult to interpret. Both the α4β2 nAChR antagonist di-hydro-β-erythroidine (DHβE) and the α7 nAChR antagonist methyllycaconitine (MLA) failed to alter task performance (Grottick and Higgins 2000), whereas the nonselective nAChR antagonist mecamylamine was found to decrease choice accuracy and increase trial omissions, correct response latencies, and perseverative responses in rats, effects opposite to those produced by nicotine (Grottick and Higgins 2000; Mirza and Stolerman 2000; Ruotsalainen et al. 2000). Increases in trial omissions and correct response latencies have also been observed in mice after mecamylamine administration (Pattij et al. 2007). To support the argument that the effects described above for mecamylamine are expressed via central nAChRs, the peripheral nAChR antagonist hexamethonium did not alter any task parameters (Blondel et al. 2000; Grottick and Higgins 2000). The information provided above might suggest that both high-affinity (α4β2) and low-affinity (α7) nAChRs (together) are required for detectable effects on 5-CSRTT performance; however, the α4β2/α7 nAChR agonist varenicline had no effect on any of the behavioral parameters assessed in the 5-CSRTT with the exception of premature responding, which was increased at low doses (Wouda et al. 2011).

The results of additional studies designed to investigate the role of the particular nAChR subtypes (α4β2 vs. α7) in mediating nicotine's response in 5-CSRTT have also been somewhat difficult to interpret. Initially, several studies in young and aged rats demonstrated that the effects of nicotine were attenuated by co-administration of either DHβE or mecamylamine, but not MLA, suggesting a more important role for the α4β2 nAChR subtype in 5-CSRTT performance (Blondel et al. 2000; Grottick and Higgins 2000; Grottick et al. 2003). However, more recently, Hahn and colleagues (2011) reported that DHβE was without effect on task accuracy produced by nicotine and that MLA co-administration counteracted the behavioral effects, thus defending the premise that α7 nAChRs have an important role 5-CSRTT performance. Other studies could be used to support either argument. For example, the important role of α4β2 nAChRs in attention can be derived from the positive observations with selective α4β2 receptor agonists (e.g., ABT-418, ABT-594, epibatidine, SIB-1765F) on 5-CSRTT performance, whereas the α7 nAChR agonist ARR-17779 lacked effects (Grottick and Higgins 2000; Grottick et al. 2003; Hahn et al. 2003; Mohler et al. 2010). Additionally, the α4β2 nAChR agonist ABT-418 was found to improve attention in adults with ADHD (Wilens et al. 1999).

To further explore the contribution of α7 nAChRs to attention, transgenic knockout mice for the α7 receptor (α7 KO) have been developed and trained in the 5-CSRTT (Hoyle et al. 2006; Young et al. 2004). Results from Young and colleagues (2004) indicated that α7 KO mice took significantly longer to acquire the task and upon reaching stable performance exhibited higher levels of omissions compared to the aged-matched wild-type mice, but that no group differences were observed for choice accuracy and correct response latency. In a more detailed examination, Hoyle and colleagues (2006) demonstrated that α7 KO mice were less accurate, had slower correct response latencies, earned fewer rewards, and exhibited higher premature responses than the wild types. Interestingly, nicotine administration failed to alter 5-CSRTT performance for either genotype. When the authors refined the original task parameters (i.e., reduced the time allowed to make a response selection and punished premature responding with a timeout period), the α7 KO mice demonstrated higher omissions and earned fewer rewards than wild types, but there were no longer any differences in accuracy, response latency, or premature responses. Again, nicotine did not alter task performance in either mouse genotype. Results from these two studies thus suggest a role for α7 nAChRs on 5-CSRTT acquisition and selective task parameters, but additional work will be required to fully determine their role in sustained attention.

Finally, other studies suggest that additional nAChR subtypes in the brain might have an important role in 5-CSRTT performance (i.e., studies where the α2-4β4 receptor agonist SIB-1553A was evaluated, Terry et al. 2002).

The evaluation of dopamine (DA) receptor agonists and antagonists in the 5-CSRTT has also revealed important roles of dopamine and its receptors in attention-related processes. For example, intra-mPFC infusions of the DA D_1 receptor agonist SKF 38393 improved choice accuracy and correct response latency in rats with low baseline accuracy (<75%) but not higher accuracies, effects blocked by intra-mPFC infusion of the DA D_1 receptor antagonist SCH 23390 (Granon et al. 2000). In contrast, infusions into the mPFC with SCH 23390 but not the DA D_2 receptor antagonist sulpiride disrupted choice accuracy in the high accuracy animals implicating a role for DA D_1 receptors. Interestingly, systemic administration of sulpiride but not the DA D_2 receptor agonists apomorphine or quinpirole counteracted the mPFC lesion decreases in choice accuracy (Passetti et al. 2003). Infusions of sulpiride into the nucleus accumbens (NAC) also resulted in improvements in choice accuracy and impulsivity in mPFC lesioned rats (Pezze et al. 2009). Moreover, infusion of SKF 38393 into the NAC improved accuracy and decreased trial omissions in normal animals, whereas intra-NAC injections with SCH 23390 disrupted task performance and quinpirole increased perseverative responding (Pezze et al. 2007). Taken together, these data suggest that dopaminergic projections to the rat PFC play a primary role in modulating choice accuracy and inhibitory control responding. Somewhat surprising, systemic administration of DA D_1 receptor agonists (e.g., SKF 38393) has not been assessed in the 5-CSRTT and, in contrast to the effects produced by intra-cerebral injection, systemic administration of DA D_2 receptor agonists or D_1 and D_2 receptor antagonists typically results in decreased premature responding, increased trial omissions, and slower response latencies with little to no effect on choice accuracy in normal animals (van Gaalen et al. 2006; Fernando et al. 2012).

Administration of indirect-acting DA agonists (e.g, amphetamine, cocaine, GBR 12909, and methylphenidate), under specific training conditions, tends to increase choice accuracy, reduce response latency, and increase premature responding in both adult and aged rats (Puumala et al. 1996; Grottick and Higgins 2002; Bizarro et al. 2004; van Gaalen et al. 2006; Paine et al. 2007; Navarra et al. 2008a; Fernando et al. 2012;) and mice (Loos et al. 2010; Yan et al. 2011). Despite their clinical utility for ADHD (Bidwell et al. 2011), amphetamine and methylphenidate produce modest effects on choice accuracy and, in general, psychostimulants appear to more significantly affect other behavioral measures in 5-CSRTT (e.g., premature responding and omissions), which may reflect activation of the striatal system opposed to cortical pathways. Interestingly, modafinil, a wake-promoting drug, seems to exert its pharmacological effects by acting as a weak DA transport inhibitor to elevate extracellular DA that, in turn, stimulates catecholamine receptors to induce arousal (Wisor 2013). The wake-promoting effects of modafinil in humans have been well documented (Westenson et al. 2002) and modafinil may have clinical utility for the treatment of cognitive impairments associated with ADHD (Turner et al. 2004a) and schizophrenia (Turner et al. 2004b; Scoriels et al. 2012). Despite the clinical interest of modafinil, very few studies have characterized its preclinical effects on sustained attention (Waters et al. 2005; Morgan et al. 2007). Using a 3-choice task, Morgan and colleagues (2007) found dose-dependent improvements in choice accuracy and premature responding with shorter response latencies in middle-aged rats, whereas Waters and colleagues (2005) using the 5-choice task under standard and high task demand conditions found no significant effects of modafinil on any behavioral measure under either task condition. Clearly, further work with modafinil on sustained attention is warranted to resolve these findings. Given the positive preclinical findings with intra-cerebrally administered DA D_1 receptor agonists and the clinical success with amphetamine, methylphenidate, and modafinil, additional research is required to understand the relationship between changes in cortical DA and attentional processing especially as it relates to attention and impulse control deficits observed in various clinical populations.

The noradrenergic (NA) system is well documented to be involved in arousal, attention, and impulsivity and pharmacological compounds modulating NA activity have been developed for the successful treatment of ADHD (Evenden 1999; Brennan and Arnsten 2008; Bidwell et al. 2011). Additionally, NA compounds may have clinical utility in other patient populations (e.g., Alzheimer's disease and schizophrenia) with similar symptoms (Egeland et al. 2003; Levinoff et al. 2005). It is thought that by increasing NA function through direct or indirect activation of α1 and α2 adrenoceptors, attention and impulsivity impairments may be diminished. Assessment of NA receptor ligands in the 5-CSRTT has indicated that α1 adrenoceptor agonists (e.g., St-587 and phenylephrine) tend to improve choice accuracy and decrease premature responding, but increase omissions and correct response latency (Puumala et al. 1997; Pattij et al. 2012). Co-administration of the α1 adrenoceptor antagonist prazosin blocked the St-587-induced improvement in accuracy, but premature responding was unaltered and prazosin alone tended to decrease task performance. Results with α2 adrenoceptor agonists and antagonists have also indicated an important role for this adrenergic receptor subtype in certain aspects of 5-CSRTT performance. For example, the α2 agonists clonidine, dexmedetomidine, and guanfacine primarily decrease premature responding (Sirvio et al. 1994; Fernando et al. 2012; Pattij et al. 2012), whereas α2 antagonists (e.g., atipamezole and yohimbine) increase premature responding (Sirvio et al. 1993; Sun et al. 2010). At higher doses, possibly reflecting sedation, α2 agonists tend to increase response latencies and trial omissions (Sirvio et al. 1994; Fernando et al. 2012). Interestingly, the α2 antagonist atipamezole was reported to improve choice accuracy under specific task parameter manipulation (i.e., reducing brightness of the visual stimulus) presumably by increasing norepinephrine (NE) levels via blockade of inhibitory autoreceptors on NA neurons leading to a state of enhanced arousal (Sirvio et al. 1993). Administration of the β2 adrenoceptor agonist clenbuterol also improved accuracy and decreased impulsivity at test doses that did not impair other task parameters (Pattij et al. 2012). Clenbuterol has previously shown beneficial effects, though modest, on working memory deficits in aged animals (Ramos et al. 2008). Based on the notion that increased NE levels leads to improvements of attentional processing (Carli et al. 1983; Dalley et al. 2001), studies have evaluated the ability of NE reuptake inhibitors (e.g., atomoxetine, desipramine, and reboxetine) to modulate 5-CSRTT performance under standard (van Gaalen et al. 2006; Paine et al. 2007; Robinson et al. 2008a; Fernando et al. 2012; Pattij et al. 2012) and high task demand conditions (Navarra et al. 2008a; Paterson et al. 2011; Robinson 2012). Overall, the findings demonstrate a consistent reduction in premature responding (impulsivity) and when the task became more challenging atomoxetine and reboxetine improved choice accuracy, though this effect was only observed in poorer performing subjects (Navarra et al. 2008a; Robinson 2012). The efficacy of NE reuptake inhibitors in modulating attention and impulsivity is thought to be related to their ability to increase extracellular catecholamine levels and simultaneously activate cortical postsynaptic DA D_1 receptors and α2 adrenoceptors. These neurochemical actions may also underlie the beneficial therapeutic effects of atomoxetine and guanfacine on core symptoms of ADHD (Bidwell et al. 2011).

Serotonin (5-HT) receptor systems are well documented to be important for learning and memory (for a review, see Seyedabadi et al. 2014) and they also have been shown to play an important role in modulating 5-CSRTT accuracy and behavioral response inhibition (Winstanley et al. 2004a). Assessment of 5-HT receptor agonists and antagonists in the 5-CSRTT reveals a complex pattern of effects on the various behavioral measures. Infusion of the 5-HT_{1A} receptor agonist 8-OH-DPAT into the mPFC enhanced choice accuracy, which was blocked by the selective 5-HT_{1A} receptor antagonist WAY 100635 implicating 5-HT_{1A} receptors on attentional processing (Winstanley et al. 2003). However, systemic administration of 8-OH-DPAT leads to impaired accuracy and increased omissions and premature responses (Carli and Samanin 2000). These divergent effects may reflect the activation of presynaptic vs. postsynaptic 5-HT_{1A} receptors (or both simultaneously). The 5-HT_2 receptor family has also been extensively studied due to its involvement in impulsivity (i.e., premature responding). Opposing roles have been described for 5-HT_{2A} and 5-HT_{2C} receptor compounds such that 5-HT_{2A} receptor agonists (e.g., DOI) and 5-HT_{2C} receptor antagonists (e.g., SB 242084) increase inhibitory control responding, whereas 5-HT_{2A} receptor antagonists (e.g., M1000907 and ketanserin) and 5-HT_{2C} receptor agonists (e.g., Ro60-0175; WAY-163909) reduce impulsivity in both rats and mice irrespective of whether the test compounds are infused directly into specific brain regions (e.g., mPFC, PrL-cortex, IL-cortex, or NAC; Koskinen and Sirvio 2001; Winstanley et al. 2003, 2004b; Robinson et al. 2008b) or administered systemically (Koskinen and Sirvio 2001; Fletcher et al. 2007; Navarra et al. 2008b; Quarta et al. 2012). Moreover, 5-HT_{2A} receptor antagonists and 5-HT_{2C} receptor agonists are efficacious in reversing psychostimulant (amphetamine and cocaine) and NMDA receptor antagonist MK-801 induced increases in 5-CSRTT premature responding (Fletcher et al. 2011) as well as cocaine seeking behavior (Neisewander and Acosta 2007) suggesting clinical utility for impulsivity and substance abuse. The recent FDA approval of the 5-HT_{2C} receptor agonist lorcaserin for obesity may afford new opportunities to evaluate this compound as well as other 5-HT_{2C} ligands for additional clinical indications (e.g., ADHD, smoking cessation, and substance abuse; see Higgins et al. 2013). To date, studies with 5-HT reuptake inhibitors (Humpston et al. 2013), 5-HT_3 receptor antagonists (Kirkby et al. 1996), and 5-HT_6 receptor antagonists (de Bruin et al. 2013) on 5-CSRTT performance have not indicated that these classes of compounds significantly affect sustained attention (or the other outcome measures assessed in this task); however, additional studies are warranted.

3 Five-Choice Continuous Performance Task

3.1 Task Description

The 5C-CPT (see Fig. 2) has recently been developed for rats (Barnes et al. 2012), mice (Young et al. 2009), and humans (Eyler et al. 2011; Young et al. 2011a). Like the 5-CSRTT, the 5C-CPT assesses the subject's ability to spatially divide its attention across five signal locations and when illumination of a single aperture (correct target) occurs the subject must choose that hole in order to receive a reward. However, the 5C-CPT adds a nontarget component in which all five aperture holes are illuminated, thus requiring the subject to inhibit choice selection. The addition of nontarget trials with correct target trials increases task similarity to the human CPT version (Young et al. 2009) and allows analysis of additional outcome measures such as hits or misses during target trials and correct rejections or false alarms during nontarget trials, thereby adding signal detection theory analysis along with the standard 5-CSRTT measurements (i.e., accuracy, omissions, premature, perseverative, and timeout responding, as well as response latency). It has been suggested that due to the absence of nontarget trials within the 5-CSRTT protocol the task may not accurately detect subtle differences that exist between vigilance and sustained attention especially as it relates to human vigilance (Robbins 1998). Consequently, the 5C-CPT was developed to overcome the limitations of the 5-CSRTT, thus increasing its translational value to human research (Young et al. 2009).

Fig. 2 Diagram of the five-choice continuous performance task (5C-CPT). In the 5C-CPT (like the 5-CSRTT), subjects are required to scan five apertures in an operant chamber for the appearance of a brief light stimulus (presented pseudorandomly) and to make a nose-poke response in the correct spatial location (i.e., the aperture where the light stimulus was presented) in order to receive a food reward (Go-Trials). However, unlike the 5-CSRTT, in the 5C-CPT, on some trials visual stimuli are presented in all five locations simultaneously and for these trials the subject must learn to withhold a response (NoGo-trials)

3.2 Neural Substrates

Given the relatively recent development of the 5C-CPT, there are no lesion- or neurochemical-based experimental data yet available to elucidate the neuroanatomical and neurobiological substrates of task performance. However, since the 5C-CPT utilizes methodological components found in both the 5-CSRTT and the operant 2-lever SDT, similar neuronal pathways (e.g., mPFC, striatum, and hippocampus) are likely involved (Robbins 2002; Sarter et al. 2005; Demeter et al. 2008). Despite this preclinical limitation, functional magnetic resonance imaging (fMRI) studies have demonstrated activation of fronto-striatal and parietal cerebral systems in humans performing the 5C-CPT (Eyler et al. 2011), which is in agreement with other CPT assessments in humans (Schneider et al. 2010).

3.3 Pharmacology

Several pharmacological studies have provided insight into the neurotransmitter systems involved in 5C-CPT performance in rats and mice (see Table 1). For example, improvements in choice accuracy of the 5C-CPT have been observed following mPFC infusions (Granon et al. 2000) as well as systemic administration (Barnes et al. 2012) of the DA D_1 receptor agonist SKF 38393. While SKF 38393 diminished task performance under baseline conditions, the drug improved attention when the task became more challenging (Barnes et al. 2012). By employing an extended variable intertrial-interval (vITI) procedure which altered standard baseline measurements (e.g., reductions in % correct, % omissions, hit rate, and sensitivity index, along with increased premature responding) SKF 38393 administration was able to reverse performance deficits such that improvements were seen in % omissions, correct response latency, premature responding, hit rate, and the sensitivity index. These results suggest that under high task demands activation of the DA D_1 receptor system can improve attention/vigilance. Further support for an important role of DA receptors in attention/vigilance is evident in the results of experiments with DA D_4 receptor knockout (D4R-KO) mice (Young et al. 2011b). D4R-KO mice exhibited an increase in false alarm responding during nontarget trials without any increases in premature responding suggesting a selective deficiency in response inhibition. The authors were able to further differentiate response inhibition and premature responding by administering the 5-HT_{2C} receptor antagonist SB 242084 to these D4R-KO mice. As expected SB 242084 increased premature responding and this effect occurred in the absence of increases in false alarm responding, providing evidence that the 5C-CPT can dissociate impulsive actions from response inhibition. Young and colleagues (2013) have also examined the effects of cholinergic manipulation on 5C-CPT performance in C57BL/6N mice. Administration of nicotine or the $\alpha4\beta2$ nicotinic receptor agonist ABT-418 alone or in combination with scopolamine improved task performance, thus providing additional support for an important role of $\alpha4\beta2$ nAChRs in sustained attention. In contrast, the $\alpha7$ nAChR agonist PNU 282987 failed to show improvements supporting previous 5-CSRTT data in rats (Grottick and Higgins 2000; Hahn et al. 2003). Thus, while pharmacological characterizations in the 5C-CPT are limited, the results obtained to date are encouraging and likely to inspire further work.

4 Signal Detection Task

4.1 Task Description

The SDT (see Fig. 3) was designed to emphasize temporal components of attention by presenting a target stimulus repeatedly in a single spatial location and varying the timing of its presentation (see Bushnell et al. 2003). The SDT thus differs from the 5-CSRTT in that the SDT requires the detection of a single centrally presented visual signal instead of detection of a target stimulus across multiple-choice locations (Bushnell 1995; McGaughy and Sarter 1995a, 1995b; Rezvani et al. 2002; Echevarria et al. 2005). Briefly, the SDT requires the subject to monitor a central panel for the presence (signal trial) or absence (non-signal or blank trial) of a visual light stimulus that varies in time and intensity. Following the trial (signal or non-signal) presentation and a short variable delay, two response levers are extended into the operant chamber and the subject must press the appropriately designated lever, based on trial type presented, to receive a reward. The SDT generates four main outcome measures: (1) hits, correct lever press for signal trial; (2) misses, incorrect lever press on the blank trial following a signal trial; (3) correct rejections, correct lever press for a blank trial; and (4) false alarms,

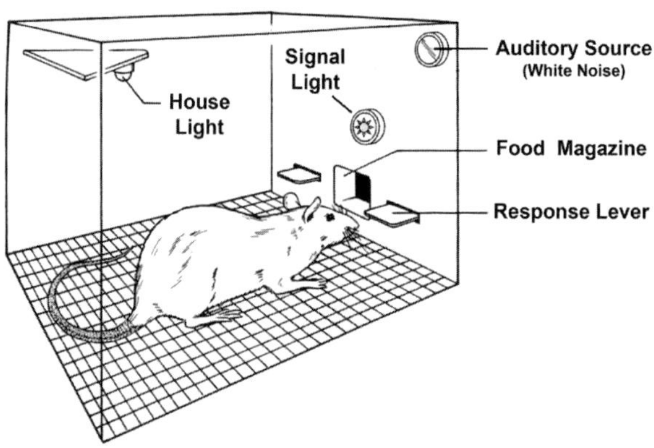

Fig. 3 Diagram of the signal detection task (SDT). In the SDT, subjects are required to monitor a central panel in an operant chamber for the presence (signal trial) or absence (non-signal or blank trial) of a light stimulus that varies in time and intensity. Following the trial (signal or non-signal) presentation and a short variable delay, two response levers are extended into the operant chamber and the subject must press the appropriately designated lever (based on the trial type presented) to receive a reward

incorrect lever press on the designated signal lever when the target signal did not occur. Response omissions (no lever selection after presentation of either trial type) and response latencies can also be assessed. The paradigm also affords the investigator with the ability to alter stimulus duration, intensity, and modality which can place greater cognitive load on the subject in order to better assess sensory function and attentiveness. Distraction (e.g., flashing house light) can be added to the protocol to further increase task demands. The SDT has been used extensively to study the neuronal networks associated with attentional processing in rats (Gill et al. 2000; Rezvani et al. 2002; Sarter et al. 2005), mice (Mohler et al. 2001; St Peters et al. 2011) and humans (Bushnell et al. 2003; Demeter et al. 2008).

4.2 Neural Substrates

Much of the work devoted to determining the neurobiological substrates of SDT performance to date has focused on basal forebrain cholinergic inputs to the PFC and other basal forebrain neurons (e.g., glutamatergic and GABAergic neurons) that are known to modulate activity of the cholinergic inputs to the PFC (for reviews, see Sarter et al. 2005; Hasselmo and Sarter 2011). The results of these experiments have revealed a complex interplay of neuronal phenotypes in modulating discrete attentional processes. For example, 192 IgG-saporin infusion into the basal forebrain (which selectively destroys cholinergic neurons while leaving GABAergic and glutamatergic neurons intact) resulted in decreases in hit rate performance without affecting correct rejection and false alarm performance (McGaughy et al. 1996). Conversely, infusion of NMDA into the basal forebrain (which leads to cholinergic overactivity specifically when a behavioral stimulus is encountered) increased false alarm performance without affecting hit rate performance. Notably, the opposite effect occurred following infusion of the NMDA receptor antagonist DL-2-amino-5-phosphonovaleric acid (APV; Turchi and Sarter 2001). Similar dissociations between the attentional functions mediated by basal forebrain cholinergic and GABAergic neurons have been observed. For example, basal forebrain lesions produced with ibotenic acid (which primarily destroy non-cholinergic, particularly GABAergic neurons) also led to a selective increase in false alarms, as opposed to affecting hit rate performance (Burk and Sarter 2001).

4.3 Pharmacology

As with the 5-CSRTT, the SDT has undergone considerable pharmacological examination to identify not only the underlying neurochemical pathways associated with performance of the task, but also to evaluate potential therapies for attention-related disorders (see Table 1). Similar to the findings with the 5-CSRTT and the 5C-CPT, administration of the muscarinic antagonist scopolamine, the NMDA glutamate antagonists ketamine and MK-801, as well as the nicotinic antagonist mecamylamine impair SDT performance by altering several behavioral measures of attention such as hit rates, correct rejections, false alarms, and omissions (McGaughy and Sarter 1995a, 1995b; Turchi et al. 1995; Bushnell et al. 1997; Presburger and Robinson 1999; Nelson et al. 2002; Rezvani et al. 2002, 2009a), thereby confirming the importance of cholinergic and glutamatergic systems in attentional processes. Performance impairments have also been observed with alcohol (Rezvani and Levin 2003), typical and atypical antipsychotics (e.g., clozapine, haloperidol, and raclopride; Rezvani and Levin 2004; Hillhouse and Prus 2013), benzodiazepines (e.g., chlordiazepoxide; McGaughy and Sarter 1995a, b; Bushnell et al. 1994, 1997), and organic chemical toxins (e.g., chlorpyrifos and toluene; Samsam et al. 2005). Deficits in attention produced by these drug classes afford the investigator with a means to (1) pharmacologically model specific disease states (e.g., Alzheimer's disease, schizophrenia, or chemical poisoning) and (2) elucidate the underlying mechanism(s) associated with investigational compounds which should, in turn, lead to the development and evaluation of new therapeutic agents. The SDT is also capable of detecting pharmacological effects in normal, aged, or genetically altered subjects (e.g., spontaneous hypertensive rats (SHR) or receptor knockout mice). Through task contingency manipulation, greater cognitive demand can be placed on the subject that decreases attentional performance and this may also mimic dysfunctions in human attention. Drugs (e.g., methylphenidate, nicotine, ABT-418) that have beneficial effects on human attention (White and Levin 1999; Wilens et al. 1999; Min et al. 2001; Bidwell et al. 2011) have also been found to improve SDT performance (McGaughy et al. 1999; Rezvani et al. 2002, 2009a; Hillhouse and Prus 2013). Methylphenidate and nicotine also attenuated scopolamine- and MK-801-induced SDT performance deficits (Rezvani et al. 2008, 2009a). Collectively, the above findings demonstrate the predictive validity as well as the versatility of the SDT model. However, it should be noted that the complementary findings in the animal studies cited above for methylphenidate and nicotine (i.e., observations of positive "stand-alone" effects and the ability to attenuate "pharmacologic-induced" impairments) have not always been observed. For example, methylphenidate, nicotine, and α4β2 nAChR agonists (e.g., ABT-418, A-82695, AZD 3480, sazetidine, and S-38232) failed to demonstrate positive "stand-alone" effects in the standard SDT version (Turchi et al. 1995; McGaughy et al. 1999; Howe et al. 2010; Rezvani et al. 2011, 2012). Yet, by decreasing attentional performance of the subject with either pharmacologic impairment (e.g., scopolamine and MK-801) or the use of the distractor version of the SDT (dSDT) α4β2 nAChR agonists were able to improve task performance (Howe et al. 2010; Rezvani et al. 2011, 2012). Howe and colleagues (2010) also suggested that some of the inconsistency with nicotine on sustained attention might be due to its activation at α7 nAChRs. In their experiments (as noted above) nicotine administered alone failed to improve performance of the dSDT, whereas nicotine combined with the nAChR antagonist MLA was beneficial. These results led the authors to suggest that activation of α4β2 nAChRs alone is sufficient to improve attention and that co-activation α7 nAChRs might actually be undesirable. Interestingly, however, the α7 nAChR agonist/5-HT3 receptor antagonist

RG3487 (MEM3454) produced "stand-alone" improvements on sustained attention (Rezvani et al. 2009b).

Acetylcholinesterase inhibitors (e.g., donepezil and physostigmine) have also been evaluated in the SDT and despite their positive clinical effects on attention (Foldi et al. 2005; Bentley et al. 2011), they failed to alter sustained attention when administered alone in animals (McGaughy and Sarter 1998; Rezvani et al. 2012). However, donepezil was efficacious in reversing MK-801 impairments (Rezvani et al. 2012).

5 Conclusion

The behavioral paradigms discussed in this chapter (5-CSRTT, 5C-CPT, and SDT) have been very useful for elucidating the neuroanatomical structures and pathways as well as the neurotransmitter systems that contribute to attentional processes and there appears to be general agreement that the construct validity of these tasks is reasonable. The tasks have also been used extensively to evaluate a variety of pharmacological agents for their potential as therapeutic agents for diseases where attention is impaired. In many cases the compounds that have a proven benefit in clinical populations are also effective in the preclinical models, supporting the argument that they may also have predictive validity. However, there are certainly exceptions to this statement and due to the number of recent clinical trials failures of novel compounds developed for neuropsychiatric disorders, the predictive validity of animal models and well as the tasks designed to assess behaviors in animal models is coming under increasing scrutiny. Thus, while it is unlikely that any behavioral task or animal model will enjoy universal acceptance as having predictive validity, it is important that additional studies be conducted to resolve conflicts in the literature where possible.

References

Amitai N, Markou A (2009) Chronic nicotine improves cognitive performance in a test of attention but does not attenuate cognitive disruption induced by repeated phencyclidine administration. Psychopharmacology (Berl) 202(1–3):275–286

Barak S, Weiner I (2011) Putative cognitive enhancers in preclinical models related to schizophrenia: the search for an elusive target. Pharmacol Biochem Behav 99:164–189

Barnes SA, Young JW, Neill JC (2012) D_1 receptor activation improves vigilance in rats as measured by the 5-choice continuous performance test. Psychopharmacology (Berl) 220:129–141

Barnett JH, Robbins TW, Leeson VC, Sahakian BJ, Joyce EM, Blackwell AD (2010) Assessing cognitive function in clinical trials of schizophrenia. Neurosci Biobehav Rev 34:1161–1177

Bentley P, Driver J, Dolan RJ (2011) Cholinergic modulation of cognition: insights from human pharmacological functional neuroimaging. Prog Neurobiol 94:360–388

Bidwell LC, McClernon FJ, Kollins SH (2011) Cognitive enhancers for the treatment of ADHD. Pharmacol Biochem Behav 99:262–274

Biederman J (2005) Attention-deficit/hyperactivity disorder: a selective overview. Biol Psychiatry 57:1215–1220

Bizarro L, Patel S, Murtagh C, Stolerman IP (2004) Differential effects of psychomotor stimulants on attentional performance in rats: nicotine, amphetamine, caffeine and methylphenidate. Behav Pharmacol 15:195–206

Blondel A, Sanger DJ, Moser PC (2000) Characterization of the effects of nicotine in the five-choice serial reaction time task in rats: antagonist studies. Psychopharmacology (Berl) 149:293–305

Brennan AR, Arnsten AFT (2008) Neuronal mechanisms underlying attention deficit hyperactivity disorder: the influence of arousal on prefrontal cortical function. Ann N Y Acad Sci 1129:236–245

Brown RG, Scott LC, Bench CJ, Dolan RJ (1994) Cognitive function in depression: its relationship to the presence and severity of intellectual decline. Psychol Med 24(4):829–847

Buccafusco JJ, Terry AV Jr (2009) A reversible model of the cognitive impairment associated with schizophrenia in monkeys: potential therapeutic effects of two nicotinic acetylcholine receptor agonists. Biochem Pharmacol 78:852–862

Burk JA, Sarter M (2001) Dissociation between the attentional functions mediated via basal forebrain cholinergic and GABAergic neurons. Neuroscience 105:899–909

Bushnell PJ (1995) Overt orienting in the rat: parametric studies of cued detection of visual targets. Behav Neurosci 109:1095–1105

Bushnell PJ, Kelly KL, Crofton KM (1994) Effects of toluene inhalation on detection of auditory signals in rats. Neurotoxicol Teratol 16:149–160

Bushnell PJ, Oshiro WM, Padnos BK (1997) Detection of visual signals by rats: effects of chlordiazepoxide and cholinergic and adrenergic drugs on sustained attention. Psychopharmacology (Berl) 134:230–241

Bushnell PJ, Benignus VA, Case MW (2003) Signal detection behavior in humans and rats: a comparison with matched tasks. Behav Processes 64:121–129

Carli M, Samanin R (2000) The 5-HT_{1A} receptor agonist 8-OH-DPAT reduces rats' accuracy of attentional performance and enhances impulsive responding in a five-choice serial reaction time task: role of presynaptic 5-HT_{1A} receptors. Psychopharmacology (Berl) 149:259–268

Carli M, Robbins TW, Evenden JL, Everitt BJ (1983) Effects of lesions to ascending noradrenergic neurons on performance of a 5-choice serial reaction task in rats: implications for theories of dorsal noradrenergic bundle function based on selective attention and arousal. Behav Brain Res 9:361–380

Chudasama Y, Robbins TW (2004) Psychopharmacological approaches to modulating attention in the five-choice serial reaction time task: implications for schizophrenia. Psychopharmacology (Berl) 174:86–98

Chudasama Y, Passetti F, Rhodes SEV, Lopian D, Desai A, Robbins TW (2003) Dissociable aspects of performance on the five-choice serial reaction time task following lesions of the dorsal anterior cingulated, infralimbic and orbitofrontal cortex in the rats: differential effects on selectivity, impulsivity and compulsivity. Behav Brain Res 146:105–119

Chudasama Y, Doobay VM, Lui Y (2012) Hippocampal-prefrontal cortical circuit mediates inhibitory response control in the rat. J Neurosci 32:10915–10924

Cole BJ, Robbins TW (1992) Forebrain norepinephrine: role in effortful but not automatic processing in the rat. Neuropsychopharmacology 7:129–142

Dalley JW, McGaughy J, O'Connell MT, Cardinal R, Levita L, Robbins TW (2001) Distinct changes in cortical acetylcholine and noradrenaline efflux during contingent and non-contingent performance of a visual attentional task. J Neurosci 21:4908–4914

Dalley JW, Theobald DE, Eagle DM, Passetti F, Robbins TW (2002) Deficits in impulse control associated with tonically elevated serotonergic function in rat prefrontal cortex. Neuropsychopharmacology 26:716–728 de Bruin NMWJ, Fransen F, Duytschaever H, Grantham C, Megens AAHP (2006) Attentional performance of (C57BL/6J x 129Sv)F2 mice in the five-choice serial reaction time task. Physiol Behav 89:692–703

de Bruin NMWJ, McCreary AC, van Loevezijn A, de Vries TJ, Venhorst J, van Drimmelen M, Kruse CG (2013) A novel highly selective 5-HT6 receptor antagonist attenuates ethanol and nicotine seeking but does not affect inhibitory response control in Wistar rats. Behav Brain Res 236:157–165

Demeter E, Sarter M, Lustig C (2008) Rats and humans paying attention: cross-species task development or translational research. Neuropsychology 22:787–799

Echevarria DJ, Brewer A, Burk JA, Brown SN, Manuzon H, Robinson JK (2005) Construct validity of an operant signal detection task for rats. Behav Brain Res 157:283–290

Egeland J, Rund BR, Sundet K, Landro NI, Asbjornsen A, Lund A, Roness A, Stordal KI, Huqdahl K (2003) Attention profile in schizophrenia compared with depression: differential effects of processing speed, selective attention and vigilance. Acta Psychiatr Scand 108:276–284

Evenden J (1999) Impulsivity: a discussion of clinical and experimental findings. J Psychopharmacol 13:180–192

Eyler LT, Dawes SE, Asgaard G, Young JW (2011) Abnormalities of brain response during vigilance and inhibition in bipolar disorder. Bipolar Disord 13(Suppl 1):42

Fernando ABP, Economidou D, Theobald DE, Zou M-F, Newman AH, Spoelder M et al (2012) Modulation of high impulsivity and attentional performance in rats by selective direct and indirect dopaminergic and noradrenergic receptor agonists. Psychopharmacology (Berl) 219:341–352

Fletcher PJ, Tampakeras M, Sinyard J, Higgins GA (2007) Opposing effects of 5-HT_{2A} and 5-HT_{2C} receptor antaonists in the rat and mouse on premature responding in the five-choice serial reaction time test. Psychopharmacology (Berl) 195:223–234

Fletcher PJ, Rizos Z, Noble K, Higgins GA (2011) Impulsive action induced by amphetamine, cocaine and MK801 is reduced by 5-HT_{2C} receptor stimulation and 5-HT_{2A} receptor blockade. Neuropharmacology 61:468–477

Fletcher PJ, Soko AD, Higgins GA (2013) Impulsive action in the 5-choice serial reaction time test in 5-HT2c receptor null mutant mice. Psychopharmacology 226:561–570

Foldi NS, White REC, Schaefer LA (2005) Detecting effects of donepezil on visual selective attention using signal detection parameters in Alzheimer's disease. Int J Geriatr Psychiatry 20:485–488

Gill TM, Sater M, Givens B (2000) Sustained visual attention performance-associated prefrontal neuronal activity: evidence for cholinergic modulation. J Neurosci 20:4745–4757

Granon S, Passetti F, Thomas KL, Dalley JW, Everitt BJ, Robbins TW (2000) Enhanced and impaired attentional performance after infusion of D1 dopaminergic receptor agents into rat prefrontal cortex. J Neurosci 20:1208–1215

Grottick AJ, Higgins GA (2000) Effect of subtype selective nicotinic compounds on attention as assessed by the five-choice serial reaction time task. Behav Brain Res 117:197–208

Grottick AJ, Higgins GA (2002) Assessing a vigilance decrement in aged rats: effects of pre-feeding, task manipulation, and psychostimulants. Psychopharmacology (Berl) 164:33–41

Grottick AJ, Haman M, Wyler R, Higgins GA (2003) Reversal of a vigilance decrement in the aged rat by subtype-selective nicotinic ligands. Neuropsychopharmacology 28:880–887

Hahn B, Sharples CG, Wonnacott S, Shoaid M, Stolerman IP (2003) Attentional effects of nicotinic agonists in rats. Neuropharmacology 44:1054–1067

Hahn B, Shoaib M, Stolerman IP (2011) Selective nicotinic receptor antagonists: effects on attention and nicotine-induced attentional enhancement. Psychopharmacology (Berl) 217:75–82

Harrison AA, Everitt BJ, Robbins TW (1997a) Central 5-HT depletion enhances impulsive responding without affecting the accuracy of attentional performance: interactions with dopaminergic mechanisms. Psychopharmacology (Berl) 133:329–342

Harrison AA, Everitt BJ, Robbins TW (1997b) Doubly dissociable effects of median and dorsal-raphe lesions on the performance of 5-choice serial reaction time test of attention in rats. Behav Brain Res 89:135–149

Hasselmo ME, Sarter M (2011) Modes and models of forebrain cholinergic neuromodulation of cognition. Neuropsychopharmacology 36:52–73

Higgins GA, Breysse N (2008) Rodent model of attention: the 5-choice serial reaction time task. Curr Protoc Pharmacol 41:5.49.1–5.49.20

Higgins GA, Sellers EM, Fletcher PJ (2013) From obesity to substance abuse: therapeutic opportunities for 5-HT_{2C} receptor agonists. Trends Pharmacol Sci 34:560–570

Hillhouse TM, Prus AJ (2013) Effects of the neurotensin NTS1 receptor agonist PD149163 on visual signal detection in rats. Eur J Pharmacol 721:201–207

Howe WM, Jinzhao J, Parikh V, Williams S, Mocaer E et al (2010) Enhancement of attentional performance by selective stimulation of α4β2* nAChRs: underlying cholinergic mechanisms. Neuropsychopharmacology 35:1391–1401

Hoyle E, Genn RF, Fernandes C, Stolerman IP (2006) Impaired performance of alpha7 nicotinic receptor knockout mice in the five-choice serial reaction time task. Psychopharmacology (Berl) 189:211–223

Humby T, Laird FM, Davies W, Wilkinson LS (1999) Visuospatial attentional orienting in mice: interactions between cholinergic manipulations and genotype. Eur J Neurosci 11:2813–2823

Humpston CS, Wood CM, Robinson ES (2013) Investigating the roles of different monoamine transmitters and impulse control using the 5-choice serial reaction time task. J Psychopharmacol 27:213–221

Jones DNC, Higgins GA (1995) Effects of scopolamine on visual attention in rats. Psychopharmacology (Berl) 120:142–149

Kirkby DL, Jones DN, Barnes JC, Higgins GA (1996) Effects of anticholinesterase drugs tacrine and E2020, the 5-HT(3) antagonist ondansetron, and the H(3) antagonist thioperamide, in models of cognition and cholinergic function. Behav Pharmacol 7:513–525

Koskinen T, Sirvio J (2001) Studies on the involvement of the dopaminergic system in the 5-HT_2 agonist (DOI)-induced premature responding in a five-choice serial reaction time task. Brain Res Bull 54:65–75

Laurent A, Saoud M, Bougerol T, d'Amato T, Anchisi AM, Biloa-Tang M, Dalery J, Rochet T (1999) Attentional deficits in patients with schizophrenia and in their non-psychotic first-degree relatives. Psychiatry Res 89(3):147–159

Lawrence AD, Sahakian BJ (1995) Alzheimer disease, attention, and the cholinergic system. Alzheimer Dis Assoc Disord 9(Suppl 2):43–49

Lee J, Park S (2006) The role of stimulus salience in CPT-AX performance of schizophrenia patients. Schizophr Res 81:191–197

Lehmann O, Grottick AJ, Cassel J-C, Higgins GA (2003) A double dissociation between serial reaction time and radial arm maze performance in rats subjected to 192 lgG-saporin- lesions of the nucleus basalis and/or the septal region. Eur J Neurosci 18:651–666

Leonard JA (1959) Five choice serial reaction apparatus. Report 326, Medical Research Council Applied Psychology Unit, Cambridge

Levin ED, Bushnell PJ, Rezvani AH (2011) Attention-modulating effects cognitive enhancers. Pharmacol Biochem Behav 99:146–154

Levinoff EJ, Saumier D, Chertkow H (2005) Focused attention deficits in patients with Alzheimer's disease and mild cognitive impairment. Brain Cogn 57:127–130

Loo SK, Hopfer C, Teale PD, Reite ML (2004) EEG correlates of methylphenidate response in ADHD: association with cognitive and behavioral measures. J Clin Neurophysiol 21:457–464

Loos M, Staal J, Schoffelmeer ANM, Smit AB, Spijker S, Pattij T (2010) Inhibitory controland response latency differences between C57BL/6J and DBA/2J mice in a Go/No-Go and 5-choice serial reaction time task and strain-specific responsivity to amphetamine. Behav Brain Res 214:216–224

McGaughy J, Sarter M (1995a) Behavioral vigilance in rats- task validation and effects of age, amphetamine and benzodiazepine receptor ligands. Psychopharmacology (Berl) 117:646–652

McGaughy J, Sarter M (1995b) Effects of chlordiazepoxide and scopolamine, but not aging, on the detection and identification of conditional visual stimuli. J Gerontol A Biol Sci Med Sci 50: B90–B96

McGaughy J, Sarter M (1998) Sustained attention performance in rats with intracortical infusions of 192 IgG-saporin-induced cortical cholinergic deafferentation: effects of physostigmine and FG 7142. Behav Neurosci 112:1519–1525

McGaughy J, Kaiser T, Sarter M (1996) Behavioral vigilance following infusions of 192 - IgG-saporin into the basal forebrain: selectivity of the behavioral impairment and relation to cortical AChE-positive fiber density. Behav Neurosci 110:247–265

McGaughy J, Decker MW, Sarter M (1999) Enhancement of sustained attention performance by the nicotinic acetylcholine receptor agonist ABT-418 in intact but not basal forebrain-lesioned rats. Psychopharmacology (Berl) 144:175–182

McGaughy J, Dalley JW, Morrison CH, Everitt BJ, Robbins TW (2002) Selective behavioral and neurochemical effects of cholinergic lesions produced by intrabasalis infusions of 192 - IgG-saporin on attentional performance in a five-choice serial reaction time task. J Neurosci 22:1905–1913

Min SK, Moon IW, Rw K, Hs S (2001) Effects of transdermal nicotine on attention and memory in healthy elderly non-smokers. Psychopharmacology (Berl) 159:83–88

Mirza N, Stolerman IP (1998) Nicotine enhances sustained attention in the rat under specific task conditions. Psychopharmacology (Berl) 138:266–274

Mirza N, Stolerman IP (2000) The role of nicotinic and muscarinic acetylcholine receptors in attention. Psychopharmacology (Berl) 107:541–550

Mohler EG, Meck WH, Williams CL (2001) Sustained attention in adult mice is modulated by prenatal choline availability. Int J Comp Psychol 14:136–150

Mohler EG, Franklin SR, Rueter LE, Fox GB, Decker MW, Browman KE (2010) ABT-594 improves performance in the 5-choice serial reaction time task under conditions of increased difficulty, sub-chronic dosing, and poorly-performing subjects. Pharmacol Biochem Behav 95:146–157

Morgan RE, Crowley JM, Smith RH, LaRoche RB, Dopheide MM (2007) Modafinil improves attention, inhibitory control, and reaction time in healthy, middle-aged rats. Pharmacol Biochem Behav 86:531–541

Muir JL, Robbins TW, Everitt BJ (1992) Disruptive effects of muscimol infused into the basal forebrain: differential interaction with cholinergic mechanisms. Psychopharmacology (Berl) 107:541–550

Muir JL, Everitt BJ, Robbins TW (1996) The cerebral cortex of the rat brain and visual attentional function: dissociable effects of mediofrontal, cingulate, anterior dorsolateral and parietal cortex lesions on a five-choice serial reaction time task. Cereb Cortex 6:470–481

Navarra R, Graf R, Huang Y, Logue S, Comery T, Hughes Z, Day M (2008a) Effects of atomoxetine and methylphenidate on attention and impulsivity in the 5-choice serial reaction time test. Prog Neuropsychopharmacol Biol Psychiatry 32:34–41

Navarra R, Comery TA, Graf R, Rosenzweig-Lipson S, Day M (2008b) The 5-HT_{2C} receptor agonist WAY-163909 decreases impulsivity in the 5-choice serial reaction time test. Behav Brain Res 188:412–415

Neisewander JL, Acosta JI (2007) Stimulation of 5-HT_{2C} receptors attenuates cue and cocaine-primed reinstatement of cocaine-seeking behavior in rats. Behav Neurosci 123:791–800

Nelson CL, Burk JA, Bruno JP, Sarter M (2002) Effects of acute and repeated systemic administration of ketamine on prefrontal acetylcholine release and sustained attention performance in rats. Psychopharmacology (Berl) 161:168–179

Nieuwenstein NR, Aleman A, de Haan EH (2001) Relationship between symptom dimensions and neurocognitive functioning in schizophrenia: a meta-analysis of WCSTand CPT studies. Wisconsin Card Sorting Test Continuous Performance Test. J Psychiatr Res 35:119–125

Paine TA, Tomasiewicz HC, Zhang K, Carlezon WA Jr (2007) Sensitivity of the five-choice serial reaction time task to the effects of various psychotropic drugs in Sprague–Dawley rats. Biol Pyschiatry 62:687–693

Parasuraman R, Warm JS, See JE (1998) Brain systems and vigilance. In: Parasuraman R (ed) The attentive brain. MIT, Cambridge, MA, pp 221–256

Passetti F, Levita L, Robbins TW (2002) The frontal cortex of the rat and visual attentional performance: dissociable functions of distinct medial prefrontal subregions. Cereb Cortex 12:1254–1268

Passetti F, Levita L, Robbins TW (2003) Sulpiride alleviates the attentional impairments of rats with medial prefrontal cortex lesions. Behav Brain Res 138:59–69

Paterson NE, Ricciardi J, Wetzler C, Hanania T (2011) Sub-optimal performance in the 5-choice serial reaction time task in rats was sensitive to methylphenidate, atomoxetine and D-amphetamine, but unaffected by the COMT inhibitor tolcapone. Neurosci Res 69:41–50

Pattij T, Janssen MCW, Loos M, Smit AB, Schoffelmeer ANM, van Gaalen MM (2007) Strain specificity and cholinergic modulation of visuospatial attention in three inbred mouse strains. Genes Brain Behav 6:579–587

Pattij T, Schetters D, Schoffelmeer ANM, van Gaalen MM (2012) On the improvement of inhibitory response control and visuospatial attention by indirect and direct adrenoceptor agonists. Psychopharmacology (Berl) 219:327–340

Pessoa L, Kastner S, Ungerleider LG (2003) Neuroimaging studies of attention: from modulation of sensory processing to top-down control. J Neurosci 23:3990–3998

Pezze M-A, Dallet JW, Robbins TW (2007) Differential roles of dopamine D1 and D2 receptors in the nucleus accumbens in attentional performance on the five-choice serial reaction time task. Neuropsychopharmacology 32:273–283

Pezze M-A, Dallet JW, Robbins TW (2009) Remediation of attentional dysfunction in rats with lesions of the medial prefrontal cortex by intra-accumbens administration of the dopamine $D_{2/3}$ receptor antagonist sulpiride. Psychopharmacology (Berl) 202:307–313

Posner MI, Petersen SE (1990) The attention system of the human brain. Annu Rev Neurosci 13:25–42

Presburger G, Robinson JK (1999) Spatial signal detection in rats is differentially disrupted by delta-9-tetrahydrocannabinol, scopolamine, and MK-801. Behav Brain Res 99:27–34

Puumala T, Sirvio J (1998) Changes in activities of dopamine and serotonin systems in the frontal cortex underlie poor choice accuracy and impulsivity of rats in an attentional task. Neuroscience 83:489–499

Puumala T, Ruotsalainen S, Jakala P, Koivisto E, Riekkinen P Jr, Sirvio J (1996) Behavioral and pharmacological studies on the validation of a new animal model for attention deficit hyperactivity disorder. Neurobiol Learn Mem 66:198–211

Puumala T, Riekkinen P Sr, Sirvio J (1997) Modulation of vigilance and behavioral activation by alpha-1 adrenoceptors in the rat. Pharmacol Biochem Behav 56:705–712

Quarta D, Naylor CG, Glennon JC, Stolerman IP (2012) Serontonin antagonists in the five-choice serial reaction time task and their interactions with nicotine. Behav Pharmacol 23:143–152

Ramos BP, Colgan LA, Nou E, Arnsten AF (2008) Beta2 adrenergic agonist, clenbuterol, enhances working memory performance in aging animals. Neurobiol Aging 29:1060–1069

Rezvani AH, Levin ED (2003) Nicotinic-alcohol interactions and attentional performance on an operant visual signal detection task in female rats. Pharmacol Biochem Behav 76:75–83

Rezvani AH, Levin ED (2004) Nicotinic-antipsychotic drug interactions and attentional performance in female rats. Eur J Pharmacol 486:175–182

Rezvani AH, Bushnell PJ, Levin ED (2002) Nicotine and mecamylamine effects on choice accuracy in an operant signal detection task. Psychopharmacology (Berl) 164:369–375

Rezvani AH, Kholdebarin E, Dawson E, Levin ED (2008) Nicotine and clozapine effects on attentional performance impaired by the NMDA antagonist dizocilpine in female rats. Int J Neuropsychopharmacol 11:63–70

Rezvani AH, Kholdebarin E, Cauley MC, Dawson E, Levin ED (2009a) Attentuation of pharmacologically-induced attentional impairment by methylphenidate in rats. Pharmacol Biochem Behav 92:141–146

Rezvani AH, Kholdebarin E, Brucato FH, Callahan PM, Lowe DA, Levin ED (2009b) Effect of R3487/MEM3454, a novel nicotinic alpha7 receptor partial agonist and 5-HT3 antagonist on sustained attention in rats. Prog Neuropsychopharmacol Biol Psychiatry 33:269–275

Rezvani AH, Cauley M, Sexton H, Xiao Y, Brown ML et al (2011) Sazetidine-A, a selective $\alpha 4\beta 2$ nicotinic acetylcholine receptor ligand: effects on dizocilpine and scopolamine-induced attentional impairments in female Sprague–Dawley rats. Psychopharmacology (Berl) 215:621–630

Rezvani AH, Cauley MC, Johnson EC, Gatto GJ, Levin ED (2012) Effects of AZD3480, a neuronal nicotinic acetylcholine receptor agonist, and donepezil on dizocilpine-induced attentional impairment in rats. Psychopharmacology (Berl) 223:251–258

Riccio CA, Reyonlds CR, Lowe P, Moore JJ (2002) The continuous performance test: a window on the neural substrates for attention. Arch Clin Neuropsychol 17:235–272

Robbins TW (1998) Arousal and attention: psychopharmacological and neuropsychological studies in experimental animals. In: Parasuraman R (ed) The attentive brain. MIT, Cambridge, MA, pp 189–219

Robbins TW (2002) The 5-choice serial reaction time task: behavioural pharmacology and functional neurochemistry. Psychopharmacology (Berl) 163:362–380

Robertson IH, Ward T, Ridgeway V, Nimmo-Smith I (1996) The structure of normal human attention: the test of everyday attention. J Int Neuropsychol Soc 2:525–534

Robinson ESJ (2012) Blockade of noradrenaline re-uptake sites improves accuracy and impulse control in rats performing a five-choice serial reaction time tasks. Psychopharmacology (Berl) 219:303–312

Robinson ESJ, Eagle DM, Mar AC, Bari A, Banerjee G, Jiang X et al (2008a) Similar effects of the selective noradrenaline reuptake inhibitor atomoxetine on three distinct forms of impulsivity in the rat. Neuropsychopharmacology 33:1028–1037

Robinson ESJ, Dalley JW, Theobald DEH, Glennon JC, Pezze MA, Murphy ER, Robbins TW (2008b) Opposing roles for 5-HT_{2A} and 5-HT_{2C} receptors in the nucleus accumbens on inhibitory response control in the 5-choice serial reaction time task. Neuropsychopharmacology 33:2398–2406

Romberg C, Mattson MP, Murgai MR, Bussey TJ, Saksida LM (2011) Impaired attention in mice in the 3XTgAD mouse model of Alzheimer's disease: rescue by donepezil (Aricept). J Neurosci 31:3500–3507

Rosvold H, Mirsky SI, Bransome ED Jr, Beck LH (1956) A continuous performance test of brain damage. J Consult Psychol 20:343–350

Ruotsalainen S, Miettinen R, MacDonald E, Koivisto E, Sirvio J (2000) Blockade of muscarinic, rather than nicotinic, receptors impairs attention, but does not interact with serotonin depletion. Psychopharmacology (Berl) 148:111–123

Sahakian BJ, Jones G, Levy R, Warburton D, Gray J (1989) The effects of nicotine on attention, information processing and short term memory in patients with dementia of the Alzheimer type. Br J Psychiatry 154:797–800

Samsam TE, Hunter DL, Bushnell PJ (2005) Effects of dietary and repeated acute exposure to chlorpyrifos on learning and sustained attention in rats. Toxicol Sci 87:460–468

Sanchez-Roige S, Pena-Oliver Y, Stephens DN (2012) Measuring impulsivity in mice: the five-choice serial reaction time task. Psychopharmacology (Berl) 219:253–270

Sarter M, Hasselmo M, Bruno J, Givens B (2005) Unraveling the attentional functions of cortical cholinergic inputs: interactions between signal-driven and cognitive modulation of a signal detection. Brain Res Rev 48:98–111

Schneider MF, Krick CM, Retz W, Hengessch G, Retz-Junginger P et al (2010) Impairment of fronto-striatal and parietal cerebral networks correlates with attentional deficit hyperactivity disorder (ADHD) psychopathology in adults-a functional magnetic resonance imaging (fMRI) study. Psychiatry Res 183:75–84

Scoriels L, Barnett JH, Soma PK, Sahakian BJ, Jones PB (2012) Effects of modafinil on cognitive functions in first episode psychosis. Psychopharmacology (Berl) 220:249–258

Seyedabadi M, Fakhfouri G, Ramezani V, Mehr SE, Rahimian R (2014) The role of serotonin in memory: interactions with neurotransmitters and downstream signaling. Exp Brain Res 232:723–738

Sirvio J, Jakala P, Mazurkiewicz M, Haapalinna A, Riekkinen P Jr, Riekkinen PJ Sr (1993) Dose- and parameter-dependent effects of atipamezole, an alpha 2-antagonist, on the performance of rats in a five-choice serial reaction time task. Pharmacol Biochem Behav 45:123–129

Sirvio J, Mazurkiewicz M, Haapalinna A, Riekkinen P Jr, Lahtinen H, Riekkinen PJ Sr (1994) The effects of selective alpha-2 adrenergic agents on the performance of rats in a 5-choice serial reaction time task. Brain Res Bull 35:451–455

Spinelli S, Pennanen L, Dettling AC, Feldon J, Higgins GA, Pryce CR (2004) Performance of the marmoset monkey on computerized tasks of attention and working memory. Cogn Brain Res 19:123–137

St Peters M, Cherian AK, Bradshaw M, Sarter M (2011) Sustained attention in mice: expanding the translational utility of the SAT by incorporating the Michigan Controlled Access Response Port (MICARP). Behav Brain Res 225:574–583

Stolerman IP, Mirza NR, Hahn B, Shoaib M (2000) Nicotine in an animal model of attention. Eur J Pharmacol 393:147–154

Sun H-S, Green TA, Theobald DEH, Birnbaum SG, Graham DL, Zeeb FD, Nestler EJ, Winstanley CA (2010) Yohimbine increases impulsivity through activation of cAMP response element binding in the orbitofrontal cortex. Biol Psychiatry 67:649–656

Terry AV Jr, Risbrough VB, Buccafusco JJ, Menzaghi F (2002) Effects of (+/−)-4-[[2-(l-methyl-2-pyrrolidinyl)ethyl]thio]phenol hydrochloride (SIB-1553-A), a selective ligand for nicotinic acetylcholine receptors, in tests of visual attention and distractibility in rats and monkeys. J Pharmacol Exp Ther 301:284–292

Terry AV Jr, Hernandez CM, Honadel EJ, Bouchard KP, Buccafusco JJ (2005) Cotinine, a neuroactive metabolite of nicotine: potential for treating disorders of impaired cognition. CNS Drug Rev 11:229–252

Terry AV Jr, Buccafusco JJ, Herman EJ, Callahan PM, Beck WD, Warner S et al (2011) The prototypical ranitidine analog JWS-USC-75-IX improves information processing and cognitive function in animal models. J Pharmacol Exp Ther 336:751–766

Terry AV Jr, Buccafusco JJ, Schade RF, Vandenhuerk L, Callahan PM, Beck WD et al (2012) The nicotine metabolite, cotinine, attenuates glutamate (NMDA) antagonist-related effects on the performance of the five choice serial reaction time task (5-CSRTT) in rats. Biochem Pharmacol 83:941–951

Turchi J, Sarter M (2001) Bidirectional modulation of basal forebrain NMDA receptor function differentially affects visual attentional but not visual discrimination performance. Neuroscience 104:407–417

Turchi J, Holley LA, Sarter M (1995) Effects of nicotinic acetylcholine receptor ligands on behavioral vigilance in rats. Psychopharmacology (Berl) 118:195–205

Turner DC, Clark L, Dowson J, Robbins TW, Sahakian BJ (2004a) Modafinil improves cognition and response inhibition in adult attention-deficit/hyperactivity disorder. Biol Psychiatry 55:1031–1040

Turner DC, Clark L, Pomarol-Clotet E, McKenna P, Robbins TW, Sahakian BJ (2004b) Modafinil improves cognition and attentional set-shifting in patients with chronic schizophrenia. Neuropsychopharmacology 29:1363–1373

van Gaalen MM, Brueggeman RJ, Bronius PFC, Schoffelmeer ANM, Vanderschuren LJMJ (2006) Behavioral disinhibition requires dopamine receptor activation. Psychopharmacology (Berl) 187:73–85

Waters KA, Burnham KE, O'Connor D, Dawson GR, Dias R (2005) Assessment of modafinil on attentional processes in a five-choice serial reaction time test in the rat. J Psychopharmacol 19:149–158

Weed MR, Taffe MA, Polis I, Roberts AC, Robbins TW, Koob GF et al (1999) Performance norms for a rhesus monkey neuropsychological testing battery: acquisition and long-term performance. Cogn Brain Res 8:185–201

Westenson NJ, Belenky G, Kautz MA, Throne DR, Reichardt RM, Balkin TJ (2002) Maintaining alertness and performance during sleep deprivation: modafinil versus caffeine. Psychopharmacology (Berl) 159:238–247

White HK, Levin ED (1999) Four-week nicotine skin patch treatment effects on cognitive performance in Alzheimer's disease. Psychopharmacology (Berl) 143:158–165

Wilens TE, Biederman J, Spencer TJ, Bostic J, Prince J, Monuteaux MC et al (1999) A pilot controlled clinical trial of ABT-418, a cholinergic agonist, in the treatment of adults with attention deficit hyperactivity disorder. Am J Psychiatry 156:1931–1937

Winstanley CA, Chidasama Y, Dalley JW, Theobald DEH, Glennon JC, Robbins TW (2003) Intra-prefrontal 8-OH-DPAT and M100907 improve visuospatial attention and decrease impulsivity on the five-choice serial reaction time task in rats. Psychopharmacology (Berl) 167:304–314

Winstanley CA, Dalley JW, Theobald DEH, Robbins TW (2004a) Fractionating impulsivity: contrasting effects of central 5-HT depletion on different measures of impulsive behavior. Neuropsychopharmacology 29:1331–1343

Winstanley CA, Theobald DEH, Dalley JW, Glennon JC, Robbins TW (2004b) 5-HT_{2A} and 5-HT_{2C} receptor antagonists have opposing effects on a measure of impulsivity: interactions with global 5-HT depletion. Psychopharmacology (Berl) 176:376–385

Wisor J (2013) Modafinil as a catecholaminergic agent: empirical evidence and unanswered questions. Front Neurology 4:1–10

Wouda JA, Riga D, De Vries W, Stegeman M, van Mourik Y, Schetters D et al (2011) Varenicline attenuates cue-induced relapse to alcohol, but not nicotine seeking, while reducing inhibitory response control. Psychopharmacology (Berl) 216:267–277

Yan TC, Dudley JA, Weir RK Grabowska EM, Pena-Oliver Y, Ripley TL et al. (2011) Performance deficits of NK1 receptor knockout mice in the 5-choice serial reaction-time task: effects of d-amphetamine, stress and time of day. PLoS one 6:el17586

Young JW, Finlayson K, Spratt C, Marston HM, Crawford N, Kelly JS, Sharkey J (2004) Nicotine improves sustained attention in mice: evidence for involvement of the α7 nicotinic acetylcholine receptor. Neuropsychopharmacology 29:891–900

Young JW, Light GA, Marston HM, Sharp R, Geyer MA (2009) The 5-choice continuous performance test: evidence for a translational test of vigilance for mice. PLoS One 4:e42

Young JW, Geyer MA, Rissling AJ, Eyler LT, Asgaard GL, Light GA (2011a) Impaired vigilance but not accuracy of patients with schizophrenia in a human version of the rodent 5-choice continuous performance test. Schizophr Bull 37:236

Young JW, Powell SB, Scott CN, Zhou X, Geyer MA (2011b) The effect of reduced dopamine D4 receptor expression in the 5-choice continuous performance task: separating response inhibition from premature responding. Behav Brain Res 222:183–192

Young JW, Meves JM, Geyer MA (2013) Nicotinic agonist-induced improvement of vigilance in mice in the 5-choice continuous performance test. Behav. Brain Res 240:119–133

Executive Function

John Talpos and Mohammed Shoaib

Contents

Abstract

Components of human executive function, like rule generation and selection in response to stimuli (attention set-shifting) or overcoming a habit (reversal learning), can be reliably modelled in rodents. The rodent paradigms are based upon tasks that assess cognitive flexibility in clinical populations and have been effective in distinguishing the neurobiological substrates and the underlying neurotransmitter systems relevant to executive function. A review of the literature on the attentional set-shifting task highlights a prominent role for the medial region of the prefrontal cortex in the ability to adapt to a new rule (extradimensional shift) while the orbitofrontal cortex has been associated

J. Talpos
Janssen R&D, Janssen Pharmaceutical Companies of Johnson & Johnson, Turnhoutseweg 30, 2340 Beerse, Belgium

M. Shoaib (✉)
Institute of Neuroscience, Newcastle University, Newcastle NE24HH, UK
e-mail: mohammed.shoaib@newcastle.ac.uk

K.M. Kantak, J.G. Wettstein (eds.), *Cognitive Enhancement*, Handbook of Experimental Pharmacology 228, DOI 10.1007/978-3-319-16522-6_6

with the reversal learning component of the task. In other paradigms specifically developed to examine reversal learning in rodents, the orbitofrontal cortex also plays a prominent role. Modulation of dopamine, serotonin, and glutamatergic receptors can disrupt executive function, a feature commonly exploited to develop concepts underlying psychiatric disorders. While these paradigms do have excellent translational construct validity, they have been less effective as predictive preclinical models for cognitive enhancers, especially for cognition in health subjects. Accordingly, a more diverse battery of tasks may be necessary to model normal human executive function in the rodent for drug development.

Keywords
Executive function • Cognitive flexibility • Reversal learning • Prefrontal cortex

1 Introduction

Executive function is one of the most widely studied aspects of human cognition. It is small wonder since the ability to evaluate a range of scenarios that we have never experienced, and adjust our behaviour to achieve our desired outcome, is in many ways at the centre of the human experience. However from a more pragmatic view many psychological disorders also express themselves in part by abnormalities in executive function. Treating these abnormalities in disorders such as schizophrenia, Alzheimer's disease, and Autism, just to name a few, should improve the quality of life for tens of millions of people worldwide. Yet, what about the remaining seven billion cognitively "normal" people, does evidence exist to suggest that executive function can be enhanced in the population at large? Clinical evidence can be found to support the first-hand reports from the hundreds of millions who find that psychostimulants improve attention and alertness under specific conditions (Wood et al. 2014). Anecdotal evidence suggests that for some people creativity can be enhanced via alcohol, marijuana, LSD, or other psychoactive substances. The fact that functioning in these other domains can be enhanced suggests that the same may be possible for executive function. However does our current understanding of the neurobiology of executive function suggest this is possible, and is there any experimental evidence suggesting such an enhancement can be achieved?

Despite being central to the human experience executive function is difficult to define. Perhaps the simplest definition is that executive function is necessary when routine behaviours will not be effective at dealing with the task in hand. In this vein, the textbook example given by Shallice and Norman may provide the best examples of when the recruitment of executive functions is crucial (Normans and Shallice 1980). These are situations that involve planning or informed decision-making: error correction and troubleshooting; situations where the responses are novel, dangerous, or technically difficult; and situations that require the overcoming of a strong habitual response. What most of these examples have in common is rapid generation of behaviour in response to a novel situation. Accordingly, for the sake of this review executive function will be defined as rule generation and selection in response to stimuli, as well as a rapid change in behaviour in response to that stimulus.

The methods used to study executive function in a clinical setting are varied. Yet relatively few of these measures can be adapted to a preclinical setting. From the outside, the experimental psychologist may seem obsessed with rodents. Despite the many differences in physiology and complexity of behaviours produced by the human and rodent brain, there are even more similarities. Accordingly the rodent can be used to "model" many aspects of human cognition and central nervous system function. Using what has become known as the "translational" approach, certain behaviours are reduced to their core elements, and if these are dependent upon overlapping neurobiology, then the preclinical species can be used to model certain aspects of human biology (for a recent review on this approach, see Talpos and Steckler 2013). The validity of the translational approach has been especially well supported for executive function and processes within the prefrontal cortex (PFC; Keeler and Robbins 2011). Despite the wide variety of methods that are used to study executive function in a clinical setting, those that are of greatest interest are the ones that have a rodent analogue. By testing rodents in a "human"-like fashion, it is possible to gain insight into how a similar manipulation might affect human cognition. Unfortunately only a handful of tests have been developed that are thought to measure executive function in the rodent. Generally, these tasks fall into three categories, those that require rule generation and selection (attentional set-shifting), those that require overcoming a habit (reversal learning), and finally those that require the rapid adaptation of a behavioural response to incoming stimuli (stop signal reaction time and the continuous performance task). Despite each being a measure of executive function, it is thought that each of these tasks is dependent upon partially dissociable brain regions and may have unique pharmacology. In this chapter, we will focus on rule generation and overcoming a habit. This is not to discount the importance of the rapid adjustment of behaviour. However rapid adjustment of behaviour is so dependent on levels of attention and impulsivity that it can be difficult to distinguish effects on executive function from other cognitive abilities.

The prefrontal cortex (PFC) is a complex structure and while there has been significant progress in understanding its function in the normal brain, greater success has been achieved in restoring compromised performance produced by either neurotoxin lesions (Dalley et al. 2004) or treatment with phencyclidine and related NMDA receptor antagonists (Floresco and Jentsch 2011; Neill et al. 2010). In a clinical setting treatments for cognitive impairments associated with Alzheimer's disease or attention deficit hyperactivity disorder are available, although these do not improve executive function per se. Moreover these treatments are generally less effective or even disruptive in a non-clinical population. Yet as we learn more about the neurobiological processes underlying executive function in normal and abnormal cognition, the possibility of enhancing these abilities by pharmacological modulation seems a scientific possibility. This will be a significant advancement from the first generation of nootropics, such as piracetam, which showed only minimal efficacy and lacked any specific pharmacological basis for their action. Since cognition can be more defined into domains such as attention, working memory, executive function, and inhibitory control, it becomes feasible to utilise animal models to develop the new generation of cognitive enhancers. Which transmitter systems should be targeted? Since the prefrontal cortex has been implicated in many of these cognitive domains (Dalley et al. 2004) and is innervated by multiple neurotransmitter systems, notably dopamine, serotonin, noradrenaline, and glutamate, opportunities to modulate these systems provide a starting point to develop novel cognitive enhancers. To facilitate this, it is imperative to understand the interplay between these neurotransmitter systems and how they influence executive control using appropriate animal models.

2 Attentional Flexibility

Attentional flexibility is commonly measured clinically using the Wisconsin card sorting test (WCST; Milner 1963), as well as intradimensional/extradimensional shifting tasks like those used as part of the Cambridge Neurological Test Battery (CANTAB). Deficits on performance in the WCST have been observed in patients with schizophrenia. Typically, the subjects must sort a series of cards dependent upon changing rules, such as pattern or colour (Berg 1948). Over the course of the test, schizophrenia patients can learn simple rules for sorting the cards, but find it difficult to adapt their behaviour once the relevant category changes (i.e. from sorting by colour to sorting by number; Egan et al. 2001). Moreover these patients also demonstrate impaired learning in the simple reversal learning tasks in which the cues signalling correct and incorrect responses are switched (Leeson et al. 2009; Waltz and Gold 2007). Many of the abnormalities in responding in patient populations have been attributed to dysfunction within the frontal cortex (Dias et al. 1996a, b; Robbins and Keeler 2011).

In rodents executive control can be modelled using the attentional set-shifting task (ASST) developed by Birrell and Brown (2000). The ASST allows investigation of the mechanisms underlying both attentional set formation and maintenance in addition to attentional flexibility and reversal learning. Furthermore, it is formally equivalent to the human and primate versions of the task (Brown and Bowman 2002; Dias et al. 1996a, b; Owen et al. 1991), but uses more species appropriate stimuli such as odour- and tactile-based discriminations. The task requires the rodent to learn to associate a food reward with a specific dimension, which is later changed during the task (extradimensional shift: EDS). Birrell and Brown (2000) observed that rodents took more trials to complete the extradimensional shift (switching to a previously irrelevant stimulus, whilst ignoring the previously relevant stimulus) component of the task than the intradimensional shift (switching within the same relevant dimension) component when rodents had bilateral lesions of the prefrontal cortex (Birrell and Brown 2000). Thus, the ASST assesses a rodent's ability to shift attention to allow investigation of the mechanisms underlying both attentional set formation and maintenance, in addition to attentional flexibility and reversal learning.

From a procedural perspective, the task requires rodents to be habituated to the test apparatus prior to the start of testing. This is typically a modified plastic home cage that has been divided into two sections by Plexiglas panels. At the end of each section is a ceramic digging bowl. During testing the bowls are filled with different digging media that are scented with different herbs and spices. The food reward (half of a cereal Cheerio™; Nestle, UK) is buried beneath the digging media in the bowl. A removable Perspex divider separates these sections from the rest of the box so access to both bowls can be controlled during testing. A smaller Perspex divider is also used when required to block access to either of the bowls individually, for example when an error was made by the rodent. As part of the "habituation" phase, rats are trained to dig in bowls of unscented wood chips for the cereal reward. Once this behaviour has been acquired rats are required to perform simple discriminations based on odour or medium to earn a food reward (for example reward is always paired with a specific odour). These simple discriminations (SD) are carried out in the same order for all rats, and in each only one of two exemplar choices is rewarded. Therefore in the "medium" SD, food reward is paired with polystyrene chips but not shredded paper, and in the "odour" SD, reward is paired with mint but not oregano. The rats are trained to a criterion performance level of six consecutive correct trials on each of the SDs and these exemplars are not used again in subsequent tests. A series of seven discriminations forms the ASST, in the order outlined in Table 1. These discriminations are presented to each rat in the same order, which encourages the formation of attention sets that can then be subjected to reversals and dimensional shifts (in the example in Table 1 this is odour, with nutmeg being the rewarded and cloves the unrewarded exemplar both in the coarse sawdust medium).

On reaching criterion on the SD phase, testing progresses to the compound discrimination (CD), where the correct and incorrect exemplars of the relevant dimension remain the same as in the SD (i.e. nutmeg vs. cloves in the example), but a second (irrelevant) dimension is introduced (i.e. fine as well as coarse sawdust medium). The CD is followed by a reversal discrimination (REV1) in which the exemplars and dimensions are unchanged from the CD, but the previously correct exemplar is now incorrect and vice versa (i.e. in our example, odour is still relevant, but it is now cloves not nutmeg which is rewarded; reversal learning will be discussed in additional detail later in the chapter). The ID shift is then carried out. New exemplars of the relevant and irrelevant dimensions are presented to the rat with the same dimension being relevant. In our example it is still odour that is the relevant dimension; however, we now have cinnamon being rewarded but not cumin. Accordingly animals are still required to ignore the element of medium, but must learn a new rule within a dimension (intradimensional shift). The ID shift is then followed by another reversal discrimination (REV2), whereas in REV1 the exemplars remain the same as in the ID shift, but the relevant and irrelevant exemplars within a dimension are reversed (so in our example, it is still odour that is the relevant dimension, but it is now cumin that is rewarded not cinnamon).

Table 1 Order of discriminations in the task and exemplar combinations used

Discrimination	Dimensions		Example paradigm	
	Relevant	Irrelevant	+	–
SD	Odour		**Nutmeg**/coarse sawdust	Cloves/coarse sawdust
CD	Odour	Medium	**Nutmeg**/coarse sawdust **Nutmeg**/fine sawdust	Cloves/coarse sawdust Cloves/fine sawdust
REV1	Odour	Medium	**Cloves**/coarse sawdust **Cloves**/fine sawdust	Nutmeg/coarse sawdust Nutmeg/fine sawdust
ID	Odour	Medium	**Cinnamon**/Darjeeling tea **Cinnamon**/Fine tea	Cumin/Darjeeling tea Cumin/Fine tea
REV2	Odour	Medium	**Cumin**/Darjeeling **Cumin**/Fine tea	Cinnamon/Darjeeling Cinnamon/Fine tea
ED	Medium	Odour	**Large pebbles**/thyme **Large Pebbles**/paprika	Small pebbles/thyme Small pebbles/paprika
REV3	Medium	Odour	**Small pebbles**/thyme **Small pebbles**/paprika	Large Pebbles/thyme Large Pebbles/paprika

The table illustrates an example of the combination of exemplar pairs in a rat shifting attentional set from odour to medium in the ED stage. Equal numbers of rats in each treatment group shifted set from odour to medium and medium to odour. In every discrimination except SD, both bowls presented differed along both perceptual dimensions. The rewarded stimulus (in bold type) was paired with either irrelevant exemplar across trials and discriminations. The order and left/right presentation of exemplar pairs was also determined pseudorandomly (adapted from Birrell and Brown 2000)

This is followed by the ED shift stage of the task. As in the ID shift, the rat is presented with completely novel exemplars of both relevant and irrelevant dimensions. However in contrast to the ID shift, the previously relevant and irrelevant dimensions are now reversed, so that for a rat initially trained on odour, medium becomes the relevant stimulus in the extradimensional, or, "ED" shift and vice versa. In the example listed in Table 1, the rat has to now attend to medium as the relevant dimension, with large pebbles being rewarded but not small pebbles—the odours of paprika and thyme are now irrelevant. The test session concludes with a final reversal discrimination (REV3) of the ED shift. Subjects are required to make six consecutive correct responses before moving on to the next stage. The ED shift is the most challenging aspect of the task and subjects typically require an increased number of trials to criterion to solve this stage compared to the ID shift.

2.1 Neural Substrates Underlying Attention Shifting

In their first report on the rodent ASST, Birrell and Brown (2000) demonstrated that different aspects of attention-shifting in the task are mediated by distinct subregions of the PFC (Birrell and Brown 2000; McAlonan and Brown 2003). Excitotoxic lesions of the medial PFC impaired ED shifting in rodents (Brown and Bowman

2002; Bissonette et al. 2008), while lesions of the orbitofrontal cortex (OFC) could also impair the reversal learning stages of the ASST (Dias et al. 1996a; Birrell and Brown 2000). The functional dissociation between these two regions has subsequently been confirmed in human studies (Hornak et al. 2004), suggesting that the rodent and non-human primate versions of the task are effective translational models of human cognitive flexibility. Furthermore, frontal lobectomy in humans (Owen et al. 1991), lesions of the primate lateral PFC (Dias et al. 1996a, b, 1997), and the equivalent prelimbic and infralimbic regions of the rat PFC (Birrell and Brown 2000) produce specific impairments in ED set-shifting, while lesions of the OFC selectively impair reversal learning in rodents and non-human primates (Dias et al. 1996a, b; Schoenbaum et al. 2002; McAlonan and Brown 2003). The neural substrates mediating the ED shift and reversal learning have been shown to be functionally dissociated since OFC is thought to be necessary for reversal learning but not ED shifting (Birrell and Brown 2000; Chase et al. 2012). In addition, lesions of the cingulate cortex can also impair ED shifting (Ng et al. 2007), although it is not clear if this type of lesion affects the formation of an attention set. The posterior parietal cortex (PPC) has also been linked with neuronal activations during an attentional set-shifting task (Asari et al. 2005) and lesions of the PPC have been shown to impair set-shifting in rodents (Fox et al. 2003). Using a different approach to identify the neural basis of ID set-shifting, one study has demonstrated significant correlations between ID performance in rats in the attentional set-shifting paradigm and immediate early gene expression in the dorsolateral striatum and the nucleus accumbens—areas that are involved in stimulus–reward associations (Egerton et al. 2005a). Therefore it is conceivable that the whole cognitive process may be more complex and involve other structures and neurotransmitter systems outside of the PFC. For example, lesions of the dorsomedial striatum have been shown to impair formation of attentional set in rats (Lindgren et al. 2013), while depletion of noradrenaline by lesions of the dorsal noradrenergic bundle has been shown to impair attentional set-shifting in the rat (Tait et al. 2007). Furthermore, the immunotoxin DβH-saporin that selectively depletes noradrenaline levels also produces a selective impairment on set-shifting in rats (McGaughy et al. 2008). These studies suggest that ASST might require inputs from a wider array of brain regions than initially described, thereby offering further clues as to neurotransmitter systems that the task may be dependent upon and which could be targeted to enhance executive function.

2.2 Pharmacological Sensitivity of the Attentional Set-Shifting Task

There have been very few published studies that have sought to enhance attentional set-shifting in normal non-compromised animals. Most investigations on pharmacological mechanisms have centred on rodent models of mental disease. To date, only a single published study has reported on the acute effects of a psychoactive substance for enhancing attentional set-shifting in rats. Nicotine injected both

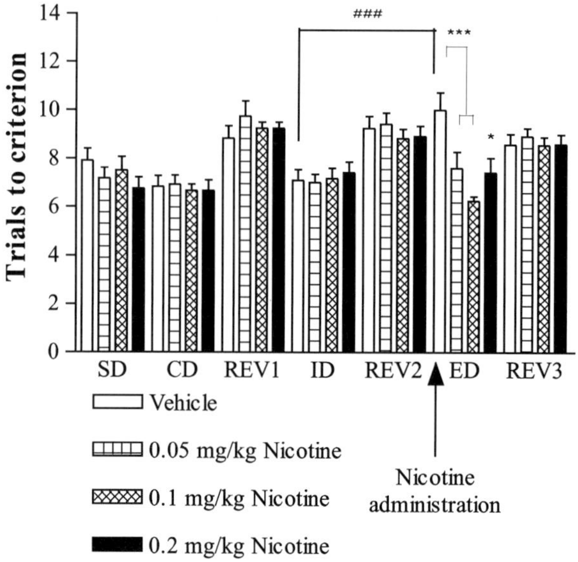

Fig. 1 Effect of administration of acute nicotine in increasing doses on extradimensional set-shifting (ED). Data expressed as mean ± SEM of the number of trials to reach criterion on the discriminations in the task. All doses of nicotine produced marked improvements in performance at the ED stage of the task. $*p < 0.05$, $***p < 0.001$ vs. vehicle-treated control. The control rats require significantly more trials to reach criterion on the ED compared to the ID stage of the task. $\#\#\#p < 0.001$ ID vs. ED in vehicle-treated group (figure reproduced from Allison and Shoaib 2013)

acutely and following repeated pre-exposure significantly improved both intradimensional and extradimensional set-shifting performance in the task (Allison and Shoaib 2013). An example of the nicotine-induced improvements on ED shifting of the dose-related improvements is illustrated in Fig. 1. These findings implicate the nicotinic receptor system in the mediation of processes underlying cognitive flexibility and suggest that nicotine improves attentional flexibility in rats, both within and between perceptual dimensions of a compound stimulus (Allison and Shoaib 2013). Further studies with more selective nicotinic agonists and positive allosteric modulators will help to confirm the therapeutic utility of targeting the cholinergic system. This is further highlighted by a recent study demonstrating that the cholinesterase inhibitor tacrine can facilitate attention shifting and age-related impairments on the reversal learning stage of the ASST (Tait et al. 2013).

The majority of the pharmacological characterisations are based around the hypo-glutamate hypothesis of schizophrenia; there have been a vast number of published studies utilising NMDA receptor antagonists to model cognitive deficits associated with schizophrenia in rodents and non-human primates (Floresco and Jentsch 2011; Barak and Weiner 2011; Frohlich and Van Horn 2014), with the key objective of restoring function. Phencyclidine has been commonly administered both acutely (Egerton et al. 2005b) and following subchronic treatment (Rodefer et al. 2008; McLean et al. 2012), which impairs ED shift performance. Similarly, ketamine, a related non-competitive NMDA receptor antagonist, has also been shown to impair ED shifting following acute (Nikiforuk et al. 2010) and subchronic (Nikiforuk and Popik 2012) treatment. Both of these treatments appear to selectively impair the ED shift in which the number of trials required to complete the ED task increases significantly, without affecting the other series of discrimination tasks (McLean et al. 2012; Nikiforuk and Popik 2012). Repeated treatment with amphetamine can also impair ED shift in rats (Fletcher et al. 2005), providing an opportunity to investigate dopaminergic mechanisms in attention shifting.

A variety of pharmacological targets have been examined with a view to reversing impairments on ED shift and reversals. These treatments range from the clinically effective neuroleptics such as clozapine to more selective compounds such as allosteric modulators at the AMPA receptor (AMPAkines). In reviewing the pharmacological sensitivity of ASST, it is apparent that there is a dearth of systematic studies that have focused on a given neurotransmitter system. For example, with serotonergic compounds, the 5HT6 receptor antagonist (SB271046) has been shown to ameliorate PCP-induced ED impairment (Rodefer et al. 2008). Restoration was also reported for sertindole (Rodefer et al. 2008), a clinically effective atypical antipsychotic that has both serotonergic and dopaminergic activities, in both PCP-treated and ketamine-treated rats (Nikiforuk et al. 2010). Similarly, risperidone and clozapine have also been shown to restore PCP-induced ED shift impairments. In a mPFC lesion model of impaired ED shift performance, asenapine, a second-generation antipsychotic, restored the deficit. Elevating noradrenaline levels by administering atomoxetine, this reuptake blocker ameliorated ED shift impairments produced by DβH-saporin NA-specific PFC lesions (Newman et al. 2008). For a more detailed review of the compounds evaluated in the ASST, Neill et al. (2010) have compiled an extensive review of psychoactive compounds tested in animal models of cognitive dysfunction associated with schizophrenia.

2.3 The Validity of the Attentional Set-Shifting Task as a Translational Model

The ASST as a rodent model of executive function appears more suited to capture cognitive deficits associated with chronic exposure to psychotomimetic drugs, rather than evaluating cognitive enhancements produced by acute exposure to psychoactive drugs. The increased number of trials to complete the ED shift produced by drugs like PCP and ketamine provides a large therapeutic window for amelioration. However, the opportunity for enhancement in normal subjects is limited and thus the ASST for this purpose may not be ideal. Furthermore, the task is very demanding to conduct, requiring a relatively large number of subjects per group ($n = 12$) for a fully counterbalanced design. There is also an element of bias that comes with scoring the performance during the trials and common recognition of what constitutes a "dig", since this constitutes a choice decision in the ASST that can vary across laboratories and thus impacts on scientific validity of the task. Thus, this model of assessing executive function is clearly not suited for purposes of drug discovery. Certainly, the ASST can be utilised for more targeted, hypothesis-driven approaches on executive control. Efforts should focus on developing automated versions of the rodent task that removes the element of subjective scoring and provides a faster throughput of data.

There are some issues relating to the validity of the ASST as a plausible measure of executive control. While the ID and ED shifts provide the key measures of attentional control, the three reversals contained within the ASST can be difficult to interpret. The sequence of discriminations as originally described by Birrell and Brown (2000) is restrictive and makes the task of evaluating acute drug effects difficult due to the varying completion times within groups of rodents and the difficulty to maintain constant drug levels throughout the ASST. The construct validity could be improved by focusing on the attention shifts by performing multiple ID and ED shifts within the ASST to infer on the attention set formations. Evaluating compounds on cognitive flexibility (see below) would be better suited to other models specifically based on reversal learning (see below). From the data generated with putative "cognitive enhancers" in the ASST thus far, very few of these molecules have not advanced to a stage where they can be tested in the equivalent version of the set-shifting task in humans; thus the predictive validity has yet to be fully appraised. Furthermore, many D2 receptor-based antipsychotics appear to have a pro-cognitive profile within ASST when tested against animal models of schizophrenia (subchronic administration of NMDA antagonists). While many of these treatments are associated with improved positive symptoms in schizophrenia, they have generally been found to have little or no effect against the cognitive deficits frequently seen in schizophrenia. This draws into question the predictive validity of the approach, especially when applied to models of schizophrenia.

3 Reversal Learning

Habitual behaviour is a huge benefit in everyday life. It allows completion of common tasks while requiring minimal mental involvement, thus freeing cognitive resources for other pursuits. However habitual behaviours are only of use as long as conditions remain constant; once contingencies change, habits can become a hindrance. For example, being able to drive to work without needing to "think" about finding your way allows you to free mental resources to focus on the traffic, plan your day, or just listen to the radio. However continuing to drive to your former place of work when you have a new job would cause some obvious inconveniences. Reversal learning is the summation of the mental process required to recognise that contingencies have changed, and update behavioural responses in accordance with these new contingencies. Classically, reversal learning is studied in the form of a discrimination. Participants (human, primate, rodent) are required to learn that a specific response is associated with a positive reward, whereas the alterative response is not. In one popular approach used in humans, monkeys, and rodents, two images are displayed upon a screen and the subject must learn which image is associated with a reward (for just a few examples, see Brigman et al. 2008; Roberts et al. 1990; Robbins et al. 1994). Once this response has been robustly acquired the contingencies are reversed, what was formerly correct is now incorrect, and vice versa. In the trials immediately following this switch in contingencies, it is expected that participants will continue to use their old response strategies. However, over time it should become apparent that the reward contingencies have changed and that behavioural responses must be altered to effectively operate in a new environment. In order to do this, it is necessary to (1) detect that the current response is no longer resulting in reward, (2) inhibit this response, (3) alter behaviour to discover the current reward contingencies, and (4) ultimately learn the new correct response. While reversal of the acquired rule will require many of the same processes required in initial rule acquisition, this need to inhibit the prepotent response should also selectively engage portions of the frontal cortex that are not necessarily required for task acquisition. Overwhelming evidence now suggests that reversal learning is dependent upon the OFC, striatum, and amygdala, but regional involvement may partially depend upon the demands unique to the task being used (Brigman et al. 2013; Clarke et al. 2008; Robbins and Keeler 2011; Boulougouris et al. 2007; and many more).

Numerous reversal learning paradigms exist for the rodent. For example procedures exist that are dependent upon the discrimination of a location, of an object/image, of an odour or texture (as described in ASST), and even of sound. Despite the different sensory modalities used, these tasks are generally thought to still be dependent upon the same brain regions except for those that are modality specific. For example, spatial reversals have been shown to engage the hippocampus, and one would expect discriminations based on odour to engage the olfactory bulb and potentially the piriform cortex. The vast majority of rodent reversal learning studies are done with a "simple" reversal learning procedure. A response to the rewarded stimulus will always result in reward, whereas a response at the non-rewarded stimulus will never result in a reward. However, this is not the case in human studies. Frequently in normal participants, a "probabilistic" reversal learning paradigm is used. In probabilistic paradigms, a correct response will in some instances result in a non-rewarded outcome, while an incorrect response will result in a rewarded outcome (Cools et al. 2007; Hornak et al. 2004) in accordance with a predetermined reward ratio (e.g. correct response results in a reward 90 % of the time and an incorrect response results in reward 10 % of the time). This causes uncertainty as to whether the reward contingencies have changed and will result in a slower "learning" of the rewarded stimulus and shift away from the former reward contingencies. As such, probabilistic reversal learning should cause an increase in responding under the old contingencies and offer a greater opportunity to detect an increase or decrease in perseverative behaviours. This distinction between simple and probabilistic reversal learning may only be relevant when studying cognitive flexibility in normal participants, as many psychiatric conditions are noted for apparent impairments in simple reversal learning negating the need of a probabilistic component. Furthermore, rodents are comparatively poor at reversal learning. "Fast" reversals, such as those dependent upon a spatial locations or odour, will often require 10s of responses prior to reversal, while "slow" reversals, like those dependent upon a visual or auditory discrimination, can take 100s of trials and span days of training. It should be noted that probabilistic reversal learning may engage additional PFC brain regions, at least in humans (Hornak et al. 2004). Therefore from a translational point of view, the use of a probabilistic reversal may still be preferable.

Another consideration that may influence the neural substrates of reversal learning is the influence of repeated reversals, or serial reversal. After subjects have been trained on an initial discrimination (A+/B−), and reversed (B+/A−), then can again be placed on the original discrimination (A+/B−). Evidence suggests that single and serial reversals alike engage the OFC; however, VLPFC is not necessary for serial reversals (Rygula et al. 2010; Clarke et al. 2007, 2008, 2011). Numerous frameworks have been proposed to explain the role of the OFC in reversal learning. A leading theory put forward by Roberts and colleagues views the OFC as being necessary for making the association between cues and the outcomes associated with these cues (Roberts et al. 1990; Roberts and Wallis 2000). In this context, the VLPFC is seen as being necessary for initial rule creation and generalisation into a new environment. While this is an extremely important process that allows us to dynamically adjust to our settings, it is, strictly speaking, not involved with the overcoming of a habitual response.

3.1 Neurobiology of Reversal Learning

A substantial body of work now exists that highlights the importance of the OFC for reversal learning in humans, monkeys, and rodents. For example, a study by Hampshire and Owen (2006) that investigated brain activity using event-related fMRI in a set-shifting task (like those previously described) found increased right and left OFC activation when reward contingencies changed within a dimension. Based in part on this finding, they concluded that the lateral OFC "plays a role in attention that is related to, although more complex than, merely processing negative feedback". A series of lesion studies in marmosets has produced data largely in line with these conclusions. For example, lesions of the OFC were found to have little effect on the learning phase of reversal (i.e. when performance was above chance), but did greatly impair performance just after a shift in reward contingencies (Clarke et al. 2008). Interestingly, an analysis of the trial-by-trial behaviour of the monkeys found that those with OFC lesions were more likely to shift responding after a correct result and less likely to shift responding after an incorrect result, as did the control group. This suggests that lesions of the OFC result in a decreased sensitivity to both positive and negative rewards, an effect that is present when the same stimuli are used for repeated reversals and if novel stimuli are used for serial reversals (Rygula et al. 2010). Lesions of the OFC have also been found to impair reversal learning in the rodent but not extradimensional shifting (McAlonan and Brown 2003; Boulougouris et al. 2007). However, unlike what has been reported in the marmosets, evidence exists to suggest that this effect is dependent upon previous task experience (Boulougouris and Robbins 2009). While these inconsistencies may be the result of task experience, it is also possible they are the result of species differences, the type of reversal task (spatial versus visual), or even the "difficulty" of the reversal (errors to criteria). Fewer efforts have been made to investigate the influence of task type (spatial, visual, olfactory) or reversal type (repeated, serial, probabilistic) on outcome within the reversal learning literature. Accordingly it is difficult to determine what are meaningful differences and what is simple variation between tasks.

Despite the well-documented importance of the OFC in reversal learning, other areas of the brain also contribute to this multifaceted process. For instance, the ventrolateral PFC has been shown to be activated during the stage of reversal when subjects have stopped perseverating and are now "looking for the solution" associated with the new reward contingencies (Hampshire and Owen 2006; Cools et al. 2002). Interestingly, marmosets that underwent lesions of the ventrolateral PFC were impaired in reversal learning only when new stimuli were used and did not show clear evidence of a perseverative phenotype. Accordingly, the ventrolateral PFC may only be of relevance under novel conditions and strictly speaking is not necessary for the overcoming of a habitual response, but will be necessary for optimal performance in a dynamic setting. A role has also been claimed for the striatum in reversal learning, and while this is very likely the case, it is unclear to what extent these effects are selective for reversal learning. For instance, a lesion of the medial striatum (marmoset) was found to impair reversal learning by inducing impairment in negative feedback learning that was not apparent during the acquisition phase of the task (Clarke et al. 2008). This result is in line with other findings where the D2 receptor preferring agonist, quinpirole, was infused directly into the ventral striatum of rats that were trained to perform a lever-based reversal (Haluk and Floresco 2009), resulting in a slower behavioural reversal. However an analysis of c-*fos* levels at different stages or visual discrimination and reversals shows high levels of activity associated with both discrimination acquisition and post-reversal performance (Brigman et al. 2013). Moreover, lesions of the striatum have also previously been shown to disrupt performance and acquisition of various discriminations (Broadbent et al. 2007. These later data would support at least some role for the striatum in task acquisition, which would erode support for a selective role of the striatum in general in reversal learning. However, this would not explain the differential effect on negative feedback in acquisition versus reversal in the study by Clarke et al. (2008).

3.2 Pharmacological Sensitivity of Reversal Learning

Many studies have been performed investigating pharmacological mechanisms that may enhance reversal learning. However, the vast majority of these have been done in conjunction with an impairment model, typically those that are thought to model aspects of schizophrenia. Far less literature exists examining the effects of neurochemical or pharmacological manipulations in non-clinical humans or unchallenged animals. Of those studies that do exist, the vast majority address the effects of either dopamine or serotonin. Some of the earliest studies investigated the effects of temporary 5-HT depletion in healthy volunteers and discovered this induced impairments in reversal learning (Park et al. 1994; Rogers et al. 1999a, b). Furthermore, chronic treatment (2 weeks administered in drinking water) with the selective serotonin reuptake inhibitor fluoxetine was found to decrease perseverative errors in mice (Brigman et al. 2010). More recently, it has been demonstrated that systemic administration of the 5-HT2a receptor antagonist M100907 will disrupt reversal learning, while the 5-HT2c receptor antagonist SB 242084 has been found to decrease perseverative behaviour in a rodent spatial reversal task (Boulougouris et al. 2007), an effect eventually attributed to the OFC (Boulougouris and Robbins 2010).

Serotonin clearly plays a role in reversal learning; however, it has also been well established that 5-HT receptors are crucial in regulating levels of dopamine. As such, it is not clear to what extent serotonergic manipulations are directly influencing behaviour versus regulating dopaminergic function. For example, administration of the D2/3 receptor antagonist raclopride was found to impair reversal learning while having little impact on discrimination acquisition in monkeys (Lee et al. 2007). In contrast, a D1/D5 receptor antagonist was not able to disrupt accuracy without also suppressing behaviour in general. The D2/3 receptor agonist quinpirole was also found to induce perseverative errors in a spatial reversal task in the rat, while in the same study, raclopride was seen to have no effect on reversal learning. These apparently conflicting results may be reconciled by claims that reversal learning is in part dependent upon striatal dopamine levels (Clatworthy et al. 2009) or striatal D2 receptor availability (Groman et al. 2011). It is possible that species or task differences could result in different optimal levels of dopamine or that D2 receptor occupancy having been obtained in each study resulted in contrasting effects after D2 receptor blockade or activation. All of these findings would support a primary role for dopamine in reversal learning. Yet this conclusion is challenged by the results of a study comparing the effects of OFC depletion of serotonin or dopamine in marmosets performing a reversal task. In this instance, the lesion of dopaminergic cells had no apparent effect on perseveration, arguing for a role of serotonin that is independent of dopamine, at least within the OFC (Clarke et al. 2007). As a whole these data suggest a complex interplay between multiple regions within the cortico-striatal loop and the dopaminergic and serotonergic systems in support of reversal learning and begin to hint at potential differences between rodents and primates in optimal levels of performance for tasks of executive function.

Despite the clear importance of the dopaminergic and serotonergic system in reversal learning, evidence also suggests that the noradrenergic system can modulate reversal learning performance. For example, the norepinephrine reuptake inhibitors methylphenidate and atomoxetine were found to enhance reversal learning in the monkey and rat (Seu and Jentsch 2009), while additionally desipramine enhanced performance in the rat alone (Seu and Jentsch 2009; Lapiz et al. 2007). Generally these compounds caused a decrease in perseverative behaviours, although the exact profile differed from drug to drug. Interestingly methylphenidate was seen to disrupt performance during the retention phase of the task in rats and monkeys, while the selective dopamine transport inhibitor GBR-12909 induced a similar impairment but offered no benefit on reversal learning (Seu and Jentsch 2009). This could explain the effects with methylphenidate, as it is a mixed NA/DA reuptake inhibitor. Finally, work by Lapiz and Morilak (2006) attributes some of the effects seen with norepinephrine reuptake inhibitors to the a2-adrenergic receptor, as stimulation with the agonist atipamezole has been shown to selectively enhance the first reversal in a set-shifting paradigm.

From the above data, it is clear that reversal learning is sensitive to a diverse pharmacology. Besides the aforementioned results, there are also data to suggest that manipulations of the cholinergic (Tait and Brown 2008; Roberts et al. 1992; Chen et al. 2004) or glutamatergic (Brigman et al. 2013) systems can influence reversal learning, as can administration of various cannabinoids (Egerton et al. 2005a). With this sensitivity to a diverse variety of mechanisms, it is important to clarify why reversal learning is of interest. The easy answer is because it is a translational measure of cognitive flexibility. While this is true, it does not mean that all aspects of reversal learning are of interest. Returning to our original definition of executive function, reversal learning is of interest because it requires overriding a prepotent response or habit. Accordingly, manipulations that improve performance at "chance" level or "late" reversal will be of little relevance, as these are not contributing to overcoming this habitual response; rather they are evidence of enhanced learning. For instance, a detailed series of studies by Fellini et al. (2014) demonstrated that acute PCP impaired the initial acquisition of a visual discrimination, the acquisition of a second visual discrimination, reversal of a visual discrimination, and reversal of a second visual discrimination. In this series of studies, PCP did disrupt reversal learning, but it would not be appropriate to attribute these results to a deficit in cognitive flexibility. While this is certainly interesting in its own right, it is not evidence of the change in executive function for which reversal learning is intended to measure. Similarly, manipulations that enhance only the first of several reversal sets may not be of interest for this specific cognitive function for two reasons. First, if the effect is only transitory, then it is possible that the task has changed from a reversal task with a distinct perseveration phase to a switching task and a lose shift strategy where animals are no longer forming true habits to overcome, but instead are adjusting behaviour in response to the most recent reward outcome. While this can look very similar to reversal learning, the lack of perseverative behaviour should mean that fundamentally different neurobiology is involved. Confirmation of this will typically require a complex behavioural analysis looking at behaviour after correct and incorrect trials similar to the approach used by Clarke et al. (2008). Second, there is no reason to think that if a manipulation influences the initial reversal phase by decreasing perseveration, assuming a habit is still being formed, that it should not do the same on subsequent reversal phases. Indeed manipulations of the dopaminergic and serotonergic systems can repeatedly influence perseverative behaviour across reversals (Boulougouris et al. 2009; Clarke et al. 2005, 2011). It is possible that since the first reversal comes after the compound discrimination stage in attentional set-shifting tasks, this reversal instead appears more like a set-shift owing to the recent addition of the irrelevant element. This may engage mPFC processes in the rodent while manipulations that influence later reversals could be having most of their effects via the OFC. While this could still be an excellent measure of executive function and cognitive flexibility, it is still removed from the original purpose of studying reversal learning—to investigate the ability to overcome a prepotent response.

3.3 The Validity of Reversal Learning as a Translational Model

As previously mentioned, a variety of approaches can be used to investigate reversal learning, and depending on the modality of the stimuli, reversal learning itself can take 10 trials (as seen in bowl digging tasks) or even 1,000 trials (as may be the case with a visual discrimination). What is unclear is the extent to which these paradigms are equivalent. For example, direct administration of the NR2B antagonist RO 25-6981 into the OFC has been shown to induce perseverative errors in a slowly learned visual discrimination-based reversal (Brigman et al. 2013), and lesions of the OFC in rodents have on multiple occasions been shown to impair "quick" reversal learning in bowl digging paradigms. However, control animals may need only ten errors to perform the reversal, while animals with lesions of the OFC may show only a minor increase in total errors that is significant in part because of the high level of baseline accuracy. This is in contrast to the remarkable effect seen in marmosets after an OFC lesion, where errors may increase by a factor of 10 as a result of the lesion. Considering the differences in trials required to complete a reversal and the magnitude of effect, it is difficult to consider all of these measures of reversal learning as equal. Yet, OFC lesions do cause impairment in all of these measures, suggesting that much of this circuitry does translate across tasks and species. To confirm this construct validity, it would be advisable to ensure that the impairment is from the same part of the learning curve (i.e. just after reversal) and that similar patterns of reward insensitivity are observed. However, this may be difficult to do practically in reversal learning assays where ten errors is the difference between a normal animal and one with an OFC lesion.

The previously mentioned marmoset data highlight another problem with reversal learning as a translational test of behavioural flexibility; primates and especially humans are very good at detecting a change in reward contingencies under normal conditions in a pathology-free state. However, this does not appear to be the case for rodents, with many compounds being shown to enhance some aspect of reversal learning. To make the task "harder" in humans, parameters are often altered to include a probabilistic element. Much of the data that have been generated in preclinical species suggest that this task manipulation results in the recruitment of other brain regions not involved with the perseverative element of reversal learning per se. While this additional mPFC involvement may still be of interest for studying cognitive flexibility in general, it is of less value for understanding how the brain overcomes a habitual response. This may represent a limit in reversal learning in the rodent as a directly translational task of normal human reversal abilities; humans appear to operate at, or close to optimal, while rodents do not. The Yerkes–Dodson principle states that varying amounts of arousal are needed to perform tasks at the highest level and that this level of arousal follows an inverted curve. Our previous example of driving a car can also be applied here. When driving a familiar route in light traffic, many individuals will prefer to listen to music or have the radio on. The increased level of arousal coming from the dashboard will cause us to focus more effectively on the task at hand and help pass the time while performing an otherwise mundane task. The additional stimulation is needed to perform at an "optimal" level. However if traffic levels suddenly change and we drive in unfamiliar territory, or perhaps it starts to rain heavily, then we will turn off the radio, which has now become a distraction, and focus completely on driving. We are now functioning at "optimal" for this task and any additional arousal (in the form of the radio) now impairs performance. Such a perspective could explain the results of a 2009 study (Clatworthy et al. 2009) investigating the relationship between D2 receptor occupancy and performance on a task of reversal learning or working memory after treatment with methylphenidate. D2 receptor occupancy in the caudate predicted performance on a reversal learning task, while performance on a task of spatial working memory was predicted by occupancy in the ventral striatum. Importantly, enhancement in one domain was often associated with impairment in the other domain. Behaving in an "optimal" fashion for *multiple* cognitive domains implies performing below *maximum* performance in *most* cognitive domains. Methylphenidate may have pharmacologically increased arousal/dopamine levels, offering a benefit to those who were under aroused but inducing a deficit in those suffering from an optimal level of arousal/D2 receptor occupancy.

Is enhancement of cognitive flexibility in the form of reversal learning possible in normal subjects? In preclinical species, the answer is undoubtedly yes. In humans, the answer is perhaps less clear. Humans are naturally very good at reversal tasks, so in instances where the task is primarily a measure of perseverative responding and primarily dependent upon the OFC, most people are probably performing too close to optimal to see much of an improvement. However, it is very possible that only a sub-selection of normal individuals will see any benefit as a result of pharmacological intervention and this will come at a cost to other cognitive functions. Few studies have investigated the effects of potentially pro-cognitive manipulations in normal humans on reversal learning, although many studies have been performed in patients suffering from schizophrenia (for a review, see Barnett et al. 2010). Those few that have been performed, however, have not resulted in positive findings. Yet the methylphenidate study by Clatworthy et al. (2009) does act as an example that some small enhancements might be possible if an individual's pre-existing neurobiology is taken into consideration. Moreover, a recent study showing a pro-cognitive effect of modafinil on working memory and a complex task of spatial planning indicates that higher order aspects of executive function can also be enhanced (Muller et al. 2013) if the task is suitably demanding. While this could limit the utility of using rodents to model normal human cognitive enhancement, it also highlights the limited availability of meaningful preclinical models of human executive function.

4 Conclusions

ASST and reversal learning in preclinical species have both been effective assays at modelling the neurobiology of human cognitive flexibility. It does appear that these tasks are dependent upon similar brain regions and neurochemical modulation as their clinical counterparts. Yet these paradigms are not without their limitations. For instance, ASST does have issues with robustness, potential cross-lab consistency, as well as being highly labour intensive. While pro-cognitive effects have been detected in this assay, none have yet translated to novel clinically effective therapeutics for individuals with or without psychiatric disorders. Moreover a number of antipsychotics have proven effective in ASST despite not generally being associated with improved cognitive ability in a clinical setting. While ASST might be very effective at understanding the fundamental neurobiology of cognitive flexibility, it appears to fall short as a predictive preclinical model of cognitive function, or at least when used in conjunction with existing impairment models in the rodent. In contrast, reversal learning can be automated, which will address some of the concerns that exist about the robustness of ASST. However, the high level of baseline performance seen in normal humans suggests that the assay may be of little translational value when used in the rodent without a validated disease model. Despite the fact that the assay can be automated, there are still many unknowns around this theoretical construct because of the wide variety of procedures that have been used to study a wide variety of manipulations. Again, as a research tool for understanding the neurobiology of cognitive flexibility reversal learning has been a powerful tool, but has fallen short as a translational model of normal functioning for the search for nootropic compounds. While it has been suggested that probabilistic reversal learning might increase the relevance of the task to a clinical setting, it is also important to consider how this might alter the underlying neurobiological constructs of the task. Like ASST, the validity of reversal learning as a translational model of cognitive flexibility is likely dependent upon the validity of the impairment model being used. Counter-intuitively, the normal rodent *might* have some utility in modelling a compromised human owing to how slowly rodents reverse on some reversal tasks and their comparatively underdeveloped frontal cortex.

Executive function is not just reversal learning and attentional set-shifting. Even if these two tasks were to completely predict a clinical setting, they certainly would not encompass all aspects of executive function. Accordingly, it is clear that this is an area in need of additional task development and creation. Recent work by Brigman and colleagues has demonstrated the potential power of the touch-screen approach for studying reversal learning. Excitingly, through the use of a touch-screen-based visual discrimination, they were able to show different structural engagement and pharmacological sensitivities at different stages within the reversal stage. This ability to detect differential neurobiological engagement that likely corresponds to different psychological processes will be extremely important for understanding cognitive flexibility and necessary to selectively target these processes with novel pharmacological agents. Moreover, an exhaustive series of studies by Fellini et al. shows how this same process can be used to compare learning, relearning, and reversal learning, and the effects of prior experience with a reversal, all in a highly comparative manner. This is a crucial step in determining what is specific to reversal learning as opposed to necessary for learning in general. Also, visual discriminations can be adapted to make use of multiple stimuli allowing further dissection of behaviour during reversal learning. Another advantage of the touch-screen approach is that through manipulating the stimuli, the rate of reversal can be increased or decreased. For example, reversal of a visual discrimination generally takes days or weeks depending on the discernibility of the stimuli, whereas the reversal of a spatial discrimination can occur several times within a session. Yet both paradigms have the same basic response requirements and yield highly comparable measurements. This improved comparability addresses one of the biggest problems with the current reversal learning literature, the extent to which results seen in once experimental setting can be extended to another.

On the surface the touch-screen approach seems like it could be ideal to address the problems with standardisation and throughput associated with the ASST. Unfortunately, attempts to address ASST to the TS environment have generally not been successful to date. Rodents have little colour vision and poor visual acuity, making the use of multidimensional complex stimuli difficult. Furthermore, rodents will often generalise across discriminations, resulting in various biases when progressing through the stages of the task. This combined with the general slow rate at which animals learn and reverse visual discriminations suggests that throughput gains made via automation may be offset by the extended duration of a given experiment. However, efforts are still ongoing to use this technology to improve upon the ASST, and it is possible that the approach may also be useful in developing other tasks of executive function, for instance a Stroop task. Adaptation of a wider variety of clinical tests of executive function into a preclinical setting will be necessary if we want to go from beginning to understand the neurobiology of executive function to actually enhancing executive function. As preclinical methods become more powerful and more "human-like", hopefully an increased variety of preclinical models of executive function will become available. As preclinical researchers make more progress in understanding executive function, it is hoped that clinical researchers will increasingly see the value of preclinical research and transition to tests and measures that can be more easily adapted to a preclinical setting. Such will be necessary to further enhance our understanding of neurobiology and develop pharmacological treatments capable of improving normal or abnormal cognitive function.

References

Allison C, Shoaib M (2013) Nicotine improves performance in an attentional set shifting task in rats. Neuropharmacology 64:314–320

Asari T, Konishi S, Jimura K, Miyashita Y (2005) Multiple components of lateral posterior parietal activation associated with cognitive set shifting. Neuroimage 26(3):694–702

Barak S, Weiner I (2011) Putative cognitive enhancers in preclinical models related to schizophrenia: the search for an elusive target. Pharmacol Biochem Behav 99(2):164–189

Barnett JH, Robbins TW, Leeson VC, Sahakian BJ, Joyce EM, Blackwell AD (2010) Assessing cognitive function in clinical trials of schizophrenia. Neurosci Biobehav Rev 34:1161–1177

Berg EA (1948) A simple objective technique for measuring flexibility in thinking. J Gen Psychol 39:15–22

Birrell JM, Brown VJ (2000) Medial frontal cortex mediates perceptual attentional set shifting in the rat. J Neurosci 20(11):4320–4324

Bissonette GB, Martins GJ, Franz TM, Harper ES, Schoenbaum G, Powell EM (2008) Double dissociation of the effects of medial and orbital prefrontal cortical lesions on attentional and affective shifts in mice. J Neurosci 28(44):11124–11130

Boulougouris V, Robbins TW (2009) Pre-surgical training ameliorates orbitofrontal-mediated impairments in spatial reversal learning. Behav Brain Res 197:469–475

Boulougouris V, Robbins TW (2010) Enhancement of spatial reversal learning by 5-HT2C receptor antagonism is neuroanatomically specific. J Neurosci 30:930–938

Boulougouris V, Dalley JW, Robbins TW (2007) Effects of orbitofrontal, infralimbic and prelimbic cortical lesions on serial spatial reversal learning in the rat. Behav Brain Res 179:219–228

Boulougouris V, Castane A, Robbins TW (2009) Dopamine D2/D3 receptor agonist quinpirole impairs spatial reversal learning in rats: investigation of D3 receptor involvement in persistent behavior. Psychopharmacology (Berl) 202:611–620

Brigman JL, Feyder M, Saksida LM, Bussey TJ, Mishina M, Holmes A (2008) Impaired discrimination learning in mice lacking the NMDA receptor NR2A subunit. Learn Mem 15:50–54

Brigman JL, Mathur P, Harvey-White J, Izquierdo A, Saksida LM, Bussey TJ, Fox S, Deneris E, Murphy DL, Holmes A (2010) Pharmacological or genetic inactivation of the serotonin transporter improves reversal learning in mice. Cereb Cortex 20:1955–1963

Brigman JL, Daut RA, Wright T, Gunduz-Cinar O, Graybeal C, Davis MI, Jiang Z, Saksida LM, Jinde S, Pease M, Bussey TJ, Lovinger DM, Nakazawa K, Holmes A (2013) GluN2B in corticostriatal circuits governs choice learning and choice shifting. Nat Neurosci 16:1101–1110

Broadbent NJ, Squire LR, Clark RE (2007) Rats depend on habit memory for discrimination learning and retention. Learn Mem 14:145–151

Brown VJ, Bowman EM (2002) Rodent models of prefrontal cortical function. Trends Neurosci 25 (7):340–343

Chase EA, Tait DS, Brown VJ (2012) Lesions of the orbital prefrontal cortex impair the formation of attentional set in rats. Eur J Neurosci 36(3):2368–2375

Chen KC, Baxter MG, Rodefer JS (2004) Central blockade of muscarinic cholinergic receptors disrupts affective and attentional set-shifting. Eur J Neurosci 20:1081–1088

Clarke HF, Walker SC, Crofts HS, Dalley JW, Robbins TW, Roberts AC (2005) Prefrontal serotonin depletion affects reversal learning but not attentional set shifting. J Neurosci 25:532–538

Clarke HF, Walker SC, Dalley JW, Robbins TW, Roberts AC (2007) Cognitive inflexibility after prefrontal serotonin depletion is behaviorally and neurochemically specific. Cereb Cortex 17:18–27

Clarke HF, Robbins TW, Roberts AC (2008) Lesions of the medial striatum in monkeys produce perseverative impairments during reversal learning similar to those produced by lesions of the orbitofrontal cortex. J Neurosci 28:10972–10982

Clarke HF, Hill GJ, Robbins TW, Roberts AC (2011) Dopamine, but not serotonin, regulates reversal learning in the marmoset caudate nucleus. J Neurosci 31:4290–4297

Clatworthy PL, Lewis SJ, Brichard L, Hong YT, Izquierdo D, Clark L, Cools R, Aigbirhio FI, Baron JC, Fryer TD, Robbins TW (2009) Dopamine release in dissociable striatal subregions predicts the different effects of oral methylphenidate on reversal learning and spatial working memory. J Neurosci 29:4690–4696

Cools R, Clark L, Owen AM, Robbins TW (2002) Defining the neural mechanisms of probabilistic reversal learning using event-related functional magnetic resonance imaging. J Neurosci 22 (11):4563–4567

Cools R, Lewis SJ, Clark L, Barker RA, Robbins TW (2007) L-DOPA disrupts activity in the nucleus accumbens during reversal learning in Parkinson's disease. Neuropsychopharmacology 32:180–189

Dalley JW, Cardinal RN, Robbins TW (2004) Prefrontal executive and cognitive functions in rodents: neural and neurochemical substrates. Neurosci Biobehav Rev 28(7):771–784

Dias R, Robbins TW, Roberts AC (1996a) Dissociation in prefrontal cortex of affective and attentional shifts. Nature 380(6569):69–72

Dias R, Robbins TW, Roberts AC (1996b) Primate analogue of the Wisconsin Card Sorting Test: effects of excitotoxic lesions of the prefrontal cortex in the marmoset. Behav Neurosci 110 (5):872–886

Dias R, Robbins TW, Roberts AC (1997) Dissociable forms of inhibitory control within prefrontal cortex with an analog of the Wisconsin Card Sort Test: restriction to novel situations and independence from "on-line" processing. J Neurosci 17(23):9285–9297

Egan MF, Goldberg TE, Kolachana BS, Callicott JH, Mazzanti CM, Straub RE, Goldman D, Weinberger DR (2001) Effect of COMT Val108/158 Met genotype on frontal lobe function and risk for schizophrenia. Proc Natl Acad Sci USA 98(12):6917–6922

Egerton A, Brett RR, Pratt JA (2005a) Acute delta9-tetrahydrocannabinol-induced deficits in reversal learning: neural correlates of affective inflexibility. Neuropsychopharmacology 30 (10):1895–1905

Egerton A, Reid L, McKerchar CE, Morris BJ, Pratt JA (2005b) Impairment in perceptual attentional set-shifting following PCP administration: a rodent model of set-shifting deficits in schizophrenia. Psychopharmacology (Berl) 179(1):77–84

Fellini L, Kumar G, Gibbs S, Steckler T, Talpos J (2014) Re-evaluating the PCP challenge as a pre-clinical model of impaired cognitive flexibility in schizophrenia. Eur Neuropsychopharmacol 24(11):1836–1849. doi:10.1016/j.euroneuro.2014.08.012

Fletcher PJ, Tenn CC, Rizos Z, Lovic V, Kapur S (2005) Sensitization to amphetamine, but not PCP, impairs attentional set shifting: reversal by a D1 agonist injected into the medial prefrontal cortex. Psychopharmacology (Berl) 183(2):190–200

Floresco SB, Jentsch JD (2011) Pharmacological enhancement of memory and executive functioning in laboratory animals. Neuropsychopharmacology 36(1):227–250

Fox MT, Barense MD, Baxter MG (2003) Perceptual attentional set-shifting is impaired in rats with neurotoxic lesions of posterior parietal cortex. J Neurosci 23(2):676–681

Frohlich J, Van Horn JD (2014) Reviewing the ketamine model for schizophrenia. J Psychopharmacol 28(4):287–302

Groman SM, Lee B, London ED, Mandelkern MA, James AS, Feiler K, Rivera R, Dahlbom M, Sossi V, Vandervoort E, Jentsch JD (2011) Dorsal striatal D2-like receptor availability covaries with sensitivity to positive reinforcement during discrimination learning. J Neurosci 31:7291–7299

Haluk DM, Floresco SB (2009) Ventral striatal dopamine modulation of different forms of behavioral flexibility. Neuropsychopharmacology 34:2041–2052

Hampshire A, Owen AM (2006) Fractionating attentional control using event-related fMRI. Cereb Cortex 16:1679–1689

Hornak J, O'Doherty J, Bramham J, Rolls ET, Morris RG, Bullock PR, Polkey CE (2004) Reward-related reversal learning after surgical excisions in orbito-frontal or dorsolateral prefrontal cortex in humans. J Cogn Neurosci 16:463–478

Keeler JF, Robbins TW (2011) Translating cognition from animals to humans. Biochem Pharmacol 81(12):1356–1366

Lapiz MD, Bondi CO, Morilak DA (2007) Chronic treatment with desipramine improves cognitive performance of rats in an attentional set-shifting test. Neuropsychopharmacology 32(5):1000–1010

Lapiz MD, Morilak DA (2006) Noradrenergic modulation of cognitive function in rat medial prefrontal cortex as measured by attentional set shifting capability. Neuroscience 137(3):1039–1049

Lee B, Groman S, London ED, Jentsch JD (2007) Dopamine D2/D3 receptors play a specific role in the reversal of a learned visual discrimination in monkeys. Neuropsychopharmacology 32:2125–2134

Leeson VC, Robbins TW, Matheson E, Hutton SB, Ron MA, Barnes TR, Joyce EM (2009) Discrimination learning, reversal, and set-shifting in first-episode schizophrenia: stability over six years and specific associations with medication type and disorganization syndrome. Biol Psychiatry 66(6):586–593

Lindgren HS, Wickens R, Tait DS, Brown VJ, Dunnett SB (2013) Lesions of the dorsomedial striatum impair formation of attentional set in rats. Neuropharmacology 71:148–153

McAlonan K, Brown VJ (2003) Orbital prefrontal cortex mediates reversal learning and not attentional set shifting in the rat. Behav Brain Res 146(1–2):97–103

McGaughy J, Ross RS, Eichenbaum H (2008) Noradrenergic, but not cholinergic, deafferentation of prefrontal cortex impairs attentional set-shifting. Neuroscience 153(1):63–71

McLean SL, Idris NF, Grayson B, Gendle DF, Mackie C, Lesage AS, Pemberton DJ, Neill JC (2012) PNU-120596, a positive allosteric modulator of α7 nicotinic acetylcholine receptors, reverses a sub-chronic phencyclidine-induced cognitive deficit in the attentional set-shifting task in female rats. J Psychopharmacol 26(9):1265–1270

Milner B (1963) Effects of different brain lesions on card sorting. Arch Neurol 9:100–110

Muller U, Rowe JB, Rittman T, Lewis C, Robbins TW, Sahakian BJ (2013) Effects of modafinil on non-verbal cognition, task enjoyment and creative thinking in healthy volunteers. Neuropharmacology 64:490–495

Neill JC, Barnes S, Cook S, Grayson B, Idris NF, McLean SL, Snigdha S, Rajagopal L, Harte MK (2010) Animal models of cognitive dysfunction and negative symptoms of schizophrenia: focus on NMDA receptor antagonism. Pharmacol Ther 128(3):419–432

Newman LA, Darling J, McGaughy J (2008) Atomoxetine reverses attentional deficits produced by noradrenergic deafferentation of medial prefrontal cortex. Psychopharmacology (Berl) 200 (1):39–50

Ng CW, Noblejas MI, Rodefer JS, Smith CB, Poremba A (2007) Double dissociation of attentional resources: prefrontal versus cingulate cortices. J Neurosci 27(45):12123–12131

Nikiforuk A, Popik P (2012) Effects of quetiapine and sertindole on subchronic ketamine-induced deficits in attentional set-shifting in rats. Psychopharmacology (Berl) 220(1):65–74

Nikiforuk A, Gołembiowska K, Popik P (2010) Mazindol attenuates ketamine-induced cognitive deficit in the attentional set shifting task in rats. Eur Neuropsychopharmacol 20(1):37–48

Normans DA, Shallice T (1980) Attention to action: willed and automatic control of behavior. In: Gazzaniga MS (ed) Cognitive neuroscience: a reader. Blackwell, Oxford

Owen AM, Roberts AC, Polkey CE, Sahakian BJ, Robbins TW (1991) Extra-dimensional versus intra-dimensional set shifting performance following frontal lobe excisions, temporal lobe excisions or amygdalo-hippocampectomy in man. Neuropsychologia 29(10):993–1006

Park SB, Coull JT, McShane RH, Young AH, Sahakian BJ, Robbins TW, Cowen PJ (1994) Tryptophan depletion in normal volunteers produces selective impairments in learning and memory. Neuropharmacology 33:575–588

Robbins TW, James M, Owen AM, Sahakian BJ, McInnes L, Rabbitt P (1994) Cambridge Neuropsychological Test Automated Battery (CANTAB): a factor analytic study of a large sample of normal elderly volunteers. Dementia 5:266–281

Roberts AC, Wallis JD (2000) Inhibitory control and affective processing in the prefrontal cortex: neuropsychological studies in the common marmoset. Cereb Cortex 10:252–262

Roberts AC, Robbins TW, Everitt BJ, Jones GH, Sirkia TE, Wilkinson J, Page K (1990) The effects of excitotoxic lesions of the basal forebrain on the acquisition, retention and serial reversal of visual discriminations in marmosets. Neuroscience 34:311–329

Roberts AC, Robbins TW, Everitt BJ, Muir JL (1992) A specific form of cognitive rigidity following excitotoxic lesions of the basal forebrain in marmosets. Neuroscience 47(2):251–264

Rodefer JS, Nguyen TN, Karlsson JJ, Arnt J (2008) Reversal of subchronic PCP-induced deficits in attentional set shifting in rats by sertindole and a 5-HT6 receptor antagonist: comparison among antipsychotics. Neuropsychopharmacology 33(11):2657–2666

Rogers RD, Everitt BJ, Baldacchino A, Blackshaw AJ, Swainson R, Wynne K, Baker NB, Hunter J, Carthy T, Booker E, London M, Deakin JF, Sahakian BJ, Robbins TW (1999a) Dissociable deficits in the decision-making cognition of chronic amphetamine abusers, opiate abusers, patients with focal damage to prefrontal cortex, and tryptophan-depleted normal volunteers: evidence for monoaminergic mechanisms. Neuropsychopharmacology 20:322–339

Rogers RD, Blackshaw AJ, Middleton HC, Matthews K, Hawtin K, Crowley C, Hopwood A, Wallace C, Deakin JF, Sahakian BJ, Robbins TW (1999b) Tryptophan depletion impairs stimulus-reward learning while methylphenidate disrupts attentional control in healthy young adults: implications for the monoaminergic basis of impulsive behaviour. Psychopharmacology (Berl) 146:482–491

Rygula R, Walker SC, Clarke HF, Robbins TW, Roberts AC (2010) Differential contributions of the primate ventrolateral prefrontal and orbitofrontal cortex to serial reversal learning. J Neurosci 30:14552–14559

Schoenbaum G, Nugent SL, Saddoris MP, Setlow B (2002) Orbitofrontal lesions in rats impair reversal but not acquisition of go, no-go odor discriminations. Neuroreport 13(6):885–890

Seu E, Jentsch JD (2009) Effect of acute and repeated treatment with desipramine or methylphenidate on serial reversal learning in rats. Neuropharmacology 57(7–8):665–672

Tait DS, Brown VJ, Farovik A, Theobald DE, Dalley JW, Robbins TW (2007) Lesions of the dorsal noradrenergic bundle impair attentional set-shifting in the rat. Eur J Neurosci 25 (12):3719–3724

Tait DS, Brown VJ (2008) Lesions of the basal forebrain impair reversal learning but not shifting of attentional set in rats. Behav Brain Res 187(1):100–108

Tait DS, Chase EA, Brown VJ (2013) Tacrine improves reversal learning in older rats. Neuropharmacology 73:284–289

Talpos J, Steckler T (2013) Touching on translation. Cell Tissue Res 354:297–308

Waltz JA, Gold JM (2007) Probabilistic reversal learning impairments in schizophrenia: further evidence of orbitofrontal dysfunction. Schizophr Res 93(1–3):296–303

Wood S, Sage JR, Shuman T, Anagnostaras SG (2014) Psychostimulants and cognition: a continuum of behavioral and cognitive activation. Pharmacol Rev 66:193–221

Declarative Memory

Wim J. Riedel and Arjan Blokland

Contents

Abstract

Declarative Memory consists of memory for events (episodic memory) and facts (semantic memory). Methods to test declarative memory are key in investigating

W.J. Riedel (✉) • A. Blokland
Department of Neuropsychology & Psychopharmacology, Faculty of Psychology & Neuroscience, Maastricht University, Universiteitssingel 40, 6229ER Maastricht, The Netherlands
e-mail: w.riedel@maastrichtuniversity.nl

K.M. Kantak, J.G. Wettstein (eds.), *Cognitive Enhancement*, Handbook of Experimental Pharmacology 228, DOI 10.1007/978-3-319-16522-6_7

effects of potential cognition-enhancing substances—medicinal drugs or nutrients. A number of cognitive performance tests assessing declarative episodic memory tapping verbal learning, logical memory, pattern recognition memory, and paired associates learning are described. These tests have been used as outcome variables in 34 studies in humans that have been described in the literature in the past 10 years. Also, the use of episodic tests in animal research is discussed also in relation to the drug effects in these tasks. The results show that nutritional supplementation of polyunsaturated fatty acids has been investigated most abundantly and, in a number of cases, but not all, show indications of positive effects on declarative memory, more so in elderly than in young subjects. Studies investigating effects of registered anti-Alzheimer drugs, cholinesterase inhibitors in mild cognitive impairment, show positive and negative effects on declarative memory. Studies mainly carried out in healthy volunteers investigating the effects of acute dopamine stimulation indicate enhanced memory consolidation as manifested specifically by better delayed recall, especially at time points long after learning and more so when drug is administered after learning and if word lists are longer. The animal studies reveal a different picture with respect to the effects of different drugs on memory performance. This suggests that at least for episodic memory tasks, the translational value is rather poor. For the human studies, detailed parameters of the compositions of word lists for declarative memory tests are discussed and it is concluded that tailored adaptations of tests to fit the hypothesis under study, rather than "off-the-shelf" use of existing tests, are recommended.

Keywords

Declarative memory • Episodic memory • Cognition-enhancing drugs

1 Introduction

Declarative memory involves the acquisition and retention of information demanding conscious or explicit learning (Lezak 2004; Squire 1987). Episodic memory is memory for events and experiences and semantic memory is memory for factual knowledge (Squire 1987; Tulving 1972). Both facts and events are encoded by Medial Temporal Lobe structures which comprise a unitary memory system (Squire 2004), as can be seen in Fig. 1. Episodic memory is responsible for storing information about events (or episodes) we have experienced, whereas semantic memory is for storing information about the world, such as the meaning of words. While semantic memory is required to understand and thus encode the words in the first place, remembering a list of words requires episodic memory.

Although historically Tulving (1972) thought that word list learning paradigms, measuring free recall of learnt words, were a typical example of tests of episodic and not semantic memory, he later revised his original definition (Tulving 2002)

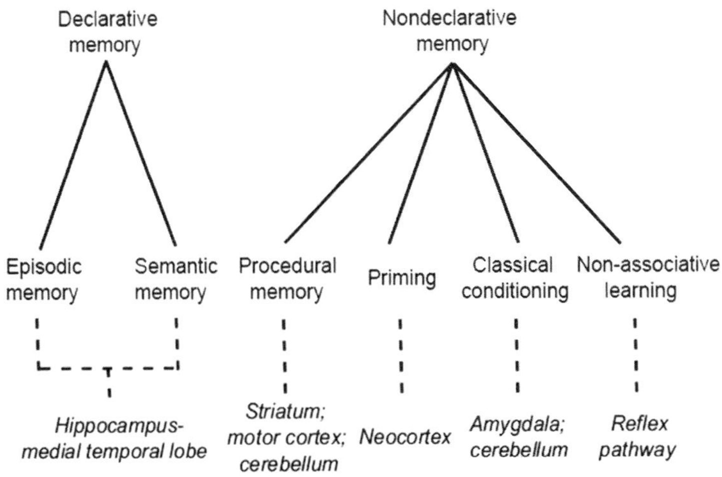

Fig. 1 Taxonomy of mammalian memory systems (after: Squire 2004)

stating that episodic memory consists of "what," "where," and "when" aspects and that free recall of word lists is almost exclusively drawing on the "what" aspect of episodic memory. The distinction between episodic and semantic memory became somewhat less clear in this respect (Tulving 2002).

Semantic memory involves the learning of factual knowledge (Squire 2004). The most common assessment of semantic memory is the verbal fluency task where participants list off as many words possible beginning with a certain letter. Verbal fluency tasks are considered a measure of strategy-driven retrieval from semantic memory stores and thus they also demand executive functions. Semantic memory is difficult to translate from/to preclinical species, so in this chapter we focus on studies of drug effects aimed at cognition enhancement primarily assessing declarative memory by means of performance tests assessing episodic memory.

2 How Declarative Episodic Memory Is Measured

The knowledge of drug effects on declarative memory in humans is largely based on studies of cognition-enhancing drugs utilizing tasks assessing episodic memory function:

- Verbal tasks such as logical memory tasks (LM) probing recall of distinct events in a short story and verbal learning tasks (VLT) probing recall of word lists
- Nonverbal tasks such as Picture or Pattern Recognition Memory (PRM) probing recognition of pictures or patterns belonging to a list and Paired Associates Learning (PAL), probing object-location recognition

Episodic memory tasks typically involve three processes: encoding, consolidation, and retrieval (Lezak 2004). During the encoding phase, the information presented is acquired and learned. Consolidation is the processing of encoded information into long-term storage for later retrieval. Stored information is then either recognized through prompting or recalled spontaneously during the retrieval phase.

2.1 Verbal Memory

2.1.1 Verbal Learning Tests

Verbal learning tasks exist in many well-known variants such as Rey's Auditory Verbal Learning Test (Lezak 2004; Rey 1958) and the California Verbal Learning Test (Delis et al. 1987). Verbal learning tasks can assess short-term memory (immediate recall) or long-term episodic memory (delayed recall and recognition). Consolidation, or conversely decay, can be inferred from the differential of immediate and delayed recall or recognition. While these tests, using word lists of respectively 15 or 16 words, may be appropriate to diagnose memory deficits, memory testing in neuropsychological and psychopharmacological research often involves normal young or healthy elderly volunteers, who might perform at ceiling level.

As can be seen in Table 1, 24 out of the 34 studies reviewed in this chapter used a VLT to assess declarative memory. List length varies considerably between studies. One study used a 12-word list, seven studies used a 15-word list as in the Rey Auditory Verbal Learning Task (Rey 1958), two studies used a 16-word list as in the California Verbal Learning Task (Delis et al. 1987), two studies used a 20-word list, nine studies used a 30-word list, and two studies used a 70-word list. The latter two probed recognition whereas all former probed free recall. It is beyond the scope of the chapter to list all details of task parameters, so we chose to provide a description of the most used VLT. Ideally a VLT should allow translation from drug effects in healthy young volunteers to drug effects in elderly suffering from memory impairment. Riedel et al. (1999) developed an adapted version of Rey's Auditory Verbal Learning test using 30-word lists (Klaassen et al. 2002; Riedel et al. 1999). In this test, 30 monosyllabic words are visually presented during 1 s at a rate of 1 per 2 s followed by a 1-min period of immediate verbal recall, a procedure that is repeated three times in total. After a 30-min delay in which other (nonverbal) cognitive tests are performed, there is another verbal recall and a recognition test.

2.1.2 Logical Memory

There is one study that uses a formal memory task that has all the hallmarks of a logical memory test (Izquierdo et al. 2008). The generic Logical Memory test, as in the Wechsler Memory Scale III (Wechsler 1997), consists of two stories (A read once and B read twice) that are read to the participant who is probed for recall of free and thematic units. Delayed free recall for both A and B is probed 30 min later. A standard cue is provided if the participant has no memory of a story. The recall

Table 1 Overview of acute and subchronic double-blind placebo-controlled studies describing cognition enhancing drug- or nutrition effects on tests of episodic memory in humans.

Study	Population sample	*n* (M/F)	Age	Tx	Mechanism	Regimen	Design	Doses	Task	Measure	Effect	Clinical area
Theunissen et al. (2013)	Healthy	11/13	31, 0	Vortioxetine	5-HT	16 d	db-co	10 mg	30 w	Recall	No drug-placebo difference	
Wingen et al. (2006)	Healthy	9/9	31, 4	Escitalopram	5-HT	15 d	db-co	10–20 mg	30 w	Recall	No drug-placebo difference	
Schmitt et al. (2005)	Healthy with PMS Sx	0/16	18–45	Alpha-lactalbumin	5-HT	Acute	db-co	20 g	PRM	Recog	Alfa-lactalbumin improved delayed pattern recognition	PMS
Sambeth et al. (2014)	Healthy	7/9	19–34	Citalopram	5-HT	Acute	db-co	20 mg	30 w	Recall	No drug-placebo difference	
Gron et al. (2005)	Healthy	30/0	23, 9	Donepezil	Ach	30 d	db-pg	5 mg	15 w	Recall	Donepezil improved immediate recall	
Gron et al. (2006)	MCI	42/0	69, 3	Galantamine	Ach	7 d	ca-ctrl	4 mg bid	16 w	Recall	Galantamine improved immediate and delayed recall	Aging, dementia
Stough et al. (2009)	Healthy	43	22–66	Procera AVH	Ach	30 d	db-pg	1,500 mg + 15 mg + 150 mg	QoM	Accuracy	The combination product improved episodic memory accuracy	Aging, dementia
Balsters et al. (2011)	Healthy	6/14	59–77	Donepezil	Ach	4 w	db-pg	5 mg	PAL	Accuracy	Donepezil impaired accuracy	Aging, dementia
Theunissen et al. (2014)	Healthy cannabis preTx	9/6	21, 23	Rivastigmine, vardenafil	Ach, PDE5	Acute	db-co	3 mg (riv), 20 mg (var)	30 w	Recall	Rivastigmine attenuated cannabis-induced impairment of delayed recall	
Izquierdo et al. (2008)	Healthy	50/55	16–82	Methylphenidate	DA	Acute	db-pg	10 mg	LM	Recall	MPH improved consolidation	Aging, dementia

(continued)

Table 1 (continued)

Study	Population sample	*n* (M/F)	Age	Tx	Mechanism	Regimen	Design	Doses	Task	Measure	Effect	Clinical area
Zeeuws and Soetens (2007)	Healthy	36/0	18–25	d-amphetamine	DA	Acute	db-co	10 mg	20 w	Recall	Improved delayed recall (consolidation)	
Zeeuws et al. (2010b)	Healthy	40/0	18–25	d-amphetamine	DA	Acute	db-co	10 mg	70 w	Recog	Improved delayed recognition (consolidation)	
Linssen et al. (2014)	Healthy	10/10	18–28	Levo/carbi-dopa	DA	Acute	db-co	125 mg	30 w	Recall	No drug-placebo difference	
Hermens et al. (2007)	Healthy	32/0	18–30	Methylphenidate	DA	Acute	db-co	5, 15, 45 mg	12 w	Recall	No drug-placebo difference	
Zeeuws et al. (2010a)	Healthy	17/0	18–30	d-amphetamine	DA	Acute	db-co	10 mg	70 w	Recog	Improved delayed recognition (consolidation)	
Apud et al. (2007)	Healthy by COMT genotype	24/23	18–55	Tolcapone	DA	1 w 6 d	db-co	100–200 tid	15 w	Recall	Trend drug x Gx, val/val Gx improved; met/met Gx worsened	
Randall et al. (2005)	Healthy	29/31	19–22	Modafinil	DA	Acute	db-pg	100, 200 mg	PRM	Recog	Modafinil both doses improved pattern recognition	
Muller et al. (2013)	Healthy	31/33	19–36	Modafinil	DA	Acute	db-pg	200 mg	PRM	Recog	Modafinil improved delayed pattern recognition	
Linssen et al. (2012)	Healthy	19/0	19–37	Methylphenidate	DA	Acute	db-co	10, 20, 40 mg	30 w	Recall	Methylphenidate improved delayed recall	
Kuypers and Ramaekers (2005)	Healthy	9/9	20–39	Methylphenidate	DA	Acute	db-co	20 mg	15 w	Recall	No drug-placebo difference	

Riby et al. (2006)	Healthy	27	20–80	Glucose	Glucose	Acute	db-co	25 g	PAL	Recall	Greater immediate recall in glucose vs. placebo condition	Aging, dementia
Christmas et al. (2014)	Healthy	32/0	18–55	Org25935	GlyTRI	Acute	db-pg	12 mg	20 w	Recall	No drug-placebo difference	
Liem-Moolenaar et al. (2010)	Healthy scopolamine preTx	43/0	18–55	R213129	GlyTRI	Acute	db-co	3, 10, 30 mg	30 w	Recall	No drug-placebo difference	
van Ruitenbeek and Mehta (2013)	Healthy	8/8	18–50	Betahistine	Histamine	Acute	db-co	2 × 48 mg	PAL	Accuracy	No drug-placebo difference	
Reneerkens et al. (2013)	Healthy	5/19	18–25	Vardenafil	PDE5	Acute	db-co	10 mg, 20 mg	30 w	Recall	No drug-placebo difference	
Benton et al. (2013)	Healthy	0/285	21, 8	DHA	PUFAs	50 d	db-pg	400 mg	30 w	Recall	DHA impaired recall at 50 days end of treatment	
Jackson et al. (2012)	Healthy	46/94	18–35	EPA, DHA	PUFAs	12 w	db-pg	90 + 450 mg; 300 + 200 mg	15 w	Recall	No drug-placebo difference	
Karr et al. (2012)	Healthy	12/29	18–35	EPA, DHA	PUFAs	4 w	db-pg	720 mg + 480 mg	15 w	Recall	Minor improvement of delayed recall after PUFA	Aging, dementia
Stonehouse et al. (2013)	Healthy low DHA in diet	83/145	18–45	EPA + DHA	PUFAs	6 mo	db-pg	170 mg + 1,160 mg	15 w	Recall	Improved speed of word recognition and improved recall in females	
Stough et al. (2012)	Healthy	74	45–80	EPA + DHA	PUFAs	90 d	db-pg	60 mg + 252 mg	QoM	Accuracy	No drug-placebo difference	Aging, dementia
Vakhapova et al. (2010)	Elderly with memory complaints	62/60	50–90	PS + EPA + DHA	PUFAs	15 w	db-pg	300 mg + 20 mg + 60 mg	15 w	Recall	PS + DHA only improved 1st trial immediate recall	Aging, dementia

(continued)

Table 1 (continued)

Study	Population sample	*n* (M/F)	Age	Tx	Mechanism	Regimen	Design	Doses	Task	Measure	Effect	Clinical area
Yurko-Mauro et al. (2010)	Healthy with ARCD	485	>=55	DHA	PUFAs	24 w	db-pg	900 mg	PAL	Accuracy	Better PAL performance after DHA supplementation	Aging, dementia
Lee et al. (2013)	MCI	8/27	>=60	EPA + DHA	PUFAs	12 mo	db-pg	150 mg + 430 mg	15 w	Recall	DHA improved immediate and delayed recall	Aging, dementia
Dangour et al. (2010)	Healthy	411/337	70–79	EPA + DHA	PUFAs	24 mo	db-pg	200 mg + 500 mg	16 w	Recall	No drug-placebo difference	Aging, dementia

MCI mild cognitive impairment, *ARCD* age-related cognitive decline, *Pre Tx* pretreatment, *Tx* treatment, *Gx* genotype, *PMS* premenstrual syndrome, *Procera AVH* acetyl-L-carnitine + vinpocetine + huperzine A, *EPA* eicosapentaenoic acid, *DHA* docosahexaenoic acid, *PS* phosphatidylserine, *PUFA* polyunsaturated fatty acid, *GlyTRI* glycine transporter reuptake inhibitor, *DA* dopamine, *Ach* acetylcholine, *5-HT* serotonin, *PDE5* phosphodiesterase-5 inhibitor, *w* weeks, *d* days, *mo* months, *db* double-blind, *pg* parallel groups, *co* crossover, *ca-ctrl* case-control, *mg* milligrams, *tid* three times daily, *bid* twice daily, *μg* micrograms, *N-w* N word list, *PAL* paired associates learning, *LM* logical memory, *PRM* pattern recognition memory and thematic unit scores are again recorded. Fifteen yes/no recognition memory questions are then asked about each story and the recognition memory scores are recorded. The formal memory task used by Izquierdo et al. involved asking the subjects to study a 15-line text with factual information on the 1954 World Cup Soccer for 10 min and then to respond to a questionnaire 2 or 7 days later on 10 of the factual items of the text (Izquierdo et al. 2008).

2.2 Nonverbal Memory

2.2.1 Pattern Recognition Memory

Studies utilized the Pattern Recognition Memory Task or similar variations in which participants identified familiar or new patterns (Muller et al. 2013; Randall et al. 2005; Schmitt et al. 2005). PRM measures the ability to encode and subsequently recognize visual information and is said to provide an assay of medial temporal lobe function (Sahakian et al. 1988). In the encoding phase, 16 patterns appear sequentially on the screen, one at a time, and participants are instructed to remember them. Immediately following the encoding phase, participants carry out a recognition test, in which each pattern from the encoding phase is presented with another pattern of similar form and color, and participants must touch the pattern they saw previously. Twenty minutes following this first recognition test, a second delayed recognition test is administered.

2.2.2 Paired Associate Learning

Boxes are displayed on the screen and open in turn to reveal a number of patterns. Participants are instructed to try to remember the location in which each pattern was shown. After all the boxes have been opened, each pattern is then shown in the center of the screen in a randomized order, and the participant touches the box in which the pattern was located. If an error is made, all the patterns are re-presented to remind the participant of their locations. As the test progresses, the stages become more difficult as the number of patterns to be remembered increases. Performance of the Paired Associates Learning (PAL) test activates the bilateral hippocampi in the Medial Temporal lobe (de Rover et al. 2011).

2.3 How to Measure Declarative Episodic Memory in Animals

By definition declarative memory assumes that memories can be declared, which in humans is the ability to verbalize or to consciously recall information. Clearly, there is no clear evidence that animals can have these abilities (Templer and Hampton 2013). However, as mentioned above, episodic memory has also been defined in terms of "what," "where," and "when" (Tulving). When considering this definition it has been shown that animals do have an episodic(like) memory (Clayton et al. 2003). Whether this is directly comparable to humans will be difficult to demonstrate, but it appears that similar brain regions do support the episodic memory in humans and animals (Allen and Fortin 2013; Morris 2001). So, there appear to be good arguments to state that animals do have an episodic memory, but there may be some differences between animals and humans.

Different tests have been used as models of episodic memory tasks in rodents (for a review, see Fouquet et al. 2010). The most used tasks are the object recognition task, and variations thereof (Dere et al. 2006; Ennaceur and Delacour 1988; Kesner et al. 2008). Other examples are specific procedures in radial mazes (Crystal 2013), odor recognition (Eichenbaum et al. 2010), and some procedures in the classic Morris water task (Zhang et al. 2008). For all these test mentioned, the "what," "where," and "when" can be applied. The PAL test has also been developed for rodents by using a touch-screen device (Talpos et al. 2009). However, it has not been claimed that this task could be considered as an episodic test for rodents.

This short overview shows that there are several episodic memory tests for rodents. Although these tests can be defined as an episodic test in animals, the comparability with human episodic tests is limited. Some tests seem not to be translatable to animal studies (i.e., VLT and LM) and some tests are considered episodic memory tests for which not such a claim has been made in animal research (i.e., PAL and spatial Morris task (only specific protocols)). The object recognition task is the most widely task in animal research to test episodic memory. However, there is not a clear human equivalent for this task in human research. Taken together, this indicates that the translation of episodic memory tasks is rather limited.

3 Overview of Studies

3.1 How Searched

For the human studies we searched for publications in the past 10 years, i.e., from 2005 to the end of 2014, about cognition-enhancing substance (either drug or nutrient) effects on at least one test of declarative memory. Most of these publications also contained reports about drug effects on other tests assessed simultaneously, but in this chapter we exclusively focus on the results concerning declarative or episodic memory. We searched PubMed using the search terms:

(episodic OR declarative) drug* cogni* enhanc*
verbal cogni* drug* enhance*
cogni* enhanc* drug* learn* paired*

In addition, publications that also fulfilled the content criterion (cognition-enhancing drug effects on tests of declarative memory) and that were already known to us or became known to us after asking around were added. For the animal studies there is a vast amount of studies in which the effects of many different drugs have been tested. Mostly, the object recognition test has been used for these purposes. Therefore, we refer to selective (review) studies in which drug effects have been examined.

3.2 Studies

A total of 34 studies were obtained and these are listed and categorized in Table 1. At large, the studies can be categorized on the basis of:

- Mechanism(s) of the substance under investigation
- Single-dose or repeated dose studies. The latter are more important when attributing the clinical meaning of the results, whereas acute single-dose studies are usually focused on a proof-of-principle type of question
- Population sample of subjects. Most studies on cognition-enhancing drugs are carried out in healthy volunteers, although in several instances these are samples described by a particular vulnerability factor, such as elderly with memory deficits or complaints, genotype, relative deficiency of omega-3 fatty acids in the diet, premenstrual syndrome, and scopolamine pretreatment
- Number of items to memorize or memory load. For example, word lists used in the different studies comprised of 12, 15, 16, 20, 30, or 70 words and the number of times they were repeatedly presented was once, twice, three times, or five times. Pattern recognition memory tests likewise present a sequence of patterns or pictures. If the pictures are concrete the test is considered verbal because pictures will be primarily verbalized. Abstract patterns, as used in the PRM and PAL, are more difficult to verbalize and are therefore considered nonverbal. Both are considered tests of episodic memory however. The memory load in PAL is increased in stepwise fashion (2, 3, 6, 8, 10, 12) until subjects reach the highest level or fail
- Number of subjects. Larger studies more often are repeated dose studies and these are deemed more important than smaller sample studies

3.2.1 PUFAs

Nine of the studies are investigating the effect of omega-3 (or n-3) polyunsaturated fatty acids (PUFAs) on declarative memory. All nine are repeated dose studies over a relatively longer period of time. However, not all treatments are the same in that the composition of PUFAs differs and also the age of the groups studied differs. From the four studies in young volunteers, one was negative (Benton et al. 2013), one found no effects (Jackson et al. 2012), and two found at best marginal positive effects (Karr et al. 2012; Stonehouse et al. 2013). In elderly, two trials were positive (Lee et al. 2013; Yurko-Mauro et al. 2010), one where phosphatidylserine was added was marginally positive (Vakhapova et al. 2010) and two found no difference (Dangour et al. 2010; Stough et al. 2012). Although different doses and different compositions of the relative contributions of EFA and DHA were used, there was no obvious relation with dose. Positive effects, if present, were mainly attributed to DHA.

3.2.2 Dopamine

Eleven studies were investigating substances that were thought to be primarily affecting dopamine (DA) of which two studies investigating modafinil are also included. All but one were single-dose studies and all but one were investigating samples of young healthy volunteers. Three were investigating single doses of d-amphetamine of which all three reported positive effects on delayed recall or recognition (Zeeuws et al. 2010a, b; Zeeuws and Soetens 2007). Four were investigating single doses of methylphenidate of which two reported positive effects on delayed recall memory (Izquierdo et al. 2008; Linssen et al. 2012) and two found no difference (Hermens et al. 2007; Kuypers and Ramaekers 2005). The studies that reported positive effects had in common that they were using considerably longer word lists (20, 30, 70 words) opposed to the studies not reporting positive effects (12 and 15 words) and were looking at late stage memory consolidation.

Two studies on modafinil in young healthy volunteers both reported positive effects on pattern memory recognition (Muller et al. 2013; Randall et al. 2005). A study looking at L-Dopa found no effects (Linssen et al. 2014). One study looking at the COMT inhibitor tolcapone found an interaction between drug and COMT (val158met) genotype such that tolcapone tended to improve word recall in healthy subjects with val/val genotype and tended to impair word recall in subjects with met/met genotype (Apud et al. 2007).

3.2.3 Acetylcholine

Five studies reported investigations of subchronic doses of substances where acetylcholine stimulation was the primary hypothesized mechanism of action, although one of them concerned a combination preparation where other mechanisms might also have played a role. Three of these were in elderly subjects, one of these concerned subjects with Mild Cognitive Impairment. Three studies reported improved episodic memory performance (Gron et al. 2006, 2005; Stough et al. 2009), whereas one reported impaired episodic memory performance (Balsters et al. 2011). The latter was in healthy elderly subjects and the only clear factor in which this study was different than the other three reporting positive effects of cholinesterase inhibitors was that Balsters et al. were using PAL, a nonverbal episodic memory task, whereas the other studies were using Word List tests. One study investigated the cholinesterase inhibitor rivastigmine after pretreatment with cannabis, which induced significant impairment of declarative memory performance and observed that a single dose of rivastigmine 3 mg attenuated cannabis-induced impairment (Theunissen et al. 2014).

3.2.4 5-HT

One study successfully demonstrated that cognitive deficits reported as symptoms, among which impaired declarative memory performance, in Premenstrual Syndrome (PMS), can be improved by treatment with a protein mixture called alpha-lactalbumin containing relatively high amounts of the serotonin precursor tryptophan (Schmitt et al. 2005). Three studies in healthy volunteers investigating acute and subchronic effects of the serotonergic antidepressants citalopram (Sambeth et al. 2014), escitalopram (Wingen et al. 2006), and vortioxetine (Theunissen et al. 2013) on declarative memory performance found no drug effects on word recall.

3.2.5 Miscellaneous (Glycine, Histamine, PDE5, Glucose)

Two studies investigating novel glycine reuptake inhibitors, hypothesized to improve glutamatergic neurotransmission in patients with schizophrenia, were described (Christmas et al. 2014; Liem-Moolenaar et al. 2010). Both were still in the early stage of human drug development investigating safety and mechanism of action. Neither improved or impaired declarative memory, and in both cases the drug may not have been in the dose window leading to an effective concentration to either confirm or reject the hypothesis.

One study hypothesized histamine augmentation to improve declarative memory by means of administration of betahistine to healthy volunteers (van Ruitenbeek and Mehta 2013). No changes were observed on declarative memory performance.

Two studies investigated the effects of acute doses of 10 and 20 mg of the phosphodiesterase-5 inhibitor (PDE5) vardenafil on declarative memory and both found no differences with placebo. One was in healthy volunteers (Reneerkens et al. 2013) and one was in healthy volunteers after pretreatment with cannabis (Theunissen et al. 2014).

One study showed that a drink containing 25 g of glucose improved paired associate learning in young and elderly volunteers. Glucose is thought to act primarily on the hippocampal region in the medial temporal cortex, known to support episodic memory (Riby et al. 2006).

3.3 Translational Issues

At large, there are three hypotheses underlying the majority of the studies reported in Table 1. These include:

- Dietary polyunsaturated fatty acid supplements improve declarative memory
  - Dietary essential fatty acids include omega-3 (n-3) and omega-6 (n-6) polyunsaturated fatty acids (PUFA). Two specific n-3 PUFAs, docosahexaenoic acid (DHA), the principal long-chain PUFA comprising 40 % of brain lipids, and eicosapentaenoic acid (EPA) have specifically been investigated. Both are important in the central nervous system, with 50 % of neuronal membranes and 70 % of myelin composed of lipids. DHA status is dependent on dietary intake, and when low, supplementation is hypothesized to improve cognitive function. The most used dietary manipulation is that of fish oil (containing a mixture of EPA and DHA) or DHA alone (Karr et al. 2012; Yurko-Mauro et al. 2010).
- Cholinesterase inhibition improves declarative memory

  - Cholinesterase inhibitors are a registered category of drugs against cognitive symptoms of Alzheimer's disease. Cholinergic decline is considered to occur relatively late in the gradual time span of age-related cognitive decline after conversion to clinical diagnosis of probable Alzheimer's disease. It is subject of considerable debate whether boosting cholinergic function by cholinesterase inhibitors is useful in earlier stages, i.e., mild cognitive impairment.
- Enhancement of dopaminergic neurotransmission improves aspects of declarative memory
  - Substances promoting dopamine may have in common that they facilitate memory consolidation. The hippocampus plays a key role in memory consolidation, which relies on novelty. The Ventral Tegmental Area (VTA) projects to the hippocampus through dopaminergic fibers acting on D1 receptors, where it enhances long-term potentiation (LTP) and learning. The hippocampal-VTA loop has been found to be important for novelty detection, for hippocampal LTP, and for memory consolidation (Izquierdo et al. 2008).

3.3.1 Prediction of Results by Preclinical Studies

When examining data from animal studies, evidence for an improvement in object recognition can be found for various drugs classes. Pro-cholinergic drugs like nicotinic agonists and acetylcholinesterase inhibitors improve performance (de Bruin and Pouzet 2006; Prickaerts et al. 2005; Sambeth et al. 2007). There are a limited number of studies using dopaminergic drugs. Most studies tested dopaminergic drugs in deficit models and transgenic animals. One study showed that especially dopamine type-1 receptor agonists can improve object memory (de Lima et al. 2011). Interestingly, acute methylphenidate treatment impaired object recognition performance in healthy animals (Chuhan and Taukulis 2006).

Also for serotonergic drugs, many studies examined the effects in deficit models. SSRIs seemed to have no effects on object recognition performance (Naudon et al. 2007; Valluzzi and Chan 2007). In animal studies there has been special interest in the 5-HT1A and the 5-HT6 receptors in cognition. Interestingly, although 5-HT1A agonists were found to improve memory performance in other tasks (Cole et al. 1994), a 5-HT1A receptor antagonist was found to improve memory performance in the object recognition task (Pitsikas et al. 2003). Also, 5-HT6 antagonists were found to improve memory performance (King et al. 2004; Lieben et al. 2005). There have been many studies that examined the effect of PDE inhibitors on memory performance and all these drugs have been found to improve object memory performance (Reneerkens et al. 2009). Finally, one study was found in which the effects of a PUFA were tested in the object recognition memory (Cutuli et al. 2014). Here an improved memory performance was found in aged animals after 8 weeks of treatment.

Although this is a restricted overview of studies in which the effects of drugs on object memory were examined, it appears that most drugs improve the performance in this task. SSRI appear to be ineffective in this task and one task showed an impaired object memory performance after methylphenidate treatment. When considering the human data, there seems to be a great discrepancy in findings between species.

3.3.2 Generalizability of Results to Clinical Therapeutic Areas

Impairments of episodic memory have been described in a wide variety of neuropsychiatric diseases: ADHD and schizophrenia (Mehta and Riedel 2006), mild cognitive impairment and dementia (Riedel 2014), as well as affective disorders (Rock et al. 2014). In Table 1, we have indicated clinical significance by referring the meaning of the results to the therapeutic areas under study. Ten studies were in elderly subjects either or not with established memory impairment or complaints. Seven of these studies noted treatment-induced improvement of declarative memory against one that noted impairment whereas two showed no change. One study in females with premenstrual symptoms including poorer performance on episodic memory tasks resulted in observed treatment-induced improvement of episodic memory performance.

Eighteen studies were in young healthy volunteers with no indication of direct clinical relevance as far as a specific population is concerned. However, results indicating treatment-induced enhancement of cognition bear the relevance that the substance is active and is therefore bearing indirect clinical relevance not specifically linked to one therapeutic area. Healthy volunteer studies are mostly relevant in uncovering principles that may be worth pursuing such as the putative manipulability of memory consolidation by means of substances boosting dopamine.

4 Conclusions

In sum, what this review of cognition-enhancing drug studies in the past 10 years shows is that tests of declarative memory are sensitive to indicate effects of cognition-enhancing treatments. There are three different areas of interest at large: (1) dietary effects of PUFAs which seem to work better in elderly groups than in young participants; (2) effects of "symptomatic" cholinergic drugs appear to be effective in supposedly prodromal or preclinical stages of Alzheimer's disease; and (3) enhancement of declarative memory consolidation by means of dopaminergic stimulation relatively late after learning, in normal young and elderly volunteers.

Enhancement of declarative memory is perhaps one of the most aspired clinical aims in a variety of recognized therapeutic areas such as dementia and its prodromal stages, ADHD, schizophrenia, and affective disorders. However, non-recognized "therapeutic" conditions such as normal aging and even young healthy volunteers aspire enhancement of declarative memory as well. It is clear that this "therapeutic goal" is wider than fulfilling only urgent medical needs. The latter may apply in the case of dementia; however, even in this area, because attention is shifting to prodromal and preclinical stages of dementia (which are characterized by absence of functional impairment and almost no, or completely no symptoms, but presence of amyloid plaques—putative neural substrate of future disease), the boundaries are fading and the classic medical model of treating disease appears to be shifting to a "prevention paradigm." As said, this primarily applies to dementia where no proven disease modifying treatment exists, yet is anticipated, so there is an urgent need for early detection of clinical symptoms of which impairment of declarative memory is the very first manifesting symptom. Measuring declarative memory can therefore be applied as a diagnostic and also as a marker of disease progression and the effects of treatments thereon. In that context, the conclusion of the present review can be: there is enough evidence that impairment and enhancement of declarative memory can be measured precisely. There is just no effective treatment yet to prevent age-related cognitive decline, mild cognitive impairment, and Alzheimer's dementia. This also can be taken to mean that the positive effects of substances such as PUFAs and cholinesterase inhibitors in elderly with mild cognitive impairment are significant, yet of marginal clinical relevance.

We observed that verbal learning tests are among the most frequently used paradigms to assess declarative episodic memory performance. There are pros and cons to each of these tests/dependent variables (LM, VLT, PRM, and PAL) reviewed in this chapter. As said, the most frequently used paradigm VLT seems to be the most sensitive, but according to Tulving (2002) could be considered too selective to be a good general index of declarative episodic memory because it addresses only the "what" aspect of memory and not the "where" and "when." This also applies to PRM. In PAL, "where" and "what" is probed. LM is the only paradigm that allows "what," "where," and "when" encoding of memory. Another objection to VLT that is sometimes heard is that VLT and therefore also LM are too language-dependent and can therefore only be applied within single language areas. It is true that there may be practical advantages to language-free tests of declarative memory such as PAL or PRM, but verbal memory encoding can be considered essential for mankind and therefore it could be a huge omission not to probe it when investigating efficacy of treatments that claim to enhance declarative memory. It can be easily argued that humans can and will verbalize even the most abstract visual information. For example, if subjects are able to employ a strategy to quickly associate each pattern in PAL with a compass heading (e.g., in clockwise direction: North, North-North-East, North-East, East-North-East, etc.), PAL performance becomes an index of verbal declarative memory encoding. Furthermore, there are numerous validated versions of VLT in different languages. In addition, even within one language area, one needs multiple comparable or parallel versions of the same word list in order to be able to apply it repeatedly within the same cohort of subjects over time or in a crossover design. More importantly, there are some other even more specific parameters, which are rarely considered that allow optimization fitted to testing a specific hypothesis. These are:

- Number of words in the list (in the present review this varies between 12 and 70)
- Number of times word list is presented and immediate recall is repeated (1–5)
- Number of syllables per word (monosyllabic, controlled multisyllabic, or uncontrolled)
- Word length (5, 5–7 letters, or uncontrolled)

– Interval between immediate and delayed recall is usually 20–30 min but can be much longer up to multiple days
– Presentation time, inter-stimulus interval, and response-stimulus interval which determine whether the test is paced or self-paced and what the pace is (1 word per second to 1 word/4 s in the studies reviewed here). These parameters are of key importance whether attentional circuits in the brain are triggered. The faster the pace, the more declarative memory performance will become dependent on attentional constraints (or in essence: time constraints)
– Associability—words within one list are usually controlled for high interrelatedness (e.g., no shoe and foot in one list)
– Recall or recognition—most studies probe free recall, but two studies by Zeeuws et al. (2010a, b) use very long lists and only probe recognition; PRM tests also only probe recognition

This review has shown that positive effects of dopaminergic stimulation on declarative memory performance seem to be best measurable using longer word lists with monosyllabic words. Drugs need to be administered not before but after learning and times of assessment of recall or recognition should be long after learning when effects of treatment on consolidation are most manifest. It remains to be shown how such effects on consolidation should be applied beyond single-dose designs at a longer timescale. What we learn from these studies is that tests need always to be tailored to the specifically predicted effects of drugs. There seems to be no off-the-shelf recipe for cognitive tests.

The prediction of animal studies with respect to episodic memory tasks appears to be very limited. First, the comparison on the level of the episodic tasks is poor. Secondly, there are clear species differences in the effects of drugs on episodic memory. Recently, some critical remarks have been made on the translational value of animal models of memory and the comparability of the effects of drugs across species (Blokland et al. 2014; Brodziak et al. 2014). At present, the translational value of episodic tasks appears to be rather poor.

Declaration of Conflict of Interest W. J. Riedel is part-time employed by Maastricht University and has in the past 12 months also carried out consultancy for Cambridge Cognition, Takeda, Lundbeck, and Keyview Labs.

References

Allen TA, Fortin NJ (2013) The evolution of episodic memory. Proc Natl Acad Sci U S A 110 (Suppl 2):10379–10386. doi:10.1073/pnas.1301199110

Apud JA, Mattay V, Chen J, Kolachana BS, Callicott JH, Rasetti R, Alce G et al (2007) Tolcapone improves cognition and cortical information processing in normal human subjects. Neuropsychopharmacology 32(5):1011–1020. doi:10.1038/sj.npp.1301227

Balsters JH, O'Connell RG, Martin MP, Galli A, Cassidy SM, Kilcullen SM, Delmonte S et al (2011) Donepezil impairs memory in healthy older subjects: behavioural, EEG and simultaneous EEG/fMRI biomarkers. PLoS One 6(9):e24126. doi:10.1371/journal.pone.0024126

Benton D, Donohoe RT, Clayton DE, Long SJ (2013) Supplementation with DHA and the psychological functioning of young adults. Br J Nutr 109(1):155–161. doi:10.1017/S0007114512000566

Blokland A, van Goethem N, Heckman P, Schreiber R, Prickaerts J (2014) Translational issues with the development of cognition enhancing drugs. Front Neurol 5:190. doi:10.3389/fneur.2014.00190

Brodziak A, Kolat E, Rozyk-Myrta A (2014) In search of memory tests equivalent for experiments on animals and humans. Med Sci Monit 20:2733–2739. doi:10.12659/MSM.891056

Christmas D, Diaper A, Wilson S, Rich A, Phillips S, Udo de Haes J, Sjogren M, Nutt D (2014) A randomised trial of the effect of the glycine reuptake inhibitor Org 25935 on cognitive performance in healthy male volunteers. Hum Psychopharmacol 29(2):163–171. doi:10.1002/hup.2384

Chuhan YS, Taukulis HK (2006) Impairment of single-trial memory formation by oral methylphenidate in the rat. Neurobiol Learn Mem 85(2):125–131. doi:10.1016/j.nlm.2005.09.001

Clayton NS, Bussey TJ, Dickinson A (2003) Can animals recall the past and plan for the future? Nat Rev Neurosci 4:685–691

Cole BJ, Jones GH, Turner JD (1994) 5-HT1A receptor agonists improve the performance of normal and scopolamine-impaired rats in an operant delayed matching to position task. Psychopharmacology (Berl) 116:135–142

Crystal JD (2013) Remembering the past and planning for the future in rats. Behav Processes 93:39–49. doi:10.1016/j.beproc.2012.11.014

Cutuli D, De Bartolo P, Caporali P, Laricchiuta D, Foti F, Ronci M, Rossi C et al (2014) n-3 polyunsaturated fatty acids supplementation enhances hippocampal functionality in aged mice. Front Aging Neurosci 6:220. doi:10.3389/fnagi.2014.00220

Dangour AD, Allen E, Elbourne D, Fasey N, Fletcher AE, Hardy P, Holder GE et al (2010) Effect of 2-y n-3 long-chain polyunsaturated fatty acid supplementation on cognitive function in older people: a randomized, double-blind, controlled trial. Am J Clin Nutr 91(6):1725–1732. doi:10.3945/ajcn.2009.29121

de Bruin N, Pouzet B (2006) Beneficial effects of galantamine on performance in the object recognition task in Swiss mice: deficits induced by scopolamine and by prolonging the retention interval. Pharmacol Biochem Behav 85:253–260

de Lima MN, Presti-Torres J, Dornelles A, Scalco FS, Roesler R, Garcia VA, Schroder N (2011) Modulatory influence of dopamine receptors on consolidation of object recognition memory. Neurobiol Learn Mem 95(3):305–310. doi:10.1016/j.nlm.2010.12.007

de Rover M, Pironti VA, McCabe JA, Acosta-Cabronero J, Arana FS, Morein-Zamir S, Hodges JR et al (2011) Hippocampal dysfunction in patients with mild cognitive impairment: a functional neuroimaging study of a visuospatial paired associates learning task. Neuropsychologia 49(7):2060–2070. doi:10.1016/j.neuropsychologia.2011.03.037

Delis DC, Kramer JH, Kaplan E, Ober BA (1987) California verbal learning test manual. The Psychological Corporation, San Antonia, TX

Dere E, Kart-Teke E, Huston JP, De Souza Silva MA (2006) The case for episodic memory in animals. Neurosci Biobehav Rev 30:1206–1224

Eichenbaum H, Fortin N, Sauvage M, Robitsek RJ, Farovik A (2010) An animal model of amnesia that uses Receiver Operating Characteristics (ROC) analysis to distinguish recollection from familiarity deficits in recognition memory. Neuropsychologia 48(8):2281–2289. doi:10.1016/j.neuropsychologia.2009.09.015

Ennaceur A, Delacour J (1988) A new one-trial test for neurobiological studies of memory in rats. 1: behavioral data. Behav Brain Res 31:47–59

Fouquet C, Tobin C, Rondi-Reig L (2010) A new approach for modeling episodic memory from rodents to humans: the temporal order memory. Behav Brain Res 215(2):172–179. doi:10.1016/j.bbr.2010.05.054

Gron G, Kirstein M, Thielscher A, Riepe MW, Spitzer M (2005) Cholinergic enhancement of episodic memory in healthy young adults. Psychopharmacology (Berl) 182(1):170–179. doi:10.1007/s00213-005-0043-2

Gron G, Brandenburg I, Wunderlich AP, Riepe MW (2006) Inhibition of hippocampal function in mild cognitive impairment: targeting the cholinergic hypothesis. Neurobiol Aging 27 (1):78–87. doi:10.1016/j.neurobiolaging.2004.12.005

Hermens DF, Cooper NJ, Clark CR, Debrota D, Clarke SD, Williams LM (2007) An integrative approach to determine the best behavioral and biological markers of methylphenidate. J Integr Neurosci 6(1):105–140

Izquierdo I, Bevilaqua LR, Rossato JI, Lima RH, Medina JH, Cammarota M (2008) Age-dependent and age-independent human memory persistence is enhanced by delayed posttraining methylphenidate administration. Proc Natl Acad Sci U S A 105 (49):19504–19507. doi:10.1073/pnas.0810650105

Jackson PA, Deary ME, Reay JL, Scholey AB, Kennedy DO (2012) No effect of 12 weeks' supplementation with 1 g DHA-rich or EPA-rich fish oil on cognitive function or mood in healthy young adults aged 18-35 years. Br J Nutr 107(8):1232–1243. doi:10.1017/S000711451100403X

Karr JE, Grindstaff TR, Alexander JE (2012) Omega-3 polyunsaturated fatty acids and cognition in a college-aged population. Exp Clin Psychopharmacol 20(3):236–242. doi:10.1037/a0026945

Kesner RP, Hunsaker MR, Warthen MW (2008) The CA3 subregion of the hippocampus is critical for episodic memory processing by means of relational encoding in rats. Behav Neurosci 122 (6):1217–1225. doi:10.1037/a0013592

King MV, Sleight AJ, Woolley ML, Topham IA, Marsden CA, Fone KC (2004) 5-HT6 receptor antagonists reverse delay-dependent deficits in novel object discrimination by enhancing consolidation—an effect sensitive to NMDA receptor antagonism. Neuropharmacology 47:195–204

Klaassen T, Riedel WJ, Deutz NEP, Van Praag HM (2002) Mood congruent memory bias induced by tryptophan depletion. Psychol Med 32(1):167–172

Kuypers KP, Ramaekers JG (2005) Transient memory impairment after acute dose of 75 mg 3.4-Methylene-dioxymethamphetamine. J Psychopharmacol 19(6):633–639

Lee LK, Shahar S, Chin AV, Yusoff NA (2013) Docosahexaenoic acid-concentrated fish oil supplementation in subjects with mild cognitive impairment (MCI): a 12-month randomised, double-blind, placebo-controlled trial. Psychopharmacology (Berl) 225(3):605–612. doi:10.1007/s00213-012-2848-0

Lezak MD (2004) Neuropsychological assessment, 3rd edn. Oxford University Press, New York

Lieben CK, Blokland A, Sik A, Sung E, van Nieuwenhuizen P, Schreiber R (2005) The selective 5-HT(6) receptor antagonist Ro4368554 restores memory performance in cholinergic and serotonergic models of memory deficiency in the rat. Neuropsychopharmacology 30:2169–2179

Liem-Moolenaar M, Zoethout RW, de Boer P, Schmidt M, de Kam ML, Cohen AF, Franson KL, van Gerven JM (2010) The effects of the glycine reuptake inhibitor R213129 on the central nervous system and on scopolamine-induced impairments in psychomotor and cognitive function in healthy subjects. J Psychopharmacol 24(11):1671–1679. doi:10.1177/0269881109106942

Linssen AM, Vuurman EF, Sambeth A, Riedel WJ (2012) Methylphenidate produces selective enhancement of declarative memory consolidation in healthy volunteers. Psychopharmacology (Berl) 221(4):611–619. doi:10.1007/s00213-011-2605-9

Linssen AM, Sambeth A, Vuurman EF, Riedel WJ (2014) Cognitive effects of methylphenidate and levodopa in healthy volunteers. Eur Neuropsychopharmacol 24(2):200–206. doi:10.1016/j.euroneuro.2013.09.009

Mehta MA, Riedel WJ (2006) Dopaminergic enhancement of cognitive function. Curr Pharm Des 12(20):2487–2500

Morris RG (2001) Episodic-like memory in animals: psychological criteria, neural mechanisms and the value of episodic-like tasks to investigate animal models of neurodegenerative disease. Philos Trans R Soc Lond B Biol Sci 356(1413):1453–1465. doi:10.1098/rstb.2001.0945

Muller U, Rowe JB, Rittman T, Lewis C, Robbins TW, Sahakian BJ (2013) Effects of modafinil on non-verbal cognition, task enjoyment and creative thinking in healthy volunteers. Neuropharmacology 64:490–495. doi:10.1016/j.neuropharm.2012.07.009

Naudon L, Hotte M, Jay TM (2007) Effects of acute and chronic antidepressant treatments on memory performance: a comparison between paroxetine and imipramine. Psychopharmacology (Berl) 191(2):353–364. doi:10.1007/s00213-006-0660-4

Pitsikas N, Rigamonti AE, Cella SG, Muller EE (2003) The 5-HT 1A receptor antagonist WAY 100635 improves rats performance in different models of amnesia evaluated by the object recognition task. Brain Res 983(1–2):215–222

Prickaerts J, Sik A, Van Der Staay FJ, De Vente J, Blokland A (2005) Dissociable effects of acetylcholinesterase inhibitors and phosphodiesterase type 5 inhibitors on object recognition memory: acquisition versus consolidation. Psychopharmacology (Berl) 177:381–390

Randall DC, Viswanath A, Bharania P, Elsabagh SM, Hartley DE, Shneerson JM, File SE (2005) Does modafinil enhance cognitive performance in young volunteers who are not sleep-deprived? J Clin Psychopharmacol 25(2):175–179

Reneerkens OA, Rutten K, Steinbusch HW, Blokland A, Prickaerts J (2009) Selective phosphodiesterase inhibitors: a promising target for cognition enhancement. Psychopharmacology (Berl) 202(1–3):419–443. doi:10.1007/s00213-008-1273-x

Reneerkens OA, Sambeth A, Ramaekers JG, Steinbusch HW, Blokland A, Prickaerts J (2013) The effects of the phosphodiesterase type 5 inhibitor vardenafil on cognitive performance in healthy adults: a behavioral-electroencephalography study. J Psychopharmacol 27(7):600–608. doi:10.1177/0269881113477747

Rey A (1958) L'examen psychologique dans les cas d'encéphalopathie traumatique. Presses Universitaires de France, Paris

Riby LM, McMurtrie H, Smallwood J, Ballantyne C, Meikle A, Smith E (2006) The facilitative effects of glucose ingestion on memory retrieval in younger and older adults: is task difficulty or task domain critical? Br J Nutr 95(2):414–420

Riedel WJ (2014) Preventing cognitive decline in preclinical Alzheimer's disease. Curr Opin Pharmacol 14:18–22. doi:10.1016/j.coph.2013.10.002

Riedel WJ, Klaassen T, Deutz NEP, Van Someren A, Van Praag HM (1999) Tryptophan depletion in normal volunteers produces selective impairment in memory consolidation. Psychopharmacology (Berl) 141(4):362–369

Rock PL, Roiser JP, Riedel WJ, Blackwell AD (2014) Cognitive impairment in depression: a systematic review and meta-analysis. Psychol Med 44(10):2029–2040. doi:10.1017/S0033291713002535

Sahakian BJ, Morris RG, Evenden JL, Heald A, Levy R, Philpot M, Robbins TW (1988) A comparative study of visuospatial memory and learning in Alzheimer-type dementia and Parkinson's disease. Brain 111(Pt 3):695–718

Sambeth A, Riedel W, Smits L, Blokland A (2007) Cholinergic drugs affect novel object recognition in rats: relation with hippocampal EEG? Eur J Pharmacol 572:151–159

Sambeth A, Riedel WJ, Klinkenberg I, Kahkonen S, Blokland A (2014) Biperiden selectively induces memory impairment in healthy volunteers: no interaction with citalopram. Psychopharmacology (Berl). doi:10.1007/s00213-014-3822-9

Schmitt JAJ, Jorissen BL, Dye L, Markus CR, Deutz NEP, Riedel WJ (2005) Memory function in women with premenstrual complaints and the effect of serotonergic stimulation by acute administration of an alpha-lactalbumin protein. J Psychopharmacol 19(4):375–384

Squire LR (1987) Memory and brain. Oxford University Press, New York

Squire LR (2004) Memory systems of the brain: a brief history and current perspective. Neurobiol Learn Mem 82(3):171–177

Stonehouse W, Conlon CA, Podd J, Hill SR, Minihane AM, Haskell C, Kennedy D (2013) DHA supplementation improved both memory and reaction time in healthy young adults: a randomized controlled trial. Am J Clin Nutr 97(5):1134–1143. doi:10.3945/ajcn.112.053371

Stough C, Kure C, Tarasuik J, Lloyd J, Downey LA, Scholey A, Wesnes KA (2009) A randomized, double-blind, placebo controlled study examining the effects of a combination nutraceutical formula on cognitive functioning and mood. JANA 12(1):12–19

Stough C, Downey L, Silber B, Lloyd J, Kure C, Wesnes K, Camfield D (2012) The effects of 90-day supplementation with the omega-3 essential fatty acid docosahexaenoic acid (DHA) on cognitive function and visual acuity in a healthy aging population. Neurobiol Aging 33(4):824.e821–e823. doi:10.1016/j.neurobiolaging.2011.03.019

Talpos JC, Winters BD, Dias R, Saksida LM, Bussey TJ (2009) A novel touchscreen-automated paired-associate learning (PAL) task sensitive to pharmacological manipulation of the hippocampus: a translational rodent model of cognitive impairments in neurodegenerative disease. Psychopharmacology (Berl) 205(1):157–168. doi:10.1007/s00213-009-1526-3

Templer VL, Hampton RR (2013) Episodic memory in nonhuman animals. Curr Biol 23(17):R801–R806. doi:10.1016/j.cub.2013.07.016

Theunissen EL, Street D, Hojer AM, Vermeeren A, van Oers A, Ramaekers JG (2013) A randomized trial on the acute and steady-state effects of a new antidepressant, vortioxetine (Lu AA21004), on actual driving and cognition. Clin Pharmacol Ther 93(6):493–501. doi:10.1038/clpt.2013.39

Theunissen EL, Heckman P, de Sousa Fernandes Perna EB, Kuypers KP, Sambeth A, Blokland A, Prickaerts J, Toennes SW, Ramaekers JG (2014) Rivastigmine but not vardenafil reverses cannabis-induced impairment of verbal memory in healthy humans. Psychopharmacology (Berl). doi:10.1007/s00213-014-3667-2

Tulving E (1972) Episodic and semantic memory. In: Tulving E, Donaldson W (eds) Organization of memory. Academic, New York

Tulving E (2002) Episodic memory: from mind to brain. Annu Rev Psychol 53:1–25. doi:10.1146/annurev.psych.53.100901.135114

Vakhapova V, Cohen T, Richter Y, Herzog Y, Korczyn AD (2010) Phosphatidylserine containing omega-3 fatty acids may improve memory abilities in non-demented elderly with memory complaints: a double-blind placebo-controlled trial. Dement Geriatr Cogn Disord 29(5):467–474. doi:10.1159/000310330

Valluzzi JA, Chan K (2007) Effects of fluoxetine on hippocampal-dependent and hippocampal-independent learning tasks. Behav Pharmacol 18(5–6):507–513. doi:10.1097/FBP.0b013e3282ee2a91

van Ruitenbeek P, Mehta MA (2013) Potential enhancing effects of histamine H(1) agonism/H(3) antagonism on working memory assessed by performance and bold response in healthy volunteers. Br J Pharmacol 170(1):144–155. doi:10.1111/bph.12184

Wechsler D (1997) Wechsler memory scale, 3rd edn. Psychological Corporation, San Antonio, TX

Wingen M, Langer S, Ramaekers JG (2006) Verbal memory performance during subchronic challenge with a selective serotonergic and a mixed action antidepressant. Hum Psychopharmacol 21(7):473–479

Yurko-Mauro K, McCarthy D, Rom D, Nelson EB, Ryan AS, Blackwell A, Salem N Jr, Stedman M, Investigators M (2010) Beneficial effects of docosahexaenoic acid on cognition in age-related cognitive decline. Alzheimers Dement 6(6):456–464. doi:10.1016/j.jalz.2010.01.013

Zeeuws I, Soetens E (2007) Verbal memory performance improved via an acute administration of D-amphetamine. Hum Psychopharmacol 22(5):279–287. doi:10.1002/hup.848

Zeeuws I, Deroost N, Soetens E (2010a) Effect of an acute d-amphetamine administration on context information memory in healthy volunteers: evidence from a source memory task. Hum Psychopharmacol 25(4):326–334. doi:10.1002/hup.1120

Zeeuws I, Deroost N, Soetens E (2010b) Verbal memory improved by D-amphetamine: influence of the testing effect. Hum Psychopharmacol 25(5):377–387. doi:10.1002/hup.1128

Zhang M, Moon C, Chan GC, Yang L, Zheng F, Conti AC, Muglia L, Muglia LJ, Storm DR, Wang H (2008) Ca-stimulated type 8 adenylyl cyclase is required for rapid acquisition of novel spatial information and for working/episodic-like memory. J Neurosci 28(18):4736–4744. doi:10.1523/JNEUROSCI.1177-08.2008

Verbal Memory

Tomiki Sumiyoshi

Contents

Abstract

Verbal memory is impaired in neurological and psychiatric conditions and provides one of the main targets of intervention. Specifically, this cognitive domain has been shown to provide a major determinant of outcome in schizophrenia and mood disorders. Therefore, verbal memory disturbances should be focused in the development of novel pharmacological and psychosocial therapeutics. Effective integration between preclinical and clinical studies should provide a key to the pursuit of drugs enhancing verbal memory.

Keywords

Verbal memory • Learning memory • Psychiatry • Antipsychotic drugs • Cognitive enhancers • Functional outcome

T. Sumiyoshi (✉)
Department of Clinical Research Promotion, National Center of Neurology and Psychiatry, 4-1-1, Ogawahigashi-cho, Kodaira, Tokyo, Japan
e-mail: sumiyot@ncnp.go.jp

K.M. Kantak, J.G. Wettstein (eds.), *Cognitive Enhancement*, Handbook of Experimental Pharmacology 228, DOI 10.1007/978-3-319-16522-6_8

1 Introduction

Memory is generally classified into short-term memory and long-term memory, with the latter divided into episodic, semantic, and procedural memory (Tulving and Fergus 2000). Sometimes, the former two aspects of long-term (or secondary) memory are jointly categorized as declarative memory (Parkin 1999). Verbal memory, representing declarative memory, is vulnerable in neurological and psychiatric conditions, and provides one of the main targets of intervention.

While impairment of verbal memory is most pronounced in dementias (e.g., Alzheimer's disease), this cognitive domain has also been shown to provide a major determinant of outcome in psychiatric diseases, such as schizophrenia (Sumiyoshi 2015) and mood disorders (Yatham et al. 2010). This chapter provides an overview on preclinical and clinical evidence for the effects of cognitive enhancers on verbal (learning) memory in these psychiatric illnesses, particularly schizophrenia.

2 Verbal Memory Impairment in Schizophrenia

Patients with schizophrenia exhibit about a 1–2.5 SD decline in a range of cognitive functions, e.g., several types of memory, executive function, verbal fluency, and attention/information processing (Keefe and Fenton 2007). A lesser degree of cognitive impairment is present in mood disorders, e.g., bipolar disorder and major depression. As demonstrated in Fig. 1, verbal memory is one of the cognitive domains most profoundly impaired in these major psychiatric diseases. This

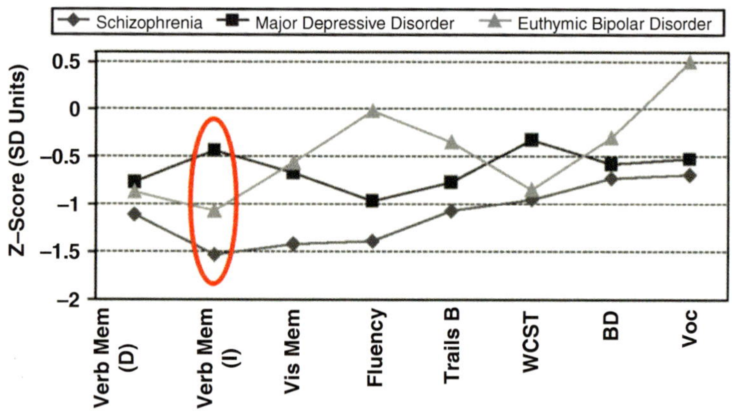

Fig. 1 Cognitive profiles in schizophrenia, major depression, and euthymic bipolar disorder. Healthy group mean = 0. Verb Mem (D), delayed verbal memory; Verb Mem (I), immediate verbal memory; Vis Mem, visual memory; Trails B, Trail Making Test, B; WCST, Wisconsin Card Sorting Test; BD, Wechsler Adult Intelligence Scale (WAIS) block design test; Voc, WAIS vocabulary. Among these cognitive domains, verbal memory is most profoundly affected in schizophrenia and bipolar disorder (Keefe et al. *Schizophr Bull* 33:912–920, 2007; permission obtained from Oxford University Press)

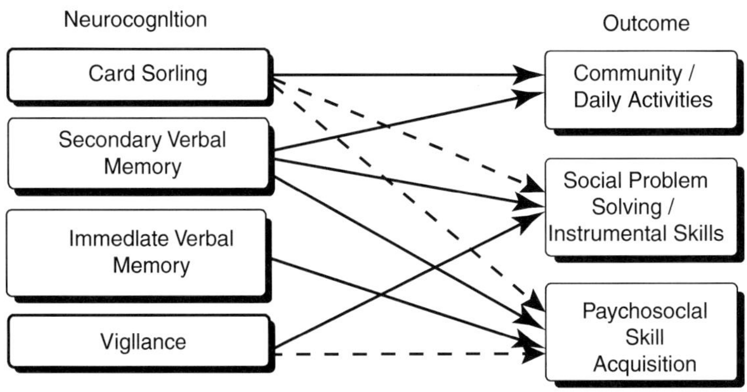

Fig. 2 Neurocognitive prediction of functional outcome. Verbal memory has been suggested to be substantially related with several key aspects of social function. A *heavy arrow* indicates that at least four separate studies found a significant relationship between the neurocognitive construct and the outcome domain. The *smaller arrows* indicate that two or three studies reported a significant relationship (Green et al. *Schizophr Bull* 26:119–136, 2000; permission obtained from Oxford University Press)

cognitive domain is included in comprehensive neurocognitive test batteries developed for the assessment of therapeutic effects, e.g., the MATRICS Comprehensive Cognitive Battery (MCCB) (Nuechterlein and Green 2006), the CogState Schizophrenia Battery (Maruff et al. 2009), and the Brief Assessment of Cognition in Schizophrenia (BACS) (Keefe et al. 2004).

Neuropsychological performance is related to community, residential, and interpersonal functioning (Harvey et al. 2011; Leifker et al. 2011). In this context, verbal memory, among several cognitive domains, has been suggested to be most strongly linked to functional outcomes, i.e., community activities, social problem solving, and psychosocial skill acquisition (Green et al. 2000) (Fig. 2).

2.1 Antipsychotic Drugs and Verbal Memory

The utility of antipsychotic drugs as a cognitive enhancer has been intensively discussed (Meltzer and Sumiyoshi 2003; Woodward et al. 2005; Goldberg et al. 2007; Sumiyoshi et al. 2013). Woodward et al. (2005) conducted a meta-analysis on the effect of atypical antipsychotic drugs (AAPDs), including clozapine, olanzapine, quetiapine, and risperidone, on cognition in schizophrenia in comparison with typical antipsychotic drugs (TAPDs), such as haloperidol. They found superiority of AAPDs, specifically for improving verbal and visual learning memory, with effect sizes $\geq$0.4 both in controlled and uncontrolled studies (Woodward et al. 2005).

On the other hand, there have been challenges to the pro-cognitive efficacy of AAPDs. Specifically, improvement of verbal memory by treatment with risperidone or olanzapine has been suggested to be no better than that of practice effect (or more precisely, test-retest effect) in normal controls (Goldberg et al. 2007). Since no data were presented in that study as to whether schizophrenia patients not receiving these AAPDs would have elicited the same degree of improvement as that in treated patients, it may not be legitimate to reject the concept that AAPDs improve cognition functions, such as verbal memory (Sumiyoshi et al. 2013).

Also, the effect of TAPDs on verbal memory has been investigated. Thus, Woodward et al. (2007) conducted a meta-analysis of cognitive change with haloperidol in schizophrenia and found small but significant effects on verbal memory tests, even restricting to studies that used alternate test forms (effect sizes 0.27–0.28). This finding challenges the notion that the cognitive improvements observed with AAPDs reflect an avoidance of a deleterious effect of haloperidol on cognitive function (Woodward et al. 2007).

The above considerations are consistent with the results from some large-scale multicenter studies to compare the effectiveness of AAPDs and TAPDs. Specifically, data from patients with first-episode schizophrenia were reported in the European First Episode Schizophrenia Trial (EUFEST) that used a randomized trial design among AAPDs (olanzapine, quetiapine, amisulpride, and ziprasidone) and haloperidol (TAPD) (Davidson et al. 2009). Verbal memory, measured by a word-list learning test, was found to be improved across the tested drugs, as shown in Fig. 3. Specifically, a *z*-score greater than 0.5 SD was indicated for olanzapine and ziprasidone (Davidson et al. 2009) (Fig. 3), two AAPDs that have been reported to also improve the semantic aspect of verbal memory (Sumiyoshi et al. 2006a). On the other hand, the benefit of AAPDs may not be evident in patients with chronic phase of schizophrenia, as in the CATIE Trial (Keefe et al. 2007).

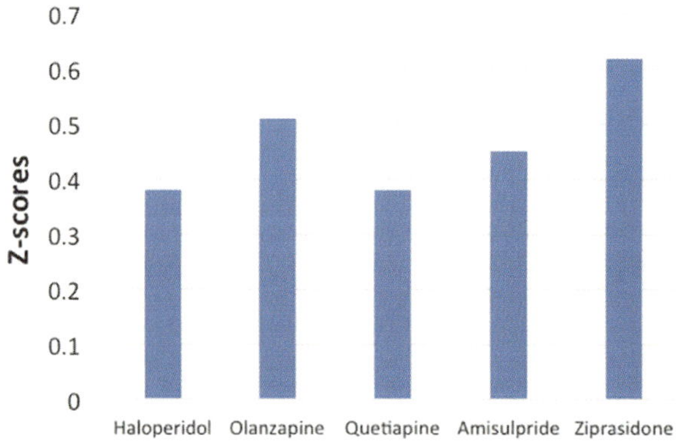

Fig. 3 Changes in verbal memory from baseline to 6 months of antipsychotic treatment in the EUFEST study (produced by T. Sumiyoshi based on data from Davidson et al. *Am J Psychiatry* 166:675–682, 2009)

The combination of neuropsychological and electrophysiological methods may be beneficial for understanding the mechanisms of enhancement of verbal memory by antipsychotic drugs (Sumiyoshi et al. 2011, 2013). Accordingly, we sought to determine the effect of olanzapine on verbal memory and P300 in patients with schizophrenia (Sumiyoshi et al. 2006b; Higuchi et al. 2008). P300 is one of the event-related potentials that have been used as a marker of attentive cognitive processes. The amplitudes of P300 have been shown to be decreased in patients with schizophrenia (Sumiyoshi et al. 2011). Specifically, we reported the effect of olanzapine on P300 current source density in discrete brain areas (Sumiyoshi et al. 2006b; Higuchi et al. 2008). At baseline, P300 current density in the left superior temporal gyrus (STG) was decreased in patients (Higuchi et al. 2008) (Fig. 4, left panel). Six-month treatment with olanzapine increased P300 current density in the left STG, accompanied with enhancement of verbal memory (Sumiyoshi et al. 2006b; Higuchi et al. 2008). In fact, this left-dominant pattern of P300 current density is similar to that for control subjects (Higuchi et al. 2008) (Fig. 4, left panel).

Importantly, the increase in P300 current density in the left STG was positively correlated with the improvement of verbal memory (Fig. 4, right panel). These

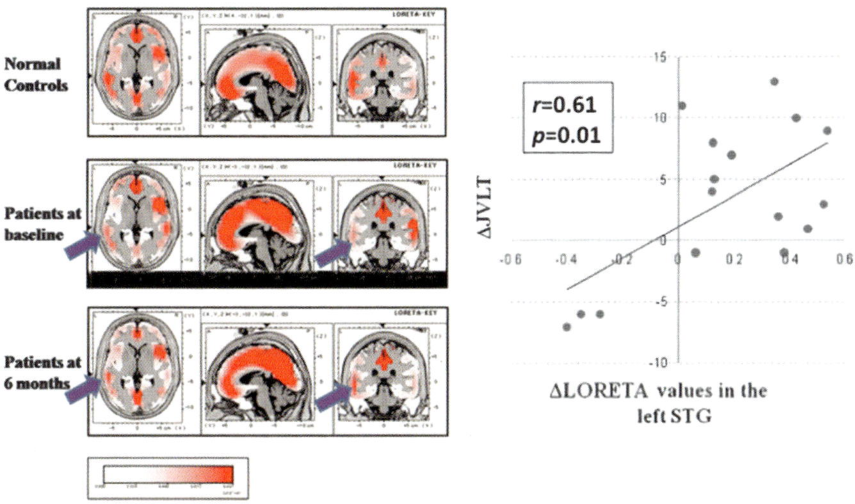

Fig. 4 (*Left*) LORETA images of P300 grand average in normal controls (*upper panel*) and patients at baseline (*middle panel*) and after 6-month treatment with olanzapine (*lower panel*). P300 current source density (as represented by LORETA value in *red*) in the left superior temporal gyrus (STG) in patients at baseline were reduced (indicated by *arrows* in the *middle panel*) compared to normal control subjects. Six-month treatment with olanzapine enhanced the current source density in the left STG of patients (indicated by *arrows* in the *lower panel*), which made the distribution pattern of P300 current density similar to that of normal controls. (*Right*) *Scatter plots* and *least squares regression lines* depicting the relationship between changes of score of the Japanese Verbal Learning Test (JVLT) vs. changes of P300 LORETA value in the left STG (Higuchi et al. *Schizophr Res* 101:320, 2008; permission obtained from Elsevier)

observations suggest that AAPDs ameliorate verbal memory impairment by correcting three-dimensional distribution of electrophysiological activity in the brain (Sumiyoshi et al. 2006b, 2009; Higuchi et al. 2008).

2.2 Preclinical Evaluation of Cognitive Enhancers

The assessment of behaviors related to cognition in animals is likely to facilitate the development of compounds to enhance cognitive performance in psychiatric illnesses. Specifically, novel objects recognition (NOR) in rodents represents recognition memory, and is considered to provide an animal model of declarative (learning) memory in humans (reviewed in Meltzer et al. 2011). NOR is non-rewarded and based on the rodent's memory for a familiar object and its natural propensity to explore novel objects to avoid harm and seek rewards. In commonly used NOR protocols, each test consists of two trials. In the first trial (acquisition), the rodents are exposed to two identical objects. In the second trial (retention), after an interval (ranging from 1 min to 2 days), the rodents are exposed to a familiar object from the first trial and a new object. Normal rodents spend more time exploring the new object in the retention trial (Meltzer et al. 2011).

Rodents treated with *N*-methyl-D-aspartate (NMDA) receptor antagonists, e.g., phencyclidine, ketamine, and MK-801, are unable to discriminate between novel and familiar objects, even after short intervals. AAPDs, such as aripiprazole (Nagai et al. 2009), perospirone (a ziprasidone-like compound) (Hagiwara et al. 2008), and lurasidone (Horiguchi et al. 2012; Horiguchi and Meltzer 2012), have been reported to ameliorate deficits in NOR induced by NMDA antagonists. These results suggest that the NOR test following treatment with an NMDA antagonist may provide a useful model for learning memory impairment in schizophrenia, and also a valuable method to screen for potential agents to treat verbal (learning) memory deficits in schizophrenia (Meltzer et al. 2011). The recent development of three-dimensional video analysis of the NOR test in rats (Matsumoto et al. 2014) (Fig. 5) is likely to add to these efforts.

2.3 Augmentation Therapy

As discussed, AAPDs have been shown to produce moderate effects on verbal memory in schizophrenia. Another approach has been the so-called augmentation therapy, in which a putative "pro-cognitive agent" is given to patients treated with antipsychotic drugs. Currently, this strategy is the mainstream in pharmacological cognitive enhancement, and a variety of compounds, derived from preclinical findings, have been tested (Harvey 2009; Michalopoulou et al. 2013).

One of the initial endeavors was a series of studies on the effects of the addition of the azapirone derivative tandospirone, a serotonin-5-HT_{1A} receptor partial agonist, to ongoing treatment with small to moderate doses of TAPDs (mainly haloperidol) (Sumiyoshi et al. 2000, 2001a, b). The addition of tandospirone (30 mg/

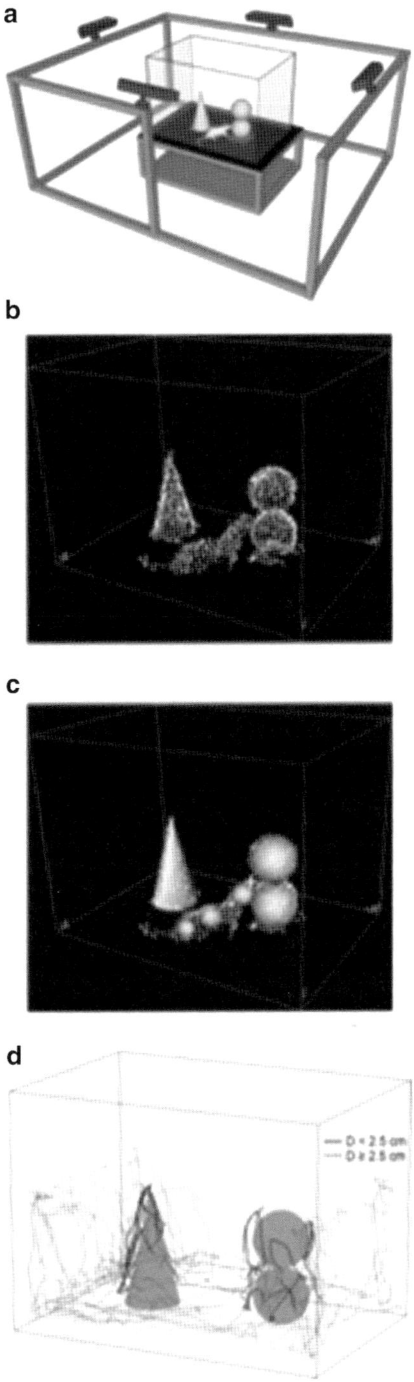

Fig. 5 3D video analysis of the novel object recognition test in rats. Experimental setup and algorithms for nose contact detection are demonstrated. (**a**) Experimental setup. (**b**) An example of a captured 3D image. Points represent the surface of objects and a rat. Two types of objects (*cone shaped* and *snowman shaped*) were used. (**c**) An example of estimated positions of a rat and objects. The rat model consists of *four connected spheres*, corresponding to head, neck, trunk, and hip. The *white cross* indicates the estimated position of the nose. (**d**) An example of a trajectory made up of estimated nose positions occurring in a trial. *Solid* and *dotted lines* indicate trajectories with and without contact with an object, respectively (Matsumoto et al. *Behav Brain Res* 272:16–24, 2014; permission obtained from Elsevier)

day), but not placebo for 4–6 weeks, was found to improve verbal learning memory (measured by the Wechsler Memory Scale-Revised) (Sumiyoshi et al. 2001a, b), as well as the semantic aspect of verbal memory (Sumiyoshi et al. 2001b), in patients with schizophrenia. Also, a subsequent study with galantamine, an acetylcholine esterase inhibitor, showed an improvement in verbal memory (Buchanan et al. 2008).

So far, few psychotropic agents have been shown to elicit apparent cognition-enhancing effects (Harvey 2009; Michalopoulou et al. 2013), and further efforts are required.

3 Verbal Memory in Mood Disorders

Cognitive disturbances in mood disorders, such as bipolar disorder and major depression, have been reported (Hellvin et al. 2012; Bourne et al. 2013). The degree of impairment in these psychiatric conditions has been demonstrated to be less than that for schizophrenia (Fig. 1) (Keefe and Fenton 2007). This concept has been confirmed by assessments with the MCCB (Burdick et al. 2011) and BACS (Hill et al. 2013), in line with the notion that the cognitive decline in mood disorders, particularly bipolar disorder, is qualitatively similar to that in schizophrenia (Hill et al. 2013).

On the other hand, there is a suggestion that these test batteries, originally developed for the assessment of cognitive disturbances of schizophrenia, may not be sensitive enough to detect impairment of patients with mood disorders (Yatham et al. 2010). With regard to verbal memory, the Hopkins Verbal Learning Test included in the MCCB has been recommended to be replaced by the California Verbal Learning Test that is more strenuous (Yatham et al. 2010).

Cognitive impairment in bipolar disorder has been associated with daily-living capacity and social function (Torres et al. 2011; Depp et al. 2012). Specifically, a longitudinal observation reports that verbal memory at baseline has been shown to correlate with functional outcome 6 months later in patients with bipolar I disorder (Torres et al. 2011). These lines of evidence indicate a need for the development of cognitive enhancers for mood disorders.

4 Conclusions

Several psychiatric illnesses, such as schizophrenia, are associated with disturbances of verbal memory. Due to its influence on social outcomes, this domain of cognition should be focused in the development of novel pharmacological and psychosocial therapeutics. Effective integration between preclinical and clinical studies, or translational research, should provide a key to the pursuit of drugs enhancing verbal memory.

Acknowledgment This work was supported by grants-in-aid for Scientific Research from Japan Society for the Promotion of Science (No. 26461761), Health and Labour Sciences Research Grants for Comprehensive Research on Disability, Health, and Welfare (H24-Seishin-Ippan-002 and H26-Seishin-Ippan-011), and Intramural Research Grant for Neurological and Psychiatric Disorders of NCNP.

The author thanks Drs. Herbert Y. Meltzer, Philip D. Harvey, Richard Keefe, and Sohee Park for the discussions.

References

Bourne C, Aydemir O, Balanza-Martinez V, Bora E, Brissos S, Cavanagh JT, Clark L, Cubukcuoglu Z, Dias VV, Dittmann S, Ferrier IN, Fleck DE, Frangou S, Gallagher P, Jones L, Kieseppa T, Martinez-Aran A, Melle I, Moore PB, Mur M, Pfennig A, Raust A, Senturk V, Simonsen C, Smith DJ, Bio DS, Soeiro-de-Souza MG, Stoddart SD, Sundet K, Szoke A, Thompson JM, Torrent C, Zalla T, Craddock N, Andreassen OA, Leboyer M, Vieta E, Bauer M, Worhunsky PD, Tzagarakis C, Rogers RD, Geddes JR, Goodwin GM (2013) Neuropsychological testing of cognitive impairment in euthymic bipolar disorder: an individual patient data meta-analysis. Acta Psychiatr Scand 128:149–162

Buchanan RW, Conley RR, Dickinson D, Ball MP, Feldman S, Gold JM, McMahon RP (2008) Galantamine for the treatment of cognitive impairments in people with schizophrenia. Am J Psychiatry 165:82–89

Burdick KE, Goldberg TE, Cornblatt BA, Keefe RS, Gopin CB, Derosse P, Braga RJ, Malhotra AK (2011) The MATRICS consensus cognitive battery in patients with bipolar I disorder. Neuropsychopharmacology 36:1587–1592

Davidson M, Galderisi S, Weiser M, Werbeloff N, Fleischhacker WW, Keefe RS, Boter H, Keet IP, Prelipceanu D, Rybakowski JK, Libiger J, Hummer M, Dollfus S, Lopez-Ibor JJ, Hranov LG, Gaebel W, Peuskens J, Lindefors N, Riecher-Rossler A, Kahn RS (2009) Cognitive effects of antipsychotic drugs in first-episode schizophrenia and schizophreniform disorder: a randomized, open-label clinical trial (EUFEST). Am J Psychiatry 166:675–682

Depp CA, Mausbach BT, Harmell AL, Savla GN, Bowie CR, Harvey PD, Patterson TL (2012) Meta-analysis of the association between cognitive abilities and everyday functioning in bipolar disorder. Bipolar Disord 14:217–226

Goldberg TE, Goldman RS, Burdick KE, Malhotra AK, Lencz T, Patel RC, Woerner MG, Schooler NR, Kane JM, Robinson DG (2007) Cognitive improvement after treatment with second-generation antipsychotic medications in first-episode schizophrenia: is it a practice effect? Arch Gen Psychiatry 64:1115–1122

Green MF, Kern RS, Braff DL, Mintz J (2000) Neurocognitive deficits and functional outcome in schizophrenia: are we measuring the "right stuff"? Schizophr Bull 26:119–136

Hagiwara H, Fujita Y, Ishima T, Kunitachi S, Shirayama Y, Iyo M, Hashimoto K (2008) Phencyclidine-induced cognitive deficits in mice are improved by subsequent subchronic administration of the antipsychotic drug perospirone: role of serotonin 5-HT1A receptors. Eur Neuropsychopharmacol 18:448–454

Harvey PD (2009) Pharmacological cognitive enhancement in schizophrenia. Neuropsychol Rev 19:324–335

Harvey PD, Raykov T, Twamley EW, Vella L, Heaton RK, Patterson TL (2011) Validating the measurement of real-world functional outcomes: phase I results of the VALERO study. Am J Psychiatry 168:1195–1201

Hellvin T, Sundet K, Simonsen C, Aminoff SR, Lagerberg TV, Andreassen OA, Melle I (2012) Neurocognitive functioning in patients recently diagnosed with bipolar disorder. Bipolar Disord 14:227–238

Higuchi Y, Sumiyoshi T, Kawasaki Y, Matsui M, Arai H, Kurachi M (2008) Electrophysiological basis for the ability of olanzapine to improve verbal memory and functional outcome in patients with schizophrenia: a LORETA analysis of P300. Schizophr Res 101:320–330

Hill SK, Reilly JL, Keefe RS, Gold JM, Bishop JR, Gershon ES, Tamminga CA, Pearlson GD, Keshavan MS, Sweeney JA (2013) Neuropsychological impairments in schizophrenia and psychotic bipolar disorder: findings from the Bipolar-Schizophrenia Network on Intermediate Phenotypes (B-SNIP) Study. Am J Psychiatry 170:1275–1284

Horiguchi M, Meltzer HY (2012) The role of 5-HT1A receptors in phencyclidine (PCP)-induced novel object recognition deficits in rats. Psychopharmacology (Berl) 221:205–215

Horiguchi M, Hannaway KE, Adelekun AE, Jayathilake K, Meltzer HY (2012) Prevention of the phencyclidine-induced impairment in novel object recognition in female rats by co-administration of lurasidone or tandospirone, a 5-HT(1A) partial agonist. Neuropsychopharmacology 37:2175–2183

Keefe RS, Fenton WS (2007) How should DSM-V criteria for schizophrenia include cognitive impairment? Schizophr Bull 33:912–920

Keefe RS, Goldberg TE, Harvey PD, Gold JM, Poe MP, Coughenour L (2004) The brief assessment of cognition in schizophrenia: reliability, sensitivity, and comparison with a standard neurocognitive battery. Schizophr Res 68:283–297

Keefe RS, Bilder RM, Davis SM, Harvey PD, Palmer BW, Gold JM, Meltzer HY, Green MF, Capuano G, Stroup TS, McEvoy JP, Swartz MS, Rosenheck RA, Perkins DO, Davis CE, Hsiao JK, Lieberman JA (2007) Neurocognitive effects of antipsychotic medications in patients with chronic schizophrenia in the CATIE Trial. Arch Gen Psychiatry 64:633–647

Leifker FR, Patterson TL, Heaton RK, Harvey PD (2011) Validating measures of real-world outcome: the results of the VALERO expert survey and RAND panel. Schizophr Bull 37:334–343

Maruff P, Thomas E, Cysique L, Brew B, Collie A, Snyder P, Pietrzak RH (2009) Validity of the CogState brief battery: relationship to standardized tests and sensitivity to cognitive impairment in mild traumatic brain injury, schizophrenia, and AIDS dementia complex. Arch Clin Neuropsychol 24:165–178

Matsumoto J, Uehara T, Urakawa S, Takamura Y, Sumiyoshi T, Suzuki M, Ono T, Nishijo H (2014) 3D video analysis of the novel object recognition test in rats. Behav Brain Res 272:16–24

Meltzer HY, Sumiyoshi T (2003) Atypical antipsychotic drugs improve cognition in schizophrenia. Biol Psychiatry 53:265–267, author reply 267–268

Meltzer HY, Horiguchi M, Massey BW (2011) The role of serotonin in the NMDA receptor antagonist models of psychosis and cognitive impairment. Psychopharmacology (Berl) 213:289–305

Michalopoulou PG, Lewis SW, Wykes T, Jaeger J, Kapur S (2013) Treating impaired cognition in schizophrenia: the case for combining cognitive-enhancing drugs with cognitive remediation. Eur Neuropsychopharmacol 23:790–798

Nagai T, Murai R, Matsui K, Kamei H, Noda Y, Furukawa H, Nabeshima T (2009) Aripiprazole ameliorates phencyclidine-induced impairment of recognition memory through dopamine D1 and serotonin 5-HT1A receptors. Psychopharmacology (Berl) 202:315–328

Nuechterlein KH, Green MF (2006) MATRICS consensus cognitive battery. MATRICS Assessment, Los Angeles, CA

Parkin AJ (1999) Memory: a guide for professionals. Wiley, West Sussex

Sumiyoshi T (2015) Cognitive impairment in schizophrenia. In: Stolerman I, Price LH (eds) Encyclopedia of Psychopharmacology, 2nd edn. Springer, New York

Sumiyoshi T, Matsui M, Yamashita I, Nohara S, Uehara T, Kurachi M, Meltzer HY (2000) Effect of adjunctive treatment with serotonin-1A agonist tandospirone on memory functions in schizophrenia [letter]. J Clin Psychopharmacol 20:386–388

Sumiyoshi T, Matsui M, Nohara S, Yamashita I, Kurachi M, Sumiyoshi C, Jayathilake K, Meltzer HY (2001a) Enhancement of cognitive performance in schizophrenia by addition of tandospirone to neuroleptic treatment. Am J Psychiatry 158:1722–1725

Sumiyoshi T, Matsui M, Yamashita I, Nohara S, Kurachi M, Uehara T, Sumiyoshi S, Sumiyoshi C, Meltzer HY (2001b) The effect of tandospirone, a serotonin(1A) agonist, on memory function in schizophrenia. Biol Psychiatry 49:861–868

Sumiyoshi C, Sumiyoshi T, Roy A, Jayathilake K, Meltzer HY (2006a) Atypical antipsychotic drugs and organization of long-term semantic memory: multidimensional scaling and cluster analyses of category fluency performance in schizophrenia. Int J Neuropsychopharmacol 9:677–683

Sumiyoshi T, Higuchi Y, Kawasaki Y, Matsui M, Kato K, Yuuki H, Arai H, Kurachi M (2006b) Electrical brain activity and response to olanzapine in schizophrenia: a study with LORETA images of P300. Prog Neuropsychopharmacol Biol Psychiatry 30:1299–1303

Sumiyoshi T, Higuchi Y, Itoh T, Matsui M, Arai H, Suzuki M, Kurachi M, Sumiyoshi C, Kawasaki Y (2009) Effect of perospirone on P300 electrophysiological activity and social cognition in schizophrenia: a three-dimensional analysis with sloreta. Psychiatry Res 172:180–183

Sumiyoshi T, Higuchi Y, Itoh T, Kawasaki Y (2011) Electrophysiological imaging evaluation of schizophrenia and treatment response. In: Risner MS (ed) Handbook of schizophrenia spectrum disorders, vol 3. Springer, New York

Sumiyoshi T, Higuchi Y, Uehara T (2013) Neural basis for the ability of atypical antipsychotic drugs to improve cognition in schizophrenia. Front Behav Neurosci 7:140

Torres IJ, DeFreitas CM, DeFreitas VG, Bond DJ, Kunz M, Honer WG, Lam RW, Yatham LN (2011) Relationship between cognitive functioning and 6-month clinical and functional outcome in patients with first manic episode bipolar I disorder. Psychol Med 41:971–982

Tulving EC, Fergus IM (2000) The Oxford handbook of memory. Oxford University Press, Oxford

Woodward ND, Purdon SE, Meltzer HY, Zald DH (2005) A meta-analysis of neuropsychological change to clozapine, olanzapine, quetiapine, and risperidone in schizophrenia. Int J Neuropsychopharmacol 8:457–472

Woodward ND, Purdon SE, Meltzer HY, Zald DH (2007) A meta-analysis of cognitive change with haloperidol in clinical trials of atypical antipsychotics: dose effects and comparison to practice effects. Schizophr Res 89:211–224

Yatham LN, Torres IJ, Malhi GS, Frangou S, Glahn DC, Bearden CE, Burdick KE, Martinez-Aran A, Dittmann S, Goldberg JF, Ozerdem A, Aydemir O, Chengappa KN (2010) The International Society for Bipolar Disorders-Battery for Assessment of Neurocognition (ISBD-BANC). Bipolar Disord 12:351–363

Emotional Memory

Karim Nader

Contents

Abstract

Research on the reconsolidation effect was greatly revitalized by the highly analytic demonstration of memory reconsolidation (Nader et al. Nature 406:722–726, 2000) in a well-defined behavioral protocol (auditory fear conditioning in the rat). Since this study, reconsolidation has been demonstrated in hundreds of studies over a range of species, tasks, and amnesic agents. Evidence for reconsolidation does not come solely from the behavioral level of analysis. Cellular and molecular correlates of reconsolidation have also been found. In this chapter, I will first define the evidence on which reconsolidation is concluded to exist. I will then discuss some of the conceptual issues facing the

K. Nader (✉)
Psychology Department, McGill University, 1205 Dr. Penfield Ave, Montreal, QC, Canada, H3A 1B1
e-mail: karim.nader@mcgill.ca

K.M. Kantak, J.G. Wettstein (eds.), *Cognitive Enhancement*, Handbook of Experimental Pharmacology 228, DOI 10.1007/978-3-319-16522-6_9

field in determining when reconsolidation does and does not occur. Lastly I will explain the clinical implications of this effect.

Keywords
Fear • Amygdala • Memory erasure • Memory enhancements

1 Introduction

This is an exciting time in the study of learning and memory. Typically, in any learning and memory study, scientists broadly differentiate between certain phases of learning and memory: learning phase in which information is acquired, stabilization phase in which specific mechanisms are engaged to stabilize initially unstable new information [referred to as synaptic consolidation] (Glickman 1961; McGaugh 1966), maintenance phase during which other mechanisms are involved to maintain the memory, and retrieval phase in which specific mechanisms will permit a memory to be retrieved (Miller and Springer 1973; Spear 1973). Prior to the year 2000, from a neurobiological perspective, only acquisition and memory stabilization (Martin et al. 2000; Kandel 2001; Dudai 2004) were considered to be active phases, in the sense that neurons had to perform certain computations or synthesize new RNA and proteins in order for these phases of memory processing to be carried out successfully. After acquisition and stabilization, all other phases were implicitly thought to be passive readout of changes in the circuits mediating the long-term memory (LTM).

Since the publication of Nader et al. (2000)'s study demonstrating that a consolidated LTM memory can become un-stored and restored, a process coined "reconsolidation" (Nader et al. 2000), about 807 research papers have been published with this term in the title. There are now cellular and molecular models of this time-dependent memory phase including exciting, rich psychopharmacological work.

This chapter will first describe the logic of the findings that brought the existence of the consolidation process into light. I will then describe how we concluded that a consolidated memory undergoes reconsolidation in a well-defined behavioral protocol involving emotional memory (auditory fear conditioning in the rat). I will then discuss the range of species, tasks, and treatments in which reconsolidation of this type has been reported. One aspect of reconsolidation that has attracted experimental attention involves the finding that there seem to be conditions that facilitate or inhibit reconsolidation from occurring. While this is an extremely exciting aspect of the phenomenon, controversy surrounds the experimental procedures that have been employed to investigate these conditions. I present a logical approach that could help to identify the veracity of such conditions. Then, I will discuss the often underappreciated data showing memories can be enhanced. Lastly, I will discuss clinical implications of reconsolidation and briefly review some of the results of the published clinical studies to date.

2 Consolidation: The Dominant Model of Memory Storage

Consolidation is defined as a time-dependent stabilization process leading eventually to the permanent stabilization of newly acquired memory (Ebbinghaus 1885; Müller and Pilzecker 1900; Glickman 1961; McGaugh 1966) (Fig. 1ai). At the level of the synapse, this process referred to as synaptic consolidation is thought to be a universal property of neurons.

The existence of the consolidation process has been shown from various lines of evidence demonstrating the presence of a post-acquisition time interval during which new memories are labile/unstable and sensitive to challenges (Fig. 1ai). First, performance can be impaired by amnesic treatments, such as electroconvulsive shock (Duncan 1949), protein synthesis inhibitors (Flexner et al. 1965), or by new learning (Gordon and Spear 1973). Second, retention can be enhanced by administration of certain compounds, such as strychnine (McGaugh and Krivanek 1970). Crucially, these manipulations are only effective when administered shortly after new learning, but not when given after a few hours. These types of results led to the conclusion that memory exists in two states: when susceptible to enhancement or impairment, memory resides in a labile state, but if it is insensitive to these treatments, memory is stable and, by definition, consolidated (McGaugh 1966; Dudai 2004).

This same logic was employed by Schafe and colleagues to test for the existence of a consolidation process in the lateral and basal amygdala (LBA) for auditory fear memory. When the protein synthesis inhibitor anisomycin is infused into the LBA shortly after training, short-term memory (STM) is intact but LTM is impaired (Schafe and LeDoux 2000) (Fig. 1aii); however, LTM remains intact when the infusion is delayed for 6 h. This pattern of results conforms to the operational definition of consolidation in the sense that the aspect of fear-conditioning memory that requires protein synthesis within the LBA is consolidated within at most 6 h after learning. In addition, we assume that the experiment manipulation induced amnesia for those computations that the LBA supposedly mediates, i.e., the association between the conditioned (the tone; CS) and the unconditioned stimulus (the foot shock; US) (Rodrigues et al. 2009).

One of the basic tenets of the cellular consolidation model is that learning induces changes in synaptic efficacy, suggesting that the physiological "unit" of cellular consolidation is the synapse. Two main candidate mechanisms that were postulated to implement these changes are long-term potentiation (LTP) and long-term depression (LTD) (Malenka and Nicoll 1999; Martin et al. 2000). In parallel to the distinction of STM and LTM, with the latter being consolidated by a protein

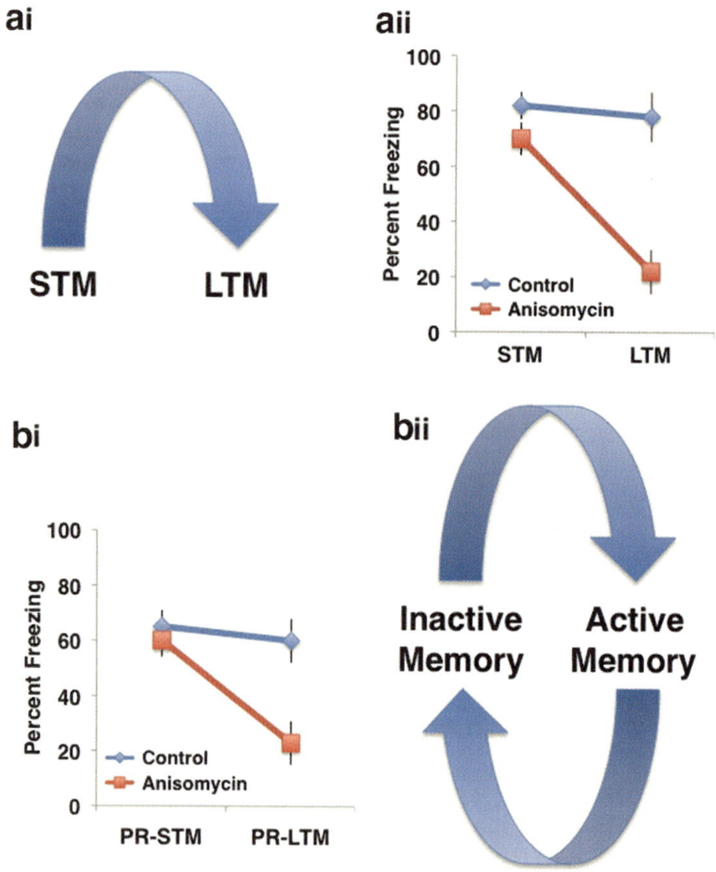

Fig. 1 (**ai**) Textbook account of consolidation demonstrating that memories consolidate over time into LTM. The critical point is to show that once a memory is in LTM it is thought to remain fixed or permanent (Glickman 1961; McGaugh 1966). (**aii**) A typical demonstration of a consolidation blockade (Schafe and LeDoux 2000). Intact STM and impaired LTM, a pattern of impairment that defines consolidation impairment (Dudai 2004; McGaugh 2004). (**bi**) A typical demonstration of a reconsolidation blockade. Intact post-reactivation STM (PR-STM) and impaired LTM, (PR-LTM), meeting the definitions for a consolidation blockade (Dudai 2004; McGaugh 2004). (**bii**) An alternate model of memory that incorporates the findings of consolidation and reconsolidation datasets proposed by Lewis (1979). Consolidation theory cannot explain the reconsolidation dataset. New and reactivated memories are in an active state and then over time they stabilize and exist in an inactive memory state. When a memory in an inactive memory state is remembered, it returns to an active memory state

synthesis-dependent process, LTP is also divided into an early transient phase (E-LTP) and a stabilized, RNA- and protein synthesis-dependent late phase (L-LTP) (Goelet et al. 1986).

3 Evidence for a Reconsolidation Process

3.1 Behavioral Evidence

The existence of a reconsolidation process in the LBA for consolidated, i.e., long-term, auditory fear memory, has been concluded from a study that in logic and design followed those for consolidation as described in Schafe and LeDoux (2000). One day after conditioning, at a time when, according to the results from Schafe and LeDoux (2000) study, memory should be fully stabilized and immune to the amnesic agent, we reminded animals of the conditioning session by exposing them again to the CS, i.e., the tone (Nader et al. 2000). Anisomycin, at the same dose, concentration, and rate as in the Schafe and colleagues' consolidation study was then either immediately or later infused into the LBA. When anisomycin was administered immediately, anisomycin-treated animals show intact post-reactivation-STM (PR-STM) but impaired PR-LTM (Fig. 1bi), a pattern of results that is identical to what is found when blocking consolidation (Schafe and LeDoux 2000) (Fig. 1aii). However, if the post-reactivation infusion was delayed by 6 h, anisomycin had no effect, demonstrating that the reactivation-induced instability/lability was transient. Importantly, animals that were not reminded of the CS prior to anisomycin infusions had intact memory.

Staying strictly within the commonly accepted consolidation framework, and applying only the definitions on which this framework is based, the following four conclusions can be drawn from the results of these experiments. First, the observation that the memory was insensitive to anisomycin when it was not reactivated demonstrates that it was "consolidated" 24 h after training—at least with regard to the specific amnesic treatment applied. Second, that only the reactivated memory was sensitive to anisomycin disruption demonstrates that memory was in a labile state after reactivation. Third, the observation that the anisomycin-treated animals showed intact STM and impaired LTM after reactivation implies that a consolidation-like process is triggered by reactivation. And finally, given the amnesic treatment was ineffective 6 h after reactivation, this post-reactivation re-stabilization process is, like consolidation, a time-dependent process. Taken together, these four conclusions yield the interpretation that reactivation of a consolidated memory returns it again to a labile state from which the memory has to undergo stabilization (i.e., reconsolidate) over time (Nader et al. 2000).

Consolidation and reconsolidation are thus both deduced from the evidence of a transient period of instability. In the case of consolidation, this window is initiated after acquisition of new information; in the case of reconsolidation, it is initiated after reactivation of an existing, consolidated memory representation. As is the case for consolidation, only during the reconsolidation phase can memory be boosted by "memory enhancers" (Gordon 1977b; Rodriguez et al. 1993; Horne et al. 1997; Rodriguez et al. 1999), or impaired by amnesic treatments (Misanin et al. 1968) and interfering new learning (Gordon 1977a). These treatments are ineffective when reconsolidation is complete, which is also the case for consolidation.

The term "reconsolidation" was introduced as early as 1973, in the context of a discussion on memory retrieval. Spear asked "...how will the dynamic aspects of memory [will] be handled, that is, with successive learning trials or related successive experiences does the entire memory reconsolidate anew or merely the new information?" (Spear 1973, p. 188). As a consequence of the perceived inability of the consolidation hypothesis to account for reconsolidation, new memory models were developed that treated new and reactivated consolidated memories in similar ways (Spear 1973; Lewis 1979) (Fig. 1bii).

Since Nader et al. (2000)'s report, reconsolidation has been shown across a variety of species, tasks, and amnesic treatments (Table 1). In light of this evidence, it is therefore postulated that reconsolidation represents a fundamental memory process (Nader and Hardt 2009).

One of the most striking findings in this literature is a study by Lee (2008), who devised specific tools to block consolidation or reconsolidation mechanisms (Lee 2008). Most students of memory would assume that presenting additional learning trials to a consolidated memory would engage consolidation mechanisms, which will make the memory stronger. However, the evidence from Lee (2008)'s study suggests that a memory has to undergo reconsolidation to be strengthened. Moreover, memory strengthening by new learning was mediated by reconsolidation and not consolidation mechanisms. This evidence suggests that a recently acquired memory will be mediated by consolidation mechanisms within a time window of approximately 5 h. However, for the rest of the memory's lifetime, the memory will engage reconsolidation mechanisms. Therefore, based on this evidence, consolidation but not reconsolidation can be considered as the atypical memory process (Lee 2009).

3.2 Alternative Interpretations

Reconsolidation, as we discussed above, has been defined by applying the very standards that define consolidation. Therefore, certain nonspecific interpretations of the reconsolidation hypothesis pose the same challenges to the consolidation hypothesis, a consequence that is rarely acknowledged. The complexity of the data poses a problem for alternative interpretations, which should not merely provide new explanations for the reconsolidation dataset, but need to allow for predictions that are different from those offered by the reconsolidation model. For this reason, we will not address all the previous alternative interpretation here. A detailed discussion of these alternative interpretations including facilitation of extinction, transient retrieval impairment, nonspecific effects, state-dependent learning, and new learning is presented in Nader and Hardt (2009).

3.3 Evidence for Reconsolidation Across Levels of Analysis

Evidence for reconsolidation does not come solely from the behavioral level of analysis. A cellular phenomenon akin to reconsolidation was shown for L-LTP

Table 1 Some of the paradigms in which reconsolidation has been reported

Experimental paradigm	Habituation (Rose and Rankin 2006)
	Auditory fear conditioning (Nader et al. 2000)
	Contextual fear conditioning (Debiec et al. 2002)
	Instrumental learning (Sangha et al. 2003), but see Hernandez and Kelley (2004)
	Inhibitory avoidance (Anokhin et al. 2002; Milekic and Alberini 2002)
	Conditioned aversion learning (Eisenberg et al. 2003)
	Motor sequence learning (Walker et al. 2003)
	Incentive learning (Wang et al. 2005)
	Object recognition (Kelly et al. 2003)
	Spatial memory (Suzuki et al. 2004; Morris et al. 2006)
	Memory for drug reward (Lee et al. 2005; Miller and Marshall 2005; Valjent et al. 2006)
	Episodic memory (Hupbach et al. 2007)
Treatment	Protein synthesis inhibition (Nader et al. 2000)
	RNA synthesis inhibition (Sangha et al. 2003)
	Inhibition of kinase activity (Kelly et al. 2003; Duvarci et al. 2005)
	Protein-knockout mice (Bozon et al. 2003)
	Anti-sense (Taubenfeld et al. 2001; Lee et al. 2004)
	Inducible knockout mice (Kida et al. 2002)
	Receptor antagonists (Przybyslawski et al. 1999; Debiec and Ledoux 2004; Suzuki et al. 2004)
	Interference by new learning (Walker et al. 2003; Hupbach et al. 2007)
	Potentiated reconsolidation by increase in kinase activity (Tronson et al. 2006)
Species	*Aplysia* (Cai et al. 2012; Lee et al. 2012)
	Nematodes (Rose and Rankin 2006)
	Honeybees (Stollhoff et al. 2005)
	Snails (Sangha et al. 2003)
	Sea slugs (Child et al. 2003)
	Fish (Eisenberg et al. 2003)
	Crabs (Pedreira et al. 2002)
	Chicks (Anokhin et al. 2002)
	Mice (Kida et al. 2002)
	Rats (Nader et al. 2000); rat pups (Gruest et al. 2004)
	Humans (Walker et al. 2003; Hupbach et al. 2007; Kindt et al. 2009; Schiller et al. 2010)

This table lists some examples from various experimental paradigms, treatments, and species for studies reporting evidence for a reconsolidation process since the year 2000

(Fonseca et al. 2006). In this study, the authors report that when anisomycin is added 2 h after the induction of L-LTP it has no effect on L-LTP maintenance. If, however, the potentiated pathway is reactivated by administering test pulses that inhibit protein synthesis, the potentiation is intact shortly after reactivation but becomes impaired over time. This suggests that reactivation of stabilized L-LTP returns its substrate to a labile state, in which it can be disrupted by inhibiting protein synthesis. This is consistent with the suggestion that the mechanisms mediating plasticity are stabilized (Finnie and Nader 2012) over time, just like consolidation (Goelet et al. 1986). Other evidence includes reports that reconsolidation blockade reverses increases in field potentials induced by fear conditioning in the LA in intact animals (Doyere et al. 2007). In sum, this evidence suggests the presence of a cellular correlate of the behaviorally demonstrated reconsolidation impairment.

More recently, two papers using classic paradigm of *Aplysia* to study sensitization and long-term facilitation (LTF) reported that reconsolidation affects these kinds of processes. Indeed, when reconsolidation was blocked, the sensory-motor synaptic enhancement typically observed after LTF was reversed (Cai et al. 2012; Lee et al. 2012).

At the molecular level, interfering with reconsolidation can, in a time-dependent manner, remove molecular correlates of memory induced by learning and subsequent consolidation. Miller and Marshall (2005) showed that place-preference learning activates the extracellular signal-regulated kinase (ERK) in the nucleus accumbens (Miller and Marshall 2005). Blocking the activated ERK in the nucleus accumbens after reactivation results in intact PR-STM but impaired PR-LTM. In these amnesic animals, this also leads to the absence of ERK and its downstream transcription factors in the nucleus accumbens [see also Valjent et al. (2006) who demonstrate reduction in ERK and GluA1 phosphorylation using a similar procedure]. Studying mechanisms of long-term habituation in *C. elegans*, Rose and Rankin (2006) showed that administering heat-shock or the non-NMDA glutamatergic antagonist, DMQX, after reactivation of a consolidated memory dramatically returns expression of AMPA receptors in the mechanosensory neuron to a level typical for naïve animals (Rose and Rankin 2006). Importantly, the reconsolidation effects in all of these studies were contingent on memory reactivation—in the absence of a reminder the amnesic treatments were ineffective.

Another study by Kaang's group described, at the level of postsynaptic α-amino-3-hydroxy-5-methyl-4-isoxazolepropionic (AMPA) receptors, the biochemical process that destabilizes a consolidated memory and the subsequent reconsolidation process. Learning is thought to lead to AMPA receptor trafficking: calcium-permeable AMPA receptors are inserted into the postsynaptic density (PSD) and then over time replaced by calcium-impermeable receptors (Rumpel et al. 2005). Kaang's group asked what the AMPA receptor dynamics would be when a memory is destabilized and then reconsolidated. These authors reported that memory destabilization is associated with calcium-permeable AMPA receptors. Indeed, blocking the introduction of calcium-permeable AMPA receptors into the PSD prevented the memory from being un-stored (Hong et al. 2013). Thus, they found that the replacement of calcium-permeable AMPA receptors by calcium-impermeable AMPA receptors mediated the process of reconsolidation.

These studies are a small sample of the datasets that provide striking evidence for the existence of a transient post-reactivation period of memory plasticity, i.e., memory reconsolidation, on the behavioral, physiological, and molecular levels of analysis.

3.4 Can Mechanisms Mediating Presynaptic Plasticity Undergo Reconsolidation?

Synapses usually have presynaptic and postsynaptic compartments. The electrical signal is conducted from the presynaptic to the postsynaptic compartment. One theory on the locus of memory posits that presynaptic changes are critical for LTM and L-LTP (Bliss and Collingridge 1993). These presynaptic changes are thought to increase the probability of vesicle release.

In all the studies that examined cellular or molecular correlates of consolidation or reconsolidation, blocking the respective memory processes reversed the learning-induced molecular/cellular correlates. For example, in a study, Bailey and colleagues (1993) reported that the blockade of consolidation in an *Aplysia* preparation with a protein synthesis inhibitor prevented the increase in the number of synapses to the point where the amount compared to levels of synapses in naïve animals. The same pattern of results has been shown in reconsolidation studies, as can be seen in the previous section.

Tsvetkov et al. (2002) have previously demonstrated that auditory fear conditioning induces predominantly presynaptic enhancements in both inputs to the lateral amygdala thought to mediate fear learning (Tsvetkov et al. 2002). Recently, this group assessed what would happen to these learning-induced presynaptic enhancements after blocking reconsolidation with rapamycin, a protein synthesis inhibitor. They reported that these presynaptic enhancements were not reduced, but that a reduction in postsynaptic AMPA receptors correlated with the behavioral impairments (Li et al. 2013). This finding suggests that the postsynaptic mechanisms must detect how much potential exists on the presynaptic terminals and reduce the postsynaptic AMPA receptors below baseline PSD levels.

There are two theoretical implications of these findings for reconsolidation. First, perhaps, presynaptic mechanisms of long-term plasticity are independent of reconsolidation. This would entail that only the postsynaptic mechanisms of long-term memory could be susceptible to reconsolidation blockade. The second possibility is that presynaptic mechanisms are affected by reconsolidation, but the amnesic treatment used, a protein synthesis inhibitor (PSI), was not appropriate to target the presynaptic mechanisms mediating reconsolidation. We know that presynaptic enhancements are not affected by PSIs. Therefore, a tool transiently challenging the mechanisms mediating long-term presynaptic efficacy would be needed to test this hypothesis.

4 Reconsolidation Is Not Universal

The fact that memory reconsolidation has been found across levels of analysis does not imply that reconsolidation is universal, i.e., observed under any circumstance. Another variation of the theme that reconsolidation is not a universal property of memory is the concept of constraints on this phenomenon, or "boundary conditions." These are situations of physiological, environmental, or psychological nature, in which memories that normally would reconsolidate no longer does. Several boundary conditions have been proposed, such as extinction consolidation (Eisenberg et al. 2003; Pederia and Maldonado 2003; Suzuki et al. 2004), memory age (Milekic and Alberini 2002; Suzuki et al. 2004), predictability of the reactivation stimulus (Pedreira et al. 2004; Morris et al. 2006), and training intensity (Suzuki et al. 2004). Others, however, have not identified similar boundary conditions in other protocols [for extinction] (Stollhoff et al. 2005; Duvarci et al. 2006), old memories (Debiec et al. 2002; Lee et al. 2005), predictability of the reactivation stimulus (Pedreira et al. 2002; Bozon et al. 2003; Sangha et al. 2003; Valjent et al. 2006), or strength of training (Debiec et al. 2002; Lee et al. 2005). Whether additional parameters moderate boundary conditions remains to be seen.

The observed inconsistencies in the identification of the boundary conditions might be due to the absence of agreed-upon, standard experimental parameters required to test the presence of such boundary conditions. For example, if memory disruption is not observed within a set of experimental parameters, then it is concluded that the memory does not undergo reconsolidation under those conditions. A number of reports, however, have demonstrated that a memory may undergo reconsolidation only under specific reactivation conditions (De Vietti and Holiday 1972; Bozon et al. 2003; Suzuki et al. 2004). The implication of these findings is that it is extremely difficult to conclude based on behavioral studies that a memory never undergoes reconsolidation. Therefore the question remains whether the negative effects upon which the boundary conditions are based imply that a given memory never undergoes reconsolidation under those conditions, or the memory is still capable of undergoing reconsolidation with another reactivation protocol (Fig. 2). Given that the parameter space of possible reactivation procedures is essentially infinite, a real boundary condition is very difficult to prove at the behavioral level. This is likely part of why there is so much inconsistency in the field of boundary conditions (Dreyfuss et al. 2009).

Wang et al. (2009) took a complementary approach to identify some of the molecular mechanisms that are induced by boundary conditions to inhibit the occurrence of reconsolidation (Wang et al. 2009). If molecular or cellular indicators of when memories stop undergoing reconsolidation were identified, then we could make strong predictions concerning when we should see these mechanisms expressed (Fig. 3).

Specifically, if strong memories, old memories, or extinction represent real boundary conditions, then the putative mechanisms mediating boundary conditions

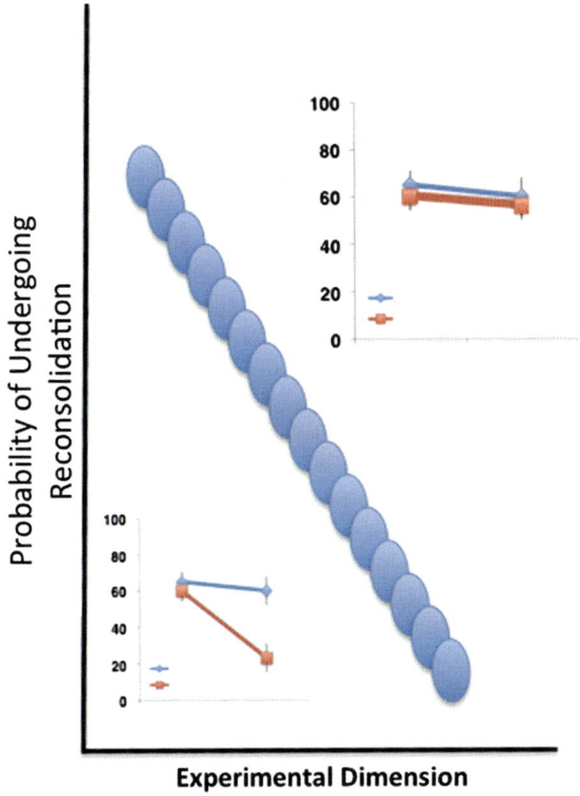

Fig. 2 Possible functions describing the constraints on reconsolidation. It is still an open question if the functions are linear or exponential. Different experimental conditions may produce different functions. The experimental space to the left of the curve is determined by examples in which the memory undergoes reconsolidation as demonstrated in the schematic behavioral impairment. The evidence for constraints on reconsolidation is derived from negative findings as shown in the schematic on the right. That is a logical limitation of the behavioral approach to this issue. Therefore, we suggested that a complementary approach to help resolve this issue would be to identify a molecular correlate for the absence of reconsolidation. This would act as positive evidence for the existence of the constraint

should be fully expressed within the respective memory system. Conversely, under conditions when a memory does undergo reconsolidation (e.g., weak training, little extinction, or young memories), then the mechanism mediating boundary conditions should be minimized. This strategy would significantly complement the behavioral studies described above in their search for true boundary conditions and help resolve some of the conflicting findings in the field.

An understanding of how boundary conditions are mediated across levels of analysis is critical because targeting reconsolidation of traumatic memories has been proposed to be a potential treatment for posttraumatic stress disorder (PTSD)

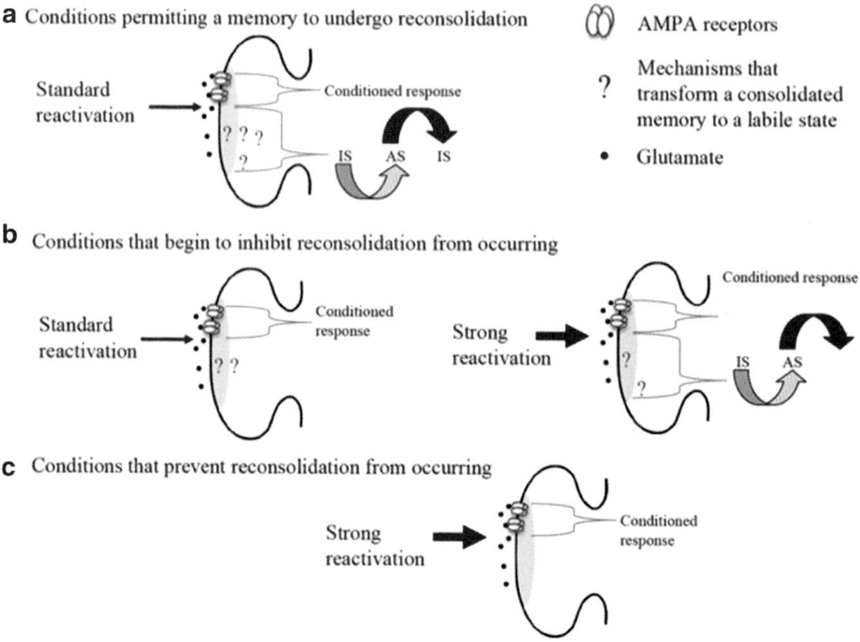

Fig. 3 Conceptual diagram demonstrating how boundary conditions could inhibit memories from undergoing reconsolidation across memories types and memory systems. (**a**) Under experimental conditions when a memory undergoes reconsolidation, the mechanisms allowing a memory to be transformed from a consolidated to a labile active state (AS) must be present and functional at the synapse ("?" in figure). These mechanisms, of course, will involve more than surface receptors and will likely include a number of molecular processes that have yet to be identified. (**b**) Experimental conditions that begin to inhibit memories from undergoing reconsolidation may lead to a partial reduction in a mechanism that is critical for the induction of reconsolidation. The partial reduction might be sufficient to prevent the induction of reconsolidation when a standard protocol is used. However, there may still be sufficient amounts of this mechanism to permit the memory to undergo reconsolidation when a stronger reactivation is used. (**c**) Under conditions when the memory does not undergo reconsolidation, a boundary condition, a necessary mechanism for the induction of reconsolidation, is reduced to the point that alternative reactivation protocols cannot induce the memory to undergo reconsolidation

(Przybyslawski and Sara 1997; Debiec et al. 2002; Schiller et al. 2010). Specifically, blocking the reconsolidation of traumatic memories might weaken the long-term maintenance of these traumatic memories, in turn, reducing PTSD pathology. However, if strong aversive experiences act as boundary conditions on reconsolidation (Suzuki et al. 2004), then this would suggest that the traumatic memories in PTSD patients may be resistant to undergoing reconsolidation, thereby negating reconsolidation as a potential therapeutic target. Therefore, understanding boundary conditions, such as strength of training, is critical to ensure we know if it is possible to target reconsolidation of very strong fear memories, and if so, what the optimal conditions are to allow an extremely strong fear memory to undergo reconsolidation.

To this end, Wang et al. (2009) found that strong auditory training produced memories that initially did not undergo reconsolidation, but they did so over time on the order of 1 month. This suggests that boundary condition induced by strong training is transient (Fig. 4a). This in itself is striking, as the implicit assumption is that once a memory stops undergoing reconsolidation it will never begin again. This was the first demonstration that a putative boundary condition could be transient (Wang et al. 2009).

Wang et al. (2009) hypothesized that one principle that could mediate boundary conditions is to downregulate the mechanisms that allow memories to undergo reconsolidation. What could be the molecular mechanism to inhibit reconsolidation of strong memories for up to 30 days after training in the LBA? Ben Mamou et al. (2006) demonstrated the NMDA receptor antagonists for the NR2B subunits are necessary in reactivation-induced destabilization, but that this destabilization does not get expressed at the behavioral level. Specifically, pre-reactivation infusion of ifenprodil (a NR2B antagonist) prevented the memory from being impaired by post-reactivation anisomycin; however, it had no effect on the expression of freezing. New strong memories show similar properties: normal expression of freezing during reactivation but insensitivity to post-reactivation anisomycin. Ben Mamou et al. (2006) reasoned that strong training may downregulate NR2B expression in the LBA, thereby making the memory insensitive to post-reactivation anisomycin infusions but capable of being expressed normally. It was hypothesized that NR2B expression in the LBA should be reduced under conditions when memories did not undergo reconsolidation but should remain normal when memories underwent reconsolidation. That was exactly what was observed. NR2B levels were normal when the memory underwent reconsolidation, but drastically reduced under the conditions in which the memory did not undergo reconsolidation (Fig. 4b). The reduction was subunit selective, with NR1 subunits constant at all time points.

The suggested role of the NR2B subunits in regulating when fear memory in the LBA will undergo reconsolidation may not generalize to all memory systems or types of memory. Currently, there are four studies that have examined the mechanisms involved in transforming a consolidated memory to a labile state. While we have demonstrated that NR2B subunit is critical for memories to return to a labile state within the LBA for fear conditioning (Ben Mamou et al. 2006), NMDA receptors in the hippocampus and within the amygdala for appetitive memories are thought to play a role in re-stabilization process (Milton et al. 2008; Suzuki et al. 2008). In the hippocampus, voltage-gated calcium channels (VGCC) (Suzuki et al. 2008) and protein degradation (Lee et al. 2008) are critical for a memory to return to a labile state.

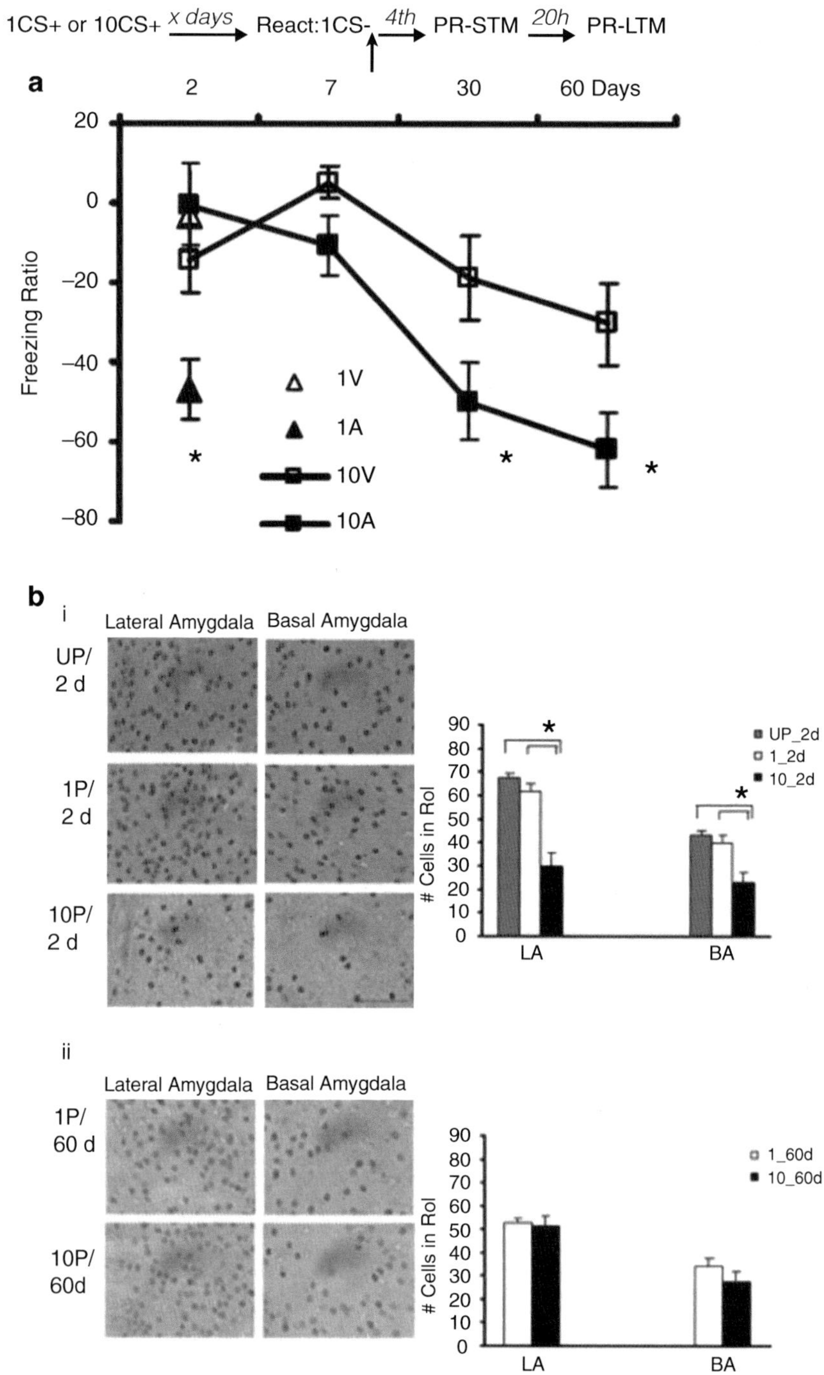

Fig. 4 (**a**) Strong memories undergo reconsolidation at 30 and 60, but not 7, days after training (taken from Wang et al. 2009). The *top panel* of each sub-figure represents the behavior protocol. Separate groups of animals were LBA cannulated and trained with 10 tone–footshock pairings.

4.1 Does Reconsolidation Imply an Exact Recapitulation of Consolidation?

An important but somewhat neglected aspect of this debate is that the protocols used to study reconsolidation are different from those used to study consolidation, which renders direct comparison of results problematic. For example, in auditory fear conditioning, both CS and US are presented, leading to the activation of afferents that relay auditory and pain information to the amygdala. Neurons that are thought to be the site of plasticity in the LBA are proposed to receive concurrent activation by these afferents (Blair et al. 2001). As a consequence, a series of second messenger systems are activated that are thought to lead to transcription and translation of proteins required for consolidation (Maren 2001; Schafe et al. 2001). In reconsolidation studies, however, typically only the CS is presented to reactivate and induce plasticity in consolidated memory. Thus, consolidation studies examine the neurobiological changes after a CS and US are presented together, while reconsolidation studies examine neurobiological changes that happen after presentation of a CS alone. For this reason, at the brain systems/circuits and molecular level, consolidation and reconsolidation must be different, as only the former directly activates the pathways that relay US information to the amygdala. Therefore, the demonstration of differences in brain regions or circuits mediating consolidation and reconsolidation may be rather trivial (Nader et al. 2005). It remains unclear which of the reported differences between consolidation and reconsolidation actually reflect genuine differences between the two processes as opposed to differences in the protocols used to induce them. A study in which differences between reconsolidation and consolidation were not attributable to differences in the protocols is the first to shed some light on this issue (Lee et al. 2004). The authors reported a double dissociation, separating the mechanisms mediating consolidation from those that mediate reconsolidation (Lee et al. 2004) [see also the work by Giese and colleagues (von Hertzen and Giese 2005)].

Fig. 4 (continued) The memory was reactivated at 7, 30, or 60 days after training. The freezing ration was computed as (PR-LTM – PR-STM)/PR-STM) × 100 %. Intra-LBA anisomycin infusion impaired the PR-LTM only when the strong memory was reactivated at 30- and 60-days after training. The *asterisk* indicates significant group differences. (**b**) NR2B-subunit levels, assessed by immunohistochemistry (IHC), are inversely related to the ability of the strong memories to undergo reconsolidation over time. (**i**) Animals received 10 tone–footshock pairings (10P), 1 pairing (1P), or 1 footshock followed by an unpaired tone (UP). They were sacrificed 2 days after training, a time when the memory does not undergo reconsolidation, and their brains were later processed for IHC. The *left panel* represents the actual staining in regions of interest (ROI) in lateral and basal amygdala (LA, BA) in individual groups ($n = 4$/group). The *graph* shows the quantification of NR2B-positive cell numbers in each ROI. While 1P and UP animals showed similar level of NR2B-immunostained cells, 10P animals showed significantly less stained cells in either LA or BA. The *asterisk* indicates significant group differences. (**ii**) Animals received either 10P or 1P. They were sacrificed 60 days after training, a time when the memory does undergo reconsolidation, and their brains were later processed for IHC. Both groups show similar level of NR2B-positive cells in LA and BA. The scalar bar represents 80 μm. All pictures in the *left panel* are in the same scale. Each data point is represented in mean +/– s.e.m

5 Clinical Implications

Consolidation and reconsolidation are processes ubiquitous to all neurons (not just those in memory systems) (Kandel 2001). The finding that consolidated memories return to a labile state and have to be restored has significant implications for a number of clinical conditions such as posttraumatic stress disorder (PTSD), addiction, obsessive–compulsive disorder (OCD), or delusions/hallucinations. An understanding of the mechanisms mediating reconsolidation could provide the basis for developing new or refining old therapeutic tools to successfully manage, if not cure, some of these conditions. As an example of how this could be applied, imagine a patient with PTSD whose symptoms were resistant to both drugs and psychotherapy. A new way of treating this condition could be to reactivate the patient's traumatic memory and block its reconsolidation. Theoretically, this should lead to a "cure" within a single session. Although finding a cure in the removal of a memory in a single session may sound worthy of a fictional reality, early studies on patients using electroconvulsive therapy (ECT) demonstrate that this possibility may not be incompatible with real life.

Franks and colleagues (Rubin et al. 1969; Rubin 1976) treated patients suffering from hallucinations, delusions, major depression, or OCD. In contrast to other studies that administered ECT when the subjects were anesthetized, Rubin and colleagues kept the patients awake and directed them to focus on the objects of their compulsions or hallucinations. This experimental procedure reactivated the neural mechanisms mediating those memories when the ECT was delivered. All of the subjects were reportedly "cured" of their condition, even though some had had up to 30 previous ECT treatments while under anesthesia. The majority remained symptom free for the 2-year period between the treatment and the publication of the manuscript. The fact that ECT was effective only when the memories were reactivated, but not when the memory reactivation was omitted (i.e., when the patient was anesthetized), suggests in principle that reconsolidation occurs in humans. Furthermore, this study provides evidence that the possibility of curing someone by removing a memory in a single session may not be so remote.

Today's treatments tend to be less intrusive than ECT. For example, beta-adrenergic antagonists such as propranolol have few side effects and are known to block reconsolidation of aversive and appetitive memories preferentially stored in the amygdala. The first attempt to target reconsolidation in patients with enduring PTSD symptoms reported a reduction in the strength of traumatic memories after a 15 min intervention (Brunet et al. 2008). It is important to note that some of these patients had been suffering from these PTSD symptoms for close to 30 years. Furthermore, it is remarkable that a single reactivation caused an old and consolidated memory to become un-stored again. Drug craving (Xue et al. 2012; Saladin et al. 2013) and PTSD (Brunet et al. 2008; Menzies 2012) are two clinical conditions in which it has been demonstrated that targeting their underlying maintenance mechanisms through reconsolidation can lead to significant clinical improvement. For a more extended discussion of these issues, please see a recent review by Nader et al. (2013).

For other clinical conditions, such as OCD/major depression, that may involve multiple maintaining mechanisms not mediated by a single brain area, a new form of ECT has been shown to block reconsolidation (Kroes et al. 2014).

5.1 Emotional Memory Enhancements via Reconsolidation

As previously discussed in this chapter, one property of new memories is that they can be enhanced. "Significance" facilitates "remembrance." We are more likely to remember significant life events than trivial ones. Evolution appears to have accomplished this feat of adaptation by natural selection of modulatory effects exerted by stress hormones on the consolidation of memory traces (McGaugh 2004). Thus, emotional memories trigger mechanisms that modulate consolidation—neutral memories do not. This effect is time dependent, i.e., the enhancement is greater when the injections are administered in temporal proximity to the training.

There have been more than a dozen of papers reporting that memory strength can be enhanced via a process of reconsolidation. This is demonstrated by observing intact PR-STM and enhanced PR-LTM. The first modern example demonstration of this effect came from the work of Tronson et al. (2006). After targeting the second messenger system protein kinase A (PKA), which is implicated in LTP, the authors showed that inhibition of PKA's activity impaired PR-LTM; facilitation of that kinase's activity enhanced PR-LTM. One recent demonstration of this equal elegance comes from Satoshi Kida' group, who identify the mechanisms of memory enhancement at the molecular and brain systems levels.

6 Conclusion

Reconsolidation as a memory process is a relatively recent entry in the domain of memory research. Reconsolidation has changed our view of memory from a passive to an active neurobiological process. The evidence for reconsolidation comes from a spectrum of species, amnesic agents, and reinforcers, spanning all levels of analysis from molecular, physiological, and behavioral levels, thereby suggesting that reconsolidation is a fundamental property of memory. What is more indicative of the status of reconsolidation in memory research is that evidence from both human and rodent studies has grown exponentially in the recent years.

Reconsolidation remains a topic of intensive research. One area of investigation that is being studied involves the identification of boundary conditions in reconsolidation. I have described a major limitation of the current approach to identify such boundary conditions and suggested a complementary approach to help resolve this important issue. Specifically, identifying a molecular or cellular indicator of when memories undergo reconsolidation represents this complementary approach.

There is a growing interest in utilizing reconsolidation blockage as a therapeutic tool in several clinical conditions, most importantly PTSD. Although most of the studies to date do not report significant clinical effects from targeting reconsolidation, evidence from these studies provides proof-of-concept demonstrations of the usefulness of reconsolidation paradigm in clinical research. Particularly, the evidence demonstrating the effectiveness of blocking reconsolidation of traumatic memories as old as 30 years is a cause for optimism. Nonetheless, future clinical research will undoubtedly benefit from advances in basic research, such as an increased understanding of the boundary conditions of reconsolidation, amongst others.

Acknowledgments K.N. was supported by the Canadian Institutes of Health Research, Natural Sciences and Engineering Research Council of Canada, Canadian Foundation for Innovation, The Volkswagen, Alfred P. Sloan, and EJL. Karim Nader is a James McGill Chair.

References

Anokhin KV, Tiunova AA, Rose SP (2002) Reminder effects—reconsolidation or retrieval deficit? Pharmacological dissection with protein synthesis inhibitors following reminder for a passive-avoidance task in young chicks. Eur J Neurosci 15:1759–1765

Bailey CH, Kandel ER (1993) Structural changes accompanying memory storage. Annu Rev Physiol 55:397–426

Ben Mamou C, Gamache K, Nader K (2006) NMDA receptors are critical for unleashing consolidated auditory fear memories. Nat Neurosci 9:1237–1239

Blair HT, Schafe GE, Bauer EP, Rodrigues SM, LeDoux JE (2001) Synaptic plasticity in the lateral amygdala: a cellular hypothesis of fear conditioning. Learn Mem 8:229–242

Bliss TV, Collingridge GL (1993) A synaptic model of memory: long-term potentiation in the hippocampus. Nature 361:31–39

Bozon B, Davis S, Laroche S (2003) A requirement for the immediate early gene zif268 in reconsolidation of recognition memory after retrieval. Neuron 40:695–701

Brunet A, Orr SP, Tremblay J, Robertson K, Nader K, Pitman RK (2008) Effect of post-retrieval propranolol on psychophysiologic responding during subsequent script-driven traumatic imagery in post-traumatic stress disorder. J Psychiatr Res 42:503–506

Cai D, Pearce K, Chen S, Glanzman DL (2012) Reconsolidation of long-term memory in aplysia. Curr Biol 22:1783–1788

Child FM, Epstein HT, Kuzirian AM, Alkon DL (2003) Memory reconsolidation in hermissenda. Biol Bull 205:218–219

De Vietti T, Holiday JH (1972) Retrograde amnesia produced by electroconvulsive shock after reactivation of a consolidated memory trace: a replication. Psychon Sci 29:137–138

Debiec J, LeDoux JE (2004) Disruption of reconsolidation of auditory fear conditioning by noradrenergic blockade in the amygdala. Neuroscience 129:267–272

Debiec J, LeDoux JE, Nader K (2002) Cellular and systems reconsolidation in the hippocampus. Neuron 36:527–538

Doyere V, Debiec J, Monfils MH, Schafe GE, LeDoux JE (2007) Synapse-specific reconsolidation of distinct fear memories in the lateral amygdala. Nat Neurosci 10:414–416

Dreyfuss JL, Regatieri CV, Jarrouge TR, Cavalheiro RP, Sampaio LO, Nader HB (2009) Heparan sulfate proteoglycans: structure, protein interactions and cell signaling. Anais da Academia Brasileira de Ciencias 81:409–429

Dudai Y (2004) The neurobiology of consolidations, or, how stable is the engram? Annu Rev Psychol 55:51–86

Duncan CP (1949) The retroactive effect of electroconvulsive shock. J Comp Physiol Psychol 42:32–44

Duvarci S, Nader K, Ledoux JE (2005) Activation of extracellular signal-regulated kinase-mitogen-activated protein kinase cascade in the amygdala is required for memory reconsolidation of auditory fear conditioning. Eur J Neurosci 21:283–289

Duvarci S, Mamou CB, Nader K (2006) Extinction is not a sufficient condition to prevent fear memories from undergoing reconsolidation in the basolateral amygdala. Eur J Neurosci 24:249–260

Ebbinghaus M (1885) Über das Gedächtnis. K. Buehler, Leipzig

Eisenberg M, Kobilo T, Berman DE, Dudai Y (2003) Stability of retrieved memory: inverse correlation with trace dominance. Science 301:1102–1104

Finnie PS, Nader K (2012) The role of metaplasticity mechanisms in regulating memory destabilization and reconsolidation. Neurosci Biobehav Rev 36:1667–1707

Flexner LB, Flexner JB, Stellar E (1965) Memory and cerebral protein synthesis in mice as affected by graded amounts of puromycin. Exp Neurol 13:264–272

Fonseca R, Nagerl UV, Bonhoeffer T (2006) Neuronal activity determines the protein synthesis dependence of long-term potentiation. Nat Neurosci 9:478–480

Glickman S (1961) Perseverative neural processes and consolidation of the memory trace. Psychol Bull 58:218–233

Goelet P, Castellucci VF, Schacher S, Kandel ER (1986) The long and short of long-term memory-a molecular framework. Nature 322:419–422

Gordon WC (1977a) Similarities of recently acquired and reactivated memories in interference. Am J Psychol 90:231–242

Gordon WC (1977b) Susceptibility of a reactivated memory to the effects of strychnine: a time-dependent phenomenon. Physiol Behav 18:95–99

Gordon WC, Spear NE (1973) Effect of reactivation of a previously acquired memory on the interaction between memories in the rat. J Exp Psychol 99:349–355

Gruest N, Richer P, Hars B (2004) Memory consolidation and reconsolidation in the rat pup require protein synthesis. J Neurosci 24:10488–10492

Hernandez PJ, Kelley AE (2004) Long-term memory for instrumental responses does not undergo protein synthesis-dependent reconsolidation upon retrieval. Learn Mem 11:748–754

Hong I, Kim J, Lee S, Ko HG, Nader K, Kaang BK, Tsien RW, Choi S (2013) AMPA receptor exchange underlies transient memory destabilization on retrieval. Proc Natl Acad Sci USA 110:8218–8223

Horne CA, Rodriguez WA, Wright TP, Padilla JL (1997) Time-dependent effects of fructose on the modulation of a reactivated memory. Prog Neuropsychopharmacol Biol Psychiatry 21:649–658

Hupbach A, Gomez R, Hardt O, Nadel L (2007) Reconsolidation of episodic memories: a subtle reminder triggers integration of new information. Learn Mem 14(1–2):47–53

Kandel ER (2001) The molecular biology of memory storage: a dialogue between genes and synapses. Science 294:1030–1038

Kelly A, Laroche S, Davis S (2003) Activation of mitogen-activated protein kinase/extracellular signal-regulated kinase in hippocampal circuitry is required for consolidation and reconsolidation of recognition memory. J Neurosci 23:5354–5360

Kida S, Josselyn SA, de Ortiz SP, Kogan JH, Chevere I, Masushige S, Silva AJ (2002) CREB required for the stability of new and reactivated fear memories. Nat Neurosci 5:348–355

Kindt M, Soeter M, Vervliet B (2009) Beyond extinction: erasing human fear responses and preventing the return of fear. Nat Neurosci 12:256–258

Kroes MC, Tendolkar I, van Wingen GA, van Waarde JA, Strange BA, Fernandez G (2014) An electroconvulsive therapy procedure impairs reconsolidation of episodic memories in humans. Nat Neurosci 17:204–206

Lee JL (2008) Memory reconsolidation mediates the strengthening of memories by additional learning. Nat Neurosci 11:1264–1266
Lee JL (2009) Reconsolidation: maintaining memory relevance. Trends Neurosci 32:413–420
Lee JL, Everitt BJ, Thomas KL (2004) Independent cellular processes for hippocampal memory consolidation and reconsolidation. Science 304:839–843
Lee JL, Di Ciano P, Thomas KL, Everitt BJ (2005) Disrupting reconsolidation of drug memories reduces cocaine-seeking behavior. Neuron 47:795–801
Lee SH, Choi JH, Lee N, Lee HR, Kim JI, Yu NK, Choi SL, Lee SH, Kim H, Kaang BK (2008) Synaptic protein degradation underlies destabilization of retrieved fear memory. Science 319 (5867):1253–1256
Lee SH, Kwak C, Shim J, Kim JE, Choi SL, Kim HF, Jang DJ, Lee JA, Lee K, Lee CH et al (2012) A cellular model of memory reconsolidation involves reactivation-induced destabilization and restabilization at the sensorimotor synapse in Aplysia. Proc Natl Acad Sci USA 109:14200–14205
Lewis DJ (1979) Psychobiology of active and inactive memory. Psychol Bull 86:1054–1083
Li Y, Meloni EG, Carlezon WA Jr, Milad MR, Pitman RK, Nader K, Bolshakov VY (2013) Learning and reconsolidation implicate different synaptic mechanisms. Proc Natl Acad Sci USA 110:4798–4803
Malenka RC, Nicoll RA (1999) Long-term potentiation—a decade of progress? Science 285:1870–1874
Maren S (2001) Neurobiology of Pavlovian fear conditioning. Annu Rev Neurosci 24:897–931
Martin SJ, Grimwood PD, Morris RG (2000) Synaptic plasticity and memory: an evaluation of the hypothesis. Annu Rev Neurosci 23:649–711
McGaugh JL (1966) Time-dependent processes in memory storage. Science 153:1351–1358
McGaugh JL (2004) The amygdala modulates the consolidation of memories of emotionally arousing experiences. Annu Rev Neurosci 27:1–28
McGaugh JL, Krivanek JA (1970) Strychnine effects on discrimination learning in mice: effects of dose and time of administration. Physiol Behav 5:1437–1442
Menzies RP (2012) Propranolol, traumatic memories, and amnesia: a study of 36 cases. J Clin Psychiatry 73:129–130
Milekic MH, Alberini CM (2002) Temporally graded requirement for protein synthesis following memory reactivation. Neuron 36(3):521–525
Miller CA, Marshall JF (2005) Molecular substrates for retrieval and reconsolidation of cocaine-associated contextual memory. Neuron 47:873–884
Miller RR, Springer AD (1973) Amnesia, consolidation, and retrieval. Psychol Rev 80:69–79
Milton AL, Lee JL, Butler VJ, Gardner R, Everitt BJ (2008) Intra-amygdala and systemic antagonism of NMDA receptors prevents the reconsolidation of drug-associated memory and impairs subsequently both novel and previously acquired drug-seeking behaviors. J Neurosci 28:8230–8237
Misanin JR, Miller RR, Lewis DJ (1968) Retrograde amnesia produced by electroconvulsive shock after reactivation of a consolidated memory trace. Science 160:203–204
Morris RG, Inglis J, Ainge JA, Olverman HJ, Tulloch J, Dudai Y, Kelly PA (2006) Memory reconsolidation: sensitivity of spatial memory to inhibition of protein synthesis in dorsal hippocampus during encoding and retrieval. Neuron 50:479–489
Müller GE, Pilzecker A (1900) Experimentelle beitrage zur lehre vom gedachtnis. Z Psychol 1:1–288
Nader K, Hardt O (2009) A single standard for memory: the case for reconsolidation. Nat Rev Neurosci 10:224–234
Nader K, Schafe GE, Le Doux JE (2000) Fear memories require protein synthesis in the amygdala for reconsolidation after retrieval. Nature 406:722–726
Nader K, Hardt O, Wang SH (2005) Response to Alberini: right answer, wrong question. Trends Neurosci 28:346–347

Nader K, Hardt O, Lanius R (2013) Memory as a new therapeutic target. Dialogues Clin Neurosci 15:475–486
Pederia ME, Maldonado H (2003) Protein synthesis subserves reconsolidation or extinction depending on reminder duration. Neuron 38:863–869
Pedreira ME, Perez-Cuesta LM, Maldonado H (2002) Reactivation and reconsolidation of long-term memory in the crab Chasmagnathus: protein synthesis requirement and mediation by NMDA-type glutamatergic receptors. J Neurosci 22:8305–8311
Pedreira ME, Perez-Cuesta LM, Maldonado H (2004) Mismatch between what is expected and what actually occurs triggers memory reconsolidation or extinction. Learn Mem 11:579–585
Przybyslawski J, Sara SJ (1997) Reconsolidation of memory after its reactivation. Behav Brain Res 84:241–246
Przybyslawski J, Roullet P, Sara SJ (1999) Attenuation of emotional and nonemotional memories after their reactivation: role of beta adrenergic receptors. J Neurosci 19:6623–6628
Rodrigues SM, LeDoux JE, Sapolsky RM (2009) The influence of stress hormones on fear circuitry. Annu Rev Neurosci 32:289–313
Rodriguez WA, Rodriguez SB, Phillips MY, Martinez JL Jr (1993) Post-reactivation cocaine administration facilitates later acquisition of an avoidance response in rats. Behav Brain Res 59:125–129
Rodriguez WA, Horne CA, Padilla JL (1999) Effects of glucose and fructose on recently reactivated and recently acquired memories. Prog Neuropsychopharmacol Biol Psychiatry 23:1285–1317
Rose JK, Rankin CH (2006) Blocking memory reconsolidation reverses memory-associated changes in glutamate receptor expression. J Neurosci 26:11582–11587
Rubin RD (1976) Clinical use of retrograde amnesia produced by electroconvulsive shock: a conditioning hypothesis. Can Psychiatr Assoc J 21:87–90
Rubin RD, Fried R, Franks CM (1969) New application of ECT. In: Rubin RD, Franks C (eds) Advances in behavior therapy, 1968. Academic, New York, pp 37–44
Rumpel S, LeDoux J, Zador A, Malinow R (2005) Postsynaptic receptor trafficking underlying a form of associative learning. Science 308:83–88
Saladin ME, Gray KM, McRae-Clark AL, Larowe SD, Yeatts SD, Baker NL, Hartwell KJ, Brady KT (2013) A double blind, placebo-controlled study of the effects of post-retrieval propranolol on reconsolidation of memory for craving and cue reactivity in cocaine dependent humans. Psychopharmacology (Berl) 226:721–737
Sangha S, Scheibenstock A, Lukowiak K (2003) Reconsolidation of a long-term memory in Lymnaea requires new protein and RNA synthesis and the soma of right pedal dorsal 1. J Neurosci 23:8034–8040
Schafe GE, LeDoux JE (2000) Memory consolidation of auditory Pavlovian fear conditioning requires protein synthesis and protein kinase A in the amygdala. J Neurosci 20:RC96
Schafe GE, Nader K, Blair HT, LeDoux JE (2001) Memory consolidation of Pavlovian fear conditioning: a cellular and molecular perspective. Trends Neurosci 24:540–546
Schiller D, Monfils MH, Raio CM, Johnson DC, Ledoux JE, Phelps EA (2010) Preventing the return of fear in humans using reconsolidation update mechanisms. Nature 463:49–53
Spear N (1973) Retrieval of memory in animals. Psychol Rev 80:163–194
Stollhoff N, Menzel R, Eisenhardt D (2005) Spontaneous recovery from extinction depends on the reconsolidation of the acquisition memory in an appetitive learning paradigm in the honeybee (Apis mellifera). J Neurosci 25:4485–4492
Suzuki A, Josselyn SA, Frankland PW, Masushige S, Silva AJ, Kida S (2004) Memory reconsolidation and extinction have distinct temporal and biochemical signatures. J Neurosci 24:4787–4795
Suzuki A, Mukawa T, Tsukagoshi A, Frankland PW, Kida S (2008) Activation of LVGCCs and CB1 receptors required for destabilization of reactivated contextual fear memories. Learn Mem 15:426–433

Taubenfeld SM, Milekic MH, Monti B, Alberini CM (2001) The consolidation of new but not reactivated memory requires hippocampal C/EBPbeta. Nat Neurosci 4:813–818

Tronson NC, Wiseman SL, Olausson P, Taylor JR (2006) Bidirectional behavioral plasticity of memory reconsolidation depends on amygdalar protein kinase A. Nat Neurosci 9:167–169

Tsvetkov E, Carlezon WA, Benes FM, Kandel ER, Bolshakov VY (2002) Fear conditioning occludes LTP-induced presynaptic enhancement of synaptic transmission in the cortical pathway to the lateral amygdala. Neuron 34:289–300

Valjent E, Aubier B, Corbille AG, Brami-Cherrier K, Caboche J, Topilko P, Girault JA, Herve D (2006) Plasticity-associated gene Krox24/Zif268 is required for long-lasting behavioral effects of cocaine. J Neurosci 26:4956–4960

von Hertzen LS, Giese KP (2005) Memory reconsolidation engages only a subset of immediate-early genes induced during consolidation. J Neurosci 25:1935–1942

Walker MP, Brakefield T, Hobson JA, Stickgold R (2003) Dissociable stages of human memory consolidation and reconsolidation. Nature 425:616–620

Wang SH, Ostlund SB, Nader K, Balleine BW (2005) Consolidation and reconsolidation of incentive learning in the amygdala. J Neurosci 25:830–835

Wang SH, de Oliveira Alvares L, Nader K (2009) Cellular and systems mechanisms of memory strength as a constraint on auditory fear reconsolidation. Nat Neurosci 12:905–912

Xue YX, Luo YX, Wu P, Shi HS, Xue LF, Chen C, Zhu WL, Ding ZB, Bao YP, Shi J et al (2012) A memory retrieval-extinction procedure to prevent drug craving and relapse. Science 336:241–245

Social Cognition

Alexandra Patin and René Hurlemann

Contents

A. Patin • R. Hurlemann (✉)
Department of Psychiatry, University of Bonn, 53105 Bonn, Germany

Division of Medical Psychology, University of Bonn, 53105 Bonn, Germany
e-mail: renehurlemann@icloud.com

K.M. Kantak, J.G. Wettstein (eds.), *Cognitive Enhancement*, Handbook of Experimental Pharmacology 228, DOI 10.1007/978-3-319-16522-6_10

Abstract

Social cognition is a major problem underlying deficiencies in interpersonal relationships in several psychiatric populations. And yet there is currently no gold standard for pharmacological treatment of psychiatric illness that directly targets these social cognitive areas. This chapter serves to illustrate some of the most innovative attempts at pharmacological modulation of social cognition in psychiatric illnesses including schizophrenia, borderline personality disorder, autism spectrum disorders, antisocial personality disorder and psychopathy, social anxiety disorder, and posttraumatic stress disorder. Pharmacological modulation includes studies administering oxytocin, ecstasy (MDMA), modafinil, methylphenidate, and D-cycloserine. Furthermore, some background on social cognition research in healthy individuals, which could be helpful in developing future treatments, is provided as well as the potential for each drug as a long-term treatment option.

Keywords

Social cognition • Schizophrenia • Anxiety disorders • Autism spectrum disorder • Oxytocin

1 Introduction

1.1 What Is Social Cognition?

Although the concepts on their own are relatively well integrated into the language of everyday literature, when put together the term social cognition is suddenly more difficult to clearly define. In a review of social behavior in humans, Ralph Adolphs describes the problem as being one of inclusion: "If the social is ubiquitous, we face the problem of including all aspects of cognition as social" (Adolphs 2003, p. 165).

At the base of social cognition traditionally lies emotion recognition, which has been argued to be the key to understanding how another person feels, what they are intending to do, or how they will react to a stimulus (Elfenbein and Ambady 2002). Included in the definition of social cognition for purposes of this chapter are also, among others, empathy and theory of mind, or the ability to infer feelings and emotions in another, cooperation, trust, and social feedback-based learning.

A further important facet of social cognition is reciprocity: it is not enough to merely perceive and understand social cues, but one must be able to give appropriate signals and reactions as well (Roepke et al. 2013).

1.2 Brain Regions Involved in Social Cognition

Although related, the neural networks involved in different social cognition domains are distinct. Face processing, for instance, involves the fusiform gyrus, or fusiform face area, for processing static features, the superior temporal sulcus for processing mimicry and dynamic changes in the face, and the amygdala (Adolphs 2003; Haxby et al. 2000). The amygdala, alongside the ventromedial prefrontal cortex, is also of great interest to social cognition researchers as they share rich functional connections and have been found to play a role in psychopathy, depression, anxiety disorders, autism, and schizophrenia (Tudusciuc and Adolphs 2013). Social cognition as a broader concept appears to have its roots in a network involving the prefrontal cortex (PFC), amygdala, cingulate gyrus, fusiform gyrus, insula and further regions in the somatosensory cortex, superior temporal sulcus, and the supramarginal gyrus (Tudusciuc and Adolphs 2013).

2 Illnesses Characterized by Low Social Cognition

A lack of social cognition is a cornerstone of several illnesses characterized by the inability to interact with others at a normal, healthy level. Specifically, emotion recognition has consistently correlated with characteristics such as social anxiety and avoidance, distress, depression, antisocial behavior, and psychopathy (McClure and Nowicki Jr 2001). Both healthy and patient populations show the importance of social cognition to interpersonal interactions. For instance, healthy participants have shown a link between fear recognition ability and an increased desire toward altruistic behavior (Marsh et al. 2007), as well as report having better relationships and a lower depression rate (Carton et al. 1999). In patient populations, social cognition has been found to have a predictive value, and autism or psychosis patients with lower social cognitive abilities statistically show lower social function (Bertrand et al. 2007; Losh et al. 2009).

There are indications in the literature that therapeutic augmentation with pharmacological modulation can be used to support social cognition in healthy participants, and one common theme among these is the emerging possibility of beneficial treatments for patients with schizophrenia, antisocial disorders, and social anxiety. Moreover, there have been several approaches toward enhancing social cognition in psychiatric illness. For one, cognitive enhancement therapy (CET) involves neurocognitive and social cognitive improvements based on a computer-based training in attention, memory, and problem solving and further exercises in perspective taking, gistfulness, social context appraisal, and other areas of social cognition (Eack 2013).

2.1 Schizophrenia

According to the Diagnostic and Statistical Manual of Mental Disorders (DSM-5), schizophrenia is a disorder within the schizophrenia spectrum and is characterized in part by negative symptoms, which make up a large part of the clinical manifestation of schizophrenia and have a direct effect on social cognition. They include diminished emotional expression, avolition (a reduced motivation for self-initiated activity), alogia (reduction in speech output), anhedonia (a reduced ability to gain pleasure from positive stimuli), and asociality (reduced interest in social interaction). Particularly salient in schizophrenia are diminished emotional expression and avolition (American Psychiatric Association 2013).

One debilitating consequence of schizophrenia is social and occupational dysfunction, although patients may well possess the cognitive ability to complete required tasks (American Psychiatric Association 2013). Initial studies directly aiming to improve social cognitive skills in individuals with schizophrenia have shown an increase in theory of mind following training (Bechi et al. 2012; Bechi et al. 2013), as well as focusing on concrete social cognition, such as in CET (Eack 2013). Patients with schizophrenia usually experience their first symptoms early on in adult life, and a large minority of patients show an onset before they reach 19 years of age (Cullen et al. 2008), and the disease affects 0.3–0.7 % of adults overall (van Os and Kapur 2009). Other findings report a much higher prevalence of schizophrenia and related categories of illness of 2–3 % (Perälä et al. 2007). Schizophrenic individuals die 12–15 years before the average population, mostly due to unhealthy lifestyles (Saha et al. 2007).

Men are affected with the disease more often than women (Roy et al. 2001). As of yet, no single cause has been pinpointed, although there is a strong evidence for a genetic predisposition (van Os and Kapur 2009). Twin studies show that there is a heritability rate of up to 80 % (van Os and Kapur 2009). The typical pharmacological treatment involves antipsychotic drugs working antagonistically at the D2 receptor, but this is usually made in addition to psychological treatment as well as social support systems (van Os and Kapur 2009).

Cognitive impairments are a common manifestation of schizophrenia, including memory, attention, and executive function (Fioravanti et al. 2005). Indeed, schizophrenia was first known as dementia praecox at its discovery precisely because of these deficits (Kraepelin 1971; van Os and Kapur 2009). Individuals with schizophrenia show a much higher use of drugs than the general population, and alcohol or nicotine use is found in half to over half the population (Mueser et al. 1992; Šagud et al. 2009). This introduces the question of whether schizophrenic individuals use substances such as nicotine as an effort to support cognitive enhancement and are perhaps lacking this support in conventional therapeutic regimes. Structural and functional neural abnormalities have been found in several cases in schizophrenic individuals, including neuroinflammation in orbitofrontal white matter and decreased astroglial density in the subgenual cingulate white matter (Najjar and Pearlman 2015), as well as abnormal functional hub density in the frontal and limbic association areas, for example, suggesting difficulty in information integration over different neural regions (Bassett et al. 2008; Zhang et al. 2012).

2.2 Borderline Personality Disorder

While borderline personality disorder (BPD) affects 1–3 % of adults (Trull et al. 2010), it affects a much larger rate of 10–20 % of psychiatric patient populations (Korzekwa et al. 2008), making it one of the leading disorders marked by social cognition deficits. The DSM-5 characterizes BPD individuals as making great effort to avoid abandonment, having feelings of emptiness, and being in unstable and intense interpersonal relationships, among other symptoms (American Psychiatric Association 2013). These are symptoms that directly reflect difficulties in social cognition, as they are illustrative of difficulties patients have in understanding and dealing with others.

From very early on in social cognition research, it became apparent that patients with BPD showed a negative bias toward judging the intentions of others as malevolent (Lerner and St. Peter 1984; Stuart et al. 1990). A lack in both cognitive and emotional empathy could also serve as a major underpinning for BPD individuals' difficulties in interactions (Roepke et al. 2013). Further studies have shown that BPD individuals have difficulties with perceiving, processing, and responding to social cues from others (Brodsky et al. 2006; Gunderson and Lyons-Ruth 2008; Stiglmayr et al. 2005).

The underlying mechanism for these deficits is difficult to discern, but there is growing support for the idea that emotional hypervigilance disrupts normal processing of emotional stimuli, specifically in terms of emotion recognition (Domes et al. 2009; Linehan 1995). This idea is in support of findings showing that BPD individuals have greater sensitivity when faced with social rejection (Staebler et al. 2011; Stiglmayr et al. 2005).

Structural and functional abnormalities have been found in the amygdala for one, suggesting that this is an important neural region for therapeutic approaches (Domes et al. 2009) focusing on cognitive and emotional empathy, emotion recognition, trust and rejection processing, and moral judgments, all of which show strong deficits in BPD individuals (Herpertz and Bertsch 2014; Roepke et al. 2013).

2.3 Autism Spectrum Disorders

Autism spectrum disorder (ASD) patients are strongly influenced by social cognitive deficits; in terms of diagnostics, the DSM-5 characterizes ASD individuals by verbal and nonverbal communication difficulties, difficulties when interacting with others, and repetitive movements or behaviors, among others (American Psychiatric Association 2014). All of these traits make normal social relationships close to impossible, but the first two are specifically concerned with social cognition per se, thus making ASD a leading candidate for social cognition treatment development.

Patients with autism suffer from a difficulty in judging the value of social signals and struggle with emotion recognition, making interpersonal interaction difficult (Gross 2004; Hill and Frith 2003). One underlying mechanism could involve reduced activation of the fusiform face gyrus, inferior occipital gyrus, superior temporal sulcus, and amygdala, and an increased response by the frontal cortex and primary visual cortex, during face processing in autism patients (Pierce et al. 2001). Early findings have indicated that autistic children lack a theory of mind (Baron-Cohen et al. 1985). Structural as well as functional deficits in autistic individuals during social cognitive tasks often include the amygdala (Baron-Cohen et al. 1999, 2000; Pierce et al. 2001), although there is currently no single identifiable neural region or network responsible for the disorder.

2.4 Antisocial Personality Disorder and Psychopathy

According to the DSM-5, antisocial personality disorder (ASPD) is at times manifested in social cognitive symptoms such as a lack of remorse, deceitfulness, and failure to conform to social norms (American Psychiatric Association 2013). The underlying mechanisms for social cognition deficiencies, however, are not entirely clear.

Psychopathy is associated with ASPD, though not interchangeable with it (Hare and Neumann 2006). One popular model of psychopathy is the Violence Inhibition Mechanism (VIM) model, which describes an inability to read and react to social signals of submission, such as facial expressions of fear, sadness, or shame (Blair 2001; Blair et al. 1997). This inability to react appropriately has the effect that individuals with ASPD or psychopathy do not empathize with victims, and the inhibiting force normally stopping violence or antisocial behavior is removed. Here, too, the underlying cause for this extreme lack of empathy is not clear, though it could depend on reduced neural responses following emotional stimuli (Meffert et al. 2013).

A meta-analysis of antisocial personality disorder showed that one of the strongest characteristics of patients is the inability to recognize fearful faces and emotional stimuli in general (Sterzer et al. 2005), which is able to be tracked to amygdala dysfunction (Birbaumer et al. 2005; Kiehl et al. 2001; Marsh and Blair 2008; Veit et al. 2013). An additional area of interest is the ventromedial PFC (Blair 2008), as both areas have shown abnormal activation during social cognitive tasks, such as emotional memory (Kiehl et al. 2001). In terms of intra-amygdala activity, individuals with psychopathy show increased activation in the central amygdala and reduced activation in the basolateral amygdala during emotional processing tasks (Moul et al. 2012).

As opposed to a specific site of trauma or lesion, the amygdala and ventromedial PFC, among other areas, make up a network in which the exact location or nature of the dysfunction is not easy to pin down (Blair 2008). Psychopathy is progressive in nature (Lynam et al. 2007), making treatment early on in life a worthwhile pursuit. Interestingly, one study found that while psychopathic individuals do not show pure executive function deficits, they do show lower levels of executive functioning when an emotional component is included (Lapierre et al. 1995). This supports the argument that social cognition, rather than pure cognition, is lacking in ASPD and psychopathic patients.

2.5 Social Anxiety Disorder

With a lifetime prevalence of approx. 12 % (Kessler et al. 2005), social anxiety disorder (SAD) affects millions of people worldwide. According to the DSM-5, SAD is characterized by symptoms such as disabling anxiety in social settings in which the individual will be under observation by others, and even complete avoidance of social situations, among other symptoms (American Psychiatric Association 2013).

Studies investigating the underlying mechanisms of these symptoms have shown a cognitive bias toward interpreting social cues to be more negative as well as toward negative self-representation (Constans et al. 1999; Hirsch and Clark 2004; Mogg et al. 2004; Rapee and Abbott 2006; Stopa and Clark 2000; Voncken et al. 2003). Furthermore, socially anxious individuals show lower theory of mind ability than non-socially anxious individuals, even independently of an interpretation bias (Hezel and McNally 2014).

2.6 Posttraumatic Stress Disorder

A highly misunderstood and prevalent disease following a traumatic event is posttraumatic stress disorder, with a lifetime prevalence of 6.8–9.2 %, depending on age (Kessler et al. 2005). Among lifetime PTSD individuals, almost half of both women and men suffer from a major depressive episode following the trauma, and over 50 % of males and almost 28 % of females will meet the criteria for alcohol abuse or alcohol dependence (Kessler et al. 1995). According to the DSM-5, PTSD individuals are also characterized by social cognition deficits, for example, persistent and exaggerated negative beliefs regarding themselves or the world around them, feelings of detachment or estrangement, hypervigilance, or an exaggerated startle response (American Psychiatric Association 2013).

3 The Effect of Pharmacological Modulation of Social Cognition

As illustrated above, deficits in social cognition are a common theme throughout psychiatric illnesses, and they are arguably the underlying reason for many of the socially debilitating consequences of disease. There is unfortunately no single pharmacological gold standard currently available for the treatment of social cognitive deficits. However, there have been several innovative efforts to develop treatments to combat these symptoms. This chapter should serve to illuminate these efforts and findings.

3.1 Oxytocin

Pharmacological properties of OT

Oxytocin (OT) research has climbed exponentially in recent years, for the most part due to the interest spawned by its effects on social cognition and social behavior as a neuromodulator. Originally named for its hormonal properties in inducing uterine contractions during birth, stemming from the Greek words ὀξύς (swift) and τόκος (birth), OT has rapidly become a drug of choice when exploring social cognition. For the most part, following synthesis in the supraoptic and paraventricular nuclei of the hypothalamus, OT proceeds via axonal transport to the neurohypophysis, where it is stored in secretory vesicles together with the OT carrier protein neurophysin I and secreted into the bloodstream (Brownstein et al. 1980).

Synthesis of OT is regulated by the OT gene located on chromosome 20 in humans. OT is mainly produced in the supraorbital (SON) and paraventricular (PVN) hypothalamic nuclei. The magnocellular cells of the PVN terminate in the neurohypophysis, amygdala, and nucleus accumbens, while the parvocellular cells of the PVN terminate in other CNS regions (Knobloch et al. 2012). OT secretion is dependent on neuronal depolarization and subsequent Ca^{2+}-dependent exocytosis of the vesicles (Brownstein et al. 1980; Gimpl and Fahrenholz 2001). OT central release also shows effects of priming (Ludwig and Leng 2006). Additionally, OT is synthesized in much smaller quantities peripherally in the uterus, placenta, amnion, corpus luteum, testis, and heart, for example (Gimpl and Fahrenholz 2001). OT was synthetically synthesized for the first time in 1953 by Vincent du Vigneaud and was thus the first polypeptide hormone to be sequenced and synthesized (Pitocin, Syntocinon; cys–tyr–ile–gln–asn–cys–pro–leu–gly–NH2) (du Vigneaud et al. 1953, 1954).

There is currently only one known OT receptor (OTR). The OTR is a G_q-protein-coupled receptor of the rhodopsin-type (class I), coded for by the OTR gene located on chromosome 3 (Gimpl and Fahrenholz 2001; Simmons Jr et al. 1995). The OTR has been found in the human brain in the central and basolateral amygdala, medial preoptic area, anterior and ventromedial hypothalamus, olfactory nucleus, vertical limb of the diagonal band, ventrolateral septum, anterior cingulate and hypoglossal and solitary nuclei, the basal nucleus of Meynert, and at times in the globus pallidus and ventral pallidum (Boccia et al. 2013; Loup et al. 1991). Social animals show high rates of OT receptor density in the nucleus accumbens and the prelimbic cortex, which

(continued)

modulate feelings of reward (Lim et al. 2004), as well as in the amygdala (Huber et al. 2005).

The exact mechanism of how or whether OT completely crosses the blood–brain barrier is unknown (Banks and Kastin 1985; Ermisch et al. 1985a, b; Meisenberg and Simmons 1983). Animal studies have shown that intravenous injection of OT results in approximately 0.01 % of OT actually crossing the blood–brain barrier (Kendrick et al. 1991). Because of this, there is most likely no correlation between endocrine OT release at the neurohypophysis and cerebrospinal fluid (CSF) levels, which are most likely influenced by neurons reaching into the third ventricle, limbic system, brain stem, and spinal cord (Altemus et al. 2004; Gimpl and Fahrenholz 2001; Kagerbauer et al. 2013; Martin et al. 2014; Striepens et al. 2013).

The exact pharmacokinetics of OT in humans are not yet completely settled. The half-life of OT ranges from roughly 2 min in plasma (Jones and Robinson 1982; Meyer et al. 1987) to 3–5 min in women in vitro (Rydén and Sjöholm 1969) and 20 min in CSF (Mens et al. 1983). OT is rapidly degraded in vitro following the addition of plasma from pregnant women, showing an 85 % reduction in OT concentration in the course of 1 h, but not in nonpregnant women or men (Leake et al. 1980). Twenty-four international units (IU) of intranasal OT has been shown to increase plasma OT levels to their highest levels 15 min following administration, whereas cerebrospinal fluid peak levels were reached 75 min later (Striepens et al. 2013).

3.1.1 OT in Healthy Individuals

Earlier studies exploring the effects of OT as a neurotransmitter focused mainly on prosocial effects of OT. However, initial groundbreaking studies soon illustrated a far more complex picture: a single dose of 24 IU of OT in healthy subjects, latency 45 min, not only enhanced prosocial behavior, but also negative emotions, such as schadenfreude and envy (Shamay-Tsoory et al. 2009). Likewise, OT has been found to enhance protective responses, evident in a potentiated acoustic startle response to negative stimuli and increased recollection of negative stimuli (Striepens et al. 2012). In this vein, the literature has strived to explore the various facets of OT in social settings, and several strides have been made in social cognitive domains. One of these domains is theory of mind, also called cognitive empathy. This ability has been suggested to increase emotional empathy, or the ability to feel what the other person is feeling, also known as putting oneself in another's shoes. Further findings show that OT increases responses to emotional faces (Shahrestani et al. 2013; Van IJzendoorn and Bakermans-Kranenburg 2012), as well as emotional empathy ratings (single dose, 24 IU, latency 45 min) (Hurlemann et al. 2010) in healthy participants.

The ability to transfer this recognition of emotion to a judgment of how the other person will likely act, a process known as "mind reading" (Siegal and Varley 2002;

Stone et al. 2003), is also increased following OT administration (single dose, 24 IU, latency 45 min) (Domes et al. 2007), suggesting a wide spectrum of areas related to emotion recognition sensitive to OT effects. At an intersection of cognitive and emotional domains, OT (single dose, 24 IU, 45 min latency) was shown to increase the effect of positive versus negative social feedback on learning during a declarative memory task in healthy males (Hurlemann et al. 2010).

In a further area of social cognition, healthy subjects given OT showed that they were more trusting during social interaction and responded less to social stress, as well as more cooperative (Bartz et al. 2011b; Heinrichs et al. 2003; Kosfeld et al. 2005). Furthermore, OT appears to increase social approach and protective behavior (Lim and Young 2006; Preckel et al. 2014; Scheele et al. 2012). The underlying mechanism for this effect could be a reduced amygdala response to vague or threatening stimuli (Baumgartner et al. 2008; Meyer-Lindenberg 2008). As such, OT appears to counteract social transmission of fear via social signals of anxiety inducing stimuli and could thus hold therapeutic potential for patients with anxiety disorders (Eckstein and Hurlemann 2013).

3.1.2 OT in Psychiatric Illness

ASD

Patient populations have shown that OT is a promising area of research in terms of treatment augmentation. For instance, ASD subjects are found to have reduced plasma OT levels (Green et al. 2001; Modahl et al. 1998; but see also Jansen et al. 2006). Furthermore, several studies have found a likely correlation between susceptibility to ASD and genetic variations in the OT receptor gene (Auranen et al. 2002; Shao et al. 2002; Wermter et al. 2010; Wu et al. 2005).

Exogenous OT administration in participants with ASD has been shown to increase comprehension and memory for the social-emotional words happy, angry, or sad (continuous infusion of 10 U/ml OT in 1.0 l of saline over 4 h per indwelling intravenous catheter; infusion rate titrated 25 ml every 15 min in the first hour, 50 ml in the second, 100 ml in the third, and infused at a constant rate of 700 ml/h in the fourth hour; testing was completed at baseline just before the infusion, 30, 60, 120, 180, and 240 min during infusion) (Hollander et al. 2007).

In an innovative study of skin conductance response (SCR) to human versus nonhuman sounds, a single dose of 24 IU OT (latency 1 h) resulted in an overall reduction in SCRs in healthy controls to all sounds, but an increase in SCRs to human sounds relative to nonhuman sounds (Lin et al. 2014). Patients under placebo showed a similar SCR pattern toward both sounds, but a similar pattern to OT-treated controls following OT administration. Thus, OT apparently worked to assimilate the patterns and levels of SCR in patients versus controls. Finally, SCR differences toward human versus nonhuman sounds following OT correlated with the patients' social function and interpersonal reactivity.

In terms of face perception, a study examining the differential neural response to emotions versus nonsocial objects found that OT enhanced neural response while processing emotions (single dose; 25 IU for participants aged 16–19 years, 18 IU for ages 12–15, 12 IU for ages 7–11; latency 45 min) (Gordon et al. 2013). A single dose of 24 IU OT, 45 min prior to testing, increased amygdala response to emotional faces (Domes et al. 2013), and a study of high-functioning ASD and Asperger patients was found to show increased social interaction and more normal face processing following a single dose of 24 IU OT 50 min prior to task completion (Andari et al. 2010). Recognition of nonverbal, information-based judgments of emotional faces was increased in ASD patients following a single dose of 24 IU OT (latency 40 min), and authors furthermore found that abnormal medial PFC activity was improved following OT (Watanabe et al. 2014).

More specifically, emotion recognition was increased following single dose of either 24 IU (participants aged 16–19) or 18 IU (ages 12–15) OT administration, latency of 45 min, in young participants with Asperger's disorder (Guastella et al. 2010). Emotion recognition, as well as emotional well-being, was also increased following a 6-week administration in ASD adults (24 IU twice daily) (Anagnostou et al. 2012). In a more long-term study, OT showed positive effects on social cognition areas including social recognition, empathic accuracy, and theory of mind in a 12-week treatment course (dosage 0.2, 0.26, 0.33, and 0.4 IU/kg/dose, twice daily) (Anagnostou et al. 2014).

Although the above studies were not successful in showing behavioral changes, a recent meta-analysis of intranasal OT as potential treatment for ASD found a moderate effect size of OT and suggested it is thus worth pursuing for therapeutic use (Bakermans-Kranenburg and van IJzendoorn 2013).

Schizophrenia

Endogenous OT in schizophrenic individuals has been found to correlate with symptom severity (Rubin et al. 2010) as well as social cognitive bias (Walss-Bass et al. 2013), and plasma OT levels in patients positively correlate with emotion recognition (Goldman et al. 2008). Additionally, there is evidence for a strong interaction between OT and other neurotransmitters such as serotonin, for instance (Lee et al. 2003; Mottolese et al. 2014).

Several studies examining the effects of exogenous OT administration in schizophrenic patients, both following a single dose as well as more long-term use, have been completed. In an augmentation study, patients received 40 IU OT for 6 weeks 30 min prior to a twice weekly social cognitive skills training sessions (Davis et al. 2014). The authors reported significant improvements in empathic accuracy both 1 week following the final training session and 1 month later. A second study employing a 6-week regimen of 24 IU OT twice daily examined participant response to social cognitive measures as well as social skills and clinical symptoms 50 min after the final morning dose at the end of week 6 (Gibson et al. 2014). Findings included improved theory of mind and fear recognition, and increased perspective taking.

Improved emotion recognition was also found in two further studies following a single dose of 24 IU OT after 45 min (fear recognition) (Fischer-Shofty et al. 2013) and after 50 min (Averbeck et al. 2012). Following a much higher single dose of 40 IU and a latency of 30 min, schizophrenic individuals showed significantly improved ability for controlled social cognition, or the ability to perceive and understand indirectly expressed emotions or intentions over long time periods (Woolley et al. 2014). In a further study using a single dose of 40 IU (latency 30 min), schizophrenic individuals increased performance on higher-level social cognition assessments, including detection of sarcasm and deception and empathy, but not on lower-level assessments, such as facial affect perception, social perception, and detection of lies (Davis et al. 2013). Thus, it appears that improvements in schizophrenia positively benefit from a higher dosage of OT, and it would be interesting to find out in further research under what conditions this correlation applies.

Interestingly, along the same vein, whereas a low single dose of 10 IU OT (latency 45 min) was detrimental to emotion recognition, a higher single dose of 20 IU (latency 45 min) improved emotion recognition in polydipsic patients, specifically reducing a bias toward fear perception (Goldman et al. 2011). Supporting these findings, 24 IU OT twice daily improved theory of mind following a 14-day treatment (latency 50 min) (Pedersen et al. 2011). Nonsignificant findings from the same study showed that schizophrenic volunteers additionally showed increased trust.

BPD

Endogenous OT levels have been found to be lower in women with BPD and negatively correlated with aggressiveness and symptom severity (Bertsch et al. 2013).

Exogenous OT administration in patients with BPD has shown mixed effects. For one, OT apparently obstructed trust and cooperation in a study of BPD individuals given a single dose of 40 IU OT and tested 35 min later, apparently due to an increased desire to punish the other player in a social dilemma game (Bartz et al. 2011a; Ebert et al. 2013).

On the other hand, BPD individuals showed a lowered stress response following 40 IU OT (latency 60 min) than placebo, manifested in a relative absence of both dysphoria and cortisol in response to the Trier Social Stress Test (Simeon et al. 2011). A study examining avoidance reactions showed that, while placebo-treated BPD individuals revealed an avoidance reaction to angry faces, BPD individuals treated with 24 IU OT (single dose, latency 45 min) did not, thus suggesting that OT abolished the hypervigilance for threatening stimuli in BPD individuals (Brüne et al. 2013). Similarly, female BPD patients given 26 IU OT (single dose, latency 45 min) were found to normalize their perception of angry faces, including reduced abnormal fixation changes to the eyes as well as the absence of hyperactive amygdala response to angry faces, indicating that the characteristic BPD hypersensitivity was abolished under OT (Bertsch et al. 2013).

ASPD and psychopathy

The literature concerning OT and psychopathy is limited, but shows some highly interesting findings. Studies of the OTR gene have found it to be influential to psychopathic traits: in one study, the rs1042778 genotype TT was linked to high levels of callous-unemotional traits in children diagnosed with disruptive behavioral problems (Dadds et al. 2014b). A further study of 4- to 16-year-old males diagnosed with oppositional-defiant or conduct disorder linked increased methylation of the OTR gene as well as with lower endogenous OT levels in older male participants to high levels of callous-unemotional traits, and reported that higher methylation correlated with low endogenous OT (Dadds et al. 2014a).

Exogenous intracerebroventricular OT in rats resulted in reduced aggressive behavior and increased social exploration (5 μl OTR peptidergic antagonist {desGly-NH_2,d$(CH_2)_5$[Tyr$(Me)^2$,Thr^4]OVT}, latency 10 min) (Calcagnoli et al. 2013). In a recent meta-analysis of the effect of OT on emotion recognition in healthy participants, OT was found to improve overall performance on recognizing facial expressions, first and foremost for happy and fearful faces (Shahrestani et al. 2013). As the authors suggest based on previous findings in this area, this could be vital to interpersonal communication, as fear recognition is key to feeling empathy for the pain of another being, blocking antisocial impulses and thus important for illnesses characterized by these (Marsh and Blair 2008; Shahrestani et al. 2013).

As of yet, no augmentation therapy has been attempted with emotional training and OT.

SAD

Endogenous OT findings in SAD patients paint an interesting picture. In one study, plasma OT was found to be similar in patients and healthy controls, but differences emerged within the patient group (Hoge et al. 2008). For one, OT levels positively correlated with symptom severity and were furthermore linked to dissatisfaction in social relationships. The authors explained this finding by suggesting that social deficits in anxiety or autistic disorders may be associated with increased levels of OT as a compensatory mechanism following OTR dysfunction.

Exogenous, intranasal administration of 24 IU OT 45–90 min prior to exposure therapy (four sessions) resulted in patients reporting a reduced negative bias toward negative metal representations of self as well as toward their speech performance and appearance (Guastella et al. 2009).

A further study showed that generalized SAD individuals given OT showed increased functional connectivity between the amygdala and bilateral insula and middle cingulate/dorsal anterior cingulate gyrus when processing fearful faces, bringing them closer to connectivity patterns shown by healthy controls (Gorka et al. 2014). OT was also shown to normalize resting state functional connectivity of the left and right amygdala with the rostral anterior cingulate cortex/medial PFC following a single dose of 24 IU OT in another generalized SAD patient population (Dodhia et al. 2014). Additionally, the study found that the more severe the social anxiety in patients, the greater amygdala-frontal connectivity was increased.

In a further study, amygdala hyperactivity in generalized SAD patients in response to fearful faces compared to healthy controls was shown to normalize following a single dose of 24 IU OT (latency 45 min) (Labuschagne et al. 2010). Additionally, an increased medial PFC and left anterior cingulate cortex response to sad faces was also improved following 24 IU OT (single dose, 50 min latency) (Labuschagne et al. 2012).

PTSD

Two studies have found that the OT receptor variation rs53576 allele is associated with posttraumatic stress and the ability to cope (Bradley et al. 2013; Lucas-Thompson and Holman 2013). Only one study has administered exogenous OT to PTSD individuals, and showed that physiologic responses during a combat imagery task were lower following 20 IU OT than placebo (single dose, latency 1 h) (Pitman et al. 1993).

3.1.3 Conclusions: Potential for OT as a Viable Long-Term Treatment Option

Although there is great promise for the success of OT as a long-term treatment option in psychiatric illness, there is still a great deal of research needed. As the authors of a comprehensive review of OT effects on social behavior report, the majority of OT studies on social cognition report that drug effects represent an interplay with stimulus or task, and that the question thus becomes, which conditions allow for an effect of OT to shine through (Bartz et al. 2011b).

For one, the pharmacodynamics of OT have yet to be explored experimentally. The optimal dosage and latency for pharmacological experiments remain unclear. This makes setting up a proper paradigm, as well as interpreting OT's effects on studies to separate from differences in paradigm more difficult. The effects are well illustrated in the finding that a lower vs. higher doses of OT can have the complete opposite effect in social cognition (Goldman et al. 2011) and can differentially influence aggressive behavior (Calcagnoli et al. 2013).

In order to measure accurately the effects of OT on social cognition, and therefore for OT to be used as a valid therapeutic method in illnesses marked by lower social cognition, there needs to be a standardization of the literature. For instance, optimal dosage and latency should be empirically determined for healthy participants (Shahrestani et al. 2013). Furthermore, response to exogenous OT should be measured in terms of standardized markers (Shahrestani et al. 2013), and intranasal administration should be standardized (Guastella et al. 2013).

In terms of emotion recognition, paradigms need to be standardized to include all basic emotions presented in a similar format to aid comparison across studies showing an effect of OT (Shahrestani et al. 2013). Currently, the literature is strongly focused on male participants. However, if OT is to become a viable treatment option, females need to be considered equally. Initial findings have shown that emotional processing is differentially affected by OT in females compared to males (Domes et al. 2010), and this remains an area to be further explored.

Additional genetic testing would be beneficial to expanding the understanding of OT in both healthy and patient groups. CD38, for example, a protein generally associated with cancer markers, has been linked in several cases to the OT receptor and the uptake of exogenous OT (Higashida et al. 2010; Higashida et al. 2011). Another therapeutically important example is the common OT receptor single nucleotide polymorphism, rs53576, which has been associated with the ability to benefit from social support under psychosocial stress (Chen et al. 2011). These are not the only potentially game-changing genetic differences that could influence the course of therapeutic OT administration, and more research is needed to determine how differences would affect different patients during treatment.

The possibility of OT for therapeutic use should not be ruled out due to these difficulties. OT remains an extremely powerful mediator of social bonds, and thus highly important for research in psychiatric illness characterized by a lack of these bonds (Scheele et al. 2013). For example, autistic children show normal response in the fusiform face gyrus when presented with their mother's face, but not when viewing other adults' faces (Pierce and Redcay 2008). OT has been long understood as a crucial element in maternal bonding (Kendrick 2004; Kendrick et al. 1997), and could thus present an important clue to the differential face processing in autistic children versus adults (Pierce et al. 2001; Pierce and Redcay 2008).

3.2 3,4-Methylenedioxymethamphetamine (Ecstasy)

Pharmacological properties of MDMA

It is not exactly clear how 3,4-methylenedioxymethamphetamine (MDMA) produces its effects, but there have been several studies documenting its potential mechanisms. Structurally, MDMA is a ring-substituted amphetamine similar to mescaline and methamphetamine (de la Torre et al. 2004). It works as an agonist to the trace amine-associated receptor 1 (TAAR1), thus working toward monoamine transporter reuptake inhibition (Miller 2011). The S(+) isomer of MDMA is a psychostimulant and has an effect on empathy, while the R isomer has hallucinogenic effects (de la Torre et al. 2004).

MDMA works mainly as a serotonin (5-HT), dopamine (DA), and noradrenaline (NA) reuptake inhibitor and/or releaser; the mechanism is one of membrane transport reversal and subsequent flow of 5-HT, DA, and NA into the synaptic cleft and to the postsynaptic membrane (de la Torre et al. 2004). Emotional excitation following 1.5 mg/kg MDMA was blocked by both a single oral dose of 50 mg of the 5-HT_{2A} receptor antagonist ketanserin and 1.4 mg of intravenous haloperidol, a D_2 receptor antagonist, implicating both serotonergic and dopaminergic influences in euphoric mood changes under MDMA (Liechti and Vollenweider 2001). MDMA also acts to increase cortisol, prolactin, ADH, and ACTH secretion (de la Torre et al. 2004).

The pharmacokinetics following a single dose of 75, 100, and 125 mg MDMA in humans are listed in Table 1. MDMA follows a nonlinear pattern, with lower doses being associated with higher urinary recovery and higher doses with lower recovery (de la Torre et al. 2004). Repeat doses show an exponential rate of plasma concentration of MDMA, with a C_{max} of 29 % following two successive doses of 100 mg MDMA over 24 h (Farre et al. 2004). Blood concentrations show a peak at 1–2 h following administration and a return to baseline after 4–6 h (Mas et al. 1999).

Table 1 Pharmacokinetics of different single oral doses of MDMA

Single dosage (mg)	C_{max} (ng/ml)	t_{max} (h)	$t_{1/2}$ (h)	k_a (h^{-1})
75 mg[a]	130.9	1.8	7.86	2.3835
100 mg[b]	225.5 ± 26.1	2.3 ± 1.1	9.0 ± 2.3	2.7 ± 1.5
125 mg[a]	236.4	2.4	8.73	2.1253

$n = 8$ for all results; C_{max}, peak plasma concentration; t_{max}, time until peak plasma concentration; $t_{1/2}$, elimination half-life; k_a, absorption constant; ng, nanograms
[a]Mas et al. (1999)
[b]de la Torre et al. (2004)

3.2.1 MDMA in Healthy Individuals

The literature regarding MDMA is far less developed than OT; however, some important observations have been documented in regard to social cognition. One study reported reduced fear recognition alongside increased self-reported loving feelings and friendliness following 1.5 mg/kg MDMA over 4 weekly sessions (latency 65 min) (Bedi et al. 2010). Similar results were reported in a study examining response to emotion recognition, which showed that 125 mg MDMA led to increased recognition of positive emotions, but decreased recognition of negative emotions (Hysek et al. 2012).

Behaviorally, 2 mg/kg MDMA (single dose, latency up to 5 h) resulted in increased friendliness, sociability, and talkativeness (Tancer and Johanson 2003). A further study found differential dose-dependent neural responses: the amygdala showed a dampened response to angry facial expressions following 1.5 mg/kg MDMA, while the ventral striatum showed an increased response to happy faces following 0.75 mg/kg MDMA (single dose, latency 45 min) (Bedi et al. 2009).

Likely mechanisms for MDMA's effects can be found in studies showing that MDMA acts as a stimulant for endogenous OT release (Dumont et al. 2009, 100 mg MDMA, single dose; Hysek et al. 2012; Kirkpatrick et al. 2014, 1.5 mg/kg MDMA, single dose, peak at 90–120 min latency; Thompson et al. 2007, injection of 5 mg/kg MDMA in rats) and could thus work as a potentiating force during prosocial behavior. Because there have been some contradictory findings (Kuypers et al. 2014), however, much more research is needed.

3.2.2 MDMA in Psychiatric Illness

Currently, there are no patient studies implementing MDMA for social cognitive improvement. However, MDMA is interesting when considering psychiatric illnesses marked by the inability to form meaningful, intimate relationships, because it increases prosocial function in healthy individuals (Bedi et al. 2009; Dumont et al. 2009). On the other hand, MDMA has been shown to significantly reduce cognitive function in many different areas in nonpsychiatric populations, as well as increase negative mood states (Parrott 2013), suggesting that it could be a difficult route of treatment development for psychiatric illness.

One area in which MDMA is surprisingly well researched relative to other illnesses is in individuals with PTSD, as an augmentation for psychotherapy. In an initial study, PTSD patients received single, initial dose of 125 mg MDMA and a supplemental, optional dose of 62.5 mg 2–2.5 h after and relaxed while participating in therapeutic discussion with a therapist over 2 sessions interspersed among 17 sessions total (latency not applicable, onset of MDMA effects 45–75 min following initial dose, peak at 2–2.5 h, duration 4–5 h following single dose, 5–6 h following supplemental dose) (Mithoefer et al. 2011). Subjects who received MDMA showed an 83 % clinical response rate following psychotherapy compared to 25 % in the placebo group. A follow-up of the same patients showed that the vast majority still showed clinical improvements (Mithoefer et al. 2013). A further study found that a single dose of 125 mg + 62.5 mg supplemental dose MDMA over three sessions of psychotherapy (alongside 12 nondrug therapy sessions) lowered self-reported symptoms of PTSD (Oehen et al. 2013). Though these studies do not show a specific improvement in social cognition per se, they do imply an improvement in social interaction, as the therapy is necessarily led by a therapist, thus dependent on a social influence.

3.2.3 Conclusions: Potential for MDMA as a Viable Long-Term Treatment Option

More than other pharmacological interventions, MDMA as a potential treatment option for psychiatric illness is made much more difficult due to the legality of its use. In addition, studies regarding its effects are lacking, and contradictory findings have been shown. That said, MDMA does show promise in several areas of emotional processing relevant to social cognition and would thus be well worth pursuing as a possible augmentation for therapy.

3.3 Modafinil

Pharmacological properties of modafinil

Modafinil (2-[(diphenylmethyl)sulfinyl] acetamide; Vigil) is a psychostimulant which works indirectly on the glutamate and GABA receptors. Additional indirect modulation of neurotransmission includes an increase in dopamine, noradrenaline, and serotonin secretion. Pharmacokinetics are t_{max} 2–3 h, $t_{1/2}$ 10–12 h, or 15 h steady state following repeated doses. Oral bioavailability is 11–52 % and plasma protein binding 62 %. Modafinil is mainly used as a waking agent for sleeping disorders. Typical dosage ranges from 200 to 400 mg/day. (Benkert et al. 2013) Further uses for modafinil include treatment for depression (Fava et al. 2007), depressive episodes in bipolar disorder (Calabrese et al. 2010; Frye et al. 2007), cocaine addiction (Dackis et al. 2004), and attention deficit/hyperactivity disorder (ADHD) (Biederman and Pliszka 2008), among other disorders.

3.3.1 Modafinil in Healthy Individuals

Healthy participants in studies investigating cognitive enhancement benefits of modafinil have shown that it improves attention, memory, spatial planning, and executive functions, but this seems to have a dose-dependent influence (Kelley et al. 2012; Repantis et al. 2010). Cognitive enhancement has also been shown in some psychiatric illnesses, such as schizophrenia (Saavedra-Velez et al. 2009; Turner et al. 2004; Wittkampf et al. 2012).

3.3.2 Modafinil in Psychiatric Illness

Findings in psychiatric participants are mixed. In terms of emotional processing, 200 mg modafinil (single dose, latency 2 h) improved recognition of emotional faces, and significantly sad faces, but did not increase sensitivity to reward or punishment or performance in cognitive tasks with emotional components or improve mood in first episode psychosis (Scoriels et al. 2011). Unfortunately, findings are few and far between for modafinil and social cognition in psychiatric illness. Potential for the drug in the future is dependent on more research and is made more difficult by contraindications such as addiction disorders (absolute contraindication) or anxiety and psychosis (relative contraindications) (Benkert et al. 2013), which plague a large portion of psychiatric patients.

3.3.3 Conclusions: Potential for Modafinil as a Viable Long-Term Treatment Option

Modafinil acts in several areas of the brain including the amygdala, lending support to the drug as a worthwhile treatment for illnesses characterized by amygdala dysfunction. However, because a single dose of 100 mg modafinil (latency 3 h) has been shown to cause increased anxiety in healthy volunteers depending on dosage (Randall et al. 2003), this would need to be closely observed in psychiatric populations characterized by hypervigilance or anxiety, for example. Furthermore, there is evidence that modafinil can create a tolerance in the user as well as have addictive properties (Volkow et al. 2009), thus creating problems as a long-term treatment option.

3.4 Methylphenidate (Ritalin)

Pharmacological properties of MPH

Methylphenidate (MPH) (methyl phenyl(piperidin-2-yl)acetate; Ritalin, Concerta, Methylin, Equasym XL, among others) is a psychostimulant used for treating ADHD in children over 6 years of age and adolescents and narcolepsy. Pharmacokinetics are as follows: t_{max} 2 h, $t_{1/2}$ 2.4 h in children, 2.1 h in adults. Oral bioavailability is 30 % and plasma protein binding approx. 20 %. Its effects are felt quickly, within 15–30 min. Following

(continued)

decay, patients sometimes report feeling the symptoms in stronger intensity (rebound phenomenon), but this disappears after further administration. Administration of MPH must be started gradually. Initial dosage is 5–10 mg/day, building up to max 60 mg/day for ADHD in children as well as narcolepsy. MPH has the potential to become addictive (Benkert et al. 2013).

3.4.1 MPH in Psychiatric Illness

MPH has been for the most part restricted to use in ADHD, and the effects of MPH in social cognition are limited to early studies showing benefits of MPH in classroom and social settings in hyperactive youth. Unfortunately, pharmacological data are not available in detail for all studies; methodological data are reported here where available. Findings include fewer negative interactions with peers in a social setting (Hinshaw et al. 1984a), greater self-control following a MPH + self-control training when confronted with a stressful and socially threatening situation (Hinshaw et al. 1984b), as well as reduced intensity of negative behaviors (both studies occurred over a 3-week period of adjunct treatment in addition to daily medication; dosages for morning administration were 5–40 mg and 0.15–1.16 mg/kg for the first study and midday administration range 5–20 mg and 0.44–0.55 mg/kg for the second) (Hinshaw et al. 1984b). Further findings supported these results and showed less disruptive behavior and improved social behavior following 10 mg MPH twice daily (Pelham et al. 1987). All studies above included adolescent boys diagnosed with ADHD, hyperactive disorder, or a similar diagnosis; unfortunately, it is not possible to provide a standard diagnosis because some studies were completed prior to current standards.

In a recent study of school children diagnosed with ADHD and comorbid social phobia, a daily dose of 0.5–1.0 mg/kg MPH per day (dose did not exceed 60 mg/day) for 12 weeks resulted in significant reductions in school-related anxiety (Golubchik et al. 2014).

In terms of aggressive behavior, MPH reduces verbal and nonverbal aggression in groups of adolescent males diagnosed with both high and low aggression levels, and, to a lesser extent, reduces aggressive response following provocation following 0.3 mg/kg MPH twice daily between 5 and 9 days over the course of 5 weeks (Murphy et al. 1992). Furthermore, MPH has been attributed to an increase in positive social interactions in ADHD patients (Hinshaw et al. 1984a). One recent study showed that MPH increased both theory of mind and empathy ratings in children with ADHD (regularly prescribed medication, latency 1–5 h) (Maoz et al. 2014). A study examining emotion recognition in ADHD children following 4-week treatment with mean 24.1 mg/day MPH (range 10–60 mg/day) 60 min prior to testing showed improved anger and fear recognition skills (Williams et al. 2008).

3.4.2 Conclusions: Potential for MPH as a Viable Long-Term Treatment Option

Unfortunately, the effects on social cognition following MPH have been researched only in a very small pool of studies and therefore are lacking in generalizability. Furthermore, a large portion of the research was completed prior to modern diagnoses, thus limiting the ability to understand results in a more modern context. Lastly, the effects of MPH on social cognition in healthy individuals could shed valuable light on the mechanisms by which this drug could help patient populations.

3.5 D-Cycloserine

Pharmacological properties of DCS

D-cycloserine (DCS) (D-4-amino-3-isoxazolidone) is a partial *N*-methyl-D-aspartate (NMDA) receptor agonist, thus expressing a glutamatergic (excitatory) influence. The NMDA receptor also holds a glycine-binding site, and which must also be co-activated to allow for NMDA receptor signaling. (Johnson and Ascher 1987; Kleckner and Dingledine 1988) Plasma concentrations are detectable within 1 h of ingestion. Peak plasma levels are 10 mg/l, reached after 3–4 h. Elimination half-life of DCS is 8–12 h. Bioavailability is excellent, and CSF levels are roughly 80–100 % of peak plasma concentrations (Holdiness 1985; Nair et al. 1956).

3.5.1 DCS in Psychiatric Illness

Traditionally an antibiotic to fight *Mycobacterium tuberculosis*, DCS has emerged as a powerful tool used in fear extinction and thus anxiety disorders, as well as cognitive functions such as memory (Onur et al. 2010). Social cognition findings, however, are limited to psychiatric populations. In individually housed mice, DCS was shown to increase social investigation and sexual behavior and decrease aggression following the introduction on an intruder (McAllister 1994). In balb/c mice reflecting behaviors mirroring those of individuals with autism, DCS led to improved sociability at a young age (Deutsch et al. 2011, 2012). Both findings suggest wide-reaching benefits of DCS in psychiatric illness in humans.

ASD

In humans, DCS has been shown to reduce withdrawal in autistic individuals, as well as generally improve clinical symptoms (Posey et al. 2004).

Schizophrenia

DCS was shown to augment cognitive remediation training (50 mg DCS administration 60 min prior to training over 8 weekly sessions) (Cain et al. 2014).

SAD

DCS has been tested as an augmentation for psychotherapy to treat SAD in initial studies. Two studies showed a greater reduction of symptoms in patients administered with 50 mg DCS 1 h prior to 4 weekly sessions employing exposure therapy than following placebo (Guastella et al. 2008; Hofmann et al. 2006). Another study, however, examining cognitive behavioral therapy (CBT) response found no benefit to 50 mg DCS 1 h prior to five exposure sessions as a part of a 12-week CBT program in terms of completion, response, or remission rate (Hofmann et al. 2013).

These findings could be explained by effects due to patient differences, as shown by two further studies. In the first, successful DCS augmentation of a 12-week CBT program, in which patients received 50 mg DCS 1 h prior to five exposure sessions, was found only for patients showing low conscientiousness and high agreeableness ratings, but not for all patients (Smits et al. 2013a). The second also used 50 mg DCS 1 h prior to five exposure sessions and found that the success of each exposure session was critical to the effect of DCS: patients who reported low fear following a session were more likely to show a greater clinical improvement at the next session if they had received DCS as opposed to placebo (Smits et al. 2013b). Likewise, the authors found that patients who received DCS and reported high fear levels following a session showed less improvement at the next session than compared to those in the placebo group. At posttreatment evaluations (week 13), patients who received DCS showed improved clinical symptoms only when they had reported low to moderate average fear levels throughout the course of treatment.

PTSD

Because of its influence on fear extinction learning, DCS has been pursued in PTSD treatment research as an augmentation to therapy. In one study, 50 mg DCS was given 30 min prior to four exposure therapy sessions, and resulted in a lower symptom reduction than patients experienced following placebo (Litz et al. 2012).

In another study, PTSD patients were given 50 mg DCS 30 min prior to a virtual reality exposure therapy over five sessions (Rothbaum et al. 2014). Primary analysis showed no difference in clinical symptoms following DCS, but when more in-depth analysis was completed, DCS was shown to increase symptom outcomes in those patients who had increased between-session learning. In a study using 50 mg DCS over 8–10 weekly exposure sessions (latency 1 h), participants showed no overall effect of having received DCS; however, DCS did show a greater reduction of symptoms in more severely affected patients (de Kleine et al. 2012). In a study examining personality differences in response to DCS, highly conscientious participants showed a better outcome following exposure therapy, as did patients with low extraversion (50 mg DCS, single dose prior to each session over ten sessions, latency n/a) (de Kleine et al. 2014).

3.5.2 Conclusions: Potential for DCS as a Viable Long-Term Treatment Option

In terms of psychiatric and psychological findings, the NMDA receptor has been implicated as having a crucial role in synaptic plasticity (Fan et al. 2014; Lee and Silva 2009; Li and Tsien 2009), and also in long-term potentiation (LTP) (Bear and Malenka 1994; Bliss and Collingridge 1993). Fear learning and fear extinction are directly dependent on LTP and thus the NMDA receptor (Blair et al. 2001; Fanselow and LeDoux 1999; Lee et al. 2001; Li et al. 1995; Walker and Davis 2002). Fear extinction is often used in therapeutic situations, and while not purely within the realm of social cognition, the social element of the therapist as a key part of treatment is supported by pharmacological modulation. A recent meta-analysis supports this and showed positive effects of DCS on exposure therapy in anxiety disorders (Rodrigues et al. 2014).

References

Adolphs R (2003) Cognitive neuroscience of human social behaviour. Nat Rev Neurosci 4(3):165–178

Altemus M, Fong J, Yang R, Damast S, Luine V, Ferguson D (2004) Changes in cerebrospinal fluid neurochemistry during pregnancy. Biol Psychiatry 56(6):386–392

American Psychiatric Association (2013) The diagnostic and statistical manual of mental disorders: DSM 5. American Psychiatric Association, Washington, DC

American Psychiatric Association (2014) Autism spectrum disorders. Retrieved 24 June 2014, from http://www.psychiatry.org/autism

Anagnostou E, Soorya L, Chaplin W, Bartz J, Halpern D, Wasserman S et al (2012) Intranasal oxytocin versus placebo in the treatment of adults with autism spectrum disorders: a randomized controlled trial. Mol Autism 3(1):16. doi:10.1186/2040-2392-3-16

Anagnostou E, Soorya L, Brian J, Dupuis A, Mankad D, Smile S, Jacob S (2014) Intranasal oxytocin in the treatment of autism spectrum disorders: a review of literature and early safety and efficacy data in youth. Brain Res 1580:188–198, http://dx.doi.org/10.1016/j.brainres.2014.01.049

Andari E, Duhamel JR, Zalla T, Herbrecht E, Leboyer M, Sirigu A (2010) Promoting social behavior with oxytocin in high-functioning autism spectrum disorders. Proc Natl Acad Sci U S A 107(9):4389–4394. doi:10.1073/pnas.0910249107

Auranen M, Vanhala R, Varilo T, Ayers K, Kempas E, Ylisaukko-Oja T et al (2002) A genomewide screen for autism-spectrum disorders: evidence for a major susceptibility locus on chromosome 3q25-27. Am J Hum Genet 71(4):777–790. doi:10.1086/342720

Averbeck BB, Bobin T, Evans S, Shergill SS (2012) Emotion recognition and oxytocin in patients with schizophrenia. Psychol Med 42(2):259–266. doi:10.1017/s0033291711001413

Bakermans-Kranenburg MJ, van IJzendoorn MH (2013) Sniffing around oxytocin: review and meta-analyses of trials in healthy and clinical groups with implications for pharmacotherapy. Transl Psychiatry 3:e258. doi:10.1038/tp.2013.34

Banks WA, Kastin AJ (1985) Permeability of the blood-brain barrier to neuropeptides: the case for penetration. Psychoneuroendocrinology 10(4):385–399, http://dx.doi.org/10.1016/0306-4530(85)90079-4

Baron-Cohen S, Leslie AM, Frith U (1985) Does the autistic child have a "theory of mind"? Cognition 21(1):37–46, http://dx.doi.org/10.1016/0010-0277(85)90022-8

Baron-Cohen S, Ring HA, Wheelwright S, Bullmore ET, Brammer MJ, Simmons A, Williams SCR (1999) Social intelligence in the normal and autistic brain: an fMRI study. Eur J Neurosci 11(6):1891–1898. doi:10.1046/j.1460-9568.1999.00621.x

Baron-Cohen S, Ring HA, Bullmore ET, Wheelwright S, Ashwin C, Williams SCR (2000) The amygdala theory of autism. Neurosci Biobehav Rev 24(3):355–364, http://dx.doi.org/10.1016/S0149-7634(00)00011-7

Bartz JA, Simeon D, Hamilton H, Kim S, Crystal S, Braun A et al (2011a) Oxytocin can hinder trust and cooperation in borderline personality disorder. Soc Cogn Affect Neurosci 6(5):556–563. doi:10.1093/scan/nsq085

Bartz JA, Zaki J, Bolger N, Ochsner KN (2011b) Social effects of oxytocin in humans: context and person matter. Trends Cogn Sci 15(7):301–309. doi:10.1016/j.tics.2011.05.002

Bassett DS, Bullmore E, Verchinski BA, Mattay VS, Weinberger DR, Meyer-Lindenberg A (2008) Hierarchical organization of human cortical networks in health and schizophrenia. J Neurosci 28(37):9239–9248

Baumgartner T, Heinrichs M, Vonlanthen A, Fischbacher U, Fehr E (2008) Oxytocin shapes the neural circuitry of trust and trust adaptation in humans. Neuron 58(4):639–650

Bear MF, Malenka RC (1994) Synaptic plasticity: LTP and LTD. Curr Opin Neurobiol 4(3):389–399

Bechi M, Riccaboni R, Ali S, Fresi F, Buonocore M, Bosia M et al (2012) Theory of mind and emotion processing training for patients with schizophrenia: preliminary findings. Psychiatry Res 198(3):371–377. doi:10.1016/j.psychres.2012.02.004

Bechi M, Spangaro M, Bosia M, Zanoletti A, Fresi F, Buonocore M et al (2013) Theory of Mind intervention for outpatients with schizophrenia. Neuropsychol Rehabil 23(3):383–400. doi:10.1080/09602011.2012.762751

Bedi G, Phan KL, Angstadt M, de Wit H (2009) Effects of MDMA on sociability and neural response to social threat and social reward. Psychopharmacology (Berl) 207(1):73–83. doi:10.1007/s00213-009-1635-z

Bedi G, Hyman D, de Wit H (2010) Is ecstasy an "empathogen"? Effects of +/-3,4-methylene-dioxymethamphetamine on prosocial feelings and identification of emotional states in others. Biol Psychiatry 68(12):1134–1140. doi:10.1016/j.biopsych.2010.08.003

Benkert O, Hippius H, Gründer G, Heiser P, Hiemke C, Himmerich H et al (2013) Kompendium der psychiatrischen Pharmakotherapie. Springer, Berlin

Bertrand MC, Sutton H, Achim AM, Malla AK, Lepage M (2007) Social cognitive impairments in first episode psychosis. Schizophr Res 95(1–3):124–133. doi:10.1016/j.schres.2007.05.033

Bertsch K, Gamer M, Schmidt B, Schmidinger I, Walther S, Kästel T et al (2013) Oxytocin and reduction of social threat hypersensitivity in women with borderline personality disorder. Am J Psychiatry 170(10):1169–1177

Biederman J, Pliszka SR (2008) Modafinil improves symptoms of attention-deficit/hyperactivity disorder across subtypes in children and adolescents. J Pediatr 152(3):394–399.e392, http://dx.doi.org/10.1016/j.jpeds.2007.07.052

Birbaumer N, Veit R, Lotze M, Erb M, Hermann C, Grodd W, Flor H (2005) Deficient fear conditioning in psychopathy: a functional magnetic resonance imaging study. Arch Gen Psychiatry 62(7):799–805. doi:10.1001/archpsyc.62.7.799

Blair RJ (2001) Neurocognitive models of aggression, the antisocial personality disorders, and psychopathy. J Neurol Neurosurg Psychiatry 71(6):727–731

Blair RJ (2008) The amygdala and ventromedial prefrontal cortex: functional contributions and dysfunction in psychopathy. Philos Trans R Soc Lond B Biol Sci 363(1503):2557–2565. doi:10.1098/rstb.2008.0027

Blair RJ, Jones L, Clark F, Smith M (1997) The psychopathic individual: a lack of responsiveness to distress cues? Psychophysiology 34(2):192–198

Blair HT, Schafe GE, Bauer EP, Rodrigues SM, LeDoux JE (2001) Synaptic plasticity in the lateral amygdala: a cellular hypothesis of fear conditioning. Learn Mem 8(5):229–242

Bliss TV, Collingridge GL (1993) A synaptic model of memory: long-term potentiation in the hippocampus. Nature 361(6407):31–39

Boccia M, Petrusz P, Suzuki K, Marson L, Pedersen C (2013) Immunohistochemical localization of oxytocin receptors in human brain. Neuroscience 253:155–164

Bradley B, Davis TA, Wingo AP, Mercer KB, Ressler KJ (2013) Family environment and adult resilience: contributions of positive parenting and the oxytocin receptor gene. Eur J Psychotraumatol 4, http://www.ejpt.net/index.php/ejpt/rt/captureCite/21659/0/ApaCitation Plugin, http://www.ncbi.nlm.nih.gov/pubmed/?term=bradley+wingo+mercer+family

Brodsky BS, Groves SA, Oquendo MA, Mann JJ, Stanley B (2006) Interpersonal precipitants and suicide attempts in borderline personality disorder. Suicide Life Threat Behav 36(3):313–322

Brownstein MJ, Russell JT, Gainer H (1980) Synthesis, transport, and release of posterior pituitary hormones. Science 207(4429):373–378. doi:10.1126/science.6153132

Brüne M, Ebert A, Kolb M, Tas C, Edel MA, Roser P (2013) Oxytocin influences avoidant reactions to social threat in adults with borderline personality disorder. Hum Psychopharmacol 28(6):552–561

Cain CK, McCue M, Bello I, Creedon T, Tang D, Laska E, Goff DC (2014) d-Cycloserine augmentation of cognitive remediation in schizophrenia. Schizophr Res 153(1–3):177–183, http://dx.doi.org/10.1016/j.schres.2014.01.016

Calabrese JR, Ketter TA, Youakim JM, Tiller JM, Yang R, Frye MA (2010) Adjunctive armodafinil for major depressive episodes associated with bipolar I disorder: a randomized, multicenter, double-blind, placebo-controlled, proof-of-concept study. J Clin Psychiatry 71 (10):1363–1370

Calcagnoli F, de Boer SF, Althaus M, den Boer JA, Koolhaas JM (2013) Antiaggressive activity of central oxytocin in male rats. Psychopharmacology (Berl) 229(4):639–651. doi:10.1007/s00213-013-3124-7

Carton JS, Kessler EA, Pape CL (1999) Nonverbal decoding skills and relationship well-being in adults. J Nonverbal Behav 23(1):91–100

Chen FS, Kumsta R, von Dawans B, Monakhov M, Ebstein RP, Heinrichs M (2011) Common oxytocin receptor gene (OXTR) polymorphism and social support interact to reduce stress in humans. Proc Natl Acad Sci U S A 108(50):19937–19942. doi:10.1073/pnas.1113079108

Constans JI, Penn DL, Ihen GH, Hope DA (1999) Interpretive biases for ambiguous stimuli in social anxiety. Behav Res Ther 37(7):643–651

Cullen KR, Kumra S, Westerman M, Schulz SC (2008) Atypical antipsychotics for treatment of schizophrenia spectrum disorders. Psychiatric Times: Schizophrenia. Retrieved 25 June 2014

Dackis CA, Kampman KM, Lynch KG, Pettinati HM, O'Brien CP (2004) A double-blind, placebo-controlled trial of modafinil for cocaine dependence. Neuropsychopharmacology 30 (1):205–211

Dadds MR, Moul C, Cauchi A, Dobson-Stone C, Hawes DJ, Brennan J, Ebstein RE (2014a) Methylation of the oxytocin receptor gene and oxytocin blood levels in the development of psychopathy. Dev Psychopathol 26(01):33–40. doi:10.1017/S0954579413000497

Dadds MR, Moul C, Cauchi A, Dobson-Stone C, Hawes DJ, Brennan J et al (2014b) Polymorphisms in the oxytocin receptor gene are associated with the development of psychopathy. Dev Psychopathol 26(01):21–31. doi:10.1017/S0954579413000485

Davis MC, Lee J, Horan WP, Clarke AD, McGee MR, Green MF, Marder SR (2013) Effects of single dose intranasal oxytocin on social cognition in schizophrenia. Schizophr Res 147 (2):393–397

Davis MC, Green MF, Lee J, Horan WP, Senturk D, Clarke AD, Marder SR (2014) Oxytocin-augmented social cognitive skills training in schizophrenia [original article]. Neuropsychopharmacology. doi:10.1038/npp.2014.68

de Kleine RA, Hendriks G-J, Kusters WJ, Broekman TG, van Minnen A (2012) A randomized placebo-controlled trial of D-cycloserine to enhance exposure therapy for posttraumatic stress disorder. Biol Psychiatry 71(11):962–968 de Kleine RA, Hendriks G-J, Smits JA, Broekman TG, van Minnen A (2014) Prescriptive variables for D-cycloserine augmentation of exposure therapy for posttraumatic stress disorder. J Psychiatr Res 48(1):40–46

de la Torre R, Farré M, Roset PN, Pizarro N, Abanades S, Segura M et al (2004) Human pharmacology of MDMA: pharmacokinetics, metabolism, and disposition. Ther Drug Monit 26(2):137–144

Deutsch SI, Burket JA, Jacome LF, Cannon WR, Herndon AL (2011) d-Cycloserine improves the impaired sociability of the Balb/c mouse. Brain Res Bull 84(1):8–11

Deutsch SI, Pepe GJ, Burket JA, Winebarger EE, Herndon AL, Benson AD (2012) d-cycloserine improves sociability and spontaneous stereotypic behaviors in 4-week old mice. Brain Res 1439:96–107, http://dx.doi.org/10.1016/j.brainres.2011.12.040

Dodhia S, Hosanagar A, Fitzgerald DA, Labuschagne I, Wood AG, Nathan PJ, Phan KL (2014) Modulation of resting-state amygdala-frontal functional connectivity by oxytocin in generalized social anxiety disorder [original article]. Neuropsychopharmacology. doi:10.1038/npp.2014.53

Domes G, Heinrichs M, Michel A, Berger C, Herpertz SC (2007) Oxytocin improves "mind-reading" in humans. Biol Psychiatry 61(6):731–733. doi:10.1016/j.biopsych.2006.07.015

Domes G, Schulze L, Herpertz SC (2009) Emotion recognition in borderline personality disorder—a review of the literature. J Pers Disord 23(1):6–19. doi:10.1521/pedi.2009.23.1.6

Domes G, Lischke A, Berger C, Grossmann A, Hauenstein K, Heinrichs M (2010) Effects of intranasal oxytocin on emotional face processing in women. Psychoneuroendocrinology 35 (1):83–93. doi:10.1016/j.psyneuen.2009.06.016

Domes G, Heinrichs M, Kumbier E, Grossmann A, Hauenstein K, Herpertz SC (2013) Effects of intranasal oxytocin on the neural basis of face processing in autism spectrum disorder. Biol Psychiatry 74(3):164–171. doi:10.1016/j.biopsych.2013.02.007

du Vigneaud V, Ressler C, Swan CJM, Roberts CW, Katsoyannis PG, Gordon S (1953) The synthesis of an octapeptide amide with the hormonal activity of oxytocin. J Am Chem Soc 75 (19):4879–4880. doi:10.1021/ja01115a553

du Vigneaud V, Ressler C, Swan JM, Roberts CW, Katsoyannis PG (1954) The synthesis of oxytocin1. J Am Chem Soc 76(12):3115–3121. doi:10.1021/ja01641a004

Dumont GJ, Sweep FC, van der Steen R, Hermsen R, Donders AR, Touw DJ et al (2009) Increased oxytocin concentrations and prosocial feelings in humans after ecstasy (3,4-methylenedioxymethamphetamine) administration. Soc Neurosci 4(4):359–366. doi:10.1080/17470910802649470

Eack SM (2013) Cognitive enhancement therapy. In: Roberts DL, Penn DL (eds) Social cognition in schizophrenia. Oxford University Press, New York

Ebert A, Kolb M, Heller J, Edel M-A, Roser P, Brüne M (2013) Modulation of interpersonal trust in borderline personality disorder by intranasal oxytocin and childhood trauma. Soc Neurosci 8 (4):305–313

Eckstein M, Hurlemann R (2013) Oxytocin: evidence for a therapeutic potential of the social neuromodulator. Nervenarzt 84(11):1321–1328. doi:10.1007/s00115-013-3832-6

Elfenbein HA, Ambady N (2002) On the universality and cultural specificity of emotion recognition: a meta-analysis. Psychol Bull 128(2):203

Ermisch A, Barth T, Rühle H, Skopkova J, Hrbas P, Landgraf R (1985a) On the blood-brain barrier to peptides: accumulation of labelled vasopressin, DesGlyNH2-vasopressin and oxytocin by brain regions. Endocrinologia Experimentalis 19(1):29–37

Ermisch A, Ruhle HJ, Landgraf R, Hess J (1985b) Blood-brain barrier and peptides. J Cereb Blood Flow Metab 5(3):350–357

Fan X, Jin WY, Wang YT (2014) The NMDA receptor complex: a multifunctional machine at the glutamatergic synapse [Review]. Front Cell Neurosci 8:160. doi:10.3389/fncel.2014.00160

Fanselow MS, LeDoux JE (1999) Why we think plasticity underlying Pavlovian fear conditioning occurs in the basolateral amygdala. Neuron 23(2):229–232

Farre M, De La Torre R, Mathúna BÓ, Roset P, Peiro A, Torrens M et al (2004) Repeated doses administration of MDMA in humans: pharmacological effects and pharmacokinetics. Psychopharmacology (Berl) 173(3–4):364–375

Fava M, Thase ME, DeBattista C, Doghramji K, Arora S, Hughes RJ (2007) Modafinil augmentation of selective serotonin reuptake inhibitor therapy in MDD partial responders with persistent fatigue and sleepiness. Ann Clin Psychiatry 19(3):153–159

Fioravanti M, Carlone O, Vitale B, Cinti ME, Clare L (2005) A meta-analysis of cognitive deficits in adults with a diagnosis of schizophrenia. Neuropsychol Rev 15(2):73–95

Fischer-Shofty M, Shamay-Tsoory SG, Levkovitz Y (2013) Characterization of the effects of oxytocin on fear recognition in patients with schizophrenia and in healthy controls. Front Neurosci 7:127

Frye M, Grunze H, Suppes T, McElroy S, Keck P, Walden J et al (2007) A placebo-controlled evaluation of adjunctive modafinil in the treatment of bipolar depression. Am J Psychiatry 164 (8):1242–1249

Gibson CM, Penn DL, Smedley KL, Leserman J, Elliott T, Pedersen CA (2014) A pilot six-week randomized controlled trial of oxytocin on social cognition and social skills in schizophrenia. Schizophr Res 156(2–3):261–265, http://dx.doi.org/10.1016/j.schres.2014.04.009

Gimpl G, Fahrenholz F (2001) The oxytocin receptor system: structure, function, and regulation. Physiol Rev 81(2):629–683

Goldman M, Marlow-O'Connor M, Torres I, Carter CS (2008) Diminished plasma oxytocin in schizophrenic patients with neuroendocrine dysfunction and emotional deficits. Schizophr Res 98(1–3):247–255. doi:10.1016/j.schres.2007.09.019

Goldman MB, Gomes AM, Carter CS, Lee R (2011) Divergent effects of two different doses of intranasal oxytocin on facial affect discrimination in schizophrenic patients with and without polydipsia. Psychopharmacology (Berl) 216(1):101–110. doi:10.1007/s00213-011-2193-8

Golubchik P, Sever J, Weizman A (2014) Methylphenidate treatment in children with attention deficit hyperactivity disorder and comorbid social phobia. Int Clin Psychopharmacol 29 (4):212–215. doi:10.1097/YIC.0000000000000029

Gordon I, Vander Wyk BC, Bennett RH, Cordeaux C, Lucas MV, Eilbott JA et al (2013) Oxytocin enhances brain function in children with autism. Proc Natl Acad Sci U S A 110 (52):20953–20958. doi:10.1073/pnas.1312857110

Gorka SM, Fitzgerald DA, Labuschagne I, Hosanagar A, Wood AG, Nathan PJ, Phan KL (2014) Oxytocin modulation of amygdala functional connectivity to fearful faces in generalized social anxiety disorder. Neuropsychopharmacology. doi:10.1038/npp.2014.168

Green L, Fein D, Modahl C, Feinstein C, Waterhouse L, Morris M (2001) Oxytocin and autistic disorder: alterations in peptide forms. Biol Psychiatry 50(8):609–613

Gross TF (2004) The perception of four basic emotions in human and nonhuman faces by children with autism and other developmental disabilities. J Abnorm Child Psychol 32(5):469–480

Guastella AJ, Richardson R, Lovibond PF, Rapee RM, Gaston JE, Mitchell P, Dadds MR (2008) A randomized controlled trial of D-cycloserine enhancement of exposure therapy for social anxiety disorder. Biol Psychiatry 63(6):544–549, http://dx.doi.org/10.1016/j.biopsych.2007.11.011

Guastella AJ, Howard AL, Dadds MR, Mitchell P, Carson DS (2009) A randomized controlled trial of intranasal oxytocin as an adjunct to exposure therapy for social anxiety disorder. Psychoneuroendocrinology 34(6):917–923, http://dx.doi.org/10.1016/j.psyneuen.2009.01.005

Guastella AJ, Einfeld SL, Gray KM, Rinehart NJ, Tonge BJ, Lambert TJ, Hickie IB (2010) Intranasal oxytocin improves emotion recognition for youth with autism spectrum disorders. Biol Psychiatry 67(7):692–694. doi:10.1016/j.biopsych.2009.09.020

Guastella AJ, Hickie IB, McGuinness MM, Otis M, Woods EA, Disinger HM et al (2013) Recommendations for the standardisation of oxytocin nasal administration and guidelines for its reporting in human research. Psychoneuroendocrinology 38(5):612–625

Gunderson JG, Lyons-Ruth K (2008) BPD's interpersonal hypersensitivity phenotype: a gene-environment-developmental model. J Pers Disord 22(1):22

Hare RD, Neumann CS (2006) The PCL-R assessment of psychopathy. In: Patrick CJ (ed) Handbook of psychopathy. Guilford, New York, pp 58–88

Haxby JV, Hoffman EA, Gobbini MI (2000) The distributed human neural system for face perception. Trends Cogn Sci 4(6):223–233

Heinrichs M, Baumgartner T, Kirschbaum C, Ehlert U (2003) Social support and oxytocin interact to suppress cortisol and subjective responses to psychosocial stress. Biol Psychiatry 54 (12):1389–1398

Herpertz SC, Bertsch K (2014) The social-cognitive basis of personality disorders. Curr Opin Psychiatry 27(1):73–77

Hezel DM, McNally RJ (2014) Theory of mind impairments in social anxiety disorder. Behav Ther 45(4):530–540, http://dx.doi.org/10.1016/j.beth.2014.02.010c

Higashida H, Lopatina O, Yoshihara T, Pichugina YA, Soumarokov AA, Munesue T et al (2010) Oxytocin signal and social behaviour: comparison among adult and infant oxytocin, oxytocin receptor and CD38 gene knockout mice. J Neuroendocrinol 22(5):373–379. doi:10.1111/j.1365-2826.2010.01976.x

Higashida H, Yokoyama S, Munesue T, Kikuchi M, Minabe Y, Lopatina O (2011) CD38 gene knockout juvenile mice: a model of oxytocin signal defects in autism. Biol Pharm Bull 34 (9):1369–1372

Hill EL, Frith U (2003) Understanding autism: insights from mind and brain. Philos Trans R Soc Lond B Biol Sci 358(1430):281–289. doi:10.1098/rstb.2002.1209

Hinshaw SP, Henker B, Whalen CK (1984a) Cognitive-behavioral and pharmacologic interventions for hyperactive boys: comparative and combined effects. J Consult Clin Psychol 52(5):739

Hinshaw SP, Henker B, Whalen CK (1984b) Self-control in hyperactive boys in anger-inducing situations: effects of cognitive-behavioral training and of methylphenidate. J Abnorm Child Psychol 12(1):55–77

Hirsch CR, Clark DM (2004) Information-processing bias in social phobia. Clin Psychol Rev 24 (7):799–825

Hofmann SG, Pollack MH, Otto MW (2006) Augmentation treatment of psychotherapy for anxiety disorders with D-cycloserine. CNS Drug Rev 12(3–4):208–217. doi:10.1111/j.1527-3458.2006.00208.x

Hofmann SG, Smits JA, Rosenfield D, Simon N, Otto MW, Meuret AE et al (2013) D-Cycloserine as an augmentation strategy with cognitive-behavioral therapy for social anxiety disorder. Am J Psychiatry 170(7):751–758. doi:10.1176/appi.ajp.2013.12070974

Hoge EA, Pollack MH, Kaufman RE, Zak PJ, Simon NM (2008) Oxytocin levels in social anxiety disorder. CNS Neurosci Ther 14(3):165–170

Holdiness MR (1985) Cerebrospinal fluid pharmacokinetics of the antituberculosis drugs. Clin Pharmacokinet 10:532–534

Hollander E, Bartz J, Chaplin W, Phillips A, Sumner J, Soorya L et al (2007) Oxytocin increases retention of social cognition in autism. Biol Psychiatry 61(4):498–503. doi:10.1016/j.biopsych.2006.05.030

Huber D, Veinante P, Stoop R (2005) Vasopressin and oxytocin excite distinct neuronal populations in the central amygdala. Science 308(5719):245–248

Hurlemann R, Patin A, Onur OA, Cohen MX, Baumgartner T, Metzler S et al (2010) Oxytocin enhances amygdala-dependent, socially reinforced learning and emotional empathy in humans. J Neurosci 30(14):4999–5007. doi:10.1523/jneurosci.5538-09.2010

Hysek CM, Domes G, Liechti ME (2012) MDMA enhances “mind reading” of positive emotions and impairs “mind reading” of negative emotions. Psychopharmacology (Berl) 222 (2):293–302. doi:10.1007/s00213-012-2645-9

Jansen LM, Gispen-de Wied CC, Wiegant VM, Westenberg HG, Lahuis BE, Van Engeland H (2006) Autonomic and neuroendocrine responses to a psychosocial stressor in adults with autistic spectrum disorder. J Autism Dev Disord 36(7):891–899

Johnson J, Ascher P (1987) Glycine potentiates the NMDA response in cultured mouse brain neurons. Nature 325(6104):529–531

Jones P, Robinson I (1982) Differential clearance of neurophysin and neurohypophysial peptides from the cerebrospinal fluid in conscious guinea pigs. Neuroendocrinology 34(4):297–302

Kagerbauer S, Martin J, Schuster T, Blobner M, Kochs E, Landgraf R (2013) Plasma oxytocin and vasopressin do not predict neuropeptide concentrations in human cerebrospinal fluid. J Neuroendocrinol 25(7):668–673

Kelley AM, Webb CM, Athy JR, Ley S, Gaydos S (2012) Cognition enhancement by modafinil: a meta-analysis. Aviat Space Environ Med 83(7):685–690

Kendrick KM (2004) The neurobiology of social bonds. J Neuroendocrinol 16(12):1007–1008

Kendrick K, Keverne E, Hinton M, Goode J (1991) Cerebrospinal fluid and plasma concentrations of oxytocin and vasopressin during parturition and vaginocervical stimulation in the sheep. Brain Res Bull 26(5):803–807

Kendrick KM, Da Costa AP, Broad KD, Ohkura S, Guevara R, Lévy F, Keverne EB (1997) Neural control of maternal behaviour and olfactory recognition of offspring. Brain Res Bull 44(4):383–395

Kessler RC, Sonnega A, Bromet E, Hughes M, Nelson CB (1995) Posttraumatic stress disorder in the National Comorbidity Survey. Arch Gen Psychiatry 52(12):1048–1060

Kessler RC, Berglund P, Demler O, Jin R, Merikangas KR, Walters EE (2005) Lifetime prevalence and age-of-onset distributions of DSM-IV disorders in the National Comorbidity Survey Replication. Arch Gen Psychiatry 62(6):593–602

Kiehl KA, Smith AM, Hare RD, Mendrek A, Forster BB, Brink J, Liddle PF (2001) Limbic abnormalities in affective processing by criminal psychopaths as revealed by functional magnetic resonance imaging. Biol Psychiatry 50(9):677–684

Kirkpatrick MG, Francis SM, Lee R, de Wit H, Jacob S (2014) Plasma oxytocin concentrations following MDMA or intranasal oxytocin in humans. Psychoneuroendocrinology 46:23–31, http://dx.doi.org/10.1016/j.psyneuen.2014.04.006

Kleckner NW, Dingledine R (1988) Requirement for glycine in activation of NMDA-receptors expressed in Xenopus oocytes. Science 241(4867):835–837

Knobloch HS, Charlet A, Hoffmann LC, Eliava M, Khrulev S, Cetin AH et al (2012) Evoked axonal oxytocin release in the central amygdala attenuates fear response. Neuron 73 (3):553–566, http://dx.doi.org/10.1016/j.neuron.2011.11.030

Korzekwa MI, Dell PF, Links PS, Thabane L, Webb SP (2008) Estimating the prevalence of borderline personality disorder in psychiatric outpatients using a two-phase procedure. Compr Psychiatry 49(4):380–386

Kosfeld M, Heinrichs M, Zak PJ, Fischbacher U, Fehr E (2005) Oxytocin increases trust in humans. Nature 435(7042):673–676. doi:10.1038/nature03701

Kraepelin E (1971) Dementia praecox and paraphrenia. Krieger, Huntington, NY

Kuypers KP, de la Torre R, Farre M, Yubero-Lahoz S, Dziobek I, Van den Bos W, Ramaekers JG (2014) No evidence that MDMA-induced enhancement of emotional empathy is related to peripheral oxytocin levels or 5-HT1a receptor activation. PLoS One 9(6):e100719

Labuschagne I, Phan KL, Wood A, Angstadt M, Chua P, Heinrichs M et al (2010) Oxytocin attenuates amygdala reactivity to fear in generalized social anxiety disorder. Neuropsychopharmacology 35(12):2403–2413

Labuschagne I, Phan KL, Wood A, Angstadt M, Chua P, Heinrichs M et al (2012) Medial frontal hyperactivity to sad faces in generalized social anxiety disorder and modulation by oxytocin. Int J Neuropsychopharmacol 15(07):883–896. doi:10.1017/S1461145711001489

Lapierre D, Braun CM, Hodgins S (1995) Ventral frontal deficits in psychopathy: neuropsychological test findings. Neuropsychologia 33(2):139–151

Leake RD, Weitzman RE, Fisher DA (1980) Pharmacokinetics of oxytocin in the human subject. Obstet Gynecol 56(6):701–704

Lee Y-S, Silva AJ (2009) The molecular and cellular biology of enhanced cognition. Nat Rev Neurosci 10(2):126–140

Lee HJ, Choi J-S, Brown TH, Kim JJ (2001) Amygdalar NMDA receptors are critical for the expression of multiple conditioned fear responses. J Neurosci 21(11):4116–4124

Lee R, Garcia F, van de Kar LD, Hauger RD, Coccaro EF (2003) Plasma oxytocin in response to pharmaco-challenge to D-fenfluramine and placebo in healthy men. Psychiatry Res 118 (2):129–136

Lerner H, St. Peter S (1984) The Roschach H response and object relations. J Pers Assess 48 (4):345–350

Li F, Tsien JZ (2009) Memory and the NMDA receptors. N Engl J Med 361(3):302

Li XF, Phillips R, LeDoux JE (1995) NMDA and non-NMDA receptors contribute to synaptic transmission between the medial geniculate body and the lateral nucleus of the amygdala. Exp Brain Res 105(1):87–100

Liechti ME, Vollenweider FX (2001) Which neuroreceptors mediate the subjective effects of MDMA in humans? A summary of mechanistic studies. Hum Psychopharmacol 16(8):589–598. doi:10.1002/hup.348

Lim MM, Young LJ (2006) Neuropeptidergic regulation of affiliative behavior and social bonding in animals. Horm Behav 50(4):506–517

Lim MM, Murphy AZ, Young LJ (2004) Ventral striatopallidal oxytocin and vasopressin V1a receptors in the monogamous prairie vole (Microtus ochrogaster). J Comp Neurol 468 (4):555–570

Lin I-F, Kashino M, Ohta H, Yamada T, Tani M, Watanabe H et al (2014) The effect of intranasal oxytocin versus placebo treatment on the autonomic responses to human sounds in autism: a single-blind, randomized, placebo-controlled, crossover design study. Mol Autism 5(1):20

Linehan M (1995) Understanding borderline personality disorder. Guildford, New York

Litz BT, Salters-Pedneault K, Steenkamp MM, Hermos JA, Bryant RA, Otto MW, Hofmann SG (2012) A randomized placebo-controlled trial of D-cycloserine and exposure therapy for posttraumatic stress disorder. J Psychiatr Res 46(9):1184–1190

Losh M, Adolphs R, Poe MD, Couture S, Penn D, Baranek GT, Piven J (2009) Neuropsychological profile of autism and the broad autism phenotype. Arch Gen Psychiatry 66(5):518–526

Loup F, Tribollet E, Dubois-Dauphin M, Dreifuss J (1991) Localization of high-affinity binding sites for oxytocin and vasopressin in the human brain. An autoradiographic study. Brain Res 555(2):220–232

Lucas-Thompson RG, Holman EA (2013) Environmental stress, oxytocin receptor gene (OXTR) polymorphism, and mental health following collective stress. Horm Behav 63(4):615–624, http://dx.doi.org/10.1016/j.yhbeh.2013.02.015

Ludwig M, Leng G (2006) Dendritic peptide release and peptide-dependent behaviours. Nat Rev Neurosci 7(2):126–136

Lynam DR, Caspi A, Moffitt TE, Loeber R, Stouthamer-Loeber M (2007) Longitudinal evidence that psychopathy scores in early adolescence predict adult psychopathy. J Abnorm Psychol 116 (1):155–165. doi:10.1037/0021-843x.116.1.155

Maoz H, Tsviban L, Gvirts HZ, Shamay-Tsoory SG, Levkovitz Y, Watemberg N, Bloch Y (2014) Stimulants improve theory of mind in children with attention deficit/hyperactivity disorder. J Psychopharmacol 28(3):212–219. doi:10.1177/0269881113492030

Marsh AA, Blair RJ (2008) Deficits in facial affect recognition among antisocial populations: a meta-analysis. Neurosci Biobehav Rev 32(3):454–465. doi:10.1016/j.neubiorev.2007.08.003

Marsh AA, Kozak MN, Ambady N (2007) Accurate identification of fear facial expressions predicts prosocial behavior. Emotion 7(2):239

Martin J, Kagerbauer SM, Schuster T, Blobner M, Kochs EF, Landgraf R (2014) Vasopressin and oxytocin in CSF and plasma of patients with aneurysmal subarachnoid haemorrhage. Neuropeptides 48(2):91–96

Mas M, Farré M, de la Torre R, Roset PN, Ortuño J, Segura J, Camí J (1999) Cardiovascular and neuroendocrine effects and pharmacokinetics of 3,4-methylenedioxymethamphetamine in humans. J Pharmacol Exp Ther 290(1):136–145

McAllister K (1994) D-cycloserine enhances social behaviour in individually-housed mice in the resident-intruder test. Psychopharmacology (Berl) 116(3):317–325

McClure EB, Nowicki S Jr (2001) Associations between social anxiety and nonverbal processing skill in preadolescent boys and girls. J Nonverbal Behav 25(1):3–19

Meffert H, Gazzola V, den Boer JA, Bartels AAJ, Keysers C (2013) Reduced spontaneous but relatively normal deliberate vicarious representations in psychopathy. Brain 136 (8):2550–2562. doi:10.1093/brain/awt190

Meisenberg G, Simmons WH (1983) Peptides and the blood-brain barrier. Life Sci 32 (23):2611–2623, http://dx.doi.org/10.1016/0024-3205(83)90352-1

Mens WB, Witter A, Van Wimersma Greidanus TB (1983) Penetration of neurohypophyseal hormones from plasma into cerebrospinal fluid (CSF): half-times of disappearance of these neuropeptides from CSF. Brain Res 262(1):143–149

Meyer C, Freund-Mercier M, Guerne Y, Richard P (1987) Relationship between oxytocin release and amplitude of oxytocin cell neurosecretory bursts during suckling in the rat. J Endocrinol 114(2):263–270

Meyer-Lindenberg A (2008) Impact of prosocial neuropeptides on human brain function. In: Inga DN, Rainer L (eds) Progress in brain research, vol 170. Elsevier, New York, pp 463–470

Miller GM (2011) The emerging role of trace amine-associated receptor 1 in the functional regulation of monoamine transporters and dopaminergic activity. J Neurochem 116 (2):164–176

Mithoefer MC, Wagner MT, Mithoefer AT, Jerome L, Doblin R (2011) The safety and efficacy of ±3, 4-methylenedioxymethamphetamine-assisted psychotherapy in subjects with chronic, treatment-resistant posttraumatic stress disorder: the first randomized controlled pilot study. J Psychopharmacol 25(4):439–452

Mithoefer MC, Wagner MT, Mithoefer AT, Jerome L, Martin SF, Yazar-Klosinski B et al (2013) Durability of improvement in posttraumatic stress disorder symptoms and absence of harmful effects or drug dependency after 3, 4-methylenedioxymethamphetamine-assisted psychotherapy: a prospective long-term follow-up study. J Psychopharmacol 27(1):28–39. doi:10.1177/0269881112456611

Modahl C, Green LA, Fein D, Morris M, Waterhouse L, Feinstein C, Levin H (1998) Plasma oxytocin levels in autistic children. Biol Psychiatry 43(4):270–277

Mogg K, Philippot P, Bradley BP (2004) Selective attention to angry faces in clinical social phobia. J Abnorm Psychol 113(1):160

Mottolese R, Redouté J, Costes N, Le Bars D, Sirigu A (2014) Switching brain serotonin with oxytocin. Proc Natl Acad Sci U S A 111(23):8637–8642. doi:10.1073/pnas.1319810111

Moul C, Killcross S, Dadds MR (2012) A model of differential amygdala activation in psychopathy. Psychol Rev 119(4):789–806. doi:10.1037/a0029342

Mueser KT, Yarnold P, Bellack A (1992) Diagnostic and demographic correlates of substance abuse in schizophrenia and major affective disorder. Acta Psychiatrica Scandinavica 85 (1):48–55

Murphy DA, Pelham WE, Lang AR (1992) Aggression in boys with attention deficit-hyperactivity disorder: methylphenidate effects on naturalistically observed aggression, response to provocation, and social information processing. J Abnorm Child Psychol 20(5):451–466

Nair KGS, Epstein IG, Baron H, Mulinos MG (1956) Absorption, distribution and excretion of cycloserine in man. Antibiot Annu 3:136–140

Najjar S, Pearlman DM (2015) Neuroinflammation and white matter pathology in schizophrenia: systematic review. Schizophr Res 161(1):102–112

Oehen P, Traber R, Widmer V, Schnyder U (2013) A randomized, controlled pilot study of MDMA (±3, 4-Methylenedioxymethamphetamine)-assisted psychotherapy for treatment of resistant, chronic Post-Traumatic Stress Disorder (PTSD). J Psychopharmacol 27(1):40–52

Onur OA, Schlaepfer TE, Kukolja J, Bauer A, Jeung H, Patin A et al (2010) The N-methyl-D-aspartate receptor co-agonist D-cycloserine facilitates declarative learning and hippocampal activity in humans. Biol Psychiatry 67(12):1205–1211, http://dx.doi.org/10.1016/j.biopsych.2010.01.022

Parrott AC (2013) Human psychobiology of MDMA or 'Ecstasy': an overview of 25 years of empirical research. Hum Psychopharmacol 28(4):289–307. doi:10.1002/hup.2318

Pedersen CA, Gibson CM, Rau SW, Salimi K, Smedley KL, Casey RL et al (2011) Intranasal oxytocin reduces psychotic symptoms and improves Theory of Mind and social perception in schizophrenia. Schizophr Res 132(1):50–53

Pelham WE, Sturges J, Hoza J, Schmidt C, Bijlsma JJ, Milich R, Moorer S (1987) Sustained release and standard methylphenidate effects on cognitive and social behavior in children with attention deficit disorder. Pediatrics 80(4):491–501

Perälä J, Suvisaari J, Saarni SI, Kuoppasalmi K, Isometsä E, Pirkola S et al (2007) Lifetime prevalence of psychotic and bipolar I disorders in a general population. Arch Gen Psychiatry 64(1):19–28

Pierce K, Redcay E (2008) Fusiform function in children with an autism spectrum disorder is a matter of "who". Biol Psychiatry 64(7):552–560

Pierce K, Müller R-A, Ambrose J, Allen G, Courchesne E (2001) Face processing occurs outside the fusiform 'face area' in autism: evidence from functional MRI. Brain 124(10):2059–2073

Pitman RK, Orr SP, Lasko NB (1993) Effects of intranasal vasopressin and oxytocin on physiologic responding during personal combat imagery in Vietnam veterans with posttraumatic stress disorder. Psychiatry Res 48(2):107–117, http://dx.doi.org/10.1016/0165-1781(93)90035-F

Posey DJ, Kem DL, Swiezy NB, Sweeten TL, Wiegand RE, McDougle CJ (2004) A pilot study of D-cycloserine in subjects with autistic disorder. Am J Psychiatry 161(11):2115–2117

Preckel K, Scheele D, Kendrick KM, Maier W, Hurlemann R (2014) Oxytocin facilitates social approach behavior in women. Front Behav Neurosci 8:191. doi:10.3389/fnbeh.2014.00191

Randall DC, Shneerson JM, Plaha KK, File SE (2003) Modafinil affects mood, but not cognitive function, in healthy young volunteers. Hum Psychopharmacol 18(3):163–173

Rapee RM, Abbott MJ (2006) Mental representation of observable attributes in people with social phobia. J Behav Ther Exp Psychiatry 37(2):113–126

Repantis D, Schlattmann P, Laisney O, Heuser I (2010) Modafinil and methylphenidate for neuroenhancement in healthy individuals: a systematic review. Pharmacol Res 62(3):187–206

Rodrigues H, Figueira I, Lopes A, Gonçalves R, Mendlowicz MV, Coutinho ESF, Ventura P (2014) Does D-cycloserine enhance exposure therapy for anxiety disorders in humans? A meta-analysis. PLoS One 9(7):e93519

Roepke S, Vater A, Preißler S, Heekeren HR, Dziobek I (2013) Social cognition in borderline personality disorder. Front Neurosci 6:195. doi:10.3389/fnins.2012.00195

Rothbaum BO, Price M, Jovanovic T, Norrholm SD, Gerardi M, Dunlop B et al (2014) A Randomized, double-blind evaluation of d-cycloserine or alprazolam combined with virtual reality exposure therapy for posttraumatic stress disorder in Iraq and Afghanistan war veterans. Am J Psychiatry 171(6):640–648

Roy M-A, Maziade M, Labbé A, Mérette C (2001) Male gender is associated with deficit schizophrenia: a meta-analysis. Schizophr Res 47(2):141–147

Rubin LH, Carter CS, Drogos L, Pournajafi-Nazarloo H, Sweeney JA, Maki PM (2010) Peripheral oxytocin is associated with reduced symptom severity in schizophrenia. Schizophr Res 124(1–3):13–21. doi:10.1016/j.schres.2010.09.014

Rydén G, Sjöholm I (1969) Half-life of oxytocin in blood of pregnant and non-pregnant women. Acta Endocrinologica 61(3):425–431. doi:10.1530/acta.0.0610425

Saavedra-Velez C, Yusim A, Anbarasan D, Lindenmayer J-P (2009) Modafinil as an adjunctive treatment of sedation, negative symptoms, and cognition in schizophrenia: a critical review. J Clin Psychiatry 70(1):104–112

Šagud M, Mihaljević-Peleš A, Mück-Šeler D, Pivac N, Vuksan-Ćusa B, Brataljenović T, Jakovljević M (2009) Smoking and schizophrenia. Psychiatria Danubina 21(3):371–375

Saha S, Chant D, McGrath J (2007) A systematic review of mortality in schizophrenia: is the differential mortality gap worsening over time? Arch Gen Psychiatry 64(10):1123–1131

Scheele D, Striepens N, Gunturkun O, Deutschlander S, Maier W, Kendrick KM, Hurlemann R (2012) Oxytocin modulates social distance between males and females. J Neurosci 32 (46):16074–16079. doi:10.1523/jneurosci. 2755-12.2012

Scheele D, Wille A, Kendrick KM, Stoffel-Wagner B, Becker B, Güntürkün O et al (2013) Oxytocin enhances brain reward system responses in men viewing the face of their female partner. Proc Natl Acad Sci U S A 110(50):20308–20313

Scoriels L, Barnett JH, Murray GK, Cherukuru S, Fielding M, Cheng F et al (2011) Effects of modafinil on emotional processing in first episode psychosis. Biol Psychiatry 69(5):457–464

Shahrestani S, Kemp AH, Guastella AJ (2013) The impact of a single administration of intranasal oxytocin on the recognition of basic emotions in humans: a meta-analysis. Neuropsychopharmacology 38(10):1929–1936. doi:10.1038/npp.2013.86

Shamay-Tsoory SG, Fischer M, Dvash J, Harari H, Perach-Bloom N, Levkovitz Y (2009) Intranasal administration of oxytocin increases envy and schadenfreude (gloating). Biol Psychiatry 66(9):864–870. doi:10.1016/j.biopsych.2009.06.009

Shao Y, Wolpert CM, Raiford KL, Menold MM, Donnelly SL, Ravan SA et al (2002) Genomic screen and follow-up analysis for autistic disorder. Am J Med Genet 114(1):99–105

Siegal M, Varley R (2002) Neural systems involved in 'theory of mind'. Nat Rev Neurosci 3(6):463–471

Simeon D, Bartz J, Hamilton H, Crystal S, Braun A, Ketay S, Hollander E (2011) Oxytocin administration attenuates stress reactivity in borderline personality disorder: a pilot study. Psychoneuroendocrinology 36(9):1418–1421. doi:10.1016/j.psyneuen.2011.03.013

Simmons CF Jr, Clancy TE, Quan R, Knoll JH (1995) The oxytocin receptor gene (OXTR) localizes to human chromosome 3p25 by fluorescence *in situ* hybridization and PCR analysis of somatic cell hybrids. Genomics 26(3):623–625

Smits JA, Hofmann SG, Rosenfield D, DeBoer LB, Costa PT, Simon NM et al (2013a) D-cycloserine augmentation of cognitive behavioral group therapy of social anxiety disorder: prognostic and prescriptive variables. J Consult Clin Psychol 81(6):1100–1112. doi:10.1037/a0034120

Smits JA, Rosenfield D, Otto MW, Marques L, Davis ML, Meuret AE et al (2013b) d-cycloserine enhancement of exposure therapy for social anxiety disorder depends on the success of exposure sessions. J Psychiatr Res 47(10):1455–1461, http://dx.doi.org/10.1016/j.jpsychires.2013.06.020

Staebler K, Renneberg B, Stopsack M, Fiedler P, Weiler M, Roepke S (2011) Facial emotional expression in reaction to social exclusion in borderline personality disorder. Psychol Med 41 (09):1929–1938

Sterzer P, Stadler C, Krebs A, Kleinschmidt A, Poustka F (2005) Abnormal neural responses to emotional visual stimuli in adolescents with conduct disorder. Biol Psychiatry 57(1):7–15. doi:10.1016/j.biopsych.2004.10.008

Stiglmayr C, Grathwol T, Linehan M, Ihorst G, Fahrenberg J, Bohus M (2005) Aversive tension in patients with borderline personality disorder: a computer-based controlled field study. Acta Psychiatr Scand 111(5):372–379

Stone VE, Baron-Cohen S, Calder A, Keane J, Young A (2003) Acquired theory of mind impairments in individuals with bilateral amygdala lesions. Neuropsychologia 41(2):209–220

Stopa L, Clark DM (2000) Social phobia and interpretation of social events. Behav Res Ther 38(3):273–283

Striepens D, Scheele D, Kendrick KM, Becker B, Schäfer L, Schwalba K et al. (2012) Oxytocin facilitates protective responses to aversive social stimuli in males. Proc Natl Acad Sci U S A 109(44):18144–18149.

Striepens N, Kendrick KM, Hanking V, Landgraf R, Wüllner U, Maier W, Hurlemann R (2013) Elevated cerebrospinal fluid and blood concentrations of oxytocin following its intranasal administration in humans. Sci Rep 3:3440

Stuart J, Western D, Lohr N, Benjamin J, Becker S, Vorus N, Silk K (1990) Object relations in borderlines, depressives, and normals: an examination of human responses on the Rorschach. J Pers Assess 55(1–2):296–318

Tancer M, Johanson C-E (2003) Reinforcing, subjective, and physiological effects of MDMA in humans: a comparison with d-amphetamine and mCPP. Drug Alcohol Depend 72(1):33–44, http://dx.doi.org/10.1016/S0376-8716(03)00172-8

Thompson MR, Callaghan PD, Hunt GE, Cornish JL, McGregor IS (2007) A role for oxytocin and 5-HT(1A) receptors in the prosocial effects of 3,4 methylenedioxymethamphetamine ("ecstasy"). Neuroscience 146(2):509–514. doi:10.1016/j.neuroscience.2007.02.032

Trull TJ, Jahng S, Tomko RL, Wood PK, Sher KJ (2010) Revised NESARC personality disorder diagnoses: gender, prevalence, and comorbidity with substance dependence disorders. J Pers Disord 24(4):412

Tudusciuc O, Adolphs R (2013) Social cognitive neuroscience: clinical foundations. In: Roberts DL, Penn DL (eds) Social cognition in schizophrenia. Oxford University Press, New York

Turner DC, Clark L, Pomarol-Clotet E, McKenna P, Robbins TW, Sahakian BJ (2004) Modafinil improves cognition and attentional set shifting in patients with chronic schizophrenia. Neuropsychopharmacology 29(7):1363–1373

Van IJzendoorn MH, Bakermans-Kranenburg MJ (2012) A sniff of trust: meta-analysis of the effects of intranasal oxytocin administration on face recognition, trust to in-group, and trust to out-group. Psychoneuroendocrinology 37(3):438–443. doi:10.1016/j.psyneuen.2011.07.008

van Os J, Kapur S (2009) Schizophrenia. Lancet 374(9690):635–645, http://dx.doi.org/10.1016/S0140-6736(09)60995-8

Veit R, Konicar L, Klinzing JG, Barth B, Yilmaz O, Birbaumer N (2013) Deficient fear conditioning in psychopathy as a function of interpersonal and affective disturbances. Front Hum Neurosci 7:706. doi:10.3389/fnhum.2013.00706

Volkow ND, Fowler JS, Logan J, Alexoff D, Zhu W, Telang F et al (2009) Effects of modafinil on dopamine and dopamine transporters in the male human brain: clinical implications. JAMA 301(11):1148–1154

Voncken MJ, Bögels SM, de Vries K (2003) Interpretation and judgmental biases in social phobia. Behav Res Ther 41(12):1481–1488

Walker DL, Davis M (2002) The role of amygdala glutamate receptors in fear learning, fear-potentiated startle, and extinction. Pharmacol Biochem Behav 71(3):379–392

Walss-Bass C, Fernandes JM, Roberts DL, Service H, Velligan D (2013) Differential correlations between plasma oxytocin and social cognitive capacity and bias in schizophrenia. Schizophr Res 147(2-3):387–392. doi:10.1016/j.schres.2013.04.003

Watanabe T, Abe O, Kuwabara H et al (2014) Mitigation of sociocommunicational deficits of autism through oxytocin-induced recovery of medial prefrontal activity: a randomized trial. JAMA Psychiatry 71(2):166–175. doi:10.1001/jamapsychiatry.2013.3181

Wermter A-K, Kamp-Becker I, Hesse P, Schulte-Körne G, Strauch K, Remschmidt H (2010) Evidence for the involvement of genetic variation in the oxytocin receptor gene (OXTR) in the etiology of autistic disorders on high-functioning level. Am J Med Genet B Neuropsychiatr Genet 153B(2):629–639. doi:10.1002/ajmg.b.31032

Williams LM, Hermens DF, Palmer D, Kohn M, Clarke S, Keage H et al (2008) Misinterpreting emotional expressions in attention-deficit/hyperactivity disorder: evidence for a neural marker and stimulant effects. Biol Psychiatry 63(10):917–926, http://dx.doi.org/10.1016/j.biopsych.2007.11.022

Wittkampf LC, Arends J, Timmerman L, Lancel M (2012) A review of modafinil and armodafinil as add-on therapy in antipsychotic-treated patients with schizophrenia. Ther Adv Psychopharmacol 2(3):115–125

Woolley J, Chuang B, Lam O, Lai W, O'Donovan A, Rankin K, Mathalon D (2014) Oxytocin administration enhances controlled social cognition in patients with schizophrenia. Psychoneuroendocrinology 47:116–125

Wu S, Jia M, Ruan Y, Liu J, Guo Y, Shuang M et al (2005) Positive association of the oxytocin receptor gene (OXTR) with autism in the Chinese Han population. Biol Psychiatry 58(1):74–77. doi:10.1016/j.biopsych.2005.03.013

Zhang Y, Lin L, Lin C-P, Zhou Y, Chou K-H, Lo C-Y et al (2012) Abnormal topological organization of structural brain networks in schizophrenia. Schizophr Res 141(2):109–118

Part III

Developmental Disorders, Alternative Approaches, and Emerging Technologies

Introduction

The final goal of this volume is to provide perspectives on multiple special topics ranging from therapeutic drug use to enhance cognition in children, emerging technologies, pharmacological disruption of maladaptive memory, and nonpharmacological approaches to cognitive enhancement. Although the range of special topics is not all-inclusive, they serve to ignite awareness of special populations that may benefit from cognitive enhancement or, in some cases, disruption and to inspire scientists to embrace innovative technologies and approaches in their quests for cognitive enhancement in neuropsychiatric and neurological disorders.

Chapter 11 (Vahabzadeh, Landino, Finger, Carlezon, and McDougle) details much of the relevant progress in the field of autism research. Autism spectrum disorder has come to the forefront in a major way over the past 10 years at least in part due to enhanced diagnosis and public awareness. A complex behavioral disorder, autism, is likely the result of numerous environmental and genetic factors—some, albeit few, are known. Of the many attributes associated with patients with autism, foremost is a deficit in social cognition. Social cognition can be enhanced by oxytocin and modulated by glutamatergic tone, and both neural systems have been studied preclinically and clinically with select pharmacological tools. Another field of focus is neuroinflammation as it may be a contributing factor to the etiology of autism. Together, learning more about the networks that control communication will be critical to understanding autism and developing potential medicines for this disorder, particularly in the area of social cognition.

Chapter 12 (Fernandez and Reeves) discusses many of the recent advances in Down syndrome research. It is quite clear that the clinical phenotypes, although overlapping somewhat, are quite variable. This is due in part to the diverse number of brain regions affected in the disorder, resulting in divergent effects on various cognitive domains in individuals with Down syndrome. The effects of the triplication of human chromosome 21 are certainly not limited to those in the central nervous system. These people often present with disorders such as hypothyroidism and congenital heart disease, and are predisposed to early aging in general. Thus, multiple biochemical and physiologic networks are altered. The expression of the various phenotypes is likely related to dysregulation of GABAergic networks, apolipoprotein E, or the processing of amyloid peptides. All these lead to certain difficulties for the conduct of clinical research studies, and thus, clinical trial design for studies with people having Down syndrome requires unique attention.

Chapter 13 (Taylor and Torregrossa) outlines the disorders of maladaptive memory and how a range of pharmaceutical agents can weaken the strength of these memories if used in conjunction with techniques that target memory reconsolidation. Disorders range from post-traumatic stress disorder to disorders of addiction, schizophrenia, and mood. Several categories of drugs have been investigated since the pioneering work showing that inhibition of protein synthesis at the time of memory reactivation can disrupt its reconsolidation and weaken its strength. The most well-studied drugs in both animal and human subjects include beta-adrenergic receptor antagonists and glucocorticoid antagonists. Preclinical research is currently discovering the relevant intracellular signaling pathways involved in memory reconsolidation, potentially giving rise to new and improved targets for treating disorders of maladaptive memory (e.g., inhibitors of PKA, ERK, mTOR, GSK3, NFkB, and histone acetylation). Parallel to the discovery of new compounds is the need to unveil the optimal reactivation conditions for reconsolidation manipulations of newer as well as older (remote) memories.

Chapter 14 (Kelly) describes the use and benefits of incorporating nonpharmacological cognitive enhancers that involve lifestyle interventions. In particular, aerobic exercise and environmental enrichment have strong empirical support, with proven efficacy in humans and in animal models. For example, in healthy humans, aerobic exercise can improve learning, response inhibition, and working memory. In the elderly, exercise has neuroprotective effects and reduces the incidence of cognitive impairment and dementia. People with depression benefit as well. Most research with environmental enrichment has been conducted in animals, and though it naturally varies to a degree in how it is implemented in people vs. animals, most investigators agree that cognitive stimulation and social stimulation offer an enriching experience in childhood through adulthood. Computerized training programs are a relatively recently developed source of targeted cognitive stimulation. Positive effects have been reported in the elderly and in Parkinson's disease patients. However, the persistence of enhancement and the transfer of the learned skills to real-life situations have not yet been demonstrated.

Chapter 15 (Kondabolu, Kowalski, Roberts, and Han) focuses on two emerging technologies that can impact, in a rather precise manner, brain networks. These are optogenetics and deep brain stimulation. Optogenetics is a revolutionary technology that allows one to control or modulate highly specific neural circuits. As it has been introduced only recently, it is still primarily in a research phase now, and clinical applications, although approachable, have not yet been realized. In practice for much longer than optogenetics, deep brain stimulation has shown therapeutic efficacy in disorders such as Parkinson's disease and depression. Although not specifically designed for cognition enhancement, the use of these technologies should allow the field to assess brain circuits that impact specific domains of cognition.

Chapter 16 (Kantak and Wettstein) provides closing thoughts for this volume on cognitive enhancement. The status of cognitive enhancement is first summarized to suggest the availability of therapeutics for improving attention and dementia, but that more research is needed to develop effective therapeutics for improving these and other cognitive domains. The advances that have been made in translational models of cognitive enhancement and in the neurobiology of learning and memory are major achievements from the past 20 years, and these achievements will help speed the availability of cognitive-enhancing (or disrupting) therapeutics for neuropsychiatric and neurological disorders. The availability of such therapeutics raises several ethical questions regarding cognitive enhancement or its disruption. Should cognitive enhancers be available for everyone, or should they be limited to special or medically defined populations? Who will be the gatekeepers dictating how and when cognitive enhancement is permissible? Scientific questions also persist, particularly the question of whether cognitive enhancement is better achieved via pharmacological vs. nonpharmacological means. There is a need to develop rational policies for the use of cognitive enhancers.

Neural Targets in the Study and Treatment of Social Cognition in Autism Spectrum Disorder

Arshya Vahabzadeh, Samantha M. Landino, Beate C. Finger, William A. Carlezon Jr., and Christopher J. McDougle

Contents

A. Vahabzadeh
Massachusetts General Hospital, 02114 Boston, MA, USA

McLean Hospital, 02478 Belmont, MA, USA

Harvard Medical School, 02115 Boston, MA, USA

S.M. Landino
Behavioral Genetics Laboratory, McLean Hospital, 02478 Belmont, MA, USA

B.C. Finger • W.A. Carlezon Jr.
Behavioral Genetics Laboratory, McLean Hospital, 02478 Belmont, MA, USA

Harvard Medical School, 02115 Boston, MA, USA

C.J. McDougle (✉)
Massachusetts General Hospital, 02114 Boston, MA, USA

Harvard Medical School, 02115 Boston, MA, USA

Lurie Center for Autism, One Maguire Road, 02421 Lexington, MA, USA
e-mail: cmcdougle@partners.org

K.M. Kantak, J.G. Wettstein (eds.), *Cognitive Enhancement*, Handbook of Experimental Pharmacology 228, DOI 10.1007/978-3-319-16522-6_11

Abstract

The purpose of this chapter is to present results from recent research on social cognition in autism spectrum disorder (ASD). The clinical phenomenology and neuroanatomical circuitry of ASD are first briefly described. The neuropharmacology of social cognition in animal models of ASD and humans is then addressed. Next, preclinical and clinical research on the neurohormone oxytocin is reviewed. This is followed by a presentation of results from preclinical and clinical studies on the excitatory amino acid glutamate. Finally, the role of neuroinflammation in ASD is addressed from the perspectives of preclinical neuroscience and research involving humans with ASD.

Keywords

Autism • Social cognition • Neuropharmacology • Preclinical • Oxytocin • Glutamate • Neuroinflammation

1 Introduction to Autism Spectrum Disorder

Autism spectrum disorder (ASD) is a group of childhood-onset neurodevelopmental disorders characterized by deficits in social communication and social interaction, along with restricted, repetitive patterns of behavior, interests, or activities (American Psychiatric Association 2013).

There has been growing scientific and societal interest in ASD, in part, as a result of its increasing prevalence. According to the Centers for Disease Control (CDC), one in 68 children born in the United States 8 years of age has an ASD (Baio 2012). The precise reasons behind this surge remain unknown, although enhanced awareness and diagnosis are likely to be key contributors.

The etiology of ASD has yet to be fully elucidated. A combination of genetic and environmental factors is thought to underlie its causes, with approximately 10 % of cases being attributed to a known genetic condition, such as fragile X syndrome (FXS) (Schaefer and Mendelsohn 2013). At present, there is no specific medical diagnostic test for ASD, with diagnosis being established through both clinical interview and observation.

Symptoms of ASD arise in early childhood and are grouped into two key symptom clusters, namely, impairments in social communication coupled with a pattern of repetitive behaviors and restricted interests. The first cluster, and arguably the most indicative of ASD, focuses on deficits in social communication.

Examples of these deficits include a lack of social–emotional reciprocity, an impaired ability to approach another person in a socially acceptable manner, and difficulty engaging in a back-and-forth conversation. A reduced capacity to initiate or respond to social interactions, including the ability to share emotions and interests, may also be demonstrated. ASD also manifests in limited nonverbal communication, including reduced eye contact and difficulty in interpreting body language. Individuals may lack imaginative play and may show reduced interest in engaging with peers.

The second primary symptom cluster of ASD focuses on restricted, repetitive behaviors and interests. Individuals with ASD may exhibit repetitive motor movements or types of play, for example, lining up toys in a particular way. Individuals may also have an insistence on adhering to a set daily routine, with even small deviations eliciting a disproportionate amount of distress. Examples of routines could include the need to get dressed in a particular sequence, or the insistence on having the same type of food for every meal. This second symptom cluster can also include the development of highly restricted but intense fixed interests that can affect an individual's daily functioning. Finally, individuals may also have unusual sensitivities to the sensory aspects of their environment, for example, a fascination with particular textures or an aversion to certain sounds (American Psychiatric Association 2013).

2 Social Cognitive Deficits in Autism Spectrum Disorder

Social interaction and communication are integral to every aspect of human life. ASD, however, is characterized by substantial impairments in social communicative processes. Research has aimed to identify specific brain mechanisms that may lead to this aberrant social cognition in ASD. Social cognition refers to a series of highly phylogenetically preserved processes that are used to interpret social interaction (Frith 2008). In humans, these cognitive processes develop rapidly from birth and include facial and facial affect recognition, identification of social sounds, orientation to biological motion, and the analysis of eye gaze (McCall and Singer 2012). These processes allow an individual to integrate large amounts of sensory data to produce a succinct, coherent understanding of the social world around them.

Central to understanding the social world is the ability to analyze the facial features of others. Facial expressions and eye movements reveal affective states and help predict the future behavior of others. Such information can be gleaned through establishing mutual eye gaze (eye contact), the most powerful method of communication between humans (Farroni et al. 2002). Eye gaze allows for the nonverbal transmission of important social communication and helps reveal an individual's intent or focus. Human infants develop the ability to identify familiar faces within a few weeks of birth (Haith et al. 1977), and they preferentially look at faces that engage them in mutual gaze (Farroni et al. 2002). Eye gaze also allows for the initiation or response to "joint attention." Joint attention is a process by which an

Table 1 Social cognition and associated brain regions

Social cognitive processes seen in autism spectrum disorder	Associated brain regions and processes
Impaired facial recognition	Fusiform gyrus
Impaired facial affect recognition	Superior Temporal sulcus
Aberrant eye gaze	Medial prefrontal cortex
Limited ability to detect biological motion	Amygdala
Difficulty in initiating/responding to joint attention	Aberrant neural connectivity

individual uses nonverbal cues to alert another individual to an object or event. Joint attention is an essential skill in social communication and learning.

A range of deficits in social cognition has been identified in ASD (see Table 1). Individuals with ASD have demonstrated impaired facial recognition (Klin et al. 1999; Weigelt et al. 2012), impaired facial affect recognition (Uljarevic and Hamilton 2013), reduced orientation to biological motion (Annaz et al. 2012; Klin et al. 2009), and aberrant eye gaze (Grice et al. 2005; Jones et al. 2008). Individuals with ASD have also demonstrated difficulty in responding to or initiating joint attention (Mundy et al. 2009).

Attempts have been made to identify the underlying neurobiology of these deficits in social cognition. Multiple interconnected brain regions are believed to be involved in ASD. Neuroimaging has identified abnormal functioning of several of these regions, including the fusiform gyrus, the superior temporal sulcus, the medial prefrontal cortex, and the amygdala (Dalton et al. 2005; Pelphrey et al. 2004; Lynch et al. 2013). Recent research has also demonstrated abnormal neural connectivity in ASD, although both hypoconnectivity and hyperconnectivity have been reported (Gotts et al. 2012; Just et al. 2012; Lynch et al. 2013; Supekar et al. 2013). Reduced connectivity of brain regions associated with social cognition has been identified in several studies of adolescents and adults with ASD (Gotts et al. 2012; von dem Hagen et al. 2013). Hyperconnectivity has also been demonstrated, albeit in children with ASD. In a large multisite study involving 110 children with ASD aged 7–13 years, widespread brain functional hyperconnectivity was reported, with greater hyperconnectivity correlated with the highest degree of social impairment (Supekar et al. 2013).

3 Social Cognitive Deficits in Other Neurodevelopmental Disorders

Impairments in social cognitive processes are not limited to ASD but are also present in disorders such as schizophrenia. Schizophrenia is a psychiatric disorder characterized by delusions, hallucinations, and disorganized behavior and speech, the so-called positive symptoms (American Psychiatric Association 2013). "Negative symptoms" may also be present and include diminished emotional expression and social engagement. Like ASD, studies of schizophrenia have identified difficulty in social interaction, communication, and emotional processing (Cheung et al. 2010). Further evidence of a link between these conditions has arisen as a result of their shared genetic underpinnings (Carroll and Owen 2009; Crespi et al. 2010). Social cognitive deficits transcend clinically determined diagnostic boundaries and can be identified in both disorders. Directly comparing ASD and schizophrenia can help uncover shared mechanisms underlying the social cognitive deficits seen in both conditions (Sasson et al. 2011).

4 Preclinical Studies of Social Cognition Related to Autism Spectrum Disorder

Behavioral findings from studies conducted in laboratory animals have provided a foundation for neuropharmacological models of social cognition and influenced human research (McCall and Singer 2012). However, preclinical models of ASD investigate social behavior in a broad context when compared to the human social cognitive processes described above. Rodent social behavior in ASD research is assessed using a variety of social paradigms such as social interaction, recognition, preference, approach, and learning, as well as social odor discrimination, social novelty recognition, and tests of ultrasonic vocalizations.

5 Neuropharmacology of Social Cognition in Animal Models of Autism Spectrum Disorder and Humans

5.1 Oxytocin

5.1.1 Preclinical Research

Oxytocin, known as the "bonding hormone," is a nonapeptide synthesized in the paraventricular (PVN) and supraoptic (SON) nuclei of the hypothalamus. Oxytocin exerts its primary action via the oxytocin receptor (OXTR), a G-protein-coupled receptor, regulating prosocial behaviors.

Oxytocin plays a modulatory role in a wide range of socially relevant behaviors, including pair bonding, anxiety, stress reactivity, aggression, maternal behavior, sexual behavior, social interaction behavior, and social recognition (Bielsky and Young 2004; Donaldson and Young 2008; Lim et al. 2005; Lukas and Neumann 2013). For these reasons, oxytocin is a target of high interest in disorders characterized by pathological symptoms in social cognition and behavior such as ASD. In rodents, social behavior can be assessed in an array of tests. Social communication and mother–infant bonding are measured in the maternal isolation test. In this procedure, pups at an early postnatal stage are removed from the nest and placed into a recording chamber to detect ultrasonic calls. In models mimicking ASD-like behavior, the number of pup vocalizations is reduced (Malkova et al. 2012). Reduced vocalizations at the pup stage in the maternal isolation test have been reported in lines of mutant mice with alterations in oxytocin systems, such as oxytocin-knockout mice, OXTR-knockout mice, and CD38 (an enzyme involved in oxytocin secretion)-knockout mice (Ferguson et al. 2000, 2001; Higashida et al. 2011, 2012; Liu et al. 2008; Takayanagi et al. 2005; Winslow and Insel 2002; Winslow et al. 2000), suggesting communication deficits at pup stage in all three mouse lines.

During adulthood, social cognition can be measured in rodents employing social preference and social recognition tasks. The social preference test measures preference for a social target (conspecific) over a nonsocial target (inanimate object), whereas the social recognition task assesses memory for a previously investigated versus a novel social target.

Oxytocin-, OXTR-, and CD38-knockout mice all exhibit deficits in social preference, as well as in social recognition, which can be reversed by acute intracerebroventricular administration of oxytocin (Ferguson et al. 2001; Jin et al. 2007; Sala et al. 2011).

Furthermore, there are inbred mouse lines with low sociability, such as the BALB/cByJ and C58/J strains, which may be useful in modeling the social deficits seen in ASD. Acute peripheral administration of oxytocin does not improve social behavior, whereas subacute dosing of oxytocin increases social behavior in the social preference task in both strains (Teng et al. 2013), indicating a modulatory role of the oxytocin system in social preference behavior in these mouse strains of low sociability. Although several causative genetic models for ASD, such as the FMR1- and MeCP2-mutant lines, show deficits in social cognition (Ey et al. 2011), there is a lack of studies addressing the direct role of oxytocin in these phenotypes.

While acute administration of oxytocin tends to yield prosocial effects, recent studies have described opposing effects following chronic administration of oxytocin. Repeated intranasal administration of oxytocin in C57BL/6 mice reduced social interaction behavior and ultrasonic vocalizations (Huang et al. 2014). Similar effects of chronic administration of oxytocin were described in the prairie vole, where chronic intranasal administration of oxytocin reduced pair bonding (Bales et al. 2013). Further studies will be necessary to evaluate the safety and efficacy of chronic oxytocin on socially relevant behavior. Considering that both of the abovementioned studies were carried out in animals with a normally functioning oxytocin system, future studies should examine the effects of chronic oxytocin administration in animal models of ASD-like behavior that carry alterations in oxytocin system function.

5.1.2 Clinical Research

Given the role of oxytocin in facilitating social behavior in animals, studies in humans have begun to explore the effects of oxytocin on human sociability and related behaviors. A recent meta-analysis of studies using healthy participants found that intranasal oxytocin enhanced emotion recognition of human faces, in particular faces that demonstrated happiness or fear (Shahrestani et al. 2013). In ASD populations, several studies have found that oxytocin not only can enhance facial affect identification but also improve eye gaze toward the eye region of faces and improve reciprocal social behavior (Andari et al. 2010; Domes et al. 2013; Hollander et al. 2007).

There have been a number of published studies of oxytocin administration to human subjects with ASD. A study of intravenous oxytocin utilized a randomized double-blind, placebo-controlled, single administration crossover design and involved 15 adults with ASD (aged 19–56 years) (Hollander et al. 2007). Participants were initially randomized to receive oxytocin or placebo infusion, followed 1 week later by administration of the other agent. The oxytocin administration consisted of an intravenous infusion of oxytocin (10 units) or normal saline over a 4-h time period. Individuals were given an affective speech comprehension task just prior to the infusion, and it was repeated at 30, 60, 120, 180, and 240 min after the infusion. Oxytocin was associated with improvement in affective speech comprehension from pre-administration levels, regardless of whether given first or second in order of the crossover (Hollander et al. 2007). Participants who received oxytocin first, when compared to those who received placebo first, were noted to have an improved ability to identify affective speech when assessed after a 1-week delay. The same study had also reported that oxytocin resulted in a significant reduction in ASD-associated repetitive behaviors compared to placebo (Hollander et al. 2003).

There have been far more studies of intranasal oxytocin administration than intravenous administration (Veening and Olivier 2013). Two primary reasons underlie this shift. First, there is a concern that oxytocin in the periphery may not cross the blood–brain barrier. Second, intranasal administration is thought to increase the half-life of oxytocin compared to intravenous administration, with oxytocin having a half-life of 28 min in the cerebrospinal fluid (CSF) and extracellular space of the brain but less than 2 min in the blood (Mens et al. 1983; Robinson and Jones 1982; Seckl and Lightman 1988).

Studies utilizing intranasal oxytocin in ASD have found that it can improve performance on the "Reading the Mind in the Eyes Task" (RMET), a measure of the ability to identify affective states from standardized pictures of eyes and their surrounding facial areas (Guastella et al. 2010). The RMET is one of the most widely used tests of affective recognition in ASD research (Haxby et al. 2002).

A randomized double-blind, placebo-controlled, single administration crossover study in 16 males with ASD (aged 12–19 years, mean age 14.9 years) demonstrated significant improvement in the RMET with intranasal oxytocin (Guastella et al. 2010). Participants were randomized to receive one dose of oxytocin nasal spray [18 or 24 international units (IU) based on age] or placebo and then the other agent at separate administration sessions. The RMET was presented 45 min after drug administration. Oxytocin was found to significantly improve RMET performance in both younger and older participants who received 18 IU or 24 IU of oxytocin, respectively. Oxytocin was well tolerated with tiredness/relaxation noted in 25 % ($n = 4$) and sweating in 6 % ($n = 1$) of participants.

A randomized double-blind, placebo-controlled study evaluated the repeated administration of intranasal oxytocin in 19 adults with ASD (16 males, 3 females; mean age 33.2 years) (Anagnostou et al. 2012). Active treatment consisted of 24 IU of oxytocin administered twice a day for a period of 6 weeks. A variety of social and behavioral outcome measures were collected. Six weeks of oxytocin led to significantly improved performance on the RMET, a secondary social cognition outcome measure, when compared to placebo. No significant improvement was observed on the primary outcome measures of repetitive behavior (Yale-Brown Obsessive Compulsive Scale and Repetitive Behavior Scale—Revised) or social cognition (Diagnostic Analysis of Nonverbal Accuracy). Oxytocin was well tolerated with no serious adverse effects. Two individuals in the oxytocin group developed mild–moderate irritability, and two participants had increased allergy symptoms. Mild fatigue, headaches, leg shaking, and increased energy were reported by single participants in the oxytocin group. No clinically significant differences were seen between the oxytocin and placebo groups in regard to laboratory or clinical tests, including complete blood count, liver and renal function, serum osmolality, and electrocardiogram.

For over a decade, abnormal visual scanning of faces has been identified in ASD (Pelphrey et al. 2002). Individuals with ASD spend less time visually attending to the eye region of others compared to non-ASD controls (Boraston et al. 2008; Dalton et al. 2005; Sterling et al. 2008). In healthy adults, oxytocin has been shown to increase gaze toward the eye region of faces, potentially improving emotion recognition and social communication (Guastella et al. 2008). Several studies have also shown that oxytocin may enhance emotion recognition and improve eye region directed gaze in adults with ASD (Domes et al. 2013).

A randomized double-blind, placebo-controlled, single administration crossover study has demonstrated that intranasal oxytocin improves gaze toward the eye region among participants with ASD. In the study, 13 adolescents and adults with ASD (aged 17–20 years, mean age 26 years) received either 24 IU of intranasal oxytocin or placebo, followed by a 1-week washout period prior to receiving the other agent. A set of facial perception tasks coupled with eye tracking was undertaken 50 min after dose administration. Compared to placebo, oxytocin significantly increased the length of time ASD participants gazed at facial stimuli, mostly toward the eye region. Despite oxytocin, significantly reduced facial and eye region gaze duration persisted in ASD compared to non-ASD control subjects (Andari et al. 2010).

Another randomized double-blind, placebo-controlled, single administration crossover study examined the effects of intranasal oxytocin in individuals with ASD using a facial emotion recognition task coupled with functional and structural brain magnetic resonance imaging (MRI) (Domes et al. 2013). A total of 14 adults with ASD (mean age 24 years) received 12 IU of intranasal oxytocin or placebo followed by a facial emotion recognition task and brain MRI 45 min after administration. Compared to controls, participants with ASD had impaired baseline emotion recognition. The use of a single intranasal oxytocin dose led to improved performance in the emotion recognition task in those with ASD. Oxytocin did not result in any significant changes in mood, wakefulness, or calmness when compared to placebo. Oxytocin was associated with an increase in amygdala reactivity and activation to facial stimuli in ASD participants, further implicating the amygdala in emotion recognition (Domes et al. 2013). Reports of amygdalar activity during facial perception tasks in ASD have been mixed, with both increased (Monk et al. 2010; Weng et al. 2011) and reduced (Kleinhans et al. 2011; Perlman et al. 2011) activity being reported.

The effects of oxytocin on brain activity could vary depending on the social significance of the stimuli being encountered. One randomized double-blind, single administration crossover study evaluated the effects of intranasal oxytocin in children with ASD using functional MRI (fMRI) (Gordon et al. 2013). The study examined the impact of a single dose of intranasal oxytocin on 17 children and adolescents with ASD (aged 8–16.5 years). The participants were given an intranasal oxytocin dose that was dependent on their age. The oxytocin dose was 12 IU for those aged 7–11 years, 18 IU for those between 12 and 15 years, and 24 IU for individuals aged 16–19 years. Oxytocin was found to modulate brain activity based on the social meaningfulness (salience) of the images presented. Socially meaningful stimuli (eyes) resulted in increased activity in the striatum, nucleus accumbens, left posterior superior temporal sulcus, and left premotor cortex. In contrast, the same brain regions demonstrated decreased activity with oxytocin exposure when nonsocially salient images (vehicles) were shown.

A randomized, double-blind, placebo-controlled, single administration crossover study in 40 adult males with ASD (mean age 28.5 years) reported improved decision-making ability in response to nonverbal information (Watanabe et al. 2013). Intranasal oxytocin (24 IU) was associated with improved performance on a social psychological task based on the interpretation of conflicting nonverbal and verbal information. Imaging data gathered using fMRI demonstrated diminished medial prefrontal cortex activity at baseline among the participants with ASD. Oxytocin was found to alter these activity patterns so that they more closely resembled those found in healthy adults. The degree of neural change seen with oxytocin was correlated with improvement on the psychological task performance.

5.2 Glutamate

5.2.1 Preclinical Research

Glutamate is the most prevalent neurotransmitter in the brain, mediating fast excitatory neurotransmission and plasticity. There are three families of ionotropic glutamate receptors (iGluRs, including AMPA [α-amino-3-hydroxy-5-methyl-4-isozolepropionic acid], NMDA [N-methyl-D-aspartate], and kainate), as well as three groups of metabotropic glutamate receptors (mGluRs I-III) (Traynelis et al. 2010). Mutations in genes encoding both AMPA and NMDA receptor subunits, which play critical roles in regulating synaptic transmission and plasticity, have been associated with ASD (Soto et al. 2014). Additionally, genetic mutations found in some cases of ASD result in alterations in postsynaptic proteins necessary for receptor scaffolding and crosstalk between mGluRs and the iGluRs AMPA and NMDA (O'Connor et al. 2014). Taken together, this genome-based research supports the hypothesis that abnormalities in genes involved in glutamate receptors and regulation of glutamate pathways may be directly involved in ASD pathology.

In preclinical models, various aspects of glutamate signaling pathways have been disrupted to investigate the role of glutamate transmission on ASD-related phenotypes. For example, mice lacking Shank2, a postsynaptic scaffolding protein, show decreases in social preference and ultrasonic vocalizations (Won et al. 2012). There are a variety of transgenic lines with disrupted glutamatergic function, via altered receptors or postsynaptic proteins linking mGluRs and iGluRs, that display disrupted social behavior analogous to the social impairment in ASD [see (Carlson 2012; O'Connor et al. 2014) for review].

Consistent with studies directly targeting glutamatergic pathways, alterations in glutamate function have also been observed in other widely used ASD models. In the rat valproic acid (VPA) model, pregnant rats are exposed to a single injection of VPA, a known teratogen. Offspring of pregnant rats exposed to a single injection of VPA display behavioral symptoms observed in ASD, including impaired social interactions, increased repetitive behaviors, and elevations in anxiety-like behaviors (Markram et al. 2008; Schneider and Przewlocki 2005). These offspring display alterations in the glutamatergic system, including increased expression of the NMDA receptor subunits NRX and NRY, as well as alterations in the expression of the second messenger CaMKII (Rinaldi et al. 2007). Prenatal injection with polyinosinic–polycytidylic acid (poly I:C), which produces an immune response similar to that caused by viral infection, reduces NR1 expression in the hippocampus (Meyer et al. 2008). Together, these studies further highlight the importance of glutamatergic neurotransmission in social cognition related to ASD.

Although dysfunction of glutamate systems can contribute to ASD-like social impairments, few preclinical studies have investigated whether glutamatergic agents can rescue social impairments. One such study utilized mice lacking the mu opioid receptor gene that display abnormal social behavior, including defective recognition of a novel social partner and decreased social preference in direct and three-chamber versions of the social interaction test (Becker et al. 2014). Chronic treatment with VU0155041, an mGluR4 positive allosteric modulator, was more effective than the atypical antipsychotic risperidone in improving the social deficits without causing sedative side effects. Specific mutation of the fragile X mental retardation 1 (FMR1) gene has been strongly linked to FXS, a leading cause of inherited intellectual disability that is associated with various social disabilities leading to a high prevalence of comorbid ASD (Kaytor and Orr 2001). FMR1-knockout mice are widely used as a model of FXS. Loss of FMR1 disrupts mRNA translation repression, which results in upregulation of mGluR5 (Bear et al. 2004; Kaytor and Orr 2001). Chronic administration of AFQ056/mavoglurant, a selective mGluR5 antagonist, restores social behavior in FMR1-knockout mice (Gantois et al. 2013). This small body of literature demonstrates that pharmacological agents targeting the glutamate system alleviate social deficits in some animal models of ASD. While these investigations support the therapeutic potential of glutamatergic agents, it is clear that the specificity and direction of glutamate dysfunction vary depending on the model. Thus, it seems likely that etiological subtypes of ASD will require different glutamate agents in order to alleviate social abnormalities.

5.2.2 Clinical Research

Excitatory glutamate and inhibitory gamma-aminobutyric acid (GABA) activity work to balance brain function in order to regulate cognitive and emotional processes, among other functions. Regional brain glutamate levels can be measured with the combined glutamate/glutamine (Glx) signal using proton magnetic resonance spectroscopy ([^{1}H]MRS). Research suggests that ASD is associated with abnormal regional levels of glutamate (using Glx). Three studies of adults with ASD have measured Glx using [^{1}H]MRS. Adults with ASD were found to have increased Glx concentration in the right amygdala–hippocampal complex (Page et al. 2006) and reduced Glx concentration in the right anterior cingulate (Bernardi et al. 2011) and basal ganglia (Horder et al. 2013). Lower Glx concentration in the basal ganglia has been significantly correlated with increased social communication deficits (Horder et al. 2013). Studies measuring Glx in children with ASD have been less remarkable, with three studies finding no significant differences compared to controls (Friedman et al. 2006; Harada et al. 2011; Hardan et al. 2008), and one demonstrating lower cortical Glx concentration (DeVito et al. 2007).

ASD has been hypothesized to be a hypoglutamatergic disorder, based on both human and animal data (Blaylock and Strunecka 2009; Carlsson 1998; Smith et al. 2011). There have been a small number of clinical trials in ASD using agents that have a glutamatergic mode of action, including D-cycloserine (DCS) and memantine (Akhondzadeh et al. 2008; Chez et al. 2007; Ghaleiha et al. 2013; Owley et al. 2006; Posey et al. 2004;). Outcome measures in these studies have included subjective clinical ratings as opposed to specific cognitive tests. However, the assessment of clinician-rated improvements in social responsiveness and interaction is thought to be indicative of improved social cognition.

DCS is a partial agonist at the glycine site of the NMDA glutamate receptor. Several studies have suggested that DCS may be beneficial in the treatment of the negative symptoms of schizophrenia, symptoms with an underlying neurobiology that may be of significance in ASD (Evins et al. 2002; Goff et al. 2008; Heresco-Levy and Javitt 2004). One prospective single-blind, placebo lead-in study investigated the role of DCS in ASD (Posey et al. 2004). A total of ten individuals with ASD (aged 5–27 years, mean age 10.0 years) were given a daily dose of DCS over 6 weeks. The dose of DCS was increased every 2 weeks, progressing from 0.7 to 1.4 mg, and then 2.8 mg/kg/day. All ten individuals completed the 6-week trial of DCS. Compared to baseline, DCS was associated with a statistically significant improvement in social withdrawal (as measured by the Aberrant Behavior Checklist (ABC) Social Withdrawal subscale). DCS at the highest dose (2.8 mg/kg/day) was associated with a 60 % decrease in social withdrawal symptom severity. DCS was well tolerated although two participants encountered adverse effects at the highest dose, with one experiencing a transient motor tic and the other having increased echolalia.

Memantine is an NMDA glutamate receptor antagonist that is used in the treatment of Alzheimer's dementia and other cognitive disorders (Peng et al. 2013). A group of open-label studies of memantine in ASD has reported improvements in memory (Owley et al. 2006) and social interaction/withdrawal (Chez et al. 2007; Erickson et al. 2007; Owley et al. 2006), although not all memantine studies have demonstrated improvements in social withdrawal (Ghaleiha et al. 2013; Niederhofer 2007).

In one large open-label add-on study, 151 patients (aged 2.58–26.33 years, mean age 9.31 years) with a diagnosis of ASD were given memantine with the dose titrated based on clinical response (range of dose 2.5–30 mg/day, mean dose 12.67 mg/day) (Chez et al. 2007). Clinically significant improvement in social interaction and language, as defined by a rating of "much improved" or "very much improved" on the Clinical Global Impression--Improvement (CGI-I) scale, was seen in 70 % of patients. Some patients were noted to have deterioration in language function or social interaction with memantine, although this occurred in only 2 % and 12 % of patients, respectively. Adverse effects encountered with memantine included agitation or manic-like behavior. Adverse effects led to 15 % of participants discontinuing memantine.

One 8-week prospective open-label study involving 14 children with ASD (aged 3–12 years, mean age 7.8 years) investigated the use of memantine (Owley et al. 2006). The dosing schedule involved a gradual stepwise increase of 5 mg of memantine per week up to a final target dose of 0.4 mg/kg/day. Study participants were noted to have significant improvement in memory as tested by the Children's Memory Scale—The Dot Locations Subtest, which is a test of visual/nonverbal memory. Significant improvements in social withdrawal and inappropriate speech, as measured by two subscales of the ABC, were also recorded. Significant improvement was also seen with memantine on measures of irritability, hyperactivity, and stereotypy, based upon the ABC.

A retrospective open-label study of memantine in children and adolescents with ASD also yielded promising results. A total of 18 participants (aged 6–19 years, mean age 11.4 years) received memantine (mean dose 10.1 mg/day, range of dose 2.5–20 mg/day) (Erickson et al. 2007). The duration of treatment varied from 1.5 to 56 weeks (mean duration 19.3 weeks). Significant symptom improvement, predominately in social withdrawal and inattention, was found in 61 % of participants. Adverse effects were seen in 39 % of participants, with 22 % requiring drug discontinuation.

5.3 Neuroinflammation

5.3.1 Preclinical Research

Findings from preclinical studies of ASD support the long-standing hypothesis that immunological factors may play a key role in a subgroup of individuals with ASD. Environmental manipulations that trigger immune responses during critical periods of development result in the core behavioral symptoms of ASD. This observation provides the foundation for animal models of ASD involving inflammatory processes. One approach often used to represent environmental manipulations in animal models of ASD is in utero immune stimulation in the maternal immune activation (MIA) model. There is now considerable evidence that maternal immune responses can induce ASD-like behaviors in the offspring, including social impairment, communication deficits, and stereotypic behavior. Administration of a variety of immune-activating agents has been utilized in MIA, including influenza virus, poly I:C, VPA, and the bacterial endotoxin lipopolysaccharide (LPS) (Malkova et al. 2012; Oskvig et al. 2012; Schneider and Przewlocki 2005; Shi et al. 2003). While these agents stimulate the immune system by varying mechanisms, they all induce social deficits in the offspring and have been used in both rat and mouse models. Many MIA protocols have demonstrated that MIA induces inflammatory alterations in the offspring, including long-lasting and region-specific changes in brain cytokines, chemokines, and microglia, and inflammatory macrophage activation and other measures of neuroinflammation (Borrell et al. 2002; Garay et al. 2013; Kannan et al. 2007; Kimura et al. 1994; Meyer et al. 2006; 2008; Onore et al. 2014). Collectively, these findings support the hypothesis that disruptions in immune function influence the behavioral abnormalities observed in MIA models of ASD.

A single maternal injection of the cytokine interleukin-6 (IL-6) has been shown to cause some of the behavioral deficits observed in poly I:C MIA offspring (Smith et al. 2007). Additionally, maternal injection of poly I:C has no effect on social interaction in the IL-6-knockout mouse (Patterson 2009). Together, these findings suggest that IL-6 is critical in mediating some of the behavioral changes observed in poly I:C MIA offspring and identify IL-6 as a potential therapeutic target. Coadministration of a neutralizing anti-IL-6 antibody with poly I:C blocks the effects of MIA on the offspring behavior, thus preventing social deficits (Patterson 2009; Smith et al. 2007). Further investigation with knockout mice can be used to investigate the role of specific cytokines in the social deficits observed in ASD.

Recent progress in preclinical ASD research provides evidence for immunological involvement in ASD, at least in a subgroup of patients. Future work will continue to delineate the specific mechanisms through which these immune-based models cause social deficits in the offspring. With these models in place, investigations of drug therapies that target the immune system can now be tested as a potential means to treat or prevent certain types of ASD.

5.3.2 Clinical Research

One area of investigation that has attracted significant attention in research in ASD, as well as other neuropsychiatric disorders, is that of neuroimmune interactions, including the processes of microglia activation and neuroinflammation (McDougle and Carlezon 2013). Results from a recent brain imaging study using positron emission tomography (PET) found evidence for greater inflammation in the brains of adults with ASD compared to matched healthy controls (Suzuki et al. 2013). Considering these data alongside a large number of previously published studies of immune function in ASD (see Stigler et al. 2009 for review) suggests that aberrant neuroimmune processes may contribute to the development and ongoing pathophysiology of a significant subgroup of those with ASDs.

In response to evidence implicating immune dysfunction and inflammation in some patients with ASD, a published literature pertaining to the use of anti-inflammatory and immune-modulating drugs in this population is beginning to emerge. As the field moves forward, it may be that some of these treatments will be studied in a rigorous double-blind, placebo-controlled manner, specifically in those patients where prominent, aberrant neuroimmune processes may be involved.

Corticosteroids are anti-inflammatory and immunosuppressive agents that inhibit pro-inflammatory cytokine production, alter T lymphocyte activity, and may also modulate microglial activation (Ros-Bernal et al. 2011; Schweingruber et al. 2012). Corticosteroids are efficacious treatments for a wide variety of autoimmune conditions, including inflammatory bowel disease, asthma, inflammatory arthritis, and neurological conditions such as multiple sclerosis. Corticosteroids are, however, associated with a diverse range of adverse effects, including neuropsychiatric symptoms such as alterations in mood, cognition, sleep, and behavior (Kenna et al. 2011). Several promising accounts of improvements in symptoms of ASD with the use of corticosteroids have been reported. Research findings are, however, limited to case reports and open-label studies in children.

One case report of a boy aged 2 years, 7 months with regressive ASD and an autoimmune lymphoproliferative condition described improved social interaction and vocalization with chronic oral prednisolone treatment (Shenoy et al. 2000). Previously, he had shown marked regression in social communication, including speech, at the age of 18 months. In order to treat the autoimmune condition, he initially received prednisolone at a dose of 2 mg/kg/day for a period of 10 weeks. Within the first month of treatment, the boy was described as having increased social interaction. The initial dose of prednisolone was associated with irritability that subsided as the prednisolone dosage was reduced to 0.5 mg/kg/day. Subsequently, his prednisolone dose was further reduced to 0.4 mg/kg every other day. However, this resulted in re-emergence of autoimmune symptoms, including diarrhea. Eventually, the dose of 0.5 mg/kg every other day was found to be an effective maintenance dose for treatment of the autoimmune condition, as well as the ASD symptoms. Continuing improvements in speech were noted over the subsequent 12 months of treatment with the emergence of a vocabulary of over 200 words. Improvements were also seen in gesturing, nonverbal communication, and language expression and comprehension as assessed by the Rossetti Infant–Toddler Language Scale.

A second case report described a 6-year-old boy with ASD who also improved with prednisolone treatment. The patient had demonstrated evidence of temporal and frontal lobe dysfunction based on a single-photon emission computed tomography (SPECT) scan and steady-state auditory evoked potential readings (Stefanatos et al. 1995). There were no electroencephalographic (EEG) abnormalities, making the diagnosis of Landau-Kleffner syndrome (LKS) unlikely. The patient was diagnosed with ASD based on prominent language and behavioral regression at the age of 22 months, with persistent impaired social interactions, motor stereotypies, and echolalia. He began a 28-week course of prednisolone, with an initial dose of 2 mg/kg/day for the first 4 weeks. Between weeks 4 and 12, the prednisolone dose was reduced by 0.5 mg/kg/day every 4 weeks. From week 12 through week 28, the prednisolone was administered on alternate days with continuing reduction of 0.25 mg/kg/day every 4 weeks. Within a few weeks of treatment, significant improvement in social communication was noted. By the end of treatment, he made relative gains of 26–36 months in expressive and receptive vocabulary in less than 18 months.

A report of two cases of children with ASD and neurologic symptoms, including seizures and motor deficits, treated with prednisolone has also been published (Mordekar et al. 2009). The children (one male and one female, both aged 4.5 years) were given a course of prednisolone (initial dose 2 mg/kg/day) for 10 weeks and 3 weeks, respectively. Both patients demonstrated a return of previously lost spoken language with a reduction in psychomotor agitation. They reportedly maintained this improvement following the discontinuation of the treatment.

An open-label study was used to investigate high-dose corticosteroid treatment in 44 children with ASD and evidence of abnormal epileptiform activity (mean age 5.6 years) (Chez et al. 1998). In these children, high-dose corticosteroid (either prednisolone or methylprednisolone at 10 mg/kg/week for 18 months) was added to ongoing treatment with divalproex sodium. Only 25 children had clinical and EEG outcomes reported after addition of a corticosteroid. Among these children, clinical improvements in speech and EEG were noted in 82 and 60 % of cases, respectively. Corticosteroid treatment was well tolerated with no significant adverse effects, including no cushingoid complications being observed even after 18 months of treatment.

Adrenocorticotropic hormone (ACTH) is produced by the anterior pituitary gland and stimulates the release of corticosteroids from adrenal cells. Children with ASD have been reported to have both abnormal circulating levels and physiologic responses to ACTH (Hamza et al. 2010; Marinovic-Curin et al. 2008). Several controlled trials have reported ACTH to be well tolerated and result in improvement in ASD symptoms in some children (Buitelaar et al. 1990, 1992a, b).

An 8-week controlled crossover trial investigated the use of synthetic ACTH (40 mg/day) in 21 children with ASD (Buitelaar et al. 1992a). The participants were aged 5–15 years, with 4 females and 17 males being enrolled. One male dropped out of the study due to gastrointestinal complaints prior to receiving ACTH. Significant improvement was seen in overall ASD severity as measured by the CGI severity (CGI-S) scale alongside improvements in specific symptoms of ASD, in particular social interaction, as measured by the ABC Social Withdrawal subscale score. While use of ACTH was generally well tolerated, six children had an increase in mood lability and "inner tension" by parent or teacher report compared to only two children experiencing the same while on placebo. Significant improvement in social interaction and eye gaze was observed. These findings supported the results of a previous double-blind, placebo-controlled crossover trial involving 14 children (aged 5–13 years) with ASD. Treatment with ACTH (20 mg/day) was noted to improve stereotypic behaviors and enhance social interaction in the subjects (Buitelaar et al. 1990, 1992b).

Celecoxib, a cyclooxygenase-2 inhibitor with anti-inflammatory effects, has been shown to have some beneficial effects on ASD symptoms, as reported in one 10-week randomized double-blind, placebo-controlled trial (Asadabadi et al. 2013). In the study, 40 children with ASD (aged 5–11 years) were assigned to risperidone combined with either celecoxib or placebo. Celecoxib was initiated at 100 mg/day and titrated to either 200 or 300 mg/day depending on body weight (limited to 200 mg/day if child weighed less than 30 kg). Risperidone was started at a dose of 0.5 mg/day with an increase of 0.5 mg/week to 2–3 mg/day, depending on body weight. Outcomes were measured using the subscales of the ABC at 2, 4, 6, and 10 weeks. By the end of the study, the risperidone plus celecoxib arm demonstrated significant improvement in irritability, social withdrawal, and stereotypic behavior when compared to risperidone plus placebo. Additionally, while risperidone plus placebo resulted in a treatment response in 20 % of patients, a significantly higher percentage, 55 %, responded to risperidone plus celecoxib. Response was defined as a 50 % reduction in the ABC Irritability subscale score. Adverse effects were similar between treatment groups, with extrapyramidal symptoms, likely due to risperidone use, being reported in half of participants (45 % placebo group, 50 % celecoxib group). Abdominal pain was reported in three patients receiving risperidone plus celecoxib compared to only one patient given risperidone plus placebo, although this difference was not statistically significant (Fig. 1).

Altered plasma immunoglobulin levels have been reported in ASD, although conflicting results have been found. For example, both IgM and IgG have been found to be increased (Trajkovski et al. 2004) and decreased (Heuer et al. 2008) in individuals with ASD.

Studies have investigated the use of oral human immunoglobulin (IGOH) and intravenous human immunoglobulin (IVIG) in the treatment of ASD. However, they have yielded divergent findings. One 8-week open-label trial of IGOH

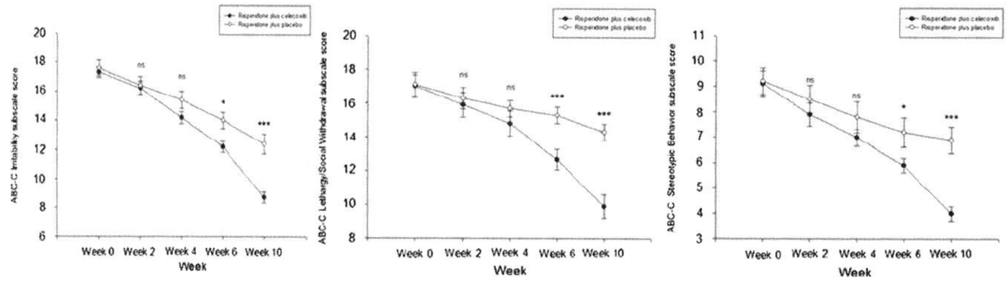

Fig. 1 Repeated measure graph for comparison of irritability, lethargy/social withdrawal, and stereotypic behavior subscales of Aberrant Behavior Checklist—Community (*ABC-C*; mean ± standard error of the mean) over time between the celecoxib and the placebo group. $*P < 0.05$; $**P < 0.01$; $***P < 0.001$. *P* values represent results of unpaired *t* test for comparison of the changes from baseline scores to that time point between the two groups. Reprinted with permission of Springer Science and Business Media

(420 mg/day) in 12 male children (aged 3–7 years) with ASD and gastrointestinal disturbances found 50 % of those treated had symptomatic improvement of both conditions (Schneider et al. 2006). Gastrointestinal outcomes were measured using the Gastrointestinal Severity Index, while symptoms of ASD were measured with the CGI-I, CGI-S, and ABC. Significant improvement in ASD symptoms, as measured by the ABC, was not only seen at 8 weeks but also at 30 days after discontinuation of medication. Three of the children withdrew from the study while receiving IGOH following the development of vomiting and fever, vomiting and nausea, and rash, respectively. There has been one double-blind, placebo-controlled, randomized trial of IGOH in the treatment of ASD (Handen et al. 2009). This study involved 125 children with ASD and chronic gastrointestinal symptoms. The treatment phase was 12 weeks long with four treatment arms, including placebo and three different dosages of IGOH (140 mg, 420 mg, and 840 mg/day). In contrast to the open-label study, IGOH was not found to be beneficial in reducing ASD or gastrointestinal symptoms in this controlled trial.

Several open-label studies examining the effects of IVIG in ASD have yielded mixed results. One study involving seven children with ASD, aged 3.5–6 years, reported on the effects of IVIG given at monthly intervals for 6 months (400 mg/kg/month) (DelGiudice-Asch et al. 1999). Two children did not finish the study as in one case a diagnosis of LKS was suspected, and in the other case, the family received the last 2 months of IVIG outside of the study. There was no significant improvement in behavior or ASD severity as measured by several scales including the Children's Yale-Brown Obsessive Compulsive Scale, the Ritvo–Freeman Real Life Rating Scale, and the CGI-I. In a study involving ten children with ASD (aged 4–17 years) and documented immunologic abnormalities, IVIG was administered every 6 weeks for a total of 4 infusions (dose 200–400 mg/kg) (Plioplys 1998). Eight of the ten children had been noted to have had ASD diagnosed following a period of regression in early life. In this study, five children did not have any change in symptoms, four children had minor improvements that the authors suggested could have been due to placebo effect, and one child had an "almost total amelioration" of ASD symptoms with treatment. The improvements were temporary, with symptoms returning to their baseline levels 5 months after treatment cessation. The authors concluded the response rate of 10 % was too low to justify the high economic costs associated with immunologic testing and IVIG administration.

There have, however, been more favorable reports of IVIG, including one by Gupta et al. (1996a) who described ten children with ASD and immunological abnormalities. The children were given IVIG on a monthly basis for 6 months (dose 400 mg/kg). Following treatment, half of the children ($n = 5$) were deemed to have had a marked ($n = 4$) or striking ($n = 1$) clinical improvement. Enhanced language use and eye contact and a reduction in agitation were specifically identified. Additionally, favorable reports were obtained from one open-label retrospective study involving 26 children with ASD who were treated with IVIG for 6 months (dose 400 mg/kg/month) (Boris et al. 2005). Treatment resulted in a significant decrease in the total ABC score alongside each of the ABC subscales (Irritability, Lethargy/Social Withdrawal, Stereotypy, Hyperactivity, and Inappropriate

Speech). Biochemical investigation of the participants revealed evidence of substantial immunological and inflammatory abnormalities, including 54 % having an elevated erythrocyte sedimentation rate (ESR), 65 % having antibodies to myelin, and 31 % having thyroid antibodies.

6 Summary

Accumulating evidence indicates that both environmental and genetic factors can play crucial roles in the etiology of ASD. Basic research on the factors that regulate communication—both between animals and, more fundamentally, nerve cells in the brain—has led to clinical trials of treatments that act by altering oxytocin or glutamate function. These approaches have been of variable success for improving social cognition in some individuals with ASD. Along other lines, it is becoming clear that many individuals with ASD have a family history of inflammation-associated conditions, often autoimmune in nature, and the core behavioral features of ASD can be produced in laboratory animals by inducing inflammatory responses during key periods of development. While it is unlikely that all cases of ASD have a common cause that would be sensitive to a single therapeutic approach, a better understanding of the ways in which environment and genetics interact may enable the development of personalized interventions that target specific etiologies.

Acknowledgment This work was supported, in part, by the Nancy Lurie Marks Family Foundation and the Robert and Donna Landreth Fund.

References

Akhondzadeh S, Tajdar H, Mohammadi MR, Mohammadi M, Nouroozinejad GH, Shabstari OL, Ghelichnia HA (2008) A double-blind placebo controlled trial of piracetam added to risperidone in patients with autistic disorder. Child Psychiatry Hum Dev 39(3):237–245. doi:10.1007/s10578-007-0084-3

American Psychiatric Association APADSMTF (2013) Diagnostic and statistical manual of mental disorders: DSM-5. Available via http://worldcat.org. http://dsm.psychiatryonline.org/book.aspx?bookid=556.

Anagnostou E, Soorya L, Chaplin W, Bartz J, Halpern D, Wasserman S, Wang AT, Pepa L, Tanel N, Kushki A, Hollander E (2012) Intranasal oxytocin versus placebo in the treatment of adults with autism spectrum disorders: a randomized controlled trial. Mol Autism 3(1):16. doi:10.1186/2040-2392-3-16

Andari E, Duhamel JR, Zalla T, Herbrecht E, Leboyer M, Sirigu A (2010) Promoting social behavior with oxytocin in high-functioning autism spectrum disorders. Proc Natl Acad Sci USA 107(9):4389–4394. doi:10.1073/pnas.0910249107

Annaz D, Campbell R, Coleman M, Milne E, Swettenham J (2012) Young children with autism spectrum disorder do not preferentially attend to biological motion. J Autism Dev Disord 42 (3):401–408

Asadabadi M, Mohammadi MR, Ghanizadeh A, Modabbernia A, Ashrafi M, Hassanzadeh E, Forghani S, Akhondzadeh S (2013) Celecoxib as adjunctive treatment to risperidone in children with autistic disorder: a randomized, double-blind, placebo-controlled trial. Psychopharmacology (Berl) 225(1):51–59. doi:10.1007/s00213-012-2796-8

Baio J, Centers for Disease Control and Prevention (2012) Prevalence of autism spectrum disorders: Autism and Developmental Disabilities Monitoring Network, 14 Sites, United States, 2008. MMWR Surveill Summ 61(3):1–19

Bales KL, Perkeybile AM, Conley OG, Lee MH, Guoynes CD, Downing GM, Yun CR, Solomon M, Jacob S, Mendoza SP (2013) Chronic intranasal oxytocin causes long-term impairments in partner preference formation in male prairie voles. Biol Psychiatry 74 (3):180–188. doi:10.1016/j.biopsych.2012.08.025

Bear MF, Huber KM, Warren ST (2004) The mGluR theory of fragile X mental retardation. Trends Neurosci 27(7):370–377. doi:10.1016/j.tins.2004.04.009

Becker JA, Clesse D, Spiegelhalter C, Schwab Y, Le Merrer J, Kieffer BL (2014) Autistic-like syndrome in mu opioid receptor null mice is relieved by facilitated mGluR4 activity. Neuropsychopharmacology 39(9):2049–2060. doi:10.1038/npp.2014.59

Bernardi S, Anagnostou E, Shen J, Kolevzon A, Buxbaum JD, Hollander E, Hof PR, Fan J (2011) In vivo 1H-magnetic resonance spectroscopy study of the attentional networks in autism. Brain Res 1380:198–205. doi:10.1016/j.brainres.2010.12.057

Bielsky IF, Young LJ (2004) Oxytocin, vasopressin, and social recognition in mammals. Peptides 25(9):1565–1574. doi:10.1016/j.peptides.2004.05.019

Blaylock RL, Strunecka A (2009) Immune-glutamatergic dysfunction as a central mechanism of the autism spectrum disorders. Curr Med Chem 16(2):157–170

Boraston ZL, Corden B, Miles LK, Skuse DH, Blakemore SJ (2008) Brief report: perception of genuine and posed smiles by individuals with autism. J Autism Dev Disord 38(3):574–580. doi:10.1007/s10803-007-0421-1

Boris M, Goldblatt A, Edelson SM (2005) Improvement in children with autism treated with intravenous gamma globulin. J Nutr Environ Med 15(4):169–176

Borrell J, Vela JM, Arevalo-Martin A, Molina-Holgado E, Guaza C (2002) Prenatal immune challenge disrupts sensorimotor gating in adult rats. Implications for the etiopathogenesis of schizophrenia. Neuropsychopharmacology 26(2):204–215. doi:10.1016/S0893-133X(01)00360-8

Buitelaar JK, van Engeland H, van Ree JM, de Wied D (1990) Behavioral effects of Org 2766, a synthetic analog of the adrenocorticotrophic hormone (4-9), in 14 outpatient autistic children. J Autism Dev Disord 20(4):467–478

Buitelaar JK, van Engeland H, de Kogel K, de Vries H, van Hooff J, van Ree J (1992a) The adrenocorticotrophic hormone (4-9) analog ORG 2766 benefits autistic children: report on a second controlled clinical trial. J Am Acad Child Adolesc Psychiatry 31(6):1149–1156

Buitelaar JK, van Engeland H, de Kogel KH, de Vries H, van Hooff JA, van Ree JM (1992b) The use of adrenocorticotrophic hormone (4-9) analog ORG 2766 in autistic children: effects on the organization of behavior. Biol Psychiatry 31(11):1119–1129

Carlson GC (2012) Glutamate receptor dysfunction and drug targets across models of autism spectrum disorders. Pharmacol Biochem Behav 100(4):850–854. doi:10.1016/j.pbb.2011.02.003

Carlsson ML (1998) Hypothesis: is infantile autism a hypoglutamatergic disorder? Relevance of glutamate—serotonin interactions for pharmacotherapy. J Neural Transm 105(4–5):525–535

Carroll LS, Owen MJ (2009) Genetic overlap between autism, schizophrenia and bipolar disorder. Genome Med 1(10):102–102.7. doi:10.1186/gm102

Cheung C, Yu K, Fung G, Leung M, Wong C, Li Q, Sham P, Chua S, McAlonan G (2010) Autistic disorders and schizophrenia: related or remote? An anatomical likelihood estimation. PLoS One 5(8):e12233

Chez ML, Loeffel M, Buchanan CP, Field-Chez M (1998) Pulse high dose steroids as combination therapy with valproic acid in epileptic aphasia patients with pervasive developmental delay or autism. Ann Neurol 44(3):53–59. doi:10.1002/ana.410440345

Chez MG, Burton Q, Dowling T, Chang M, Khanna P, Kramer C (2007) Memantine as adjunctive therapy in children diagnosed with autistic spectrum disorders: an observation of initial clinical response and maintenance tolerability. J Child Neurol 22(5):574–579. doi:10.1177/0883073807302611

Crespi B, Stead P, Elliot M (2010) Comparative genomics of autism and schizophrenia. Proc Natl Acad Sci 107(suppl 1):1736–1741

Dalton KM, Nacewicz BM, Johnstone T, Schaefer HS, Gernsbacher MA, Goldsmith HH, Alexander AL, Davidson RJ (2005) Gaze fixation and the neural circuitry of face processing in autism. Nat Neurosci 8(4):519–526. doi:10.1038/nn1421

DelGiudice-Asch G, Simon L, Schmeidler J, Cunningham-Rundles C, Hollander E (1999) Brief report: a pilot open clinical trial of intravenous immunoglobulin in childhood autism. J Autism Dev Disord 29(2):157–160

DeVito TJ, Drost DJ, Neufeld RW, Rajakumar N, Pavlosky W, Williamson P, Nicolson R (2007) Evidence for cortical dysfunction in autism: a proton magnetic resonance spectroscopic imaging study. Biol Psychiatry 61(4):465–473. doi:10.1016/j.biopsych.2006.07.022

Domes G, Kumbier E, Heinrichs M, Herpertz SC (2013) Oxytocin promotes facial emotion recognition and amygdala reactivity in adults with asperger syndrome. Neuropsychopharmacology 39(3):698–706. doi:10.1038/npp.2013.254

Donaldson ZR, Young LJ (2008) Oxytocin, vasopressin, and the neurogenetics of sociality. Science 322(5903):900–904. doi:10.1126/science.1158668

Erickson CA, Posey DJ, Stigler KA, Mullett J, Katschke AR, McDougle CJ (2007) A retrospective study of memantine in children and adolescents with pervasive developmental disorders. Psychopharmacology (Berl) 191(1):141–147. doi:10.1007/s00213-006-0518-9

Evins AE, Amico E, Posever TA, Toker R, Goff DC (2002) D-Cycloserine added to risperidone in patients with primary negative symptoms of schizophrenia. Schizophr Res 56(1–2):19–23

Ey E, Leblond CS, Bourgeron T (2011) Behavioral profiles of mouse models for autism spectrum disorders. Autism Res 4(1):5–16. doi:10.1002/aur.175

Farroni T, Csibra G, Simion F, Johnson MH (2002) Eye contact detection in humans from birth. Proc Natl Acad Sci U S A 99(14):9602–9605. doi:10.1073/pnas.152159999

Ferguson JN, Young LJ, Hearn EF, Matzuk MM, Insel TR, Winslow JT (2000) Social amnesia in mice lacking the oxytocin gene. Nat Genet 25(3):284–288. doi:10.1038/77040

Ferguson JN, Aldag JM, Insel TR, Young LJ (2001) Oxytocin in the medial amygdala is essential for social recognition in the mouse. J Neurosci 21(20):8278–8285

Friedman SD, Shaw DW, Artru AA, Dawson G, Petropoulos H, Dager SR (2006) Gray and white matter brain chemistry in young children with autism. Arch Gen Psychiatry 63(7):786–794. doi:10.1001/archpsyc.63.7.786

Frith CD (2008) Social cognition. Philos Trans R Soc Lond B Biol Sci 363(1499):2033–2039. doi:10.1098/rstb.2008.0005

Gantois I, Pop AS, de Esch CE, Buijsen RA, Pooters T, Gomez-Mancilla B, Gasparini F, Oostra BA, D'Hooge R, Willemsen R (2013) Chronic administration of AFQ056/Mavoglurant restores social behaviour in Fmr1 knockout mice. Behav Brain Res 239:72–79. doi:10.1016/j.bbr.2012.10.059

Garay PA, Hsiao EY, Patterson PH, McAllister AK (2013) Maternal immune activation causes age- and region-specific changes in brain cytokines in offspring throughout development. Brain Behav Immun 31:54–68. doi:10.1016/j.bbi.2012.07.008

Ghaleiha A, Asadabadi M, Mohammadi MR, Shahei M, Tabrizi M, Hajiaghaee R, Hassanzadeh E, Akhondzadeh S (2013) Memantine as adjunctive treatment to risperidone in children with autistic disorder: a randomized, double-blind, placebo-controlled trial. Int J Neuropsychopharmacol 16(4):783–789. doi:10.1017/S1461145712000880

Goff DC, Cather C, Gottlieb JD, Evins AE, Walsh J, Raeke L, Otto MW, Schoenfeld D, Green MF (2008) Once-weekly D-cycloserine effects on negative symptoms and cognition in schizophrenia: an exploratory study. Schizophr Res 106(2–3):320–327. doi:10.1016/j.schres.2008.08.012

Gordon I, Vander Wyk BC, Bennett RH, Cordeaux C, Lucas MV, Eilbott JA, Zagoory-Sharon O, Leckman JF, Feldman R, Pelphrey KA (2013) Oxytocin enhances brain function in children with autism. Proc Natl Acad Sci USA 110(52):20953–20958. doi:10.1073/pnas.1312857110

Gotts SJ, Simmons WK, Milbury LA, Wallace GL, Cox RW, Martin A (2012) Fractionation of social brain circuits in autism spectrum disorders. Brain 135(Pt 9):2711–2725. doi:10.1093/brain/aws160

Grice SJ, Halit H, Farroni T, Baron-Cohen S, Bolton P, Johnson MH (2005) Neural correlates of eye-gaze detection in young children with autism. Cortex 41(3):342–353

Guastella AJ, Mitchell PB, Dadds MR (2008) Oxytocin increases gaze to the eye region of human faces. Biol Psychiatry 63(1):3–5. doi:10.1016/j.biopsych.2007.06.026

Guastella AJ, Einfeld SL, Gray KM, Rinehart NJ, Tonge BJ, Lambert TJ, Hickie IB (2010) Intranasal oxytocin improves emotion recognition for youth with autism spectrum disorders. Biol Psychiatry 67(7):692–694. doi:10.1016/j.biopsych.2009.09.020

Gupta S, Aggarwal S, Heads C (1996) Dysregulated immune system in children with autism: beneficial effects of intravenous immune globulin on autistic characteristics. J Autism Dev Disord 26(4):439–452

Haith M, Bergman T, Moore M (1977) Eye contact and face scanning in early infancy. Science 198(4319):853–855. doi:10.1126/science.918670

Hamza RT, Hewedi DH, Ismail MA (2010) Basal and adrenocorticotropic hormone stimulated plasma cortisol levels among Egyptian autistic children: relation to disease severity. Ital J Pediatr 36:71. doi:10.1186/1824-7288-36-71

Handen BL, Melmed RD, Hansen RL, Aman MG, Burnham DL, Bruss JB, McDougle CJ (2009) A double-blind, placebo-controlled trial of oral human immunoglobulin for gastrointestinal dysfunction in children with autistic disorder. J Autism Dev Disord 39(5):796–805. doi:10.1007/s10803-008-0687-y

Harada M, Taki MM, Nose A, Kubo H, Mori K, Nishitani H, Matsuda T (2011) Non-invasive evaluation of the GABAergic/glutamatergic system in autistic patients observed by MEGA-editing proton MR spectroscopy using a clinical 3 T instrument. J Autism Dev Disord 41 (4):447–454. doi:10.1007/s10803-010-1065-0

Hardan AY, Minshew NJ, Melhem NM, Srihari S, Jo B, Bansal R, Keshavan MS, Stanley JA (2008) An MRI and proton spectroscopy study of the thalamus in children with autism. Psychiatry Res 163(2):97–105. doi:10.1016/j.pscychresns.2007.12.002

Haxby JV, Hoffman EA, Gobbini MI (2002) Human neural systems for face recognition and social communication. Biol Psychiatry 51(1):59–67

Heresco-Levy U, Javitt DC (2004) Comparative effects of glycine and D-cycloserine on persistent negative symptoms in schizophrenia: a retrospective analysis. Schizophr Res 66(2–3):89–96. doi:10.1016/S0920-9964(03)00129-4

Heuer L, Ashwood P, Schauer J, Goines P, Krakowiak P, Hertz-Picciotto I, Hansen R, Croen LA, Pessah IN, Van de Water J (2008) Reduced levels of immunoglobulin in children with autism correlates with behavioral symptoms. Autism Res 1(5):275–283. doi:10.1002/aur.42

Higashida H, Yokoyama S, Munesue T, Kikuchi M, Minabe Y, Lopatina O (2011) CD38 gene knockout juvenile mice: a model of oxytocin signal defects in autism. Biol Pharm Bull 34 (9):1369–1372

Higashida H, Yokoyama S, Kikuchi M, Munesue T (2012) CD38 and its role in oxytocin secretion and social behavior. Horm Behav 61(3):351–358. doi:10.1016/j.yhbeh.2011.12.011

Hollander E, Novotny S, Hanratty M, Yaffe R, DeCaria CM, Aronowitz BR, Mosovich S (2003) Oxytocin infusion reduces repetitive behaviors in adults with autistic and Asperger's disorders. Neuropsychopharmacology 28(1):193–198. doi:10.1038/sj.npp.1300021

Hollander E, Bartz J, Chaplin W, Phillips A, Sumner J, Soorya L, Anagnostou E, Wasserman S (2007) Oxytocin increases retention of social cognition in autism. Biol Psychiatry 61 (4):498–503. doi:10.1016/j.biopsych.2006.05.030

Horder J, Lavender T, Mendez MA, O'Gorman R, Daly E, Craig MC, Lythgoe DJ, Barker GJ, Murphy DG (2013) Reduced subcortical glutamate/glutamine in adults with autism spectrum disorders: a [(1)H]MRS study. Transl Psychiatry 3:e279. doi:10.1038/tp.2013.53

Huang H, Michetti C, Busnelli M, Manago F, Sannino S, Scheggia D, Giancardo L, Sona D, Murino V, Chini B, Scattoni ML, Papaleo F (2014) Chronic and acute intranasal oxytocin produce divergent social effects in mice. Neuropsychopharmacology 39(5):1102–1114. doi:10.1038/npp.2013.310

Jin D, Liu HX, Hirai H, Torashima T, Nagai T, Lopatina O, Shnayder NA, Yamada K, Noda M, Seike T, Fujita K, Takasawa S, Yokoyama S, Koizumi K, Shiraishi Y, Tanaka S, Hashii M, Yoshihara T, Higashida K, Islam MS, Yamada N, Hayashi K, Noguchi N, Kato I, Okamoto H, Matsushima A, Salmina A, Munesue T, Shimizu N, Mochida S, Asano M, Higashida H (2007) CD38 is critical for social behaviour by regulating oxytocin secretion. Nature 446 (7131):41–45. doi:10.1038/nature05526

Jones W, Carr K, Klin A (2008) Absence of preferential looking to the eyes of approaching adults predicts level of social disability in 2-year-old toddlers with autism spectrum disorder. Arch Gen Psychiatry 65(8):946–954. doi:10.1001/archpsyc.65.8.946

Just MA, Keller TA, Malave VL, Kana RK, Varma S (2012) Autism as a neural systems disorder: a theory of frontal-posterior underconnectivity. Neurosci Biobehav Rev 36(4):1292–1313. doi:10.1016/j.neubiorev.2012.02.007

Kannan S, Saadani-Makki F, Muzik O, Chakraborty P, Mangner TJ, Janisse J, Romero R, Chugani DC (2007) Microglial activation in perinatal rabbit brain induced by intrauterine inflammation: detection with 11C-(R)-PK11195 and small-animal PET. J Nucl Med 48(6):946–954. doi:10.2967/jnumed.106.038539

Kaytor MD, Orr HT (2001) RNA targets of the fragile X protein. Cell 107(5):555–557

Kenna HA, Poon AW, de los Angeles CP, Koran LM (2011) Psychiatric complications of treatment with corticosteroids: review with case report. Psychiatry Clin Neurosci 65 (6):549–560. doi:10.1111/j.1440-1819.2011.02260.x

Kimura M, Toth LA, Agostini H, Cady AB, Majde JA, Krueger JM (1994) Comparison of acute phase responses induced in rabbits by lipopolysaccharide and double-stranded RNA. Am J Physiol 267(6 Pt 2):R1596–R1605

Kleinhans NM, Richards T, Johnson LC, Weaver KE, Greenson J, Dawson G, Aylward E (2011) fMRI evidence of neural abnormalities in the subcortical face processing system in ASD. Neuroimage 54(1):697–704. doi:10.1016/j.neuroimage.2010.07.037

Klin A, Sparrow SS, de Bildt A, Cicchetti DV, Cohen DJ, Volkmar FR (1999) A normed study of face recognition in autism and related disorders. J Autism Dev Disord 29(6):499–508

Klin A, Lin DJ, Gorrindo P, Ramsay G, Jones W (2009) Two-year-olds with autism orient to non-social contingencies rather than biological motion. Nature 459(7244):257–261

Lim MM, Bielsky IF, Young LJ (2005) Neuropeptides and the social brain: potential rodent models of autism. Int J Dev Neurosci 23(2–3):235–243. doi:10.1016/j.ijdevneu.2004.05.006

Liu HX, Lopatina O, Higashida C, Tsuji T, Kato I, Takasawa S, Okamoto H, Yokoyama S, Higashida H (2008) Locomotor activity, ultrasonic vocalization and oxytocin levels in infant CD38 knockout mice. Neurosci Lett 448(1):67–70. doi:10.1016/j.neulet.2008.09.084

Lukas M, Neumann ID (2013) Oxytocin and vasopressin in rodent behaviors related to social dysfunctions in autism spectrum disorders. Behav Brain Res 251:85–94. doi:10.1016/j.bbr.2012.08.011

Lynch CJ, Uddin LQ, Supekar K, Khouzam A, Phillips J, Menon V (2013) Default mode network in childhood autism: posteromedial cortex heterogeneity and relationship with social deficits. Biol Psychiatry 74(3):212–219. doi:10.1016/j.biopsych.2012.12.013

Malkova NV, Yu CZ, Hsiao EY, Moore MJ, Patterson PH (2012) Maternal immune activation yields offspring displaying mouse versions of the three core symptoms of autism. Brain Behav Immun 26(4):607–616. doi:10.1016/j.bbi.2012.01.011

Marinovic-Curin J, Marinovic-Terzic I, Bujas-Petkovic Z, Zekan L, Skrabic V, Dogas Z, Terzic J (2008) Slower cortisol response during ACTH stimulation test in autistic children. Eur Child Adolesc Psychiatry 17(1):39–43. doi:10.1007/s00787-007-0632-1

Markram K, Rinaldi T, La Mendola D, Sandi C, Markram H (2008) Abnormal fear conditioning and amygdala processing in an animal model of autism. Neuropsychopharmacology 33 (4):901–912. doi:10.1038/sj.npp.1301453

McCall C, Singer T (2012) The animal and human neuroendocrinology of social cognition, motivation and behavior. Nat Neurosci 15(5):681–688. doi:10.1038/nn.3084

McDougle CJ, Carlezon WA Jr (2013) Neuroinflammation and autism: toward mechanisms and treatments. Neuropsychopharmacology 38(1):241–242. doi:10.1038/npp.2012.174

Mens WB, Witter A, van Wimersma Greidanus TB (1983) Penetration of neurohypophyseal hormones from plasma into cerebrospinal fluid (CSF): half-times of disappearance of these neuropeptides from CSF. Brain Res 262(1):143–149

Meyer U, Nyffeler M, Engler A, Urwyler A, Schedlowski M, Knuesel I, Yee BK, Feldon J (2006) The time of prenatal immune challenge determines the specificity of inflammation-mediated brain and behavioral pathology. J Neurosci 26(18):4752–4762. doi:10.1523/JNEUROSCI.0099-06.2006

Meyer U, Murray PJ, Urwyler A, Yee BK, Schedlowski M, Feldon J (2008) Adult behavioral and pharmacological dysfunctions following disruption of the fetal brain balance between pro-inflammatory and IL-10-mediated anti-inflammatory signaling. Mol Psychiatry 13 (2):208–221. doi:10.1038/sj.mp.4002042

Monk CS, Weng SJ, Wiggins JL, Kurapati N, Louro HM, Carrasco M, Maslowsky J, Risi S, Lord C (2010) Neural circuitry of emotional face processing in autism spectrum disorders. J Psychiatry Neurosci 35(2):105–114

Mordekar SR, Prendergast M, Chattopadhyay AK, Baxter PS (2009) Corticosteroid treatment of behaviour, language and motor regression in childhood disintegrative disorder. Eur J Paediatr Neurol 13(4):367–369. doi:10.1016/j.ejpn.2008.06.001

Mundy P, Sullivan L, Mastergeorge AM (2009) A parallel and distributed-processing model of joint attention, social cognition and autism. Autism Res 2(1):2–21. doi:10.1002/aur.61

Niederhofer H (2007) Glutamate antagonists seem to be slightly effective in psychopharmacologic treatment of autism. J Clin Psychopharmacol 27(3):317–318. doi:10.1097/01.jcp.0000270082.30500.69

O'Connor EC, Bariselli S, Bellone C (2014) Synaptic basis of social dysfunction: a focus on postsynaptic proteins linking group-I mGluRs with AMPARs and NMDARs. Eur J Neurosci 39(7):1114–1129. doi:10.1111/ejn.12510

Onore CE, Schwartzer JJ, Careaga M, Berman RF, Ashwood P (2014) Maternal immune activation leads to activated inflammatory macrophages in offspring. Brain Behav Immun 38:220–226. doi:10.1016/j.bbi.2014.02.007

Oskvig DB, Elkahloun AG, Johnson KR, Phillips TM, Herkenham M (2012) Maternal immune activation by LPS selectively alters specific gene expression profiles of interneuron migration and oxidative stress in the fetus without triggering a fetal immune response. Brain Behav Immun 26(4):623–634. doi:10.1016/j.bbi.2012.01.015

Owley T, Salt J, Guter S, Grieve A, Walton L, Ayuyao N, Leventhal BL, Cook EH Jr (2006) A prospective, open-label trial of memantine in the treatment of cognitive, behavioral, and memory dysfunction in pervasive developmental disorders. J Child Adolesc Psychopharmacol 16(5):517–524. doi:10.1089/cap.2006.16.517

Page LA, Daly E, Schmitz N, Simmons A, Toal F, Deeley Q, Ambery F, McAlonan GM, Murphy KC, Murphy DG (2006) In vivo 1H-magnetic resonance spectroscopy study of amygdala-hippocampal and parietal regions in autism. Am J Psychiatry 163(12):2189–2192. doi:10.1176/appi.ajp.163.12.2189

Patterson PH (2009) Immune involvement in schizophrenia and autism: etiology, pathology and animal models. Behav Brain Res 204(2):313–321. doi:10.1016/j.bbr.2008.12.016

Pelphrey KA, Sasson NJ, Reznick JS, Paul G, Goldman BD, Piven J (2002) Visual scanning of faces in autism. J Autism Dev Disord 32(4):249–261
Pelphrey K, Adolphs R, Morris JP (2004) Neuroanatomical substrates of social cognition dysfunction in autism. Ment Retard Dev Disabil Res Rev 10(4):259–271. doi:10.1002/mrdd.20040
Peng D, Yuan X, Zhu R (2013) Memantine hydrochloride in the treatment of dementia subtypes. J Clin Neurosci 20(11):1482–1485. doi:10.1016/j.jocn.2013.02.041
Perlman SB, Hudac CM, Pegors T, Minshew NJ, Pelphrey KA (2011) Experimental manipulation of face-evoked activity in the fusiform gyrus of individuals with autism. Soc Neurosci 6 (1):22–30. doi:10.1080/17470911003683185
Plioplys AV (1998) Intravenous immunoglobulin treatment of children with autism. J Child Neurol 13(2):79–82
Posey DJ, Kem DL, Swiezy NB, Sweeten TL, Wiegand RE, McDougle CJ (2004) A pilot study of D-cycloserine in subjects with autistic disorder. Am J Psychiatry 161(11):2115–2117. doi:10.1176/appi.ajp.161.11.2115
Rinaldi T, Kulangara K, Antoniello K, Markram H (2007) Elevated NMDA receptor levels and enhanced postsynaptic long-term potentiation induced by prenatal exposure to valproic acid. Proc Natl Acad Sci U S A 104(33):13501–13506. doi:10.1073/pnas.0704391104
Robinson IC, Jones PM (1982) Oxytocin and neurophysin in plasma and CSF during suckling in the guinea-pig. Neuroendocrinology 34(1):59–63
Ros-Bernal F, Hunot S, Herrero MT, Parnadeau S, Corvol JC, Lu L, Alvarez-Fischer D, Carrillo-de Sauvage MA, Saurini F, Coussieu C, Kinugawa K, Prigent A, Hoglinger G, Hamon M, Tronche F, Hirsch EC, Vyas S (2011) Microglial glucocorticoid receptors play a pivotal role in regulating dopaminergic neurodegeneration in Parkinsonism. Proc Natl Acad Sci USA 108 (16):6632–6637. doi:10.1073/pnas.1017820108
Sala M, Braida D, Lentini D, Busnelli M, Bulgheroni E, Capurro V, Finardi A, Donzelli A, Pattini L, Rubino T, Parolaro D, Nishimori K, Parenti M, Chini B (2011) Pharmacologic rescue of impaired cognitive flexibility, social deficits, increased aggression, and seizure susceptibility in oxytocin receptor null mice: a neurobehavioral model of autism. Biol Psychiatry 69 (9):875–882. doi:10.1016/j.biopsych.2010.12.022
Sasson NJ, Pinkham AE, Carpenter KL, Belger A (2011) The benefit of directly comparing autism and schizophrenia for revealing mechanisms of social cognitive impairment. J Neurodev Disord 3(2):87–100. doi:10.1007/s11689-010-9068-x
Schaefer GB, Mendelsohn NJ (2013) Clinical genetics evaluation in identifying the etiology of autism spectrum disorders: 2013 guideline revisions. Genet Med 15(5):399–407. doi:10.1038/gim.2013.32
Schneider T, Przewlocki R (2005) Behavioral alterations in rats prenatally exposed to valproic acid: animal model of autism. Neuropsychopharmacology 30(1):80–89. doi:10.1038/sj.npp.1300518
Schneider CK, Melmed RD, Barstow LE, Enriquez FJ, Ranger-Moore J, Ostrem JA (2006) Oral human immunoglobulin for children with autism and gastrointestinal dysfunction: a prospective, open-label study. J Autism Dev Disord 36(8):1053–1064. doi:10.1007/s10803-006-0141-y
Schweingruber N, Reichardt SD, Luhder F, Reichardt HM (2012) Mechanisms of glucocorticoids in the control of neuroinflammation. J Neuroendocrinol 24(1):174–182. doi:10.1111/j.1365-2826.2011.02161.x
Seckl J, Lightman S (1988) Cerebrospinal fluid neurohypophysial peptides in benign intracranial hypertension. J Neurol Neurosurg Psychiatry 51(12):1538–1541
Shahrestani S, Kemp AH, Guastella AJ (2013) The impact of a single administration of intranasal oxytocin on the recognition of basic emotions in humans: a meta-analysis. Neuropsychopharmacology 38(10):1929–1936. doi:10.1038/npp.2013.86
Shenoy S, Arnold S, Chatila T (2000) Response to steroid therapy in autism secondary to autoimmune lymphoproliferative syndrome. J Pediatr 136(5):682–687. doi:10.1067/mpd.2000.105355

Shi L, Fatemi SH, Sidwell RW, Patterson PH (2003) Maternal influenza infection causes marked behavioral and pharmacological changes in the offspring. J Neurosci 23(1):297–302

Smith SE, Li J, Garbett K, Mirnics K, Patterson PH (2007) Maternal immune activation alters fetal brain development through interleukin-6. J Neurosci 27(40):10695–10702. doi:10.1523/JNEUROSCI. 2178-07.2007

Smith SE, Zhou YD, Zhang G, Jin Z, Stoppel DC, Anderson MP (2011) Increased gene dosage of Ube3a results in autism traits and decreased glutamate synaptic transmission in mice. Sci Transl Med 3(103):103ra97. doi:10.1126/scitranslmed.3002627

Soto D, Altafaj X, Sindreu C, Bayes A (2014) Glutamate receptor mutations in psychiatric and neurodevelopmental disorders. Commun Integr Biol 7(1):e27887. doi:10.4161/cib.27887

Stefanatos GA, Grover W, Geller E (1995) Case study: corticosteroid treatment of language regression in pervasive developmental disorder. J Am Acad Child Adolesc Psychiatry 34 (8):1107–1111. doi:10.1097/00004583-199508000-00022

Sterling L, Dawson G, Webb S, Murias M, Munson J, Panagiotides H, Aylward E (2008) The role of face familiarity in eye tracking of faces by individuals with autism spectrum disorders. J Autism Dev Disord 38(9):1666–1675. doi:10.1007/s10803-008-0550-1

Stigler KA, Sweeten TL, Posey DJ, McDougle CJ (2009) Autism and immune factors: a comprehensive review. Res Autism Spectrum Disord 3(4):840–860. doi:10.1016/j.rasd. 2009.01-007

Supekar K, Uddin LQ, Khouzam A, Phillips J, Gaillard WD, Kenworthy LE, Yerys BE, Vaidya CJ, Menon V (2013) Brain hyperconnectivity in children with autism and its links to social deficits. Cell Rep 5(3):738–747. doi:10.1016/j.celrep.2013.10.001

Suzuki K, Sugihara G, Ouchi Y, Nakamura K, Futatsubashi M, Takebayashi K, Yoshihara Y, Omata K, Matsumoto K, Tsuchiya K, Iwata Y, Tsujii M, Sugiyama T, Mori N (2013) Microglial activation in young adults with autism spectrum disorders. JAMA Psychiatry 70 (1):49–58. doi:10.1001/jamapsychiatry.2013.272

Takayanagi Y, Yoshida M, Bielsky IF, Ross HE, Kawamata M, Onaka T, Yanagisawa T, Kimura T, Matzuk MM, Young LJ, Nishimori K (2005) Pervasive social deficits, but normal parturition, in oxytocin receptor-deficient mice. Proc Natl Acad Sci USA 102 (44):16096–16101. doi:10.1073/pnas.0505312102

Teng BL, Nonneman RJ, Agster KL, Nikolova VD, Davis TT, Riddick NV, Baker LK, Pedersen CA, Jarstfer MB, Moy SS (2013) Prosocial effects of oxytocin in two mouse models of autism spectrum disorders. Neuropharmacology 72:187–196. doi:10.1016/j.neuropharm.2013.04.038

Trajkovski V, Ajdinski L, Spiroski M (2004) Plasma concentration of immunoglobulin classes and subclasses in children with autism in the Republic of Macedonia: retrospective study. Croat Med J 45:746–749

Traynelis SF, Wollmuth LP, McBain CJ, Menniti FS, Vance KM, Ogden KK, Hansen KB, Yuan H, Myers SJ, Dingledine R (2010) Glutamate receptor ion channels: structure, regulation, and function. Pharmacol Rev 62(3):405–496. doi:10.1124/pr.109.002451

Uljarevic M, Hamilton A (2013) Recognition of emotions in autism: a formal meta-analysis. J Autism Dev Disord 43(7):1517–1526. doi:10.1007/s10803-012-1695-5

Veening JG, Olivier B (2013) Intranasal administration of oxytocin: behavioral and clinical effects, a review. Neurosci Biobehav Rev 37(8):1445–1465. doi:10.1016/j.neubiorev.2013.04.012

von dem Hagen EA, Stoyanova RS, Baron-Cohen S, Calder AJ (2013) Reduced functional connectivity within and between 'social' resting state networks in autism spectrum conditions. Soc Cogn Affect Neurosci 8(6):694–701. doi:10.1093/scan/nss053

Watanabe T, Abe O, Kuwabara H, Yahata N, Takano Y, Iwashiro N, Natsubori T, Aoki Y, Takao H, Kawakubo Y, Kamio Y, Kato N, Miyashita Y, Kasai K, Yamasue H (2013) Mitigation of sociocommunicational deficits of autism through oxytocin-induced recovery of medial prefrontal activity: a randomized trial. JAMA Psychiatry 71(2):166–175. doi:10.1001/jamapsychiatry.2013.3181

Weigelt S, Koldewyn K, Kanwisher N (2012) Face identity recognition in autism spectrum disorders: a review of behavioral studies. Neurosci Biobehav Rev 36(3):1060–1084

Weng SJ, Carrasco M, Swartz JR, Wiggins JL, Kurapati N, Liberzon I, Risi S, Lord C, Monk CS (2011) Neural activation to emotional faces in adolescents with autism spectrum disorders. J Child Psychol Psychiatry 52(3):296–305. doi:10.1111/j.1469-7610.2010.02317.x

Winslow JT, Insel TR (2002) The social deficits of the oxytocin knockout mouse. Neuropeptides 36(2–3):221–229

Winslow JT, Hearn EF, Ferguson J, Young LJ, Matzuk MM, Insel TR (2000) Infant vocalization, adult aggression, and fear behavior of an oxytocin null mutant mouse. Horm Behav 37 (2):145–155. doi:10.1006/hbeh.1999.1566

Won H, Lee HR, Gee HY, Mah W, Kim JI, Lee J, Ha S, Chung C, Jung ES, Cho YS, Park SG, Lee JS, Lee K, Kim D, Bae YC, Kaang BK, Lee MG, Kim E (2012) Autistic-like social behaviour in Shank2-mutant mice improved by restoring NMDA receptor function. Nature 486 (7402):261–265. doi:10.1038/nature11208

Assessing Cognitive Improvement in People with Down Syndrome: Important Considerations for Drug-Efficacy Trials

Fabian Fernandez and Roger H. Reeves

Contents

F. Fernandez (✉) • R.H. Reeves
Department of Physiology, Johns Hopkins University School of Medicine, Baltimore, MD 21205, USA

Institute for Genetic Medicine, Johns Hopkins University School of Medicine, Baltimore, MD 21205, USA
e-mail: ffernan6@jhmi.edu

K.M. Kantak, J.G. Wettstein (eds.), *Cognitive Enhancement*, Handbook of Experimental Pharmacology 228, DOI 10.1007/978-3-319-16522-6_12

Abstract

Experimental research over just the past decade has raised the possibility that learning deficits connected to Down syndrome (DS) might be effectively managed by medication. In the current chapter, we touch on some of the work that paved the way for these advances and discuss the challenges associated with translating them. In particular, we highlight sources of phenotypic variability in the DS population that are likely to impact performance assessments. Throughout, suggestions are made on how to detect meaningful changes in cognitive–adaptive function in people with DS during drug treatment. The importance of within-subjects evaluation is emphasized.

Keywords

Down syndrome • Intellectual disability • Clinical trial • Cognition • Adaptive behavior • Drug treatment

1 Introduction

1.1 Down Syndrome: Background and Etiology

Down syndrome results from the gametic triplication of human chromosome 21 (meiotic nondisjunction, Hsa21; Sherman et al. 2007). From this event, it is reasonable to assume that the physical presence of an extra nuclear chromosome and dosage imbalance of approximately 170 protein-coding trisomic genes change the constitution and regulatory environment of the genome (Hattori et al. 2000; Patterson 2007; Reeves 2001). Elevated levels of categorically diverse Hsa21 transcripts and finer alterations in disomic allelic expression likely influence developmental trajectories from conception, reprogramming rates of symmetric and asymmetric cell division and undermining the temporal stability of transient populations that guide cellular migration or cue terminal differentiation (Chakrabarti et al. 2010; Tyler and Haydar 2013). Ultimately, the accumulated effects of trisomy reshape the size and properties of every tissue system in the body in a way that is inwardly and outwardly recognizable. In mature systems, approximately 1.5-fold steady-state overexpression of Hsa21 transcripts and fluctuations in Hsa21 proteins and signaling pathways further modify the autonomous/nonautonomous function of adult cells and their response to experience (Begenisic et al. 2011; Chakrabarti et al. 2011; Kahlem et al. 2004; Roper and Reeves 2006; Sultan et al. 2007).

Upon birth, infants with Down syndrome (DS) exhibit a panoply of several dozen clinical features that vary, sometimes widely, in severity or expressivity (Steingass et al. 2011; Weijerman and de Winter 2010). The penetrance for these phenotypes has been thought to be equally variable, but technical diagnostic advances capable of discerning more subtle pathology suggest that trisomy

21 produces a few "common" features. The typical child with DS will show a flattened facial profile due to brachycephaly of the skull, upslanting palpebral fissures and ring-like brushfield spots of connective tissue across the eyes, smaller dysplastic ears, and characteristic changes in form to the upper and lower extremities including a palmar crease on the hands, exaggerated curvature of the fifth finger, and a wide space between the first and second toes known as "sandal gap" (Baum et al. 2008). Further examination of the muscles will reveal hypotonia and joint laxity. More involved procedures will reveal impairments in the visual and auditory systems (Al-Bagdady et al. 2011; Blaser et al. 2006; Courage et al. 1997; Hassmann et al. 1998; Raut et al. 2011; Shott et al. 2001), septation defects in the heart (Dennis et al. 2010; Park et al. 1977), and pulmonary hypoplasia of the lungs (Bertrand et al. 2003; McDowell and Craven 2011) in a significant fraction of individuals.

The hallmark of the DS phenotype is intellectual disability or cognitive impairment (Edgin 2013; Nadel 2003), which refers to a suite of deficits in how information is sought, attended or oriented to, and manipulated for short and long periods of time to effect learning and memory and ascendant functions such as reasoning, problem-solving, and planning. Perhaps nonintuitively, many anatomical aspects of the brain and its function appear normal soon after birth. As development of the nervous system ensues, young children with DS exhibit slowing trajectories of developmental quotient (DQ) and intelligence quotient (IQ) growth that compromise language acquisition and expansion of cognitive skill sets (Abbeduto et al. 2007; Carr 1970, 1988; Chapman 1997; Patterson et al. 2013). Lacking these proper tools, people with DS are hindered in important aspects of their adaptive function (e.g., self-help skills) and participation in society (e.g., social-interpretative skills) (Sigman et al. 1999). The endpoint of having three copies of Hsa21 instead of two is, arguably, the absence of a sense of personal agency that is uncomfortably felt over the entire lifespan in those with DS and their families.

The period within the first few years of life during which children with DS begin a terminal IQ decline provides a window into the brain mechanisms responsible for chronic problems in declarative-episodic memory (i.e., those processes that encode the contextual and factual details of everyday experiences). In a long-running series of elegant investigations, Wishart and colleagues found that children with trisomy 21 could perform four sequentially more difficult levels of an object concept task, but over time, performance was erratic (1986, 1993, 2001). Longitudinal profiles of the same children showed that early success at the lower- or higher-level variations of the test could be followed up by failures at older ages. Within individual sessions, curious patterns of performance also emerged where the children could execute more difficult parts of the test but not less difficult ones. These findings suggested that the development of the various sub-domains comprising episodic memory was not *integrated* in people with DS the way that they are in the typically developing population. In Sect. 2.1 below, we lay out evidence to suggest that this unstable integration is a byproduct of altered connectivity loops between parts of the brain known to be strongly affected by DS, namely, the cerebellum and frontotemporal lobes.

The last decade has seen steady progress toward characterizing pharmacological interventions that might improve phenotypes related to the brain and cognition of individuals with DS. We describe the origins of these drug treatments, starting with some background discussion of the advent of the Ts65Dn mouse, an animal model of trisomy 21 that has catalyzed a deeper understanding of DS biology and left an indelible mark on DS translational research. We go on to systematically document the variability of several clinical and psychological features of DS that would be predicted to interfere with assessments of cognitive improvement in drug-efficacy trials and stipulate that the least imperfect methodological design for such trials is a within-subjects assessment. Further considerations are made with regard to the US Food and Drug Administration (FDA) approval process and some lessons that have been gleaned from ongoing efforts to bring drugs to market for the treatment of autism spectrum disorders. People born with trisomy 21 display a tremendous amount of phenotypic variation stemming from the genetic background in which aneuploidy occurs and environmental factors at work in caregiver and educational support. Despite expectations to the contrary, they are not a homogeneous group, and efforts to treat the physical and mental health of individuals with DS should give careful thought to this diversity.

1.2 Use of Trisomic Mouse Models as Platforms for Therapeutics Discovery

Can a mouse be used as an informative model of DS, a *human* condition caused by the presence of a supernumerary *human* chromosome and typified by deficits in quintessential *human qualities* such as cognition and language? Is there a genetic state that could be engineered in a mouse so that it, in turn, logically reproduces the trajectory and range of genetic, cellular, and larger phenotypic effects observed in people with DS? These were contentious questions over three decades ago as modern techniques in molecular biology and murine genetics were beginning to be implemented in DS research (Epstein et al. 1982, 1985). To put them in some perspective, little comparative mapping had been done between the two species to indicate how material from Hsa21 was distributed in the mouse genome. If, for instance, segments of Hsa21 were scattered in small blocks across several different mouse chromosomes (Mmu's), then the likelihood that an animal model could be devised was poor to nonexistent. If these segments were better consolidated to a few Mmu's, then problems might still remain because of disparities in how individual gene-coding regions were ordered or how they were subject to regulation and transcription. There was not even a guarantee that the biological function of each Hsa21 gene would be conserved in its mouse ortholog. By the mid 1980s, several genes had been located to both Hsa21 and Mmu16, improving the prospects that a large degree of synteny was shared between Hsa21 and a single Mmu (Reeves et al. 1986). Later efforts employing linkage analysis of restriction fragment length polymorphisms (RFLPs) and high-resolution physical mapping using contigs constructed from yeast artificial chromosome (YAC) libraries showed particular stretches of homology between Hsa21 and the distal portions of Mmu16 (Cabin et al. 1998; Pletcher et al. 2001; Reeves et al. 1997). Against this backdrop, it became increasingly clear—from both published and unpublished work—that the q-arm of Mmu16 housed many Hsa21 orthologs, had preserved their organization and spacing (i.e., the genes were found in the same sequential order and, presumably, with the same regulatory elements in between), and could be targeted to engineer a mouse counterpart to DS.

In a *tour de force* project to develop the first postnatally viable mammalian model of DS, Davisson, Reeves, and colleagues screened hundreds of irradiated male mice for translocations involving Mmu16 and used these animals to generate mice segmentally trisomic for most of the Hsa21 homologous region (Davisson et al. 1993). These mice are now referred to as Ts65Dn (Reeves et al. 1995). Though they do not comprehensively reproduce the range of human symptoms in people with DS, their study has elucidated many impressive parallels in how Hsa21 gene dosage impacts the three-dimensional construction of the craniofacial skeleton (Richtsmeier et al. 2000), septation of the heart (Moore 2006; Williams et al. 2008), solid tumor growth (Sussan et al. 2008; Yang and Reeves 2011), the function of sensory-motor systems (Hampton et al. 2004; Han et al. 2009; Scott-McKean et al. 2010), and sleep (Colas et al. 2008). This fidelity has extended to the prediction of brain phenotypes in people with DS that were first discovered in Ts65Dn mice (Baxter et al. 2000). The corpus of work in Ts65Dn has suggested that selective overexpression of Hsa21 orthologs under physiological control can *phenocopy* significant aspects of DS in animals and that the mice themselves might be platforms for the discovery of drugs that mitigate intellectual disability in children and adults born with trisomy (Fernandez and Garner 2008; Hyde et al. 2001; Smith et al. 2014). Since the mid 2000s, several different drug classes have shown an ability to improve rudimentary behavioral indices of declarative learning and memory in the Ts65Dn model, including GABA antagonists (Fernandez et al. 2007) and inverse agonists (Braudeau et al. 2011a, b), open-channel antagonists of the NMDA receptor (Costa et al. 2008), Smoothened agonists (SAG; Das et al. 2013), compounds that stimulate norepinephrine (Dang et al. 2014; Salehi et al. 2009), and dietary choline (Moon et al. 2010). It remains to be seen whether any of these agents will prove to be similarly efficacious in people, though some early signs are encouraging. For the rest of the chapter, we discuss the finer points of these translational endeavors.

2 Variability of Clinical Phenotypes in Individuals with Down Syndrome

2.1 Neurocognitive Development

A modern understanding of neurocognitive function in DS was established in the 1986 work "Down syndrome in Neurobiological Perspective" by Nadel, who suggested that intellectual disabilities gradually arose in early childhood from developmental arrest of late-maturing brain structures, including the prefrontal cortex, hippocampus, and cerebellum. Prenatal brain development is characterized by a series of explosive additive events, where large populations of cells undergo brisk division, migration, and neural or glial differentiation. Upon birth, the brain undergoes a more deliberative mix of additive events (such as myelination) and *regressive* events, where many neurons are lost through apoptosis or refined in their connectivity via activity-dependent pruning of the dendritic tree and elimination of synaptic contacts (Gogtay et al. 2004; Huttenlocher 1979; Huttenlocher et al. 1982; Lebel et al. 2008; Petanjek et al. 2011). Several large-scale analyses of DS fetal brain tissue suggest that prenatal growth is similar to that of the typically developing population. There are no differences in weight or lipid metabolism (Brooksbank et al. 1989; Mito et al. 1991; Schmidt-Sidor et al. 1990), and cells appear to have the same morphology, dendritic branching, spine numbers (Cragg 1975; Takashima et al. 1981), and apposition length and ratios of asymmetric/symmetric junctions (Petit et al. 1984). The situation shifts postnatally as changes in phylogenetically older motor-sensory areas of the brain can be observed within the first year of age in infants with DS. The changes signify an overindulgence in mechanisms that generate regressive events. Trisomic neurons in layers of the visual, auditory, and somatosensory cortex are culled more than euploid neurons (Ross et al. 1984; Schmidt-Sidor et al. 1990; Wisniewski et al. 1984), and the dendritic perimeter of the remaining cells is pared back to a greater extent (Becker et al. 1986; Golden and Hyman 1994; Takashima et al. 1981). Readouts of connectivity in older children and adults with DS indicate that, in light of this, local circuits are organized normally, but longer range networks are less synchronized (Ábrahám et al. 2012; Anderson et al. 2013; Elul et al. 1975).

Exaggerated regressive events hit the hippocampus and cerebellum especially hard because important functional pockets within them are still proliferating after birth. In the case of the hippocampus, granule cells of the dentate gyrus (or *fascia dentata*) are still dividing to sculpt a niche that will act as the main portal for information flow from areas of adult neocortex to the hippocampal formation. In the case of the cerebellum, a much larger group of granule neurons is actively dividing (and migrating) to establish the fundamental wiring of the structure and its interface with circuit loops to and from the frontotemporal lobes. Both proliferating areas eventually become impoverished in the brains of individuals with DS (Baxter et al. 2000; Contestabile et al. 2007; Guidi et al. 2008, 2011; Insausti et al. 1998; Lorenzi and Reeves 2006; Roper et al. 2006).

The prefrontal cortex (PFC)—the single largest region in the human brain comprising up to one-third of the entire cerebral cortex—remains an ornate chrysalis over the first three decades of life, undergoing repeated waves of developmental remodeling that fine-tune its properties (Petanjek et al. 2011; Spencer-Smith and Anderson 2009). Given the conditions of the postnatal DS brain, it would be prudent to assume that the nature and sophistication of this remodeling are permanently altered (Jernigan et al. 1993; Lögdberg and Brun 1993; Raz et al. 1995; Wang 1996). Reduced complexity is initially evident within the shape and outer surface of the frontal lobes. The fronto-occipital diameter shortens, as if growth on the leading edges of the brain has grounded to a halt (Schmidt-Sidor et al. 1990). The cortical surface, usually elaborated with tiles of sulci and gyri, also flattens with shallower primary and fewer secondary sulci. Later, myelination of the PFC is poor and interferes with timely communication in a regional network that is the most interconnected of any in the adult central nervous system (Fuster 1989; Wisniewski and Schmidt-Sidor 1989).

Incomplete development of the PFC, hippocampus, and cerebellum has a variety of effects on DS cognition. In a bit of a departure from the Nadel doctrine, we argue that these effects are best conceptualized from the standpoint of *interactive specialization*—the idea that although cognitive modules can work semiautonomously if pressed, in an intact state, they do not (Johnson 2001). As such, most cognitive skill sets emerge not just from activity in one brain region or another but *also* from patterns of connectivity between them. Vis-à-vis DS, intellectual disabilities might stem from the failure to build an interregional axis between the PFC–hippocampus–cerebellum during the larger progression of general development that sweeps from the posterior aspects of the brain to the anterior. The suggestion that the Nadel doctrine presents an untenable dichotomy (i.e., mental development arises from either the elaboration of a single capacity or the specification of several "separate" and "autonomous" modules) has ramifications for how we approach treating intellectual disabilities in people with DS.

2.1.1 Prefrontal Cortex

Historically, the PFC has been thought to oversee a hierarchical array of cognitive abilities collectively referred to as executive function (EF; Miller and Cohen 2001). The earliest developing component of EF and a necessary prerequisite for other EF skills is (1) attentional control, which is the ability to monitor and filter sensory information over an extended period of time without distraction (Anderson 2002; Ridderinkhof and Van der Stelt 2000). A closely related subcomponent to attentional control is behavioral inhibition—the practice of suppressing an otherwise knee-jerk or preconditioned response within an inappropriate context (Rueda et al. 2005). Corollaries to it include self-monitoring and self-regulation (i.e., adaptive management of thoughts, feelings, and actions; Kochanska et al. 2001). Younger children with DS demonstrate problems with behavioral inhibition that are overcome with age. While 3- to 4-year-olds show significantly shorter latencies to touch a prohibited toy relative to age-matched controls (Kopp et al. 1983), teenagers with DS perform at levels commensurate with their overall intellectual development in a digital "stopping" task where they are asked to respond to the visual presentation of a letter only in the presence/absence of an auditory tone (Pennington et al. 2003). These psychometrics suggest that the PFC might develop to a point where it can subserve some indices of attentional control. However, maturation stops here (Campbell et al. 2013).

Attentional control lays the foundation for more complex domains like (2) cognitive flexibility and (3) problem-solving (Campbell et al. 2013; Klenberg et al. 2001). Cognitive flexibility, itself, incorporates the concepts of working memory and set-shifting behavior (Garon et al. 2008; Oh et al. 2014). Working memory permits the transient evaluation and synthesis of multimodal material from strings of information being acquired in real time with knowledge that has been stored long term (Baddeley and Jarrold 2007; Laws 2002). Individuals engage in set-shifting behavior when they switch from one adaptive response strategy to another depending on the demands of a particular task (Rowe et al. 2006; Weed et al. 2008). Together, these "flexible" processes allow for planning and problem-solving (Lee et al. 2011).

Individuals with DS experience significant deficits in verbal working or serial order memory as revealed in assessments where they are asked to repeat the sequence of a list of words that are spoken to them. Over two decades of careful study, Jarrold has found that the deficits occur selectively in verbal short-term memory (Brock and Jarrold 2004, 2005; Jarrold and Baddeley 1997, 2001; Jarrold et al. 2000, 2002), with less impact in visuospatial processing (though see Spanò, Lanfranchi and colleagues for further clarification of this distinction; Lanfranchi et al. 2009). The working memory difficulties of those with DS are accompanied by difficulties in set-shifting attention as exemplified in variations of the Dimensional Change Card Sorting (DCCS) task (Pennington et al. 2003; Zelazo et al. 1996). Here, children or adults are asked to sort cards across various rules, for instance, by color vs. by object category. People with DS can often execute the first session of the task when sorting requires an understanding of one novel rule. Once they have grown accustomed to this requirement, however, they show an inability to segregate the cards in observation of a second rule. This underlying perseveration or cognitive rigidity surfaces in problem-solving situations where toddlers with DS cannot adapt unsuccessful strategies into successful ones so as to achieve a desired outcome (Fidler et al. 2005b). Overall, the profile of the PFC in DS suggests that there is broken hierarchical elaboration of attentional control to fuller expressions of EF skills like cognitive flexibility and that this arrest precludes real-world abilities to creatively troubleshoot and overcome problems.

2.1.2 The Hippocampus and Episodic Memory

The hippocampus functions at the core of a web of neural circuitry in the medial temporal lobe that processes and encodes the contextual and factual details of everyday life for later conscious recollection (Squire et al. 2004). Historically, it has been associated with the concepts of explicit long-term memory and information storage, but pioneering work throughout the 1970s by O'Keefe and Nadel demonstrated that it also constructs internal two-dimensional representations of three-dimensional space (1978). Research has unequivocally shown that individuals with DS are impaired in verbal and nonverbal assessments of intermediate or long-term memory (e.g., in the verbal domain, deficits in delayed word recognition and prose recall; in the non-verbal domain, deficits in picture, object, or visuospatial recognition; Carlesimo et al. 1997; Ellis et al. 1989; Katz and Ellis 1991; Vicari et al. 2000, 2005). Teenagers and adults also have trouble navigating in real environments when forced to use geometric and layout information (Edgin et al. 2012) and cannot use digital cues to improvise on previously learned routes through a virtual environment (i.e., wayfinding behavior; Courbois et al. 2013). In other words, while they maintain a semblance of direction and can pick up information related to space, they are impaired in the use of these facilities to create a consolidated map of the world around them.

Another conceptualization of the hippocampus is that of a powerful associative mnemonic device that is capable of binding various temporal and contextual features of a momentary episode and overlaying them with internal states related to motivation and affect (Cox et al. 2014; Retailleau et al. 2012; Wolosin et al. 2012; Yonelinas 2013). The hippocampus takes snapshots of moment-by-moment episodes (Jacobs et al. 2013), fuses them to construct a uniform "past" experience (Olsen et al. 2012), and records the subjective qualities of these experiences—for instance, whether they were reinforcing or aversive (Wimmer and Shohamy 2012). This conceptualization might be more particular to its function than just that of a reservoir from which information is categorized and stored. Many of the activities associated with memory encoding and retrieval that have been ascribed to the hippocampus have been empirically shown to tap the PFC in humans. Clinicians have noted memory deficits in patients with PFC lesions (Shimamura 1995), and ongoing investigations by Ranganath and colleagues have reported significant activation of dorsolateral and ventrolateral PFC during performance of tasks requiring long-term memory for faces or words (Blumenfeld and Ranganath 2007; Jenkins and Ranganath 2010; Murray and Ranganath 2007; Ranganath et al. 2003).

In light of these data, one should consider the possibility that childhood development is marked by a carefully orchestrated "melding" of the frontal and temporal lobes into a uniform episodic memory system. Under this model, the hippocampus's extraordinary associative properties are subjugated by the PFC as the latter structure continues its decades-long refinement. The PFC carries some oversight capacity that is able to modulate the valence of some of the information that the episodic system is encoding (De Saint et al. 2013; Dolcos et al. 2004), string episodes *temporally* together when there are quick or unexpected changes to context (Kesner et al. 1994; Gutchess et al. 2007; Murty and Adcock 2013), and prioritize overlapping associations between memory representations (i.e., offset proactive interference; Shimamura et al. 1995). This view of frontotemporal development suggests that the emergence of cognition in people born with trisomy 21 could be turbulent. At first, skills in early childhood would accrue as individual regions of the brain mature, but as integration becomes necessary (i.e., via some genetically predetermined timetable), these skills might disappear or be significantly compromised. What follows is a situation not unlike what Wishart described in her longitudinal studies on object permanence in toddlers with DS (1986, 1993, 2001).

2.1.3 The Cerebellum and Interactive Specialization

Per conventional thinking, the cerebellum acts as a filter for aggregated sensorimotor information that is used to maintain coordination and counteract fine changes to vestibular balance (Massaquoi 2012). Difficulties with gross motor coordination are nuanced in individuals with DS, despite overt structural pathology in the cerebellum including low density of granule cells and reduced volume (Aylward et al. 1997; Baxter et al. 2000). While older children and adults with DS display atypical patterns of movement and problems with handwriting and other tasks that require use of fine digits (Galli et al. 2010; Latash et al. 2002), it has not been determined whether these problems result from altered cerebellar function or originate from poor functioning of, perhaps, the motor cortex (Dmitriev 2001). How stunted cerebellar growth manifests clinically in those with DS beyond diminished optokinetic and vestibular reflexes remains an open question (Costa 2011a, b), although recent work in mice and humans suggests critical contributions to navigation and verbal working memory (Burguière et al. 2005, 2010; Marvel and Desmond 2010; Paulesu et al. 1993; Passot et al. 2012; Rochefort et al. 2011, 2013).

Understanding the role of the cerebellum in hippocampal spatial operation has evolved over just the past several years. An elegant series of experiments has been conducted with L7-PKCI, a mouse model in which synaptic plasticity at the parallel fiber–Purkinje cell circuit is specifically ablated without disrupting the electrophysiological properties of individual granule or Purkinje cells, the structural integrity of the cerebellum, or the behavioral sensorimotor reflexes of the animals. Rochefort, Rondi-Reig, and their colleagues found that L7-PKCI cannot maintain stable spatial or "place field" maps in a water maze task (2011). The authors have suggested that when learning a new environment, the contextual self-motion information that the cerebellum normally collects to coordinate movement is co-opted by the hippocampus to perfect cognitive representations of space. This way, maps can be continually compared with online ambulatory experience and primed bidirectionally with activity in either brain region. The use of individual movement to compute spatial relationships has been described in the human literature and is termed "spatial updating" (Burgess 2006). A role for the cerebellum and vestibular system in this phenomenon has been well characterized (Jahn et al. 2012; Smith 1997).

Though several polysynaptic pathways are likely, the establishment of more direct connections between the cerebellum and hippocampus has proven challenging. A small corpus of anatomical work and electrophysiology in awake animals during the 1960s–1970s suggests projection of the fastigial nucleus bilaterally to the dentate gyrus and CA3 subfield of rhesus monkeys, cats, and rats (Harper and Heath 1973; Heath 1973; Heath et al. 1978, 1980; Heath and Harper 1974; Newman and Reza 1979; Paul et al. 1973; Snider and Maiti 1975, 1976). In one report, stimulation of the fastigial nucleus produced evoked potentials in the hippocampus with a latency of as little as 1–2 ms (Heath and Harper 1974). Since no limbic connections of any sort are evident from the lateral cerebellar hemispheres or other deep output nuclei such as the dentate nucleus, these data suggest that the fastigial nucleus might represent a unique pathway by which the cerebellum instructs hippocampal function, though this supposition is only theoretical at the moment. What has been made clear is that the vestibular system makes significant contributions to human spatial processing (Smith 1997); loss of the system results in selective hippocampal atrophy and spatial memory deficits (Brandt et al. 2005).

Lackluster development of the cerebellum and hippocampus implies a double hit on how people with DS gain knowledge of space and contextualize (i.e., bind) objects and faces within memory. It might not be a coincidence that researchers have charted IQ declines in children with DS that begin around the time toddlers learn to walk. However, a larger dysfunctional axis is also implicated between the cerebellum and frontotemporal lobes as cognition stagnates. The ventral cerebellum normally exhibits protracted maturation in lobules VI/Crus1, VIIB, and VIII and has been shown to subserve inner speech processes essential for verbal working memory (Marvel and Desmond 2010; Menghini et al. 2011; Tiemeier et al. 2010). fMRI (functional magnetic resonance imaging) activation of VI occurs when individuals see verbal materials like alphabetical letters relative to nonverbal symbols (Ravizza et al. 2004, 2006). Activation of VIIB/VIII occurs when this information has to be maintained across a delay and retrieved (Chen and Desmond 2005; Ghosh et al. 2008). Several histological studies have demonstrated cerebellar connections to multiple divisions of the PFC; the circuitry is thought to solidify gradually over the course of adolescence (Middleton and Strick 2001; for review, please reference Ramnani 2006). As individuals age, they likely depend more on frontotemporal–cerebellum networks for verbal working memory (Diamond 2000). The fact that people with DS struggle so significantly with this aspect of cognition might be explained by failures to bridge nodes within these networks.

2.1.4 Posterior–Anterior Shift with Aging

Cognition deteriorates to varying degrees in typically developing individuals as they enter advanced age. Older adults who maintain successful performance—even with accumulating pathology—will often display increased bilateral recruitment of the left and right hemispheres and engage additional neural circuitry immediately adjacent to that utilized in younger individuals for task performance (Grady 2012; Reuter-Lorenz and Lustig 2005). They will also show "migration" of activity from perceptual and medial-temporal regions of the brain to the frontal lobe (Dalton et al. 2014; Eyler et al. 2011). This phenomenon, termed the posterior–anterior shift with aging (PASA; Davis et al. 2008), has been reported across several behavioral measures of working memory, object recognition, and visuospatial processing (Ansado et al. 2012; Grady et al. 2005; Gutchess et al. 2005; Park et al. 2003). It likely reflects attempts by the episodic memory system to compensate for hippocampal damage usually inflicted through decades of living. Synaptic loss, atrophy, and decreased efficiency of activation are all observed in the hippocampus of normally aging individuals as they get older.

Given that the hippocampus is impaired much earlier in people born with trisomy 21, one might predict to see premature signs of PASA in neuroimaging and psychometric studies. Early data suggest this might be the case. One investigation found that teenagers and twentysomethings with DS strongly activated the frontal cortex during an object semantic task that recruited primarily the occipital and parietal lobes in chronological age-matched (CA) controls (Jacola et al. 2011). Another, using a similarly comprised DS cohort, documented increased metabolic rates of glucose in the frontal lobe at rest relative to CA young adults (age range, 20–35 years; Horwitz et al. 1990). The prospect of regional activity compensation in the DS brain has been noted by Edgin and colleagues who characterized inappropriate developmental patterns of contextual object recognition in individuals with the condition (2014). In concert, these findings point to the likelihood that some with DS might appear to have steady "hippocampal" function even in the presence of severe temporal lobe disease. They further underscore two thematic points of this section: (1) drug development for intellectual disabilities should not be solely based on modular-centric approaches but also take into account interplays between the PFC, hippocampus, and cerebellum and (2) brain imaging biomarkers in DS might shift with age in ways that are hard to reconcile with behavioral measures and might not bear any relationship to drug efficacy or lack thereof.

2.2 Early-Onset Hypothyroidism

Inclusion and exclusion criteria and group-matching designs reported in the DS clinical trials literature codify a number of obvious factors that might introduce performance variability among people with DS. Conspicuous or "loud" factors that could interfere with the assessment of cognitive improvement during drug treatment include differences in baseline global or linguistic function (Abbeduto et al. 2008; Chapman et al. 1998) and sufficiency of hearing and vision (Austeng et al. 2013; Cregg et al. 2003; Felius et al. 2014; John et al. 2004; Libb et al. 1985; Suttle and Lloyd 2005; Woodhouse et al. 2000), along with any personal history of neurological events such as infantile spasms (Eisermann et al. 2003; Silva et al. 1996), epilepsy (Arya et al. 2011; Goldberg-Stern et al. 2001), or sleep apnea (Rosen 2011). In the sections that follow, we will discuss a few "silent" factors that are not usually considered. The first of these, congenital hypothyroidism and heart disease, cannot be adequately controlled for in assessments of the adult population. The latter, including APOE (apolipoprotein E) status and secretion of amyloid-beta peptides from *a*myloid *p*recursor *p*rotein (APP) cleavage, can be screened and evaluated for their effects on treatment response. The complex matrix of known and unknown phenotypes concealed in the average person with DS has the potential to create a false sense of security in clinical trial settings where matched-group assignments are based on loud variability factors alone and might interfere with efforts to make informative cross-group comparisons of drug efficacy.

People with DS show biochemical markers of thyroid dysfunction across the lifespan, with an adult prevalence of up to 54 % (Hawli et al. 2009; King et al. 2014; Pueschel et al. 1991). The thyroid gland is a butterfly-shaped organ that develops within the anterior side of the neck, with physical attachment to underlying tracheal cartilage. It responds to pituitary secretion of *t*hyroid-*s*timulating *h*ormone (TSH) by augmenting the synthesis and release of two chemicals, thyroxine (T_4) and triiodothyronine (T_3), which go on to regulate anabolic energy metabolism and hormone sensitivity throughout the body (Stenzel and Huttner 2013). The brain is one of the primary sites of action for T_4/T_3 during development and adulthood (Kester et al. 2004). The clinical significance of adult- or acute-onset hypothyroidism on cognitive function (e.g., episodic memory) and mental health has been established in the general population, and exclusion criteria for the participation of individuals with DS in drug-efficacy trials often specify removing those with active thyroid problems (e.g., Boada et al. 2012). However, the impact of congenital hypothyroidism on later adult cognition and behavior—and its bearing on DS clinical trial enrollment—has received less attention.

Animal research has long shown that maternal or postnatal thyroid deficiency decreases neurogenesis, synaptogenesis, and myelination in the fetal–neonatal brain. In humans, maternal hypothyroxinemia caused by low intake of dietary iodine or untreated *c*ongenital *h*ypothyroidism (CH) produces a severe intellectual disability syndrome. Despite *American Academy of Pediatrics* guidelines advising thyroid screening in newborns and young children with DS (i.e., as embodied in their policy statement "Health Supervision for Children with Down syndrome"), a recent study estimated that only about 40 % of children receive all required exams (Fergeson et al. 2009). Medical detection is often complicated by the overlap of features shared by impaired thyroid function and DS, such as poor intellectual or physical growth and muscle fatigue (Hardy et al. 2004).

Around birth, many infants with DS will exhibit a mild hypothyroid state defined by compensatory elevations of TSH and decreased concentrations of circulating T_4 that are left-shifted relative to the normal population distribution (Luton et al. 2012; Sarici et al. 2012; Van Trotsenburg et al. 2003, 2006). This form of CH is insidious because it cannot be easily diagnosed and generates ambiguity as to whether it should be formally treated (Graber et al. 2012). To evaluate whether thyroid supplementation would improve basic indices of mental and motor development in DS *without* a diagnosis of CH, Van Trotsenburg et al. conducted a randomized, double-blind trial in which the investigators treated ~200 neonates with DS daily with T_4 or placebo (2005). By 2 years of age, toddlers given T_4 from birth displayed significantly smaller delays in motor development and smaller delays in mental development that reached borderline statistical significance (see Kowalczyk et al. 2013 for additional data on thyroxine-mediated growth improvement). These results suggest the existence of an underlying hypothyroid disorder that might variably affect a wide swath of the DS population within the first few years of brain maturation (Tenenbaum et al. 2012). It may or may not be addressed, and its effects on later developmental trajectories have not been comprehensively studied.

2.3 Congenital Heart Disease

Congenital heart disease (CHD) occurs in nearly half of people born with trisomy 21 (Freeman et al. 1998; Li et al. 2012; Mogra et al. 2011; Paladini et al. 2000; Vis et al. 2009; Wells et al. 1994). The term refers to a suite of developmental defects that arise in the septation of the heart into four distinct chambers: the right atrium/right ventricle and the left atrium/left ventricle. These chambers normally function at the intersection of a systemic and pulmonary circuit that conveys low-oxygen blood from general circulation to the right side of the heart to the lungs, and oxygen-rich blood from the lungs to the left side of the heart back out to the rest of the body. Incomplete closure of the septal walls can allow mixing of deoxygenated and oxygenated blood, forming a left-to-right shunt that will increase pulmonary artery pressure (McDowell and Craven 2011; Suzuki et al. 2000) and heart failure if not surgically corrected (Fudge et al. 2010; Majdalany et al. 2010).

Children with DS are screened for CHD soon after birth using Doppler echocardiography and electrocardiography; these noninvasive diagnostics yield anatomical and functional information (Dennis et al. 2010; Ghaffar et al. 2005). Echocardiography uses ultrasounds to image the structural integrity of the heart tissue and its blood vessels and, via the Doppler effect, estimates the velocity and direction of blood flow (Turan et al. 2014). Electrocardiography (EKG) uses a system of surface electrodes placed on the torso and limbs to record electrical impulses created when the heart beats (Vetter et al. 2011). The resulting waveforms visualize whether the activity is rhythmic (healthy) or arrhythmic (unhealthy) and can be analyzed in ways that will provide estimates of heart rate and the size or relative position of each of the four chambers.

A few studies suggest that individuals with DS exhibit patterns of cardiovascular pathology that cannot be fully ascertained by one-time echocardiography or EKG. In one, Al-Biltagi et al. found that children with DS who had been previously determined to have no anatomical changes to the heart still showed left and right ventricular systolic/diastolic blood pressure differences relative to controls under closer scrutiny of Doppler ultrasounds (2013). Higher pulmonary artery pressure was noted. In another, Narchi found that some 2-day-old infants with DS could have normal EKG readings, but show anatomical evidence of CHD with echocardiography (1999). These findings suggest that there are individuals in the DS population who might be "asymptomatic" for CHD at birth, but still carry structural defects that slowly create conditions favoring a left-to-right shunt with increases in pulmonary hypertension as development proceeds. Presumably, pulmonary vascular damage and poor transport of oxygenated blood to the periphery and brain could ensue—a chilling proposition given that ~40 % of those with DS are already at much greater risk of respiratory distress from airway anomalies, lung hypoplasia, and subpleural cysts (Biko et al. 2008; Bloemers et al. 2007, 2010a, b). Any idiosyncratic variability introduced by altered developmental trajectories in the heart, lungs, and brain would be uncharted at the time people with DS are recruited for clinical trial participation (Hilton et al. 1999; Mogra et al. 2011; Shah et al. 2004; Shapiro et al. 2000). This variability could influence cognitive outcome measures as overt, surgically corrected CHD in DS has been shown to account for some individual differences in expressive language delay (Visootsak et al. 2013).

2.4 Premature Aging

2.4.1 Triplication of APP Drives Abeta Production

People with DS are trisomic for *APP*, a gene at the center of a major model for how neuropathology and cognitive impairment occur in Alzheimer's disease (AD). The amyloid hypothesis posits that elevated cleavage of APP by β- and γ-secretase produces toxic accumulation of beta amyloid (Aβ) peptides in the extracellular space between neurons. APP fragments, like $A\beta_{1-42}$, polymerize into diffusible oligomers that can self-aggregate into fibrils and thick plaque deposits that then interfere with synaptic transmission and, over time, cause localized brain damage. Support for this hypothesis has been bookended by two landmark observations published in 1991 and 2012, in which researchers uncovered genetic mutations around the protease β-site in APP that either promoted Aβ production and early-onset dementia or impeded generation of both in the elderly (Chartier-Harlin et al. 1991; Jonsson et al. 2012; Murrell et al. 1991).

By virtue of APP overexpression, individuals with DS secrete larger quantities of Aβ. Immunoreactivity and metabolic signatures for the presence of soluble Aβ have appeared in the brains of the youngest children with DS surveyed, from 5 months gestation to 8 years of age (Cataldo et al. 2000; Leverenz and Raskind 1998; Mori et al. 2002; Teller et al. 1996). At about 10 years of age, plasma levels of $A\beta_{1-42}$ are significantly increased relative to slightly older siblings (32 pg/ml versus 23 pg/ml), as are levels of $A\beta_{1-40}$, which are nearly doubled (277 pg/ml versus 155 pg/ml) (Mehta et al. 2007). These trends in plasma concentration are remarkably well maintained throughout the first five decades of life ($A\beta_{1-42}$, DS median range [20–35 pg/ml], control median range [15–17 pg/ml]; $A\beta_{1-40}$, DS median range [150–200 pg/ml], control median range [75–125 pg/ml]) (Jones et al. 2009; Mehta et al. 2003; Prasher et al. 2010; Schupf et al. 2001).

Stepping back from the numbers, a few things stand out about the development of the DS brain. First, Aβ is generated during the fetal period at a time when available data suggest that it should not be. Though full-length APP has defined roles in neurodifferentiation and synaptogenesis (Apelt et al. 1997), secretory cleavage of it is practically undetectable in euploid fetal neurons, astrocytes, and microglia (Haass et al. 1991; Hung et al. 1992). Aβ serves a homeostatic function in adult neural circuits. The peptide gets processed for exocytotic release from activity-driven increases in APP internalization (Bero et al. 2011; Cirrito et al. 2005) and, in turn, decreases excitatory synaptic currents via AMPA and NMDA receptor endocytosis (Hsieh et al. 2006; Snyder et al. 2005). To effect this, oligomers of Aβ bind leukocyte immunoglobulin-like receptor B2, leading to a signaling cascade that breaks down the actin cytoskeleton encasing the postsynaptic density (Kim et al. 2013). Amid the latter part of gestation and within the first few years after birth, the brain wires local and distal assortments of neurons observing a number of activity- and NMDA-dependent rules. Precocious accumulation of Aβ in DS might therefore handicap formation of neural circuits and have a hand in some of the exaggerated regressive events (i.e., apoptosis, dendritic atrophy) routinely noted during this developmental window.

The second item worthy of attention is the observation that the episodic memory system in the DS brain begins to weather AD-like neuropathology at much earlier time points than the typically maturing euploid one. Plaque deposits materialize during adolescence and the early twenties in the trisomic PFC and hippocampus when connectivity between these regions should be consolidated, not divided (Hof et al. 1995; Lemere et al. 1996; Motte and Williams 1989). This scenario clouds the historical distinction made between developmental trajectories associated with growing neurocognitive sophistication in school-aged children and disease trajectories associated with cognitive decline starting in middle age. Both collide in younger people with DS. It is likely that individual differences in Aβ peptide secretion (e.g., disomic genetic modifiers of APP metabolism) scale the degree of overlap between these events (Patel et al. 2011). For example, higher adolescent production of Aβ could predispose subsets of individuals with DS to an earlier version of mild cognitive impairment that would be difficult to distinguish from preexisting intellectual disabilities. Given the uncertainties regarding how AD processes impinge on development, clinical trial organizers might be advised to consider screening participants for Aβ levels prior to enrollment or group assignment. Whether amyloid-modifying treatments should be prescribed from childhood on to normalize plasma Aβ and, possibly, improve cognitive outcomes is another provocative question.

2.4.2 APOE Status

APOE is a constituent of *l*ow-*d*ensity *l*ipoproteins (LDLs), which bind and shuttle excess cholesterol and other triglycerides from the blood to the liver. In the brain, it is mainly secreted by astrocytes and can be found circulating in the interstitial space and cerebrospinal fluid. Further roles in restorative processes linked to phospholipid redistribution and cell membrane repair (including maintenance of the blood–brain barrier) have been hypothesized (for review, see Zlokovic 2013). Allelic variants of APOE differing by single-amino acid substitutions at two residues exhibit divergences in conformational stability that impact vascular clearance of Aβ from the brain. While APOE ε2 and APOE ε3 interact strongly with LDL receptor-related protein 1, which directs rapid clearance of Aβ peptides, APOE ε4 interacts only weakly with the protein (Deane et al. 2008). As a result, APOE ε4 will often shunt Aβ removal to other disposal mechanisms that work 2- to 3-fold less efficiently (Castellano et al. 2011; Deane et al. 2008). These dynamics have real-world consequences on AD progression. Carriers of two ε4 alleles are at increased risk of developing sporadic, age-related AD by 1,000 % (Tanzi 2012).

In the typical population, APOE status correlates with brain differences early in life. Infants carrying ε4 alleles have less white and gray matter in areas of the parietal and temporal lobes, but more frontal lobe matter (a phenomenon showing some thematic resemblance to PASA; Dean et al. 2014). Despite its long-held connection to dementia onset in the DS population (Coppus et al. 2008; Deb et al. 2000; Forte et al. 2007; Hardy et al. 1994; Lai et al. 1999; Lambert et al. 1996; Prasher et al. 1997, 2008; Royston et al. 1994, 1996), APOE status might also affect the course of neurocognitive growth and decline in those with trisomy 21. Linear regression analysis of performance IQ from 5 to 30 years of age suggests that subgroups of individuals with DS positive for only ε2/ε3 are protected from worsening function compared to subgroups with an ε4 allele (Del Bo et al. 1997). These data argue that the APOE AD susceptibility locus might shape structural and functional outcomes in the brain to a degree meriting attention irrespective of the age of the cohort being tested. This could be particularly relevant for clinical trials in people with DS, in whom development and aging is conflated by genetic triplication of *APP* and Aβ overexpression.

3 Trisomy 21: A Syndrome-Specific Behavioral Profile That Limits Learning Opportunities

People with DS will perform differently from one another in structured assessments of learning and memory. Whatever their scores, there is a general expectation that test performance accurately reflects underlying competence. This assumption might be strained by psychological factors that are inextricably tied to the DS behavioral profile. Infants and toddlers born with trisomy 21 start life with deficits in mastery motivation (MM), an intrinsic quality that compels very young children to explore and gain control over the surrounding environment (Niccols et al. 2003). In older children, MM drives the individual to build new and ever more complicated skill sets and to solve challenging tasks that require perseverance. Behaviors associated with MM are inherently satisfying and "pleasurable," as they occur independent of any immediate external reinforcement. Ultimately, they allow for the child's mental construction of the world and their sense of personal efficacy within it (Cuskelly et al. 1998; Vlachou and Farrell 2000).

Experiments done by Wishart in the 1980s provide an elegant case study in how MM never seems to materialize in 0- to 2-year-olds with DS (1993). The children were trained on a contingency in which a footkick across a light sensor activated rotation of a colorful baby mobile—an event that children generally find appealing. When noncontingent activation of the mobile was available but at a very reduced frequency relative to what could have been provided by a footkick, infants with DS were content to passively watch random turns. It was only when the contingency had been fully reinstated that they rushed to pair their movements once again to the movement of the toy. This situation contrasted to that observed in typically developing infants, who sought to gain maximal activation of the mobile irrespective of whether "free" turns were offered or not. On the surface, Wishart's findings suggest that infants with DS might be satisfied or satiated by lower levels of stimulation. More careful insight would suggest that they derived the same pleasure as typically developing children from seeing the mobile move, but did not experience the added satisfaction of being the agent to prod it (i.e., to be an agent of change in their environment). Passivity at this early age has been documented in the play behavior of toddlers with DS as well. When maternal attention is directed to a particular object, they will often just hold the object rather than manipulate it to see how it works (Landry and Chapieski 1989).

Deficits in MM cripple the emergence of means–end, or instrumental learning, in those with DS (Fidler 2006; Fidler et al. 2005b). Instrumental learning occurs whenever a child links a series of behaviors together so as to achieve a desired outcome. It serves as the developmental foundation for adult strategic thinking. Infants with DS take longer than CA controls to move from shorter chains of continuous goal-directed behaviors to longer chains and are less "happy" when performing more complex chain linking (Dunst 1988; Ruskin et al. 1994). Gradually, motivational issues exacerbate the cognitive disabilities that arise from poor brain development (Cicchetti and Sroufe 1976).

Three- to five-year-olds with DS begin to realize that they do not have the faculties to support smooth execution of learning and memory tasks. Wishart observed the endpoints of this budding awareness in her tests of object permanence, in which children with DS would avoid harder portions of the task by refusing to comply with them, electing to fail by default by routinely choosing the same answer on each trial, or distracting attention away from the examination using social ploys (e.g., amused hand clapping or other party gestures, feigning interest in other things, attempts at charming discussion, etc). Left to their own devices, children and young adults with DS eventually avoid putting themselves in strenuous situations that involve learning difficult-to-grasp concepts or engaging new environments so as to obviate the possibility of failure (Fidler 2006; Gilmore et al. 2003; Schwethelm and Mahoney 1986; Wishart 2001). Instead, they grow to rely on caregivers and others to take personal initiatives for them, asking for help even when assistance might not be of use (Berry and Gunn 1984; Pitcairn and Wishart 1994). Kasari and Freeman provided a compelling laboratory demonstration of this neediness when they asked older children and preteens with DS to assemble puzzles of different complexity (2001). Regardless of how easy the puzzle could be solved, those with DS looked to the experimenter more often than children with other forms of intellectual disability and requested more help.

Individuals with DS participate in special education or mainstream schooling. Despite efforts to the contrary, they are exposed to significant cognitive difficulties in these settings that will end in only some limited success and skills achievement. Less-than-favorable learning histories and dependence on caregivers might create the impression of diminishing returns on further educational activities. The result is seclusion from peers and negative self-perceptions that prime feelings of inadequacy and depression (Ali et al. 2012; Capone et al. 2006; Dykens et al. 2002; Fidler et al. 2005a, 2006). In those subgroups of individuals with DS that become discouraged, preexisting problems with speech intelligibility and autobiographical memory can complicate identifying mental health conditions (i.e., diagnostic overshadowing; Reiss et al. 1982). People with DS have trouble conveying their emotional struggles and will tend to exhibit negative affective symptoms such as psychomotor retardation and withdrawal (Dykens et al. 2002; Stein et al. 2013; Visootsak and Sherman 2007). For some with DS, the psychological impact of intellectual disability weighs heavily and might influence how they approach formal testing.

In light of this behavioral orientation and the challenging learning style that accompanies it, clinical trial organizers might be advised to consider not only the suitability of their cognitive outcome measures but the manner in which their neuropsychological assessments are administered. Drawing on almost two decades worth of experience conducting clinical trials in the DS population, Heller, Kishnani, and colleagues stipulate that performance changes in people with DS will be best detected in regimented testing sessions that (1) maintain the same examiner, who is introduced to the participants in advance of any formal assessment; (2) are held under the same environmental conditions, in the same orientation within a familiar room with a fixed decor; (3) are identically scheduled, with organized times for breaks, snacks, and meals; (4) limit distractions from noise, interruptions, or secondary observers; and (5) employ a clinical research team with experience testing people with intellectual abilities and who are capable of maintaining positive encouragement throughout the study so as to avoid artificial plateaus in subject performance (2006). People with DS are three-dimensional individuals who happen to have been born with a debilitating cognitive disorder. Their reactions to this condition are on par with what might be expected in those with early stage AD or anyone suffering from sudden onset cognitive impairment due to stroke or other trauma. Sensitivity is recommended when partnering with this population in experimental clinical trials for a new therapeutic drug.

4 Clinical Trials in the Down Syndrome Population

4.1 Within-Subjects Design

In many ways, DS represents a soft-spectrum disorder. The DS community is comprised of an eclectic mix of individuals whose neurocognitive development has been variably steered by interactions between several tissue systems, including the brain, skeleton, thyroid, and heart. These interactions translate into unique but greatly overlapping intellectual outcomes. Despite the surface similarity in what the average person with DS can and cannot do, it is likely that a particular level of performance is arrived at by different developmental trajectories from one individual to the next (D'Souza and Karmiloff-Smith 2011; Karmiloff-Smith 1998). One subgroup of those with DS might have overt or silent CHD that is compensated for by disomic modifiers that restrain APP metabolism. Another might see precocious elevations of Aβ in the brain that are not further augmented by obstructive sleep apnea or respiratory distress. Still, a third subgroup may buffer the regressive loss of neurons in the hippocampus and cerebellum with better-than-expected thyroid function. What this suggests, in some cases, is that people with DS who share a particular range of IQ scores cannot be directly compared to one another. Corollary to this, they might not be predicted to show the same performance responses to the same pharmacological treatment.

One of the standard designs for establishing drug efficacy in clinical trials involves matching individuals for various factors that are highly likely to influence performance in their own right (Kover and Atwood 2013). Some of these factors are obvious but, even here, are not always recognized when making group assignments. An example of the matched-pairs design in a clinical trial of people with DS is highlighted in a study led by Costa (Boada et al. 2012). Here, the researchers evaluated whether the NMDA open-channel blocker memantine could improve measures of episodic memory relative to placebo. Groups receiving drug or mock placebo tablets were balanced according to age, gender, baseline intelligence scores, verbal ability, *s*ocio*e*conomic *s*tatus (SES), hypothyroidism, body-mass index, diabetes, and the incidence of sleep apnea. The Costa trial illustrates the lengths that a research team can go to minimize nuisance variables that will interfere with an interpretation of drug efficacy. Yet, in doing so, it ironically makes clear that not all *possible* confounds could ever be controlled for in a population with DS, where the amount of known medical information is decisively outweighed by unknowns regarding developmental trajectories (Mervis and Robinson 1999).

An alternative approach that reduces variance associated with individual differences is the within-subjects design (for textbook description, see Kim 2010). Rather than assigning separate pools of individuals with DS to a control or treatment group, all of the people in the study go through a phase where they are given the placebo pill *and* a phase where they are allotted the drug. The same cognitive outcomes are assessed repeatedly under both conditions. Under this design, those participating in the clinical trial serve as their own controls. Group changes in performance, thus, cannot be attributed to nebulous differences that could potentially exist in the larger pool of subjects that happened to be recruited for the trial or how they were randomized; all unexplained baseline variables not relating to the treatment response are now removed. The magnitude of the response can also be comfortably dissected from person to person as a function of their own medical history.

Two additional strengths of the within-subjects approach are universally noted and particularly relevant for conducting clinical trials in the DS population. First, subtle therapeutic effects that might have been statistically hidden in a between-groups comparison are now magnified to improve detection, requiring less post hoc analysis and facilitating decisions about whether a clinical investigation should advance to subsequent stages or expand to more domestic or international sites. Secondly, statistical power is increased. Enrollment of individuals with DS into clinical trials is complicated by the total number of individuals available for long-term study and cultural attitudes in the DS community that might shun the idea of cognitive pharmacotherapy (Inglis et al. 2014). Because participants serve as their own controls in a within-subjects assessment, the total number of people required for the trial is cut by half.

The within-subjects design has some disadvantages. Repeated testing of subjects might create practice effects that contribute to better performance in the latter half of the trial relative to previous sessions (e.g., Acosta et al. 2011; Edwards et al. 1996). Trend analysis can be used to isolate these effects and to model whether they occur linearly or quadratically (i.e., whether performance improves by some fixed increment with each exposure or whether the magnitude of the improvement grows with more testing). Fatigue effects that impair performance are also possible due to the fact that subjects are required to double their time commitments to the trial. People with DS suffer from muscle weakness, have problems getting quality sleep, and are known to have a hard time staying on task. When employing a within-subjects design, it might be necessary to break up testing sessions into shorter periods and to motivate continuing participation when subjects are at home in between visits (Berry-Kravis et al. 2008).

Crossover strategies that randomize the order in which the participants receive treatment can mitigate practice and fatigue effects. In the initial phase of a trial, for instance, half the individuals might be randomly chosen to receive the drug, while the other half receives placebo. After a sizable wash-out period, the groups then switch to the other treatment condition. The most ideal structure for a clinical trial involving people with DS should assume, however, that practice and fatigue effects will occur (naturally) even with attempts at controlling for them via counterbalancing. To completely unmask them for statistical analysis and correction, one might design a within-subjects assessment with four discrete phases. The first phase would act as an introductory baseline to allow all of the individuals to familiarize themselves with the testing routine and to collect control data. The second and third phases would encompass the traditional crossover period, in which the subjects alternate between the placebo and drug treatments. The fourth would act as a "debriefing" baseline condition. Here, fatigue effects can be assessed in the subgroup that was initially allotted drug so that trends in performance *sans* pharmacotherapy can be assessed across two sequential phases like in those participants initially receiving placebo. Carryover effects specific to transitioning from the study medication to placebo can now be assessed in both subgroups as well.

Implicit in the within-subjects design is a sampling bias. Individuals do not get recruited for lengthy clinical trials; *families do*. Caregivers take on extensive responsibilities when participating in drug-efficacy studies, often tantamount to a second job. They attend multiple onsite visits that might require long-distance travel, are responsible for the appropriate timing and dosing of medication, and are dispatched as surrogate examiners during the study to record elements of the subject's behavior that will shed light on FDA-relevant outcome measures like adaptive function (see Sect. 4.2.2). These requirements, codified in most trial inclusion criteria, select for families high on the SES scale, who likewise tend to be more highly educated. Not much, if anything, can be done about the selection process and what it might portend for the generalizability of drug effects to other cohorts of people with DS with lower IQ's or those who reside in less affluent families (Fernald 2010; Fernald et al. 2013; Hoff 2003; Mani et al. 2013; Tomalski et al. 2013). A positive demonstration that a drug can improve cognitive function in any person with DS nevertheless establishes an important proof of principle that would materially benefit the entire community.

We have suggested the possibility that clinical trial organizers adopt a within-subjects rather than a randomized or paired between-subjects design to assess cognitive interventions in people with DS. The within-subjects approach eliminates variability arising from individual differences in development, but introduces performance artifacts arising from repeated learning and memory assessments. We contend that this trade-off is worthwhile, because it exchanges unknown factors that cannot be controlled for factors that are acknowledged and open to systematic correction. No drug will be a panacea for every individual with DS, and as such, the within-subjects design affords the added benefit of scrutinizing personalized medical information that can inform future therapeutic opportunities.

4.2 Cognitive Outcome Measures: General Principles

4.2.1 Fundamental Learning and Memory Systems as Early Indicators of Cognitive Improvement

A fine line separates whether an experimental drug can improve cognitive outcomes for people with DS in a regimented "laboratory setting" and whether these performance gains are indicative of real-world efficacy. The compound might raise scores on a particular memory assessment and do so by a magnitude predicted by clinical trial organizers, but does this endpoint mean anything for the ability of the treated individuals to live more independently—to acquire additional cognitive–adaptive and communication skills that will permit them to hold down a job and have a better quality of life? Away from the academic debate about suitable response measures for those with intellectual disability and what facets of cognition might be changed with pharmacotherapeutic intervention, establishing *clinical relevance* is the overarching, FDA-mandated goal of any clinical study. It allows for the results of the trial to be authenticated, so that the drug can now be prescribed by a primary care or specialty physician as a medication indicated for the management of learning disabilities in those born with trisomy 21.

Pharmaceutical companies typically address clinical relevance in one of two ways. They can elect to use (1) normed psychometric batteries like full-scale IQ tests (Facon 2008), which have been historically linked to practical educational and achievement outcomes (e.g., assimilating language). These assessments are standardized in the typically developing population and factor out validity and reliability issues that crop up during psychological testing (Dacey et al. 1999), yet are difficult to interpret in people with intellectual disabilities because they invariably fall on the extreme end of the performance distribution below the first percentile (Couzens et al. 2004; Hessl et al. 2009; Nelson and Dacey 1999). At the tail end of the scales, estimates of performance are significantly less accurate, and there is narrower resolution to detect meaningful change (Couzens et al. 2004; Silverman et al. 2010; Spinks et al. 2009). Normative testing misses a key point in understanding how cognitive improvement might occur in an individual with DS;

this improvement might be better compared to other people with DS, less so to the function of the everyday person.

Medication can do little to compensate for the widespread developmental abnormalities wrought by an extra Hsa21. Seen in that light, comparisons between individuals with and without trisomy 21 might not be informative. Instead, a drug likely to be approved by the FDA will enhance daily function in *a treated person with DS* beyond that which could reasonably be expected for the *average untreated person with the condition.* (2) Anchoring is the other strategy employed by trial organizers to establish clinical relevance (Crosby et al. 2003). It associates raw scores on learning and memory assessments to ratings on adaptive behavior scales, thus allowing the scores to be used as endpoints with external validity. Because errors or the number of items correct on a memory test will vary across a normal distribution (i.e., without appreciable floor or ceiling effects), the performance of an individual with DS can be tracked with respect to past performance and to that of peers in a larger cohort under study (see Hall et al. 2008 for an elegant demonstration of this approach). Whole distribution shifts for a group participating in a clinical trial can also be ascertained if a test has been validated well enough in a representative sample of the general DS population.

The ability to use raw scores from cognitive assessments as an outcome measure for drug efficacy gives a research team flexibility when crafting an up-to-date DS neuropsychological battery. Many norm-referenced tests were standardized with sample groups that may no longer be indicative of the typically developing population, let alone any specific individual with DS. Pending proper characterization, trial organizers have the advantage of being able to modify their assessments over time to reflect what knowledge has been gained about trisomy 21 and its effects on neurocognitive development. Thought, in turn, can be allocated to crafting an instrument that reflects the unique cognitive strengths and weaknesses of those in the DS community.

Edgin et al. designed what is arguably the first cognitive test battery to capture snapshots of performance in older children and young adults with DS (2010a) as part of the Down Syndrome Cognition Project (DSCP), a multi-institutional effort seeking to identify genotypic and phenotypic factors that influence the course of intellectual disability in DS. The Arizona Cognitive Test Battery (ACTB) was compiled from tasks that tap either the PFC (i.e., set shifting, modified *DOTS* task), hippocampus (paired-associates learning, virtual navigation), or cerebellum (i.e., reaction time, finger tapping). While individuals with DS showed deficits in each cognitive domain relative to mentally aged-matched controls, the DSCP study found that performance across the domains was interrelated. How an individual with DS fared on paired-associates learning predicted how they would score on EF tasks, such as set shifting and *DOTS*, and on tests of cerebellar function. Correlations occurred among several measures but could also be quite specific. In particular, a unique association was found between navigation success in the computer arena and levels of coordination achieved during finger tapping—a hippocampal–cerebellar link that might be expected given animal work with L7-PKCI mice in the Morris water maze (2010a).

The gestalt of these findings is consistent with the interpretation of an episodic memory axis from cerebellum to the frontotemporal lobes and with reports describing co-activation of the PFC and hippocampus in tasks originally thought to isolate contributions from either brain region. Human fMRI studies indicate BOLD responses from the PFC–hippocampus during the encoding phase of the paired-associates test (De Rover et al. 2011) as well as during performance of the Wisconsin Card Sorting Task when the dimensional rule for organizing the cards is first identified (Graham et al. 2009). The ACTB articulates an important theoretical principle in devising performance criteria for people with DS in a clinical trial, namely, that the memory assessments chosen should triangulate activity as best as possible from all elements of the episodic memory system. Edgin couches the point astutely in a recent treatise on the cognitive neuroscience of DS, in which the author emphasizes that modular or brain region-specific theories of cognitive delay have not held up to scrutiny upon assessment in those with DS and that the condition is more accurately described by computational inefficiencies in dialogue between the frontal and temporal lobes (2013).

The endowment of neuropsychological information that has been built from decades of academic study of people with DS strongly suggests that these individuals cannot adequately process verbal short-term memory (Baddeley and Jarrold 2007; Næss et al. 2011), an ability that hinges on proper and timely feedback between the PFC (Andreasen et al. 1995; Bunge et al. 2000; Cabeza et al. 1997; Cohen et al. 1994; Hashimoto and Sakai 2002; Paulesu et al. 1993), hippocampus (Ekstrom 2014; Karlsgodt et al. 2005; Miller et al. 2013; Mueller et al. 2011; Schmidt-Wilcke et al. 2009; Travis et al. 2014), and cerebellum (Andreasen et al. 1995; Chen and Desmond 2005; Ghosh et al. 2008; Paulesu et al. 1993; Ravizza et al. 2004, 2006). Deficient verbal memory might be, in fact, one of the signature cognitive deficits of DS, and work by several investigators has clarified that performance of individuals with trisomy is not an artifact of impaired hearing, phonetic discrimination/speech perception, language competence, cultural background, general forgetting, or word learning capacity (Brock and Jarrold 2004; Jarrold and Baddeley 1997; Frenkel and Bourdin 2009; Jarrold et al. 2002; Laws 2002; Marcell and Cohen 1992; Marcell and Weeks 1988; Mosse and Jarrold 2010; Purser and Jarrold 2005, 2013).

Considering this strong historical record and the fact that these tasks tap networks encompassing the cerebellum–frontotemporal lobes, clinical trial organizers might seek to use visual or auditory verbal short-term memory (VSTM) batteries as early litmus tests for cognitive therapeutic efficacy. Normed assessments that evaluate verbal performance, like components of the Repeatable Battery for the Assessment of Neuropsychological Status (RBANS; Berry-Kravis et al. 2008; Randolph 1998), are available but lock researchers into tests that cannot be modified as experience dictates. Chronic rehabilitation of verbal memory function in people with DS will likely contribute to hard-won but stable gains in language assimilation that will relate to measures of adaptive behavior (Laws and Gunn 2004; Miles and Chapman 2002; Miolo et al. 2005). Via anchoring, clinical researchers might be free, thus, to devise paradigms that look at various aspects of

VSTM. Brock and Jarrold have suggested that one particular VSTM weakness is verbal serial order reconstruction, a form of memory that is accessed when participants are asked to remember the correct *sequence* of a list of words or numbers (2005). Vicari and colleagues have also documented disproportionately lower verbal *backward* spans in teenagers with DS relative to a mentally and chronologically aged-matched group with intellectual disability of different etiology (1995). Doubtless, other VSTM mechanics could be evaluated such as word-length, repetition, and context or cross-modality effects (e.g., Duarte et al. 2011; Edgin et al. 2010b; Kanno and Ikeda 2002). To our knowledge, a comprehensive VSTM battery—comprised of several informative subtests with derivatives that scale in executive difficulty—has not been devised for people with DS. This might be a good place to start when devising an "early warning" detection system that will signal the real-world efficacy of a new medication.

4.2.2 Adaptive Behavior: Parent Tailored Measures

Adaptive behavior is a deceptively simple term that embodies a deeper operational meaning. It is an integral part of functional intelligence, the ability to translate underlying competencies into consistent daily habits that enable self-sufficiency. Skill sets associated with adaptive behavior are organized along three domains: (1) conceptual skills, which refer to language-oriented abilities like reading and writing and computational ones like counting money or telling time; (2) practical skills, which refer to everyday activities such as using a computer, navigating public transit, or preparing a meal; and (3) social skills, which involve interpersonal awareness and communication, social reasoning, and an understanding of customs and laws (for review, see Tassé et al. 2012). Estimates of functional intelligence factor into formal legal definitions of learning disability and, in many countries, determine whether an individual is eligible for governmental health and educational services. Because of this and its obvious relationship to quality of life, performance on adaptive behavior scales is a credible primary endpoint and "anchor" for a clinical trial. Scores on memory assessments often correlate significantly with adaptive performance as measured by instruments such as the Vineland or the Scales of Independent Behavior-Revised (Edgin et al. 2010a; Hessl et al. 2009; Sparrow et al. 1984).

People with DS climb a wall of function to achieve some level of independence by young adulthood (Carr 1994; Van Gameren-Oosterom et al. 2013). While IQ relative to the typically developing population drops steadily throughout life, evidence suggests that those with DS continue to learn and improve on extant cognitive abilities until about 50 years of age (Berry et al. 1984; Hawkins et al. 2003). There is tremendous room to maneuver from baselines that often leave many teenage adults with DS unable to live autonomously outside of highly scripted settings—to bathe, dress, eat meals, read correspondence, pay bills, and interact with the community (Sloper and Turner 1996). What is slowing them during this journey? Are the issues similar from one person to the next?

The FDA has emphasized that regulatory approval for a new drug must be accompanied by outcome measures in a clinical trial that are germane to the condition to be treated. In other words, endpoints with external validity, such as adaptive behavior, should be specific to those problems that keep people with DS from having a better quality of life. This perspective deserves a moment of pause. As we have discussed in the current chapter, individuals born with trisomy 21 exhibit significant heterogeneity in how they develop and adapt to their social and educational environments. They do not comprise a uniform block, and it might be heuristic to assume that one general-purpose diagnostic instrument will accurately chronicle the daily intellectual struggles of every single person with DS and the therapeutic influence of medication. This being the case, clinical and academic researchers should consider the possibility of designing a circumscribed set of adaptive behavior scales (e.g., 2–3 separate inventories) built around pragmatic observations that have been made in children with DS that relate to core problems likely to affect most of the DS population. From a practical standpoint, this effort would be paid off in the scales' subsequent ease of use, minimal expense, and reliability across sites in an international trial. Parents would be instructed on what behavioral variability looks like across many children and young adults with DS, not just their own son or daughter, and then work with investigators to identify divisions within the questionnaires that are especially relevant to the issues they see in their child's daily function. All three informant-based questionnaires would be completed but, for the purposes of evaluating drug efficacy, would be rolled into a single composite primary outcome measure that is weighted differently in each participant in accordance to those weaknesses that were prespecified.

What skills or constructs might a DS adaptive behavior battery cover? One instrument could assess "cognitive behavior" in a similar fashion to the Behavior Rating Inventory of Executive Function (BRIEF; Gioia et al. 2000), a caregiver report that surveys real-world instances of EF skills such as emotional control/inhibition (Is the person impulsive? Do they overreact to minor setbacks?), set shifting (Does the person acclimate well to new friends, schools, or changes in routine?), working memory (Does the person finish tasks that they start, especially ones with multiple steps?), and planning (Is the person organized? Is the person punctual? Can they easily find possessions?). The BRIEF has been validated in the typically developing population and has versions that are age-appropriate for young children, teenagers, and adults but might not be directly applicable to those with DS, who show a wide range of abilities that could be better described in the preschool version of the instrument, the school-age version, or in a mixture of the two (Lee et al. 2011). Some work is needed to refine a more universally relevant EF scale for DS.

Another section of the adaptive battery could operationalize the concept of mastery motivation by creating an inventory of goal-directed activities or behaviors that one might expect to see practiced in a home or school setting. Pulling from many of the studies canvassed in Sect. 3 (Fidler et al. 2005b; Kasari and Freeman 2001; Landry and Chapieski 1989), one might ask whether children with DS show more sustained bouts of object exploration and tinkering with new toys, whether they take the initiative to start puzzles or games or whether they are more effusive in their questioning about the world (Do their questions strike caregivers as being more inquisitive?). In teenagers and young adults with DS, one might further ask how self-reliant they are in starting and maintaining a homework schedule, taking on new hobbies, or sticking with tough problems and improvising on them. A key common thread in this survey would be to track the degree to which a person with DS will persevere with daily tasks and interests even when their ability to accomplish or profit from them might not necessarily be assured.

A final element of the adaptive battery could monitor instances of collaborative problem-solving, a skill which is becoming increasingly significant in modern professional life. Despite the widespread perception that people born with trisomy 21 are extroverted and socially adept, empirical studies suggest that children with the condition are reserved in their interactions with peers, have difficulty interpreting the social rules and cues that accompany these engagements, and need coaxing to get involved with group projects (Guralnick et al. 2009, 2011). Wishart and colleagues have examined how older children with DS approach collaborative learning situations. In contrast to the learning facilitation that occurred in pairings between two typically developing children or in children with nonspecific intellectual disability, the researchers found that dyads that included a student with DS did not benefit from collaborating on a set of object-sorting tasks (Wishart 2007; Wishart et al. 2007). From viewing interactions among ~100 children with DS and those with other kinds of disability, it was noted that the students with DS simply did not engage their partners, preferring to work in parallel with little "chitchat" or task-related communication. This style of interaction, if addressed in some way with drug treatment, might positively foreshadow additional benefits on language, which is a phenomenon that requires joint attention and willful attempts at shared understanding.

5 Conclusions

Drug efficacy for any intellectual disability is a very personalized matter that should be assessed over a lifetime. Clinical researchers can orient group outcome measures toward highly responsive phenotypes with real-world applicability, but the tangible benefits of medication on a specific person's ongoing daily life are hard to predict a priori. In fact, this might be an impossible feat for conditions that are heterogeneous in both their causes (i.e., developmental trajectories) and symptoms. The FDA authenticates the efficacy of an experimental drug through the lens of one and only one primary endpoint. As we have stated, there is a dangerous assertion embedded in this approach that assumes that a single performance metric can be devised that will apply equally to most participants in a clinical trial. This assumption, phrased alternatively, is that if a psychotropic drug "works," then it should work the same exact way in each and every individual with a defined learning disability. Unfortunately, treating a learning disorder is not as straightforward a proposition as treating high blood pressure. It is possible that the FDA's regulatory practices will need to be reconsidered in light of advances in neuroscience and psychiatry that have laid the groundwork for efforts to address cognitive impairment in spectrum conditions like autism or DS.

The conundrum that ensues is illustrated by recent efforts to bring a medication to market for some of the core symptoms of Fragile X syndrome. During a phase II–III study of arbaclofen in people with Fragile X (Berry-Kravis et al. 2012), investigators from Roche and Seaside Therapeutics noted significant improvements in adaptive behavior measures on the Vineland II and Aberrant Behavior Checklist (ABC) Socialization scales, but no significant treatment advantage over placebo on the trial's primary endpoint, the ABC's Irritability subscale. Only mild adverse effects were reported for the subjects taking the drug. By many personal accounts, the study was successful. Children and teenagers receiving arbaclofen exhibited *social awakenings*, sharing moments and requesting hugs from their caregivers for the first time, becoming interested in making friends, and displaying eagerness to go on social outings (Pollack 2013; Rubin 2013). Yet, clinical trials seeking to purpose the drug for neurobehavioral deficits in Fragile X were halted due to failures to meet the stated primary objective (*ibid.*; ClinicalTrials.gov Identifier NCT01013480).

The aftermath of the Roche–Seaside arbaclofen affair posits a closing consideration for drug-efficacy trials of cognitive rehabilitation in DS. How does one package a primary outcome measure to withstand the phenotypic and response variability that is bound to exist in the subject pool? Though no simple answers are available, some of the suggestions made in the current chapter, such as using a within-subjects design and primary endpoint composites that are weighted differently among the participants depending on problem areas nominated by parents, might be employed to mitigate the risk that a real treatment effect will not be sufficiently documented to garner FDA approval. The challenges of running clinical trials in the DS population should remind us just how far we have come in the discussion on how to improve the everyday quality of life in those suffering from the ill effects of having three copies of Hsa21. We are witnessing exciting times that might soon herald better integration of people with DS as independent members of society. Now is the time to see these challenges through.

Acknowledgements We acknowledge Ms. Lucy Han for her assistance in preparing the manuscript. Special thanks are extended to the LuMind Foundation, the Anna and John J. Sie Foundation, and Research Down syndrome for their tireless support. This work was also funded under R01 HD38384 from the National Institute of Child Health and Human Development. F.F. is indebted to J.O. Edgin for her wellspring of enthusiasm, inspiration, and strength.

References

Abbeduto L, Warren SF, Conners FA (2007) Language development in Down syndrome: from the prelinguistic period to the acquisition of literacy. Ment Retard Dev Disabil Res Rev 13:247–261

Abbeduto L, Murphy MM, Kover ST, Giles ND, Karadottir S, Amman A, Bruno L, Kim JS, Schroeder S, Anderson JA, Nollin KA (2008) Signaling noncomprehension of language: a comparison of fragile X syndrome and Down syndrome. Am J Ment Retard 113:214–230

Ábrahám H, Vincze A, Veszprémi B, Kravják A, Gömöri É, Kovács GG, Seress L (2012) Impaired myelination of the human hippocampal formation in Down syndrome. Int J Dev Neurosci 30:147–158

Acosta MT, Kardel PG, Walsh KS, Rosenbaum KN, Gioia GA, Packer RJ (2011) Lovastatin as treatment for neurocognitive deficits in neurofibromatosis type 1: phase I study. Pediatr Neurol 45:241–245

Al-Bagdady M, Murphy PJ, Woodhouse JM (2011) Development and distribution of refractive error in children with Down's syndrome. Br J Ophthalmol 95:1091–1097

Al-Biltagi M, Serag AR, Hefidah MM, Mabrouk MM (2013) Evaluation of cardiac functions with Doppler echocardiography in children with Down syndrome and anatomically normal heart. Cardiol Young 23:174–180

Ali A, Hassiotis A, Strydom A, King M (2012) Self stigma in people with intellectual disabilities and courtesy stigma in family carers: a systematic review. Res Dev Disabil 33:2122–2140

Anderson P (2002) Assessment and development of executive function (EF) during childhood. Child Neuropsychol 8:71–82

Anderson JS, Nielsen JA, Ferguson MA, Burback MC, Cox ET, Dai L, Gerig G, Edgin JO, Korenberg JR (2013) Abnormal brain synchrony in Down syndrome. NeuroImage: Clinical 2:703–715

Andreasen NC, O'Leary DS, Arndt S, Cizadlo T, Hurtig R, Rezai K, Watkins GL, Ponto LL, Hichwa RD (1995) Short-term and long-term verbal memory: a positron emission tomography study. Proc Natl Acad Sci USA 92:5111–5115

Ansado J, Monchi O, Ennabil N, Faure S, Joanette Y (2012) Load-dependent posterior-anterior shift in aging in complex visual selective attention situations. Brain Res 1454:14–22

Apelt J, Schliebs R, Beck M, Rossner S, Bigl V (1997) Expression of amyloid precursor protein mRNA isoforms in rat brain is differentially regulated during postnatal maturation and by cholinergic activity. Int J Dev Neurosci 15:95–112

Arya R, Kabra M, Gulati S (2011) Epilepsy in children with Down syndrome. Epileptic Disord 13:1–7

Austeng ME, Akre H, Falkenberg ES, Øverland B, Abdelnoor M, Kværner KJ (2013) Hearing level in children with Down syndrome at the age of eight. Res Dev Disabil 34:2251–2256

Aylward EH, Habbak R, Warren AC, Pulsifer MB, Barta PE, Jerram M, Pearlson GD (1997) Cerebellar volume in adults with Down syndrome. Arch Neurol 54:209–212

Baddeley A, Jarrold C (2007) Working memory and Down syndrome. J Intellect Disabil Res 51:925–931

Baum RA, Nash PL, Foster JE, Spader M, Ratliff-Schaub K, Coury DL (2008) Primary care of children and adolescents with Down syndrome: an update. Curr Probl Pediatr Adolesc Health Care 38:241–261

Baxter LL, Moran TH, Richtsmeier JT, Troncoso J, Reeves RH (2000) Discovery and genetic localization of Down syndrome cerebellar phenotypes using the Ts65Dn mouse. Hum Mol Genet 9:195–202

Becker LE, Armstrong DL, Chan F (1986) Dendritic atrophy in children with Down's syndrome. Ann Neurol 20:520–526

Begenisic T, Spolidoro M, Braschi C, Baroncelli L, Milanese M, Pietra G, Fabbri ME, Bonanno G, Cioni G, Maffei L, Sale A (2011) Environmental enrichment decreases GABAergic inhibition and improves cognitive abilities, synaptic plasticity, and visual functions in a mouse model of Down syndrome. Front Cell Neurosci 5:29

Bero AW, Yan P, Roh JH, Cirrito JR, Stewart FR, Raichle ME, Lee JM, Holtzman DM (2011) Neuronal activity regulates the regional vulnerability to amyloid-β deposition. Nat Neurosci 14:750–756

Berry P, Gunn P (1984) Maternal influence on the task behaviour of young Down's syndrome children. J Ment Defic Res 28:269–274

Berry P, Groeneweg G, Gibson D, Brown RI (1984) Mental development of adults with Down syndrome. Am J Ment Defic 89:252–256

Berry-Kravis E, Sumis A, Kim OK, Lara R, Wuu J (2008) Characterization of potential outcome measures for future clinical trials in fragile X syndrome. J Autism Dev Disord 38:1751–1757

Berry-Kravis EM, Hessl D, Rathmell B, Zarevics P, Cherubini M, Walton-Bowen K, Mu Y, Nguyen DV, Gonzalez-Heydrich J, Wang PP, Carpenter RL, Bear MF, Hagerman RJ (2012) Effects of STX209 (arbaclofen) on neurobehavioral function in children and adults with fragile X syndrome: a randomized, controlled, phase 2 trial. Sci Transl Med 4:152ra127

Bertrand P, Navarro H, Caussade S, Holmgren N, Sánchez I (2003) Airway anomalies in children with Down syndrome: endoscopic findings. Pediatr Pulmonol 36:137–141

Biko DM, Schwartz M, Anupindi SA, Altes TA (2008) Subpleural lung cysts in Down syndrome: prevalence and association with coexisting diagnoses. Pediatr Radiol 38:280–284

Blaser S, Propst EJ, Martin D, Feigenbaum A, James AL, Shannon P, Papsin BC (2006) Inner ear dysplasia is common in children with Down syndrome (trisomy 21). Laryngoscope 116:2113–2119

Bloemers BL, van Furth AM, Weijerman ME, Gemke RJ, Broers CJ, van den Ende K, Kimpen JL, Strengers JL, Bont LJ (2007) Down syndrome: a novel risk factor for respiratory syncytial virus bronchiolitis—a prospective birth-cohort study. Pediatrics 120:e1076–e1081

Bloemers BL, Broers CJ, Bont L, Weijerman ME, Gemke RJ, van Furth AM (2010a) Increased risk of respiratory tract infections in children with Down syndrome: the consequence of an altered immune system. Microbes Infect 12:799–808

Bloemers BL, van Furth AM, Weijerman ME, Gemke RJ, Broers CJ, Kimpen JL, Bont L (2010b) High incidence of recurrent wheeze in children with Down syndrome with and without previous respiratory syncytial virus lower respiratory tract infection. Pediatr Infect Dis J 29:39–42

Blumenfeld RS, Ranganath C (2007) Prefrontal cortex and long-term memory encoding: an integrative review of findings from neuropsychology and neuroimaging. Neuroscientist 13:280–291

Boada R, Hutaff-Lee C, Schrader A, Weitzenkamp D, Benke TA, Goldson EJ, Costa AC (2012) Antagonism of NMDA receptors as a potential treatment for Down syndrome: a pilot randomized controlled trial. Transl Psychiatry 2:e141

Brandt T, Schautzer F, Hamilton DA, Brüning R, Markowitsch HJ, Kalla R, Darlington C, Smith P, Strupp M (2005) Vestibular loss causes hippocampal atrophy and impaired spatial memory in humans. Brain 128:2732–2741

Braudeau J, Dauphinot L, Duchon A, Loistron A, Dodd RH, Hérault Y, Delatour B, Potier MC (2011a) Chronic treatment with a Promnesiant GABA-A α5-selective inverse agonist increases immediate early genes expression during memory processing in mice and rectifies their expression levels in a Down Syndrome Mouse Model. Adv Pharmacol Sci 2011:153218

Braudeau J, Delatour B, Duchon A, Pereira PL, Dauphinot L, de Chaumont F, Olivo-Marin JC, Dodd RH, Hérault Y, Potier MC (2011b) Specific targeting of the GABA-A receptor α5 subtype by a selective inverse agonist restores cognitive deficits in Down syndrome mice. J Psychopharmacol 25:1030–1042

Brock J, Jarrold C (2004) Language influences on verbal short-term memory performance in Down syndrome: item and order recognition. J Speech Lang Hear Res 47:1334–1346

Brock J, Jarrold C (2005) Serial order reconstruction in Down syndrome: evidence for a selective deficit in verbal short-term memory. J Child Psychol Psychiatry 46:304–316

Brooksbank BW, Walker D, Balázs R, Jørgensen OS (1989) Neuronal maturation in the foetal brain in Down's syndrome. Early Hum Dev 18:237–246

Bunge SA, Klingberg T, Jacobsen RB, Gabrieli JD (2000) A resource model of the neural basis of executive working memory. Proc Natl Acad Sci USA 97:3573–3578

Burgess N (2006) Spatial memory: how egocentric and allocentric combine. Trends Cogn Sci 10:551–557

Burguière E, Arleo A, MR H, Elgersma Y, De Zeeuw CI, Berthoz A, Rondi-Reig L (2005) Spatial navigation impairment in mice lacking cerebellar LTD: a motor adaptation deficit? Nat Neurosci 8:1292–1294

Burguière E, Arabo A, Jarlier F, De Zeeuw CI, Rondi-Reig L (2010) Role of the cerebellar cortex in conditioned goal-directed behavior. J Neurosci 30:13265–13271
Cabeza R, Mangels J, Nyberg L, Habib R, Houle S, McIntosh AR, Tulving E (1997) Brain regions differentially involved in remembering what and when: a PET study. Neuron 19:863–870
Cabin DE, McKee-Johnson JW, Matesic LE, Wiltshire T, Rue EE, Mjaatvedt AE, Huo YK, Korenberg JR, Reeves RH (1998) Physical and comparative mapping of distal mouse chromosome 16. 5 p5. Genome Res 8:940–950
Campbell C, Landry O, Russo N, Flores H, Jacques S, Burack JA (2013) Cognitive flexibility among individuals with Down syndrome: assessing the influence of verbal and nonverbal abilities. Am J Intellect Dev Disabil 118:193–200
Capone G, Goyal P, Ares W, Lannigan E (2006) Neurobehavioral disorders in children, adolescents, and young adults with Down syndrome. Am J Med Genet C Semin Med Genet 142C:158–172
Carlesimo GA, Marotta L, Vicari S (1997) Long-term memory in mental retardation: evidence for a specific impairment in subjects with Down's syndrome. Neuropsychologia 35:71–79
Carr J (1970) Mental and motor development in young mongol children. J Ment Defic Res 14:205–220
Carr J (1988) Six weeks to twenty-one years old: a longitudinal study of children with Down's syndrome and their families. Third Jack Tizard memorial lecture. J Child Psychol Psychiatry 29:407–431
Carr J (1994) Long-term-outcome for people with Down's syndrome. J Child Psychol Psychiatry 35:425–439
Castellano JM, Kim J, Stewart FR, Jiang H, DeMattos RB, Patterson BW, Fagan AM, Morris JC, Mawuenyega KG, Cruchaga C, Goate AM, Bales KR, Paul SM, Bateman RJ, Holtzman DM (2011) Human apoE isoforms differentially regulate brain amyloid-β peptide clearance. Sci Transl Med 3:89ra57
Cataldo AM, Peterhoff CM, Troncoso JC, Gomez-Isla T, Hyman BT, Nixon RA (2000) Endocytic pathway abnormalities precede amyloid beta deposition in sporadic Alzheimer's disease and Down syndrome: differential effects of APOE genotype and presenilin mutations. Am J Pathol 157:277–286
Chakrabarti L, Best TK, Cramer NP, Carney RS, Isaac JT, Galdzicki Z, Haydar TF (2010) Olig1 and Olig2 triplication causes developmental brain defects in Down syndrome. Nat Neurosci 13:927–934
Chakrabarti L, Scafidi J, Gallo V, Haydar TF (2011) Environmental enrichment rescues postnatal neurogenesis defect in the male and female Ts65Dn mouse model of Down syndrome. Dev Neurosci 33:428–441
Chapman RS (1997) Language development in children and adolescents with Down syndrome. Ment Retard Dev Disabil Res Rev 3:307–312
Chapman RS, Seung HK, Schwartz SE, Kay-Raining Bird E (1998) Language skills of children and adolescents with Down syndrome: II. Production deficits. J Speech Lang Hear Res 41:861–873
Chartier-Harlin MC, Crawford F, Houlden H, Warren A, Hughes D, Fidani L, Goate A, Rossor M, Roques P, Hardy J, Mullan M (1991) Early-onset Alzheimer's disease caused by mutations at codon 717 of the beta-amyloid precursor protein gene. Nature 353:844–846
Chen SH, Desmond JE (2005) Cerebrocerebellar networks during articulatory rehearsal and verbal working memory tasks. Neuroimage 24:332–338
Cicchetti D, Sroufe LA (1976) The relationship between affective and cognitive development in Down's Syndrome infants. Child Dev 47:920–929
Cirrito JR, Yamada KA, Finn MB, Sloviter RS, Bales KR, May PC, Schoepp DD, Paul SM, Mennerick S, Holtzman DM (2005) Synaptic activity regulates interstitial fluid amyloid-beta levels in vivo. Neuron 48:913–922

Cohen JD, Forman SD, Braver TS, Casey BJ, Servan-Schreiber D, Noll DC (1994) Activation of the prefrontal cortex in a nonspatial working memory task with functional MRI. Hum Brain Mapp 1:293–304

Colas D, Valletta JS, Takimoto-Kimura R, Nishino S, Fujiki N, Mobley WC, Mignot E (2008) Sleep and EEG features in genetic models of Down syndrome. Neurobiol Dis 30:1–7

Contestabile A, Fila T, Ceccarelli C, Bonasoni P, Bonapace L, Santini D, Bartesaghi R, Ciani E (2007) Cell cycle alteration and decreased cell proliferation in the hippocampal dentate gyrus and in the neocortical germinal matrix of fetuses with Down syndrome and in Ts65Dn mice. Hippocampus 17:665–678

Coppus AM, Evenhuis HM, Verberne GJ, Visser FE, Arias-Vasquez A, Sayed-Tabatabaei FA, Vergeer-Drop J, Eikelenboom P, van Gool WA, van Duijn CM (2008) The impact of apolipoprotein E on dementia in persons with Down's syndrome. Neurobiol Aging 29:828–835

Costa AC (2011a) An assessment of the vestibulo-ocular reflex (VOR) in persons with Down syndrome. Exp Brain Res 214:199–213

Costa AC (2011b) An assessment of optokinetic nystagmus (OKN) in persons with Down syndrome. Exp Brain Res 214:381–391

Costa AC, Scott-McKean JJ, Stasko MR (2008) Acute injections of the NMDA receptor antagonist memantine rescue performance deficits of the Ts65Dn mouse model of Down syndrome on a fear conditioning test. Neuropsychopharmacology 33:1624–1632

Courage ML, Adams RJ, Hall EJ (1997) Contrast sensitivity in infants and children with Down syndrome. Vision Res 37:1545–1555

Courbois Y, Farran EK, Lemahieu A, Blades M, Mengue-Topio H, Sockeel P (2013) Wayfinding behaviour in Down syndrome: a study with virtual environments. Res Dev Disabil 34:1825–1831

Couzens D, Cuskelly M, Jobling A (2004) The Stanford Binet Fourth Edition and its use with individuals with Down syndrome: cautions for clinicians. Int J Disabil Dev Educ 51:39–56

Cox R, Tijdens RR, Meeter MM, Sweegers CC, Talamini LM (2014) Time, not sleep, unbinds contexts from item memory. PLoS One 9:e88307

Cragg BG (1975) The density of synapses and neurons in normal, mentally defective ageing human brains. Brain 98:81–90

Cregg M, Woodhouse JM, Stewart RE, Pakeman VH, Bromham NR, Gunter HL, Trojanowska L, Parker M, Fraser WI (2003) Development of refractive error and strabismus in children with Down syndrome. Invest Ophthalmol Vis Sci 44:1023–1030

Crosby RD, Kolotkin RL, Williams GR (2003) Defining clinically meaningful change in health-related quality of life. J Clin Epidemiol 56:395–407

Cuskelly M, Zhang A, Gilmore L (1998) The importance of self-regulation in young children with Down syndrome. Int J Disabil Dev Educ 45:331–341

Dacey CM, Nelson WM 3rd, Stoeckel J (1999) Reliability, criterion-related validity and qualitative comments of the Fourth Edition of the Stanford-Binet Intelligence Scale with a young adult population with intellectual disability. J Intellect Disabil Res 43:179–184

Dalton MA, Tu S, Hornberger M, Hodges JR, Piguet O (2014) Medial temporal lobe contributions to intra-item associative recognition memory in the aging brain. Front Behav Neurosci 7:222

Dang V, Medina B, Das D, Moghadam S, Martin KJ, Lin B, Naik P, Patel D, Nosheny R, Wesson Ashford J, Salehi A (2014) Formoterol, a long-acting β2 adrenergic agonist, improves cognitive function and promotes dendritic complexity in a mouse model of Down syndrome. Biol Psychiatry 75:179–188

Das I, Park JM, Shin JH, Jeon SK, Lorenzi H, Linden DJ, Worley PF, Reeves RH (2013) Hedgehog agonist therapy corrects structural and cognitive deficits in a Down syndrome mouse model. Sci Transl Med 5:201ra120

Davis SW, Dennis NA, Daselaar SM, Fleck MS, Cabeza R (2008) Que PASA? The posterior-anterior shift in aging. Cereb Cortex 18:1201–1209

Davisson MT, Schmidt C, Reeves RH, Irving NG, Akeson EC, Harris BS, Bronson RT (1993) Segmental trisomy as a mouse model for Down syndrome. Prog Clin Biol Res 384:117–133

De Rover M, Pironti VA, McCabe JA, Acosta-Cabronero J, Arana FS, Morein-Zamir S, Hodges JR, Robbins TW, Fletcher PC, Nestor PJ, Sahakian BJ (2011) Hippocampal dysfunction in patients with mild cognitive impairment: a functional neuroimaging study of a visuospatial paired associates learning task. Neuropsychologia 49:2060–2070

De Saint BP, Hok V, Save E, Poucet B, Chaillan FA (2013) Differential role of the dorsal hippocampus, ventro-intermediate hippocampus, and medial prefrontal cortex in updating the value of a spatial goal. Hippocampus 23:342–351

Dean DC 3rd, Jerskey BA, Chen K, Protas H, Thiyyagura P, Roontiva A, O'Muircheartaigh J, Dirks H, Waskiewicz N, Lehman K, Siniard AL, Turk MN, Hua X, Madsen SK, Thompson PM, Fleisher AS, Huentelman MJ, Deoni SC, Reiman EM (2014) Brain differences in infants at differential genetic risk for late-onset Alzheimer disease: A cross-sectional imaging study. JAMA Neurol 71:11–22

Deane R, Sagare A, Hamm K, Parisi M, Lane S, Finn MB, Holtzman DM, Zlokovic BV (2008) apoE isoform-specific disruption of amyloid beta peptide clearance from mouse brain. J Clin Invest 118:4002–4013

Deb S, Braganza J, Norton N, Williams H, Kehoe PG, Williams J, Owen MJ (2000) APOE epsilon 4 influences the manifestation of Alzheimer's disease in adults with Down's syndrome. Br J Psychiatry 176:468–472

Del Bo R, Comi GP, Bresolin N, Castelli E, Conti E, Degiuli A, Ausenda CD, Scarlato G (1997) The apolipoprotein E epsilon4 allele causes a faster decline of cognitive performances in Down's syndrome subjects. J Neurol Sci 145:87–91

Dennis J, Archer N, Ellis J, Marder L (2010) Recognising heart disease in children with Down syndrome. Arch Dis Child Educ Pract Ed 95:98–104

Diamond A (2000) Close interrelation of motor development and cognitive development and of the cerebellum and prefrontal cortex. Child Dev 71:44–56

Dmitriev V (2001) Early intervention for children with Down syndrome: time to begin. Pro-Ed, Austin, TX

Dolcos F, LaBar KS, Cabeza R (2004) Dissociable effects of arousal and valence on prefrontal activity indexing emotional evaluation and subsequent memory: an event-related fMRI study. Neuroimage 23:64–74

D'Souza D, Karmiloff-Smith A (2011) When modularization fails to occur: a developmental perspective. Cogn Neuropsychol 28:276–287

Duarte CP, Covre P, Braga AC, de Macedo EC (2011) Visuospatial support for verbal short-term memory in individuals with Down syndrome. Res Dev Disabil 32:1918–1923

Dunst CJ (1988) Stage transitioning in the sensorimotor development of Down's syndrome infants. J Ment Defic Res 32:405–410

Dykens EM, Shah B, Sagun J, Beck T, King BH (2002) Maladaptive behaviour in children and adolescents with Down's syndrome. J Intellect Disabil Res 46:484–492

Edgin JO (2013) Cognition in Down syndrome: a developmental cognitive neuroscience perspective. WIREs Cogn Sci 4:307–317

Edgin JO, Mason GM, Allman MJ, Capone GT, Deleon I, Maslen C, Reeves RH, Sherman SL, Nadel L (2010a) Development and validation of the Arizona Cognitive Test Battery for Down syndrome. J Neurodev Disord 2:149–164

Edgin JO, Pennington BF, Mervis CB (2010b) Neuropsychological components of intellectual disability: the contributions of immediate, working, and associative memory. J Intellect Disabil Res 54:406–417

Edgin JO, Mason GM, Spanò G, Fernández A, Nadel L (2012) Human and mouse model cognitive phenotypes in Down syndrome: implications for assessment. Prog Brain Res 197:123–151

Edgin JO, Spanò G, Kawa K, Nadel L (2014) Remembering things without context: development matters. Child Dev 85:1491–1502

Edwards S, Brice C, Craig C, Penri-Jones R (1996) Effects of caffeine, practice, and mode of presentation on Stroop task performance. Pharmacol Biochem Behav 54:309–315

Eisermann MM, DeLaRaillère A, Dellatolas G, Tozzi E, Nabbout R, Dulac O, Chiron C (2003) Infantile spasms in Down syndrome—effects of delayed anticonvulsive treatment. Epilepsy Res 55:21–27

Ekstrom AD (2014) Cognitive neuroscience: navigating human verbal memory. Curr Biol 24: R167–R168

Ellis NR, Woodley-Zanthos P, Dulaney CL (1989) Memory for spatial location in children, adults, and mentally retarded persons. Am J Ment Retard 93:521–526

Elul R, Hanley J, Simmons JQ (1975) Non-Gaussian behavior of the EEG in Down's syndrome suggests decreased neuronal connections. Acta Neurol Scand 51:21–28

Epstein CJ, Epstein LB, Weil J, Cox DR (1982) Trisomy 21: mechanisms and models. Ann N Y Acad Sci 396:107–118

Epstein CJ, Cox DR, Epstein LB (1985) Mouse trisomy 16: an animal model of human trisomy 21 (Down syndrome). Ann N Y Acad Sci 450:157–168

Eyler LT, Sherzai A, Kaup AR, Jeste DV (2011) A review of functional brain imaging correlates of successful cognitive aging. Biol Psychiatry 70:115–122

Facon B (2008) A cross-sectional test of the similar-trajectory hypothesis among adults with mental retardation. Res Dev Disabil 29:29–44

Felius J, Beauchamp CL, Stager DR Sr (2014) Visual acuity deficits in children with nystagmus and Down syndrome. Am J Ophthalmol 157:458–463

Fergeson MA, Mulvihill JJ, Schaefer GB, Dehaai KA, Piatt J, Combs K, Bright BC, Neas BR (2009) Low adherence to national guidelines for thyroid screening in Down syndrome. Genet Med 11:548–551

Fernald A (2010) Getting beyond the "convenience sample" in research on early cognitive development. Behav Brain Sci 33:91–92

Fernald A, Marchman VA, Weisleder A (2013) SES differences in language processing skill and vocabulary are evident at 18 months. Dev Sci 16:234–248

Fernandez F, Garner CC (2008) Episodic-like memory in Ts65Dn, a mouse model of Down syndrome. Behav Brain Res 188:233–237

Fernandez F, Morishita W, Zuniga E, Nguyen J, Blank M, Malenka RC, Garner CC (2007) Pharmacotherapy for cognitive impairment in a mouse model of Down syndrome. Nat Neurosci 10:411–413

Fidler DJ (2006) The emergence of a syndrome-specific personality profile in young children with Down syndrome. Downs Syndr Res Pract 10:53–60

Fidler DJ, Barrett KC, Most DE (2005a) Age-related differences in smiling and personality in Down syndrome. J Dev Phys Disabil 17:263–280

Fidler DJ, Philofsky A, Hepburn SL, Rogers SJ (2005b) Nonverbal requesting and problem-solving by toddlers with Down syndrome. Am J Ment Retard 110:312–322

Fidler DJ, Most DE, Booth-LaForce C, Kelly JF (2006) Temperament and behaviour problems in young children with Down syndrome at 12, 30, and 45 months. Downs Syndr Res Pract 10:23–29

Forte GI, Piccione M, Scola L, Crivello A, Galfano C, Corsi MM, Chiappelli M, Candore G, Giuffrè M, Verna R, Licastro F, Corsello G, Caruso C, Lio D (2007) Apolipoprotein E genotypic frequencies among Down syndrome patients imply early unsuccessful aging for ApoE4 carriers. Rejuvenation Res 10:293–299

Freeman SB, Taft LF, Dooley KJ, Allran K, Sherman SL, Hassold TJ, Khoury MJ, Saker DM (1998) Population-based study of congenital heart defects in Down syndrome. Am J Med Genet 80:213–217

Frenkel S, Bourdin B (2009) Verbal, visual, and spatio-sequential short-term memory: assessment of the storage capacities of children and teenagers with Down's syndrome. J Intellect Disabil Res 53:152–160

Fudge JC Jr, Li S, Jaggers J, O'Brien SM, Peterson ED, Jacobs JP, Welke KF, Jacobs ML, Li JS, Pasquali SK (2010) Congenital heart surgery outcomes in Down syndrome: analysis of a national clinical database. Pediatrics 126:315–322

Fuster J (1989) The prefrontal cortex. Raven, New York, NY
Galli M, Cimolin V, Patti P, Ferrario D, Heaney G, Albertini G, Freedland R (2010) Quantifying established clinical assessment measures using 3D-movement analysis in individuals with Down syndrome. Disabil Rehabil 32:1768–1774
Garon N, Bryson SE, Smith IM (2008) Executive function in preschoolers: a review using an integrative framework. Psychol Bull 134:31–60
Ghaffar S, Lemler MS, Fixler DE, Ramaciotti C (2005) Trisomy 21 and congenital heart disease: effect of timing of initial echocardiogram. Clin Pediatr (Phila) 44:39–42
Ghosh SS, Tourville JA, Guenther FH (2008) A neuroimaging study of premotor lateralization and cerebellar involvement in the production of phonemes and syllables. J Speech Lang Hear Res 51:1183–1202
Gilmore L, Cuskelly M, Hayes A (2003) Self-regulatory behaviors in children with Down syndrome and typically developing children measured using the Goodman Lock Box. Res Dev Disabil 24:95–108
Gioia GA, Isquith PK, Guy SC, Kenworthy L (2000) Behavior rating inventory of executive function. Child Neuropsychol 6:235–238
Gogtay N, Giedd JN, Lusk L, Hayashi KM, Greenstein D, Vaituzis AC, Nugent TF 3rd, Herman DH, Clasen LS, Toga AW, Rapoport JL, Thompson PM (2004) Dynamic mapping of human cortical development during childhood through early adulthood. Proc Natl Acad Sci USA 101:8174–8179
Goldberg-Stern H, Strawsburg RH, Patterson B, Hickey F, Bare M, Gadoth N, Degrauw TJ (2001) Seizure frequency and characteristics in children with Down syndrome. Brain Dev 23:375–378
Golden JA, Hyman BT (1994) Development of the superior temporal neocortex is anomalous in trisomy 21. J Neuropathol Exp Neurol 53:513–520
Graber E, Chacko E, Regelmann MO, Costin G, Rapaport R (2012) Down syndrome and thyroid function. Endocrinol Metab Clin North Am 41:735–745
Grady C (2012) The cognitive neuroscience of ageing. Nat Rev Neurosci 13:491–505
Grady CL, McIntosh AR, Craik FI (2005) Task-related activity in prefrontal cortex and its relation to recognition memory performance in young and old adults. Neuropsychologia 43:1466–1481
Graham S, Phua E, Soon CS, Oh T, Au C, Shuter B, Wang SC, Yeh IB (2009) Role of medial cortical, hippocampal and striatal interactions during cognitive set-shifting. Neuroimage 45:1359–1367
Guidi S, Bonasoni P, Ceccarelli C, Santini D, Gualtieri F, Ciani E, Bartesaghi R (2008) Neurogenesis impairment and increased cell death reduce total neuron number in the hippocampal region of fetuses with Down syndrome. Brain Pathol 18:180–197
Guidi S, Ciani E, Bonasoni P, Santini D, Bartesaghi R (2011) Widespread proliferation impairment and hypocellularity in the cerebellum of fetuses with Down syndrome. Brain Pathol 21:361–373
Guralnick MJ, Connor RT, Johnson LC (2009) Home-based peer social networks of young children with Down syndrome: a developmental perspective. Am J Intellect Dev Disabil 114:340–355
Guralnick MJ, Connor RT, Johnson LC (2011) The peer social networks of young children with Down syndrome in classroom programmes. J Appl Res Intellect Disabil 24:310–321
Gutchess AH, Welsh RC, Hedden T, Bangert A, Minear M, Liu LL, Park DC (2005) Aging and the neural correlates of successful picture encoding: frontal activations compensate for decreased medial-temporal activity. J Cogn Neurosci 17:84–96
Gutchess AH, Hebrank A, Sutton BP, Leshikar E, Chee MW, Tan JC, Goh JO, Park DC (2007) Contextual interference in recognition memory with age. Neuroimage 35:1338–1347
Haass C, Hung AY, Selkoe DJ (1991) Processing of beta-amyloid precursor protein in microglia and astrocytes favors an internal localization over constitutive secretion. J Neurosci 11:3783–3793
Hall SS, Burns DD, Lightbody AA, Reiss AL (2008) Longitudinal changes in intellectual development in children with Fragile X syndrome. J Abnorm Child Psychol 36:927–939

Hampton TG, Stasko MR, Kale A, Amende I, Costa AC (2004) Gait dynamics in trisomic mice: quantitative neurological traits of Down syndrome. Physiol Behav 82:381–389

Han F, Yu H, Zhang J, Tian C, Schmidt C, Nava C, Davisson MT, Zheng QY (2009) Otitis media in a mouse model for Down syndrome. Int J Exp Pathol 90:480–488

Hardy J, Crook R, Perry R, Raghavan R, Roberts G (1994) ApoE genotype and Down's syndrome. Lancet 343:979–980

Hardy O, Worley G, Lee MM, Chaing S, Mackey J, Crissman B, Kishnani PS (2004) Hypothyroidism in Down syndrome: screening guidelines and testing methodology. Am J Med Genet A 124A:436–437

Harper JW, Heath RG (1973) Anatomic connections of the fastigial nucleus to the rostral forebrain in the cat. Exp Neurol 39:285–292

Hashimoto R, Sakai KL (2002) Specialization in the left prefrontal cortex for sentence comprehension. Neuron 35:589–597

Hassmann E, Skotnicka B, Midro AT, Musiatowicz M (1998) Distortion products otoacoustic emissions in diagnosis of hearing loss in Down syndrome. Int J Pediatr Otorhinolaryngol 45:199–206

Hattori M, Fujiyama A, Taylor TD, Watanabe H, Yada T, Park HS, Toyoda A, Ishii K, Totoki Y, Choi DK, Groner Y, Soeda E, Ohki M, Takagi T, Sakaki Y, Taudien S, Blechschmidt K, Polley A, Menzel U, Delabar J, Kumpf K, Lehmann R, Patterson D, Reichwald K, Rump A, Schillhabel M, Schudy A, Zimmermann W, Rosenthal A, Kudoh J, Schibuya K, Kawasaki K, Asakawa S, Shintani A, Sasaki T, Nagamine K, Mitsuyama S, Antonarakis SE, Minoshima S, Shimizu N, Nordsiek G, Hornischer K, Brant P, Scharfe M, Schon O, Desario A, Reichelt J, Kauer G, Blocker H, Ramser J, Beck A, Klages S, Hennig S, Riesselmann L, Dagand E, Haaf T, Wehrmeyer S, Borzym K, Gardiner K, Nizetic D, Francis F, Lehrach H, Reinhardt R, Yaspo ML, Chromosome 21 mapping and sequencing consortium (2000) The DNA sequence of human chromosome 21. Nature 405:311–319

Hawkins BA, Eklund SJ, James DR, Foose AK (2003) Adaptive behavior and cognitive function of adults with Down syndrome: modeling change with age. Ment Retard 41:7–28

Hawli Y, Nasrallah M, El-Hajj Fuleihan G (2009) Endocrine and musculoskeletal abnormalities in patients with Down syndrome. Nat Rev Endocrinol 5:327–334

Heath RG (1973) Fastigial nucleus connections to the septal region in monkey and cat: a demonstration with evoked potentials of a bilateral pathway. Biol Psychiatry 6:193–196

Heath RG, Harper JW (1974) Ascending projections of the cerebellar fastigial nucleus to the hippocampus, amygdala, and other temporal lobe sites: evoked potential and histological studies in monkeys and cats. Exp Neurol 45:268–287

Heath RG, Dempesy CW, Fontana CJ, Myers WA (1978) Cerebellar stimulation: effects on septal region, hippocampus, and amygdala of cats and rats. Biol Psychiatry 13:501–529

Heath RG, Dempesy CW, Fontana CJ, Fitzjarrell AT (1980) Feedback loop between cerebellum and septal-hippocampal sites: its role in emotion and epilepsy. Biol Psychiatry 15:541–556

Heller JH, Spiridigliozzi GA, Crissman BG, Sullivan-Saarela JA, Li JS, Kishnani PS (2006) Clinical trials in children with Down syndrome: issues from a cognitive research perspective. Am J Med Genet C Semin Med Genet 142C:187–195

Hessl D, Nguyen DV, Green C, Chavez A, Tassone F, Hagerman RJ, Senturk D, Schneider A, Lightbody A, Reiss AL, Hall S (2009) A solution to limitations of cognitive testing in children with intellectual disabilities: the case of fragile X syndrome. J Neurodev Disord 1:33–45

Hilton JM, Fitzgerald DA, Cooper DM (1999) Respiratory morbidity of hospitalized children with Trisomy 21. Paediatr Child Health 35:383–386

Hof PR, Bouras C, Perl DP, Sparks DL, Mehta N, Morrison JH (1995) Age-related distribution of neuropathologic changes in the cerebral cortex of patients with Down's syndrome. Quantitative regional analysis and comparison with Alzheimer's disease. Arch Neurol 52:379–391

Hoff E (2003) The specificity of environmental influence: socioeconomic status affects early vocabulary development via maternal speech. Child Dev 74:1368–1378

Horwitz B, Schapiro MB, Grady CL, Rapoport SI (1990) Cerebral metabolic pattern in young adult Down's syndrome subjects: altered intercorrelations between regional rates of glucose utilization. J Ment Defic Res 34:237–252

Hsieh H, Boehm J, Sato C, Iwatsubo T, Tomita T, Sisodia S, Malinow R (2006) AMPAR removal underlies Abeta-induced synaptic depression and dendritic spine loss. Neuron 52:831–843

Hung AY, Koo EH, Haass C, Selkoe DJ (1992) Increased expression of beta-amyloid precursor protein during neuronal differentiation is not accompanied by secretory cleavage. Proc Natl Acad Sci U S A 89:9439–9443

Huttenlocher PR (1979) Synaptic density in human frontal cortex—developmental changes and effects of aging. Brain Res 163:195–205

Huttenlocher PR, de Courten C, Garey LJ, Van der Loos H (1982) Synaptogenesis in human visual cortex—evidence for synapse elimination during normal development. Neurosci Lett 33:247–252

Hyde LA, Frisone DF, Crnic LS (2001) Ts65Dn mice, a model for Down syndrome, have deficits in context discrimination learning suggesting impaired hippocampal function. Behav Brain Res 118:53–60

Inglis A, Lohn Z, Austin JC, Hippman C (2014) A "cure" for Down syndrome: What do parents want? Clin Genet 86:310–317

Insausti AM, Megías M, Crespo D, Cruz-Orive LM, Dierssen M, Vallina IF, Insausti R, Flórez J (1998) Hippocampal volume and neuronal number in Ts65Dn mice: a murine model of Down syndrome. Neurosci Lett 253:175–178

Jacobs NS, Allen TA, Nguyen N, Fortin NJ (2013) Critical role of the hippocampus in memory for elapsed time. J Neurosci 33:13888–13893

Jacola LM, Byars AW, Chalfonte-Evans M, Schmithorst VJ, Hickey F, Patterson B, Hotze S, Vannest J, Chiu CY, Holland SK, Schapiro MB (2011) Functional magnetic resonance imaging of cognitive processing in young adults with Down syndrome. Am J Intellect Dev Disabil 116:344–359

Jahn G, Wendt J, Lotze M, Papenmeier F, Huff M (2012) Brain activation during spatial updating and attentive tracking of moving targets. Brain Cogn 78:105–113

Jarrold C, Baddeley AD (1997) Short-term memory for verbal and visuospatial information in Down's syndrome. Cogn Neuropsychiatry 2:101–122

Jarrold C, Baddeley AD (2001) Short-term memory in Down syndrome: applying the working memory model. Downs Syndr Res Pract 7:17–23

Jarrold C, Baddeley AD, Hewes AK (2000) Verbal short-term memory deficits in Down syndrome: a consequence of problems in rehearsal? J Child Psychol Psychiatry 41:233–244

Jarrold C, Baddeley AD, Phillips CE (2002) Verbal short-term memory in Down syndrome: a problem of memory, audition, or speech? J Speech Lang Hear Res 45:531–544

Jenkins LJ, Ranganath C (2010) Prefrontal and medial temporal lobe activity at encoding predicts temporal context memory. J Neurosci 30:15558–15565

Jernigan TL, Bellugi U, Sowell E, Doherty S, Hesselink JR (1993) Cerebral morphologic distinctions between Williams and Down syndromes. Arch Neurol 50:186–191

John FM, Bromham NR, Woodhouse JM, Candy TR (2004) Spatial vision deficits in infants and children with Down syndrome. Invest Ophthalmol Vis Sci 45:1566–1572

Johnson MH (2001) Functional brain development in humans. Nat Rev Neurosci 2:475–483

Jones EL, Hanney M, Francis PT, Ballard CG (2009) Amyloid beta concentrations in older people with Down syndrome and dementia. Neurosci Lett 451:162–164

Jonsson T, Atwal JK, Steinberg S, Snaedal J, Jonsson PV, Bjornsson S, Stefansson H, Sulem P, Gudbjartsson D, Maloney J, Hoyte K, Gustafson A, Liu Y, Lu Y, Bhangale T, Graham RR, Huttenlocher J, Bjornsdottir G, Andreassen OA, Jönsson EG, Palotie A, Behrens TW, Magnusson OT, Kong A, Thorsteinsdottir U, Watts RJ, Stefansson K (2012) A mutation in APP protects against Alzheimer's disease and age-related cognitive decline. Nature 488:96–99

Kahlem P, Sultan M, Herwig R, Steinfath M, Balzereit D, Eppens B, Saran NG, Pletcher MT, South ST, Stetten G, Lehrach H, Reeves RH, Yaspo ML (2004) Transcript level alterations reflect gene dosage effects across multiple tissues in a mouse model of Down syndrome. Genome Res 14:1258–1267

Kanno K, Ikeda Y (2002) Word-length effect in verbal short-term memory in individuals with Down's syndrome. J Intellect Disabil Res 46:613–618

Karlsgodt KH, Shirinyan D, van Erp TG, Cohen MS, Cannon TD (2005) Hippocampal activations during encoding and retrieval in a verbal working memory paradigm. Neuroimage 25:1224–1231

Karmiloff-Smith A (1998) Development itself is the key to understanding developmental disorders. Trends Cogn Sci 2:389–398

Kasari C, Freeman SF (2001) Task-related social behavior in children with Down syndrome. Am J Ment Retard 106:253–264

Katz ER, Ellis NR (1991) Memory for spatial location in retarded and nonretarded persons. J Ment Defic Res 35:209–220

Kesner RP, Hopkins RO, Fineman B (1994) Item and order dissociation in humans with prefrontal cortex damage. Neuropsychologia 32:881–891

Kester MH, Martinez de Mena R, Obregon MJ, Marinkovic D, Howatson A, Visser TJ, Hume R, Morreale de Escobar G (2004) Iodothyronine levels in the human developing brain: major regulatory roles of iodothyronine deiodinases in different areas. J Clin Endocrinol Metab 89:3117–3128

Kim J (2010) Within-subjects design. In: Salkind N (ed) Encyclopedia of research design. Sage, Thousand Oaks, CA, pp 1639–1645

Kim T, Vidal GS, Djurisic M, William CM, Birnbaum ME, Garcia KC, Hyman BT, Shatz CJ (2013) Human LilrB2 is a β-amyloid receptor and its murine homolog PirB regulates synaptic plasticity in an Alzheimer's model. Science 341:1399–1404

King K, O'Gorman C, Gallagher S (2014) Thyroid dysfunction in children with Down syndrome: a literature review. Ir J Med Sci 183:1–6

Klenberg L, Korkman M, Lahti-Nuuttila P (2001) Differential development of attention and executive functions in 3- to 12-year-old Finnish children. Dev Neuropsychol 20:407–428

Kochanska G, Coy KC, Murray KT (2001) The development of self-regulation in the first four years of life. Child Dev 72:1091–1111

Kopp CB, Krakow JB, Johnson KL (1983) Strategy production by young Down syndrome children. Am J Ment Defic 88:164–169

Kover ST, Atwood AK (2013) Establishing equivalence: methodological progress in group-matching design and analysis. Am J Intellect Dev Disabil 118:3–15

Kowalczyk K, Pukajło K, Malczewska A, Król-Chwastek A, Barg E (2013) L-thyroxine therapy and growth processes in children with Down syndrome. Adv Clin Exp Med 22:85–92

Lai F, Kammann E, Rebeck GW, Anderson A, Chen Y, Nixon RA (1999) APOE genotype and gender effects on Alzheimer disease in 100 adults with Down syndrome. Neurology 53:331–336

Lambert JC, Pérez-Tur J, Dupire MJ, Delacourte A, Frigard B, Chartier-Harlin MC (1996) Analysis of the APOE alleles impact in Down's syndrome. Neurosci Lett 220:57–60

Landry SH, Chapieski ML (1989) Joint attention and infant toy exploration: effects of Down syndrome and prematurity. Child Dev 60:103–118

Lanfranchi S, Carretti B, Spanò G, Cornoldi C (2009) A specific deficit in visuospatial simultaneous working memory in Down syndrome. J Intellect Disabil Res 53:474–483

Latash ML, Kang N, Patterson D (2002) Finger coordination in persons with Down syndrome: atypical patterns of coordination and the effects of practice. Exp Brain Res 146:345–355

Laws G (2002) Working memory in children and adolescents with Down syndrome: evidence from a colour memory experiment. J Child Psychol Psychiatry 43:353–364

Laws G, Gunn D (2004) Phonological memory as a predictor of language comprehension in Down syndrome: a five-year follow-up study. J Child Psychol Psychiatry 45:326–337

Lebel C, Walker L, Leemans A, Phillips L, Beaulieu C (2008) Microstructural maturation of the human brain from childhood to adulthood. Neuroimage 40:1044–1055

Lee NR, Fidler DJ, Blakeley-Smith A, Daunhauer L, Robinson C, Hepburn SL (2011) Caregiver report of executive functioning in a population-based sample of young children with Down syndrome. Am J Intellect Dev Disabil 116:290–304

Lemere CA, Blusztajn JK, Yamaguchi H, Wisniewski T, Saido TC, Selkoe DJ (1996) Sequence of deposition of heterogeneous amyloid beta-peptides and APO E in Down syndrome: implications for initial events in amyloid plaque formation. Neurobiol Dis 3:16–32

Leverenz JB, Raskind MA (1998) Early amyloid deposition in the medial temporal lobe of young Down syndrome patients: a regional quantitative analysis. Exp Neurol 150:296–304

Li H, Cherry S, Klinedinst D, DeLeon V, Redig J, Reshey B, Chin MT, Sherman SL, Maslen CL, Reeves RH (2012) Genetic modifiers predisposing to congenital heart disease in the sensitized Down syndrome population. Circ Cardiovasc Genet 5:301–308

Libb JW, Dahle A, Smith K, McCollister FP, McLain C (1985) Hearing disorder and cognitive function of individuals with Down syndrome. Am J Ment Defic 90:353–356

Lögdberg B, Brun A (1993) Prefrontal neocortical disturbances in mental retardation. J Intellect Disabil Res 37:459–468

Lorenzi HA, Reeves RH (2006) Hippocampal hypocellularity in the Ts65Dn mouse originates early in development. Brain Res 1104:153–159

Luton D, Azria E, Polak M, Carré A, Vuillard E, Delezoide AL, Guibourdenche J (2012) Thyroid function in fetuses with Down syndrome. Horm Res Paediatr 78:88–93

Majdalany DS, Burkhart HM, Connolly HM, Abel MD, Dearani JA, Warnes CA, Schaff HV (2010) Adults with Down syndrome: safety and long-term outcome of cardiac operation. Congenit Heart Dis 5:38–43

Mani A, Mullainathan S, Shafir E, Zhao J (2013) Poverty impedes cognitive function. Science 341:976–980

Marcell MM, Cohen S (1992) Hearing abilities of Down syndrome and other mentally handicapped adolescents. Res Dev Disabil 13:533–551

Marcell MM, Weeks SL (1988) Short-term memory difficulties and Down's syndrome. J Ment Defic Res 32:153–162

Marvel CL, Desmond JE (2010) Functional topography of the cerebellum in verbal working memory. Neuropsychol Rev 20:271–279

Massaquoi SG (2012) Physiology of clinical dysfunction of the cerebellum. Handb Clin Neurol 103:37–62

McDowell KM, Craven DI (2011) Pulmonary complications of Down syndrome during childhood. J Pediatr 158:319–325

Mehta PD, Mehta SP, Fedor B, Patrick BA, Emmerling M, Dalton AJ (2003) Plasma amyloid beta protein 1–42 levels are increased in old Down syndrome but not in young Down syndrome. Neurosci Lett 342:155–158

Mehta PD, Capone G, Jewell A, Freedland RL (2007) Increased amyloid beta protein levels in children and adolescents with Down syndrome. J Neurol Sci 254:22–27

Menghini D, Costanzo F, Vicari S (2011) Relationship between brain and cognitive processes in Down syndrome. Behav Genet 41:381–393

Mervis CB, Robinson BF (1999) Methodological issues in cross-syndrome comparisons: matching procedures, sensitivity (Se), and specificity (Sp). Monogr Soc Res Child Dev 64:115–130

Middleton FA, Strick PL (2001) Cerebellar projections to the prefrontal cortex of the primate. J Neurosci 21:700–712

Miles S, Chapman RS (2002) Narrative content as described by individuals with Down syndrome and typically developing children. J Speech Lang Hear Res 45:175–189

Miller EK, Cohen JD (2001) An integrative theory of prefrontal cortex function. Annu Rev Neurosci 24:167–202

Miller JF, Neufang M, Solway A, Brandt A, Trippel M, Mader I, Hefft S, Merkow M, Polyn SM, Jacobs J, Kahana MJ, Schulze-Bonhage A (2013) Neural activity in human hippocampal formation reveals the spatial context of retrieved memories. Science 342:1111–1114

Miolo G, Chapman RS, Sindberg HA (2005) Sentence comprehension in adolescents with Down syndrome and typically developing children: role of sentence voice, visual context, and auditory-verbal short-term memory. J Speech Lang Hear Res 48:172–188

Mito T, Pereyra PM, Becker LE (1991) Neuropathology in patients with congenital heart disease and Down syndrome. Pediatr Pathol 11:867–877

Mogra R, Zidere V, Allan LD (2011) Prenatally detectable congenital heart defects in fetuses with Down syndrome. Ultrasound Obstet Gynecol 38:320–324

Moon J, Chen M, Gandhy SU, Strawderman M, Levitsky DA, Maclean KN, Strupp BJ (2010) Perinatal choline supplementation improves cognitive functioning and emotion regulation in the Ts65Dn mouse model of Down syndrome. Behav Neurosci 124:346–361

Moore CS (2006) Postnatal lethality and cardiac anomalies in the Ts65Dn Down syndrome mouse model. Mamm Genome 17:1005–1012

Mori C, Spooner ET, Wisniewsk KE, Wisniewski TM, Yamaguch H, Saido TC, Tolan DR, Selkoe DJ, Lemere CA (2002) Intraneuronal Abeta42 accumulation in Down syndrome brain. Amyloid 9:88–102

Mosse EK, Jarrold C (2010) Searching for the Hebb effect in Down syndrome: evidence for a dissociation between verbal short-term memory and domain-general learning of serial order. J Intellect Disabil Res 54:295–307

Motte J, Williams RS (1989) Age-related changes in the density and morphology of plaques and neurofibrillary tangles in Down syndrome brain. Acta Neuropathol 77:535–546

Mueller SG, Chao LL, Berman B, Weiner MW (2011) Evidence for functional specialization of hippocampal subfields detected by MR subfield volumetry on high resolution images at 4 T. Neuroimage 56:851–857

Murray LJ, Ranganath C (2007) The dorsolateral prefrontal cortex contributes to successful relational memory encoding. J Neurosci 27:5515–5522

Murrell J, Farlow M, Ghetti B, Benson MD (1991) A mutation in the amyloid precursor protein associated with hereditary Alzheimer's disease. Science 254:97–99

Murty VP, Adcock RA (2013) Enriched encoding: Reward motivation organizes cortical networks for hippocampal detection of unexpected events. Cereb Cortex 24:2160–2168

Nadel L (1986) Down syndrome in neurobiological perspective. In: Epstein CJ (ed) The neurobiology of Down syndrome. Raven, New York, NY, pp 239–251

Nadel L (2003) Down's syndrome: a genetic disorder in biobehavioral perspective. Genes Brain Behav 2:156–166

Næss KA, Lyster SA, Hulme C, Melby-Lervåg M (2011) Language and verbal short-term memory skills in children with Down syndrome: a meta-analytic review. Res Dev Disabil 32:2225–2234

Narchi H (1999) Neonatal ECG screening for congenital heart disease in Down syndrome. Ann Trop Paediatr 19:51–54

Nelson WM 3rd, Dacey CM (1999) Validity of the Stanford-Binet Intelligence Scale-IV: its use in young adults with mental retardation. Ment Retard 37:319–325

Newman PP, Reza H (1979) Functional relationships between the hippocampus and the cerebellum: an electrophysiological study of the cat. J Physiol 287:405–426

Niccols A, Atkinson L, Pepler D (2003) Mastery motivation in young children with Down's syndrome: relations with cognitive and adaptive competence. J Intellect Disabil Res 47:121–133

O'Keefe J, Nadel L (1978) The hippocampus as a cognitive map. Oxford University Press, USA

Oh A, Vidal J, Taylor MJ, Pang EW (2014) Neuromagnetic correlates of intra- and extra-dimensional set-shifting. Brain Cogn 86:90–97

Olsen RK, Moses SN, Riggs L, Ryan JD (2012) The hippocampus supports multiple cognitive processes through relational binding and comparison. Front Hum Neurosci 6:146

Paladini D, Tartaglione A, Agangi A, Teodoro A, Forleo F, Borghese A, Martinelli P (2000) The association between congenital heart disease and Down syndrome in prenatal life. Ultrasound Obstet Gynecol 15:104–108

Park SC, Mathews RA, Zuberbuhler JR, Rowe RD, Neches WH, Lenox CC (1977) Down syndrome with congenital heart malformation. Am J Dis Child 131:29–33

Park DC, Welsh RC, Marshuetz C, Gutchess AH, Mikels J, Polk TA, Noll DC, Taylor SF (2003) Working memory for complex scenes: age differences in frontal and hippocampal activations. J Cogn Neurosci 15:1122–1134

Passot JB, Sheynikhovich D, Duvelle É, Arleo A (2012) Contribution of cerebellar sensorimotor adaptation to hippocampal spatial memory. PLoS One 7:e32560

Patel A, Rees SD, Kelly MA, Bain SC, Barnett AH, Thalitaya D, Prasher VP (2011) Association of variants within APOE, SORL1, RUNX1, BACE1 and ALDH18A1 with dementia in Alzheimer's disease in subjects with Down syndrome. Neurosci Lett 487:144–148

Patterson D (2007) Genetic mechanisms involved in the phenotype of Down syndrome. Ment Retard Dev Disabil Res Rev 13:199–206

Patterson T, Rapsey CM, Glue P (2013) Systematic review of cognitive development across childhood in Down syndrome: implications for treatment interventions. J Intellect Disabil Res 57:306–318

Paul SM, Heath RG, Ellison JP (1973) Histochemical demonstration of a direct pathway from the fastigial nucleus to the septal region. Exp Neurol 40:798–805

Paulesu E, Frith CD, Frackowiak RS (1993) The neural correlates of the verbal component of working memory. Nature 362:342–345

Pennington BF, Moon J, Edgin J, Stedron J, Nadel L (2003) The neuropsychology of Down syndrome: evidence for hippocampal dysfunction. Child Dev 74:75–93

Petanjek Z, Judaš M, Šimic G, Rasin MR, Uylings HB, Rakic P, Kostovic I (2011) Extraordinary neoteny of synaptic spines in the human prefrontal cortex. Proc Natl Acad Sci USA 108:13281–13286

Petit TL, LeBoutillier JC, Alfano DP, Becker LE (1984) Synaptic development in the human fetus: a morphometric analysis of normal and Down's syndrome neocortex. Exp Neurol 83:13–23

Pitcairn TK, Wishart JG (1994) Reactions of young children with Down's syndrome to an impossible task. Br J Dev Psychol 12:485–489

Pletcher MT, Wiltshire T, Cabin DE, Villanueva M, Reeves RH (2001) Use of comparative physical and sequence mapping to annotate mouse chromosome 16 and human chromosome 21. Genomics 74:45–54

Pollack A (2013) An experimental drug's bitter end. The New York Times, June 6th

Prasher VP, Chowdhury TA, Rowe BR, Bain SC (1997) ApoE genotype and Alzheimer's disease in adults with Down syndrome: meta-analysis. Am J Ment Retard 102:103–110

Prasher VP, Sajith SG, Rees SD, Patel A, Tewari S, Schupf N, Zigman WB (2008) Significant effect of APOE epsilon 4 genotype on the risk of dementia in Alzheimer's disease and mortality in persons with Down syndrome. Int J Geriatr Psychiatry 23:1134–1140

Prasher VP, Sajith SG, Mehta P, Zigman WB, Schupf N (2010) Plasma beta-amyloid and duration of Alzheimer's disease in adults with Down syndrome. Int J Geriatr Psychiatry 25:202–207

Pueschel SM, Jackson IM, Giesswein P, Dean MK, Pezzullo JC (1991) Thyroid function in Down syndrome. Res Dev Disabil 12:287–296

Purser HR, Jarrold C (2005) Impaired verbal short-term memory in Down syndrome reflects a capacity limitation rather than atypically rapid forgetting. J Exp Child Psychol 91:1–23

Purser HR, Jarrold C (2013) Poor phonemic discrimination does not underlie poor verbal short-term memory in Down syndrome. J Exp Child Psychol 115:1–15

Ramnani N (2006) The primate cortico-cerebellar system: anatomy and function. Nat Rev Neurosci 7:511–522

Randolph C (1998) Repeatable battery for the assessment of neuropsychological status. Pearson, San Antonio, TX

Ranganath C, Johnson MK, D'Esposito M (2003) Prefrontal activity associated with working memory and episodic long-term memory. Neuropsychologia 41:378–389

Raut P, Sriram B, Yeoh A, Hee KY, Lim SB, Daniel ML (2011) High prevalence of hearing loss in Down syndrome at first year of life. Ann Acad Med Singapore 40:493–498

Ravizza SM, Delgado MR, Chein JM, Becker JT, Fiez JA (2004) Functional dissociations within the inferior parietal cortex in verbal working memory. Neuroimage 22:562–573
Ravizza SM, McCormick CA, Schlerf JE, Justus T, Ivry RB, Fiez JA (2006) Cerebellar damage produces selective deficits in verbal working memory. Brain 129:306–320
Raz N, Torres IJ, Briggs SD, Spencer WD, Thornton AE, Loken WJ, Gunning FM, McQuain JD, Driesen NR, Acker JD (1995) Selective neuroanatomic abnormalities in Down's syndrome and their cognitive correlates: evidence from MRI morphometry. Neurology 45:356–366
Reeves RH (2001) Down's syndrome. A complicated genetic insult. Lancet 358:S23
Reeves RH, Gearhart JD, Littlefield JW (1986) Genetic basis for a mouse model of Down syndrome. Brain Res Bull 16:803–814
Reeves RH, Irving NG, Moran TH, Wohn A, Kitt C, Sisodia SS, Schmidt C, Bronson RT, Davisson MT (1995) A mouse model for Down syndrome exhibits learning and behaviour deficits. Nat Genet 11:177–184
Reeves RH, Rue EE, Citron MP, Cabin DE (1997) High-resolution recombinational map of mouse chromosome 16. Genomics 43:202–208
Reiss S, Levitan GW, Szyszko J (1982) Emotional disturbance and mental retardation: diagnostic overshadowing. Am J Ment Defic 86:567–574
Retailleau A, Etienne S, Guthrie M, Boraud T (2012) Where is my reward and how do I get it? Interaction between the hippocampus and the basal ganglia during spatial learning. J Physiol Paris 106:72–80
Reuter-Lorenz PA, Lustig C (2005) Brain aging: reorganizing discoveries about the aging mind. Curr Opin Neurobiol 15:245–251
Richtsmeier JT, Baxter LL, Reeves RH (2000) Parallels of craniofacial maldevelopment in Down syndrome and Ts65Dn mice. Dev Dyn 217:137–145
Ridderinkhof KR, Van der Stelt O (2000) Attention and selection in the growing child: views derived from developmental psychophysiology. Biol Psychol 54:55–106
Rochefort C, Arabo A, André M, Poucet B, Save E, Rondi-Reig L (2011) Cerebellum shapes hippocampal spatial code. Science 334:385–389
Rochefort C, Lefort JM, Rondi-Reig L (2013) The cerebellum: a new key structure in the navigation system. Front Neural Circuits 7:35
Roper RJ, Reeves RH (2006) Understanding the basis for Down syndrome phenotypes. PLoS Genet 2:e50
Roper RJ, Baxter LL, Saran NG, Klinedinst DK, Beachy PA, Reeves RH (2006) Defective cerebellar response to mitogenic Hedgehog signaling in Down syndrome mice. Proc Natl Acad Sci U S A 103:1452–1456
Rosen D (2011) Management of obstructive sleep apnea associated with Down syndrome and other craniofacial dysmorphologies. Curr Opin Pulm Med 17:431–436
Ross MH, Galaburda AM, Kemper TL (1984) Down's syndrome: is there a decreased population of neurons? Neurology 34:909–916
Rowe J, Lavender A, Turk V (2006) Cognitive executive function in Down's syndrome. Br J Clin Psychol 45:5–17
Royston MC, Mann D, Pickering-Brown S, Owen F, Perry R, Raghavan R, Khin-Nu C, Tyrer S, Day K, Crook R et al (1994) Apolipoprotein E epsilon 2 allele promotes longevity and protects patients with Down's syndrome from dementia. Neuroreport 5:2583–2585
Royston MC, Mann D, Pickering-Brown S, Owen F, Perry R, Ragbavan R, Khin-Nu C, Tyner S, Day K, Crook R, Hardy J, Roberts GW (1996) ApoE2 allele, Down's syndrome, and dementia. Ann N Y Acad Sci 777:255–259
Rubin BM (2013) Dreams dashed as drug trial canceled. Chicago Tribune, August 4th
Rueda MR, Posner MI, Rothbart MK (2005) The development of executive attention: contributions to the emergence of self-regulation. Dev Neuropsychol 28:573–594
Ruskin EM, Kasari C, Mundy P, Sigman M (1994) Attention to people and toys during social and object mastery in children with Down syndrome. Am J Ment Retard 99:103–111

Salehi A, Faizi M, Colas D, Valletta J, Laguna J, Takimoto-Kimura R, Kleschevnikov A, Wagner SL, Aisen P, Shamloo M, Mobley WC (2009) Restoration of norepinephrine-modulated contextual memory in a mouse model of Down syndrome. Sci Transl Med 1:7ra17

Sarici D, Akin MA, Kurtoglu S, Gunes T, Ozturk MA, Akcakus M (2012) Thyroid functions of neonates with Down syndrome. Ital J Pediatr 38:44

Schmidt-Sidor B, Wisniewski KE, Shepard TH, Sersen EA (1990) Brain growth in Down syndrome subjects 15 to 22 weeks of gestational age and birth to 60 months. Clin Neuropathol 9:181–190

Schmidt-Wilcke T, Poljansky S, Hierlmeier S, Hausner J, Ibach B (2009) Memory performance correlates with gray matter density in the ento-/perirhinal cortex and posterior hippocampus in patients with mild cognitive impairment and healthy controls—a voxel based morphometry study. Neuroimage 47:1914–1920

Schupf N, Patel B, Silverman W, Zigman WB, Zhong N, Tycko B, Mehta PD, Mayeux R (2001) Elevated plasma amyloid beta-peptide 1–42 and onset of dementia in adults with Down syndrome. Neurosci Lett 301:199–203

Schwethelm B, Mahoney G (1986) Task persistence among organically impaired mentally retarded children. Am J Ment Defic 90:432–439

Scott-McKean JJ, Chang B, Hurd RE, Nusinowitz S, Schmidt C, Davisson MT, Costa AC (2010) The mouse model of Down syndrome Ts65Dn presents visual deficits as assessed by pattern visual evoked potentials. Invest Ophthalmol Vis Sci 51:3300–3308

Shah PS, Hellmann J, Adatia I (2004) Clinical characteristics and follow up of Down syndrome infants without congenital heart disease who presented with persistent pulmonary hypertension of newborn. J Perinat Med 32:168–170

Shapiro NL, Huang RY, Sangwan S, Willner A, Laks H (2000) Tracheal stenosis and congenital heart disease in patients with Down syndrome: diagnostic approach and surgical options. Int J Pediatr Otorhinolaryngol 54:137–142

Sherman SL, Allen EG, Bean LH, Freeman SB (2007) Epidemiology of Down syndrome. Ment Retard Dev Disabil Res Rev 13:221–227

Shimamura AP (1995) Memory and the prefrontal cortex. Ann N Y Acad Sci 769:151–159

Shimamura AP, Jurica PJ, Mangels JA, Gershberg FB, Knight RT (1995) Susceptibility to memory interference effects following frontal lobe damage: findings from tests of paired-associate learning. J Cogn Neurosci 7:144–152

Shott SR, Joseph A, Heithaus D (2001) Hearing loss in children with Down syndrome. Int J Pediatr Otorhinolaryngol 61:199–205

Sigman M, Ruskin E, Arbeile S, Corona R, Dissanayake C, Espinosa M, Kim N, López A, Zierhut C (1999) Continuity and change in the social competence of children with autism, Down syndrome, and developmental delays. Monogr Soc Res Child Dev 64:1–114

Silva ML, Cieuta C, Guerrini R, Plouin P, Livet MO, Dulac O (1996) Early clinical and EEG features of infantile spasms in Down syndrome. Epilepsia 37:977–982

Silverman W, Miezejeski C, Ryan R, Zigman W, Krinsky-McHale S, Urv T (2010) Stanford-Binet & WAIS IQ differences and their implications for adults with intellectual disability (aka mental retardation). Intelligence 38:242–248

Sloper P, Turner S (1996) Progress in social-independent functioning of young people with Down's syndrome. J Intellect Disabil Res 40:39–48

Smith PF (1997) Vestibular-hippocampal interactions. Hippocampus 7:465–471

Smith GK, Kesner RP, Korenberg JR (2014) Dentate gyrus mediates cognitive function in the Ts65Dn/DnJ mouse model of Down syndrome. Hippocampus 24:354–362

Snider RS, Maiti A (1975) Septal afterdischarges and their modification by the cerebellum. Exp Neurol 49:529–539

Snider RS, Maiti A (1976) Cerebellar contributions to the Papez circuit. J Neurosci Res 2:133–146

Snyder EM, Nong Y, Almeida CG, Paul S, Moran T, Choi EY, Nairn AC, Salter MW, Lombroso PJ, Gouras GK, Greengard P (2005) Regulation of NMDA receptor trafficking by amyloid-beta. Nat Neurosci 8:1051–1058

Sparrow SS, Balla DA, Cicchetti DV (1984) Vineland adaptive behavior scales: interview edition (Survey Form). American Guidance Services, Circle Pines, MN

Spencer-Smith M, Anderson V (2009) Healthy and abnormal development of the prefrontal cortex. Dev Neurorehabil 12:279–297

Spinks R, McKirgan LW, Arndt S, Caspers K, Yucuis R, Pfalzgraf CJ (2009) IQ estimate smackdown: comparing IQ proxy measures to the WAIS-III. J Int Neuropsychol Soc 15:590–596

Squire LR, Stark CE, Clark RE (2004) The medial temporal lobe. Annu Rev Neurosci 27:279–306

Stein DS, Munir KM, Karweck AJ, Davidson EJ, Stein MT (2013) Developmental regression, depression, and psychosocial stress in an adolescent with Down syndrome. J Dev Behav Pediatr 34:216–218

Steingass KJ, Chicoine B, McGuire D, Roizen NJ (2011) Developmental disabilities grown up: Down syndrome. J Dev Behav Pediatr 32:548–558

Stenzel D, Huttner WB (2013) Role of maternal thyroid hormones in the developing neocortex and during human evolution. Front Neuroanat 7:19

Sultan M, Piccini I, Balzereit D, Herwig R, Saran NG, Lehrach H, Reeves RH, Yaspo ML (2007) Gene expression variation in Down's syndrome mice allows prioritization of candidate genes. Genome Biol 8:R91

Sussan TE, Yang A, Li F, Ostrowski MC, Reeves RH (2008) Trisomy represses Apc(Min)-mediated tumours in mouse models of Down's syndrome. Nature 451:73–75

Suttle CM, Lloyd R (2005) Chromatic and achromatic transient VEPs in adults with Down syndrome. Ophthalmic Physiol Opt 25:501–513

Suzuki K, Yamaki S, Mimori S, Murakami Y, Mori K, Takahashi Y, Kikuchi T (2000) Pulmonary vascular disease in Down's syndrome with complete atrioventricular septal defect. Am J Cardiol 86:434–437

Takashima S, Becker LE, Armstrong DL, Chan F (1981) Abnormal neuronal development in the visual cortex of the human fetus and infant with down's syndrome. A quantitative and qualitative Golgi study. Brain Res 225:1–21

Tanzi RE (2012) The genetics of Alzheimer disease. Cold Spring Harb Perspect Med 2:a006296

Tassé MJ, Schalock RL, Balboni G, Bersani H Jr, Borthwick-Duffy SA, Spreat S, Thissen D, Widaman KF, Zhang D (2012) The construct of adaptive behavior: its conceptualization, measurement, and use in the field of intellectual disability. Am J Intellect Dev Disabil 117:291–303

Teller JK, Russo C, DeBusk LM, Angelini G, Zaccheo D, Dagna-Bricarelli F, Scartezzini P, Bertolini S, Mann DM, Tabaton M, Gambetti P (1996) Presence of soluble amyloid beta-peptide precedes amyloid plaque formation in Down's syndrome. Nat Med 2:93–95

Tenenbaum A, Lebel E, Malkiel S, Kastiel Y, Abulibdeh A, Zangen DH (2012) Euthyroid submedian free T4 and subclinical hypothyroidism may have a detrimental clinical effect in Down syndrome. Horm Res Paediatr 78:113–118

Tiemeier H, Lenroot RK, Greenstein DK, Tran L, Pierson R, Giedd JN (2010) Cerebellum development during childhood and adolescence: a longitudinal morphometric MRI study. Neuroimage 49:63–70

Tomalski P, Moore DG, Ribeiro H, Axelsson EL, Murphy E, Karmiloff-Smith A, Johnson MH, Kushnerenko E (2013) Socioeconomic status and functional brain development—associations in early infancy. Dev Sci 16:676–687

Travis SG, Huang Y, Fujiwara E, Radomski A, Olsen F, Carter R, Seres P, Malykhin NV (2014) High field structural MRI reveals specific episodic memory correlates in the subfields of the hippocampus. Neuropsychologia 53:233–245

Turan S, Turan OM, Desai A, Harman CR, Baschat AA (2014) A prospective study of first trimester fetal cardiac examination using spatiotemporal image correlation, tomographic ultrasound and color Doppler imaging for the diagnosis of complex congenital heart disease in high-risk patients. Ultrasound Obstet Gynecol 44:562–567

Tyler WA, Haydar TF (2013) Multiplex genetic fate mapping reveals a novel route of neocortical neurogenesis, which is altered in the Ts65Dn mouse model of Down syndrome. J Neurosci 33:5106–5119

Van Gameren-Oosterom HB, Fekkes M, Reijneveld SA, Oudesluys-Murphy AM, Verkerk PH, Van Wouwe JP, Buitendijk SE (2013) Practical and social skills of 16–19-year-olds with Down syndrome: independence still far away. Res Dev Disabil 34:4599–4607

Van Trotsenburg AS, Vulsma T, van Santen HM, Cheung W, de Vijlder JJ (2003) Lower neonatal screening thyroxine concentrations in Down syndrome newborns. J Clin Endocrinol Metab 88:1512–1515

Van Trotsenburg AS, Vulsma T, van Rozenburg-Marres SL, van Baar AL, Ridder JC, Heymans HS, Tijssen JG, de Vijlder JJ (2005) The effect of thyroxine treatment started in the neonatal period on development and growth of two-year-old Down syndrome children: a randomized clinical trial. J Clin Endocrinol Metab 90:3304–3311

Van Trotsenburg AS, Kempers MJ, Endert E, Tijssen JG, de Vijlder JJ, Vulsma T (2006) Trisomy 21 causes persistent congenital hypothyroidism presumably of thyroidal origin. Thyroid 16:671–680

Vetter VL, Dugan N, Guo R, Mercer-Rosa L, Gleason M, Cohen M, Vogel RL, Iyer R (2011) A pilot study of the feasibility of heart screening for sudden cardiac arrest in healthy children. Am Heart J 161:1000–1006

Vicari S, Carlesimo A, Caltagirone C (1995) Short-term memory in persons with intellectual disabilities and Down's syndrome. J Intellect Disabil Res 39:532–537

Vicari S, Bellucci S, Carlesimo GA (2000) Implicit and explicit memory: a functional dissociation in persons with Down syndrome. Neuropsychologia 38:240–251

Vicari S, Bellucci S, Carlesimo GA (2005) Visual and spatial long-term memory: differential pattern of impairments in Williams and Down syndromes. Dev Med Child Neurol 47:305–311

Vis JC, Duffels MG, Winter MM, Weijerman ME, Cobben JM, Huisman SA, Mulder BJ (2009) Down syndrome: a cardiovascular perspective. J Intellect Disabil Res 53:419–425

Visootsak J, Sherman S (2007) Neuropsychiatric and behavioral aspects of trisomy 21. Curr Psychiatry Rep 9:135–140

Visootsak J, Hess B, Bakeman R, Adamson LB (2013) Effect of congenital heart defects on language development in toddlers with Down syndrome. J Intellect Disabil Res 57:887–892

Vlachou M, Farrell P (2000) Object mastery motivation in pre-school children with and without disabilities. Educ Psychol 20:167–176

Wang PP (1996) A neuropsychological profile of Down syndrome: Cognitive skills and brain morphology. Ment Retard Dev Disabil Res Rev 2:102–108

Weed MR, Bryant R, Perry S (2008) Cognitive development in macaques: attentional set-shifting in juvenile and adult rhesus monkeys. Neuroscience 157:22–28

Weijerman ME, de Winter JP (2010) Clinical practice. The care of children with Down syndrome. Eur J Pediatr 169:1445–1452

Wells GL, Barker SE, Finley SC, Colvin EV, Finley WH (1994) Congenital heart disease in infants with Down's syndrome. South Med J 87:724–727

Williams AD, Mjaatvedt CH, Moore CS (2008) Characterization of the cardiac phenotype in neonatal Ts65Dn mice. Dev Dyn 237:426–435

Wimmer GE, Shohamy D (2012) Preference by association: how memory mechanisms in the hippocampus bias decisions. Science 338:270–273

Wishart JG (1986) The effects of step-by-step training on cognitive performance in infants with Down's syndrome. J Ment Defic Res 30:233–250

Wishart JG (1993) The development of learning difficulties in children with Down's syndrome. J Intellect Disabil Res 37:389–403

Wishart JG (2001) Motivation and learning styles in young children with Down syndrome. Downs Syndr Res Pract 7:47–51

Wishart JG (2007) Socio-cognitive understanding: a strength or weakness in Down's syndrome? J Intellect Disabil Res 51:996–1005

Wishart JG, Willis DS, Cebula KR, Pitcairn TK (2007) Collaborative learning: comparison of outcomes for typically developing children and children with intellectual disabilities. Am J Ment Retard 112:361–374

Wisniewski KE, Schmidt-Sidor B (1989) Postnatal delay of myelin formation in brains from Down syndrome infants and children. Clin Neuropathol 8:55–62

Wisniewski KE, Laure-Kamionowska M, Wisniewski HM (1984) Evidence of arrest of neurogenesis and synaptogenesis in brains of patients with Down's syndrome. N Engl J Med 311:1187–1188

Wolosin SM, Zeithamova D, Preston AR (2012) Reward modulation of hippocampal subfield activation during successful associative encoding and retrieval. J Cogn Neurosci 24:1532–1547

Woodhouse JM, Cregg M, Gunter HL, Sanders DP, Saunders KJ, Pakeman VH, Parker M, Fraser WI, Sastry P (2000) The effect of age, size of target, and cognitive factors on accommodative responses of children with Down syndrome. Invest Ophthalmol Vis Sci 41:2479–2485

Yang A, Reeves RH (2011) Increased survival following tumorigenesis in Ts65Dn mice that model Down syndrome. Cancer Res 71:3573–3581

Yonelinas AP (2013) The hippocampus supports high-resolution binding in the service of perception, working memory and long-term memory. Behav Brain Res 254:34–44

Zelazo PD, Burack JA, Benedetto E, Frye D (1996) Theory of mind and rule use in individuals with Down's syndrome: a test of the uniqueness and specificity claims. J Child Psychol Psychiatry 37:479–484

Zlokovic BV (2013) Cerebrovascular effects of apolipoprotein E: implications for Alzheimer disease. JAMA Neurol 70:440–444

Pharmacological Disruption of Maladaptive Memory

Jane R. Taylor and Mary M. Torregrossa

Contents

J.R. Taylor
Department of Psychiatry, Yale University, New Haven, CT, USA

M.M. Torregrossa (✉)
Department of Psychiatry, University of Pittsburgh, Bridgeside Point II, Room 223, 450 Technology Drive, 15219 Pittsburgh, PA, USA
e-mail: torregrossam@upmc.edu

K.M. Kantak, J.G. Wettstein (eds.), *Cognitive Enhancement*, Handbook of Experimental Pharmacology 228, DOI 10.1007/978-3-319-16522-6_13

Abstract

Many psychiatric disorders are characterized by intrusive, distracting, and disturbing memories that either perpetuate the illness or hinder successful treatment. For example, posttraumatic stress disorder (PTSD) involves such strong reemergence of memories associated with a traumatic event that the individual feels like the event is happening again. Furthermore, drug addiction is characterized by compulsive use and repeated relapse that is often driven by internal memories of drug use and/or by exposure to external stimuli that were associated with drug use. Therefore, identifying pharmacological methods to weaken the strength of maladaptive memories is a major goal of research efforts aimed at finding new treatments for these disorders. The primary mechanism by which memories could be pharmacologically disrupted or altered is through manipulation of memory reconsolidation. Reconsolidation occurs when an established memory is remembered or reactivated, reentering a labile state before again being consolidated into long-term memory storage. Memories are subject to disruption during this labile state. In this chapter we will discuss the preclinical and clinical studies identifying potential pharmacological methods for disrupting the integrity of maladaptive memory to treat mental illness.

Keywords

Maladaptive memory • Reconsolidation • Posttraumatic stress disorder • Addictive disorders • Memory-disrupting drugs

1 Introduction: Fundamentals of Memory

The ability to learn, retain, and retrieve information is critical to survival. Individual organisms must learn where to find food, how to prepare it, how to build shelters, and what predators to avoid. The formation of all types of memory involves multiple distinct phases (Fig. 13.1). First, the memory must be acquired or encoded. Most learning paradigms involve an acquisition phase where the animal learns a behavior over the course of multiple "trials" or encounters with the situation. A variety of factors can affect how effectively an individual acquires new information including availability of attentional resources, motivation, arousal, etc. During learning, new information is held in short-term or working memory to maintain ongoing behavior. Once new information or skills are acquired, they must be stored in long-term memory so that the information can be retrieved when needed at a later date. The process of storing newly learned information is known as consolidation (Abel and Lattal 2001). Memory consolidation occurs over the course of several hours, requires glutamatergic receptor activity, RNA transcription, and protein synthesis. Disruption of these signaling events or molecular processes immediately following acquisition can prevent consolidation of the memory into long-term storage, as evidenced by subsequent impairments in

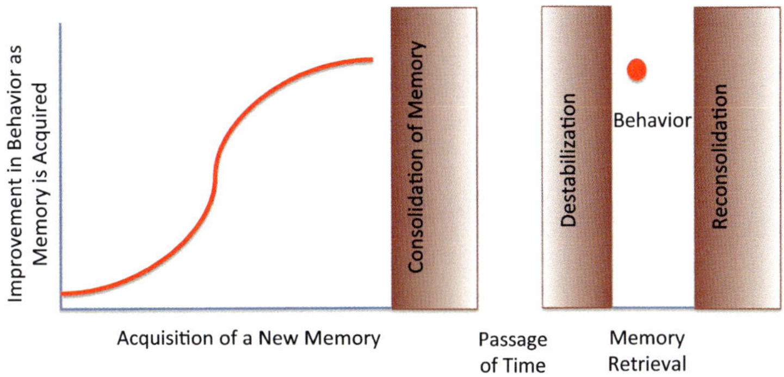

Fig. 13.1 Illustration of the different phases of learning and memory. The figure shows how behavior improves with learning during memory acquisition and that the memory then undergoes consolidation into long-term storage. The memory can then be retrieved at a later date to guide behavior, and as it is retrieved, the memory is destabilized. Once the memory is no longer needed to guide behavior, it is stored again in a process called reconsolidation

memory retrieval (Wang et al. 2006). Retrieval is the last stage of the learning and memory process where information is brought back into working memory from long-term storage so that it can be used to guide behavior. The retrieval process can also be disrupted by lesioning or inactivating specific brain regions or by pharmacological manipulations.

Traditionally, acquisition, consolidation, and retrieval have been considered the three cornerstones of the learning and memory process. However, in the past couple of decades, a lot of attention has been focused on what happens to memories during and after retrieval. First, recent research has established that in the process of retrieving information from long-term storage, the memory becomes "destabilized." That is, the molecular mechanisms supporting the memory are reactivated, and the memory becomes labile and subject to disruption. Once the memory is destabilized, it is then restored or "restabilized" in long-term memory in a process termed reconsolidation, which requires many of the same molecular mechanisms essential for initial consolidation (Nader et al. 2000b; Duvarci and Nader 2004; Tronson and Taylor 2007; Taylor et al. 2009). In general, when memories are consolidated or reconsolidated and are currently not actively being used, they are considered stable and not subject to disruption by pharmacological manipulations. Though, it should be noted that a non-reactivated memory can still be lost if the brain region responsible for long-term storage is lesioned or otherwise damaged. Pharmacological manipulations can affect all phases of memory to either enhance or strengthen the memory or to make it weaker or even forgotten. However, treating disorders of maladaptive memory with pharmacological manipulations is likely going to be most realistic clinically by focusing on memory reconsolidation processes. In this chapter we will primarily focus on the preclinical and clinical evidence for using pharmacological methods to disrupt reconsolidation, though the possibility of affecting other phases of memory will be discussed.

2 Disorders of Maladaptive Memory

In order to understand why one might want to use pharmacological tools to disrupt maladaptive memory, we first need to discuss the diseases or disorders that are characterized by maladaptive memories. For the purposes of this chapter, we are concentrating on psychiatric disorders that are perpetuated by disruptive memories such as posttraumatic stress disorder and addiction. We are not going to discuss disorders of memory loss such as Alzheimer's disease, dementia, or amnesia. There may be ways to pharmacologically prevent memory loss or improve memory in these disorders, but the point of this chapter is not to discuss memory loss, but rather how one might disrupt *maladaptive* memory.

2.1 Posttraumatic Stress Disorder

Posttraumatic stress disorder or PTSD is the prototypical example of a psychiatric memory disorder characterized by pathological reoccurrence of fear and fear memories after a traumatic event. PTSD is associated with deficits in maintaining extinction of the traumatic memories, generalization of fear to safe contexts, and enhancement rather than degradation of fear and anxiety as time elapses after the traumatic event (Pitman et al. 2012; Parsons and Ressler 2013). People suffering from PTSD also often report that unwanted memories of the traumatic event enter their thoughts unbidden and that the memories can be so strong that they feel like they are re-experiencing the traumatic event. From these descriptions it is reasonable to conclude that PTSD may involve disruptions at multiple phases of learning and memory. Due to genetic and/or environmental factors, the individual may have acquired the traumatic memory more readily. The memory may have undergone stronger consolidation into long-term storage, creating a more stable memory trace. Retrieval mechanisms certainly also appear to be dysfunctional as the traumatic memory is spuriously retrieved at inappropriate times and in inappropriate contexts. Finally, reconsolidation mechanisms also seem to be enhanced in PTSD, with repeated memory reactivations leading to a progressively stronger memory. Indeed, health professionals now recommend that individuals not be asked to recount the details of a traumatic event immediately after so as not to invoke overly robust reconsolidation mechanisms. Thus, finding ways to weaken the intensity of traumatic memories so the memories are either less frightening or less intrusive is a major goal of PTSD treatment research.

2.2 Other Anxiety Disorders

In addition to PTSD, most other anxiety disorders, including generalized anxiety disorder (GAD), phobias, and obsessive-compulsive disorder (OCD), potentially involve maladaptive learning and memory processes (Zlomuzica et al. 2014). GAD and phobias are associated with an unnaturally high fear response in situations that should be interpreted as safe. In other words, normal fear responses are expressed in situations where the individual has never explicitly learned that they should be frightened. Exposing these individuals to the triggers that induce fear and then pharmacologically interfering with reconsolidation has the potential to short-circuit the induction of feelings of anxiety when the trigger is encountered again. OCD is also characterized by feelings of anxiety induced by certain triggers that require the individual to perform specific compulsive behaviors. These obsessions and the compulsive behaviors required to deal with the obsessions can make it very difficult for these individuals to function well within society. As with the other anxiety disorders described, it may be possible to expose these individuals to the triggers that "reactivate" the obsessive memory and interfere with the reconsolidation process so that the obsessive thoughts are less likely to be retrieved again in the future. Most studies suggest that if you could get rid of the obsessions, then the compulsive behaviors would be unnecessary and the individual would be less anxious and better able to perform daily activities. One potential caveat to this theory is that the compulsive behavior may be based on implicit procedural learning; that is, the behavior is habitual. Less research has explicitly investigated mechanisms for disrupting reconsolidation of procedural memories that are associated with stimulus-driven habitual behavior, and it is possible that different treatment strategies may be necessary in these situations. However, addictive disorders (see below) are also thought to have a compulsive and/or habitual component, and there is evidence that reconsolidation disruption can be beneficial in these disorders.

2.3 Addictive Disorders

When we discuss addictive disorders, we will generally be talking about drug addiction or alcoholism. However, the principles that we apply to drug addiction should also be true for other disorders that have an addictive component such as compulsive gambling or binge-eating disorder. All addictive disorders are characterized by compulsive use or consumption despite adverse consequences to health or well-being. The majority of individuals suffering from an addiction recognize that they have a problem and will quit and abstain from use for some time. Unfortunately, most individuals eventually relapse and ultimately endure multiple cycles of abstinence and relapse throughout their lives. One of the primary drivers of relapse is exposure to the environmental stimuli or cues (i.e., the people, places, and things) that were associated with drug use. These cues are able to induce relapse because of the strong memories that were formed over long periods of associating the stimulus with the effects of the drug. The cues can initiate craving or drug "wanting" and/or initiate the subconscious habitual behaviors associated with obtaining and taking drugs (Torregrossa et al. 2011; Tronson and Taylor 2013). Like PTSD, addictive disorders are thought to involve disruptions in multiple aspects of learning and memory. First, the initial acquisition and consolidation of memories associated with drug use are thought to be more robust or stronger than the acquisition of nondrug, reward-associated memories. Due to the fact that drugs of abuse potently increase dopamine release and the activity of brain circuits responsible for normal learning and memory, new learning while under the influence of most drugs of abuse tends to occur more quickly and may be consolidated more strongly (Torregrossa et al. 2011; Milton and Everitt 2012; Gipson et al. 2013). Indeed, in a study comparing brain response of cocaine-using individuals to controls when confronted with a sexually arousing stimulus versus a cocaine stimulus, the cocaine-using individuals showed a greatly diminished response to the sexual stimulus relative to controls but showed robust brain activation when presented with a cocaine-associated stimulus (Garavan et al. 2000). Thus, cocaine-associated memories may be so strong that they overshadow those of natural rewards. In addition, the ability to retrieve drug-associated memories may also be enhanced, such that thoughts of drug use are also likely to come to mind more readily than other memories. Finally, every time a drug-associated memory is retrieved and reconsolidated into long-term memory, the reconsolidation process may also be stronger resulting in progressively more invasive drug-associated memories with repeated use (Sara 2000; Tronson and Taylor 2007; Sorg 2012). Ultimately, this may make drug-associated memories particularly difficult to disrupt but amenable to pharmacological methods for inhibiting the reconsolidation process.

2.4 Schizophrenia and Mood Disorders

Psychiatric disorders like schizophrenia, major depression, and bipolar disorder may also include dysregulated learning and memory systems. In general, maladaptive memory in these disorders has been studied to a lesser extent than PTSD and addiction, but there are compelling theoretical reasons, and in some cases direct experimental evidence, to suggest that these disorders are influenced by maladaptive memories that could be targets for therapeutic intervention. Schizophrenia is characterized by positive, negative, and cognitive symptoms. The cognitive symptoms are unquestionably characterized by disruptions in learning and memory (e.g., working memory) that could benefit from cognitive-enhancing agents. However, the positive symptoms, specifically the formation and persistence of elaborate delusions, may also arise from the formation of maladaptive memories that we have argued involve aberrant reconsolidation mechanisms (Corlett et al. 2009, 2010). For example, several studies have found that the psychotomimetic drug ketamine can disrupt processing of learning-based prediction errors in frontocortical regions and that the degree of activation to prediction errors under placebo conditions predicted the severity of delusional thoughts experienced under ketamine (e.g., Corlett et al. 2006; Corlett and Fletcher 2012). Specifically, the persistence of delusions may be due to overactive memory reconsolidation systems that persistently strengthen and reinforce the bizarre beliefs despite little evidence for the accuracy of these beliefs. Indeed, evidence for this hypothesis was supported by the finding that when fear memories are reactivated under the influence of ketamine, they are stronger when recalled the following day, suggesting that memories may be more likely to undergo reconsolidation in the psychotic state (Corlett et al. 2013). Therefore, there is a possibility that unpleasant delusions may in fact represent maladaptive memories that could be subject to pharmacological disruption, which might improve the social function and quality of life of those with schizophrenia.

Depressive disorders, on the other hand, are characterized by negative affect, lack of motivation, anhedonia, and persistent rumination, defined as repeated negative thoughts about one's self (Nejad et al. 2013). The persistence of these cycles of negative thought could be interpreted as repeated memory reactivation and reconsolidation events and that with repetition negative memories become stronger and stronger, perpetuating the depressive episode. While this hypothesis of rumination as aberrant reconsolidation has not been directly tested, there is evidence that depressed subjects show heightened amygdala reactivity to negative stimuli (Jaworska et al. 2014). The amygdala has consistently been shown to be the locus of fear memory reconsolidation (Duvarci and Nader 2004; Tronson and Taylor 2007); thus, heightened reactivity of this region may represent an increased likelihood that thoughts and images with negative valence will be strongly reconsolidated. Moreover, memories for negative life events may initially be more robustly consolidated and thus have greater cognitive representation than in healthy subjects.

Finally, bipolar disorder is the cycling of depressive symptoms, as just described, and symptoms of mania. Manic phases include enhanced hedonic responses, sleeplessness, impulsivity/risk-taking, and delusions of grandeur (Phillips and Swartz 2014). Many potential mechanisms could lead to this cluster of symptoms, though the extreme focus on positive outcomes could also be due to disruptions in amygdala/ventral striatal consolidation/reconsolidation mechanisms that enhance the expression of memories with positive valence. While this hypothesis is highly speculative, it may be possible to test clinically. The prediction would be that disruption of reconsolidation of grandiose memories may help someone cycle out of mania, or at least be less likely to act on their emotions. In general, many psychiatric disorders involve potential disruptions in learning and memory process and often result in the formation of maladaptive memories. Thus, identifying pharmacological methods for specifically interfering with maladaptive memories could lead to improved therapeutic outcomes for many, hard-to-treat psychiatric disorders.

3 Anatomy of Reconsolidation

The ultimate goal of studies aimed at identifying pharmacological tools for disrupting maladaptive memories in psychiatric disorders is to find targets that can be manipulated safely and effectively after oral administration of a drug. Of course, such a drug would affect both the periphery and the whole brain. Unfortunately, many of the studies that will be discussed in the next section that have identified potential mechanisms of memory disruption have only been examined in

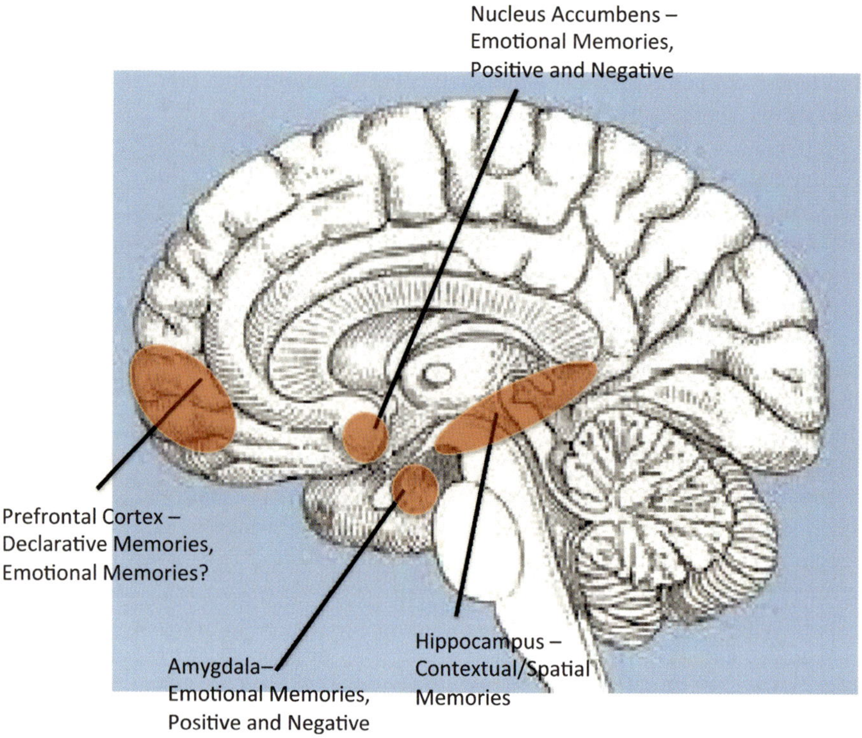

Fig. 13.2 Schematic illustrating the approximate anatomical locations of brain regions required for the reconsolidation of different types of memories

a specific brain region. Therefore, it is unclear if any of the manipulations that will be discussed below are realistic targets for clinical intervention. Brain region-specific studies do provide insight into relevant neurobiological mechanisms and consequently what targets might be clinically relevant. Notably, there are several examples of drugs that are effective by both central and systemic routes of administration. In addition, complex maladaptive memories may be represented by a network ensemble of brain circuits that may even be more easily disrupted through systemic manipulations (Kandel et al. 2014). Here we will discuss generally what is known about the brain loci required for reconsolidation of different types of memory (Fig 13.2).

3.1 Hippocampus

The hippocampus has long been established as a critical region for acquiring and consolidating multiple types of memory. The hippocampus is responsible for spatial and contextual-based learning. In addition, the hippocampus is required for the acquisition of new episodic and declarative memories. Interestingly, once enough time passes from the formation of an episodic or declarative memory, the memory is no longer stored in the hippocampus, but is "transferred" to the frontal cortex. In addition, there is evidence that contextual memories no longer engage the hippocampus once sufficient time has passed (Hall et al. 2001). Therefore, memories for past traumatic events may be more likely to be disrupted by amygdala/cortical manipulations as opposed to the hippocampus. Nevertheless, there have been numerous studies demonstrating that interfering with reconsolidation processes in the hippocampus can disrupt contextual and spatial memories. Moreover, several maladaptive memories have a contextual component. For example, the places, or contexts, where drugs of abuse are commonly ingested are strong drivers of relapse and are thought to be stored in the hippocampus. In animal models of drug addiction, reconsolidation of a contextual memory associated with drug self-administration or feelings of drug reward can be disrupted by manipulations of the hippocampus (cf. Brown et al. 2007; Raybuck and Lattal 2014; Shi et al. 2014). However, while reconsolidation of a memory for a context associated with cocaine self-administration can be disrupted by inactivation with tetrodotoxin (TTX), this form of memory was not affected by protein synthesis inhibition, suggesting that dorsal hippocampal activity may drive protein synthesis-dependent reconsolidation in another region, such as the amygdala (Ramirez et al. 2009; Wells et al. 2011). Finally, it should be noted that many studies of hippocampal mediation of reconsolidation have focused on the dorsal hippocampus, but the ventral hippocampus is much more intricately linked with the parts of the brain that regulated emotion, like the amygdala and nucleus accumbens, and, thus, would be more likely to regulate maladaptive memories. Fewer studies have manipulated the ventral hippocampus specifically because it is a difficult target, but future research is likely to find an important role for the ventral hippocampus in the reconsolidation of maladaptive memory. Nevertheless, many of the potential mechanisms for disrupting memory reconsolidation in general were discovered using contextual fear conditioning and manipulations of the dorsal hippocampus. Thus, maladaptive fear memories in PTSD or other anxiety disorders may have their root in the hippocampal dysfunction.

3.2 Amygdala

The amygdala is critical for the reconsolidation of many types of emotional memory (Duvarci and Nader 2004). Manipulations of the amygdala disrupt reconsolidation of negative fear-related memories and positive cue/context memories associated with drug use. Generally speaking the amygdala is critical for the reconsolidation of any cued associative learning-based memory. So, if a cue is paired with foot shock or drug reward, reconsolidation of this memory will require activity in the amygdala. In addition, both contextual fear and drug-associated memories that necessitate activity in the hippocampus for reconsolidation also require at least some forms of signaling in the amygdala. There are several examples of signaling molecules that are important for reconsolidation of memory in one brain region, but not in another. For example, Ras-related C3 botulinum toxin substrate (Rac) signaling is critical in the amygdala for reconsolidation of a cued fear conditioning memory, but not a contextual memory, while the CA1 of the hippocampus is required for contextual but not cued fear memory reconsolidation (Wu et al. 2014a). On the other hand, contextual fear memory reconsolidation does appear to involve cAMP response element-binding protein (CREB) activation and protein synthesis in the amygdala (Mamiya et al. 2009). Moreover, several studies have found that reconsolidation of drug-associated contextual memories can be disrupted by manipulations in the amygdala, including inhibition of protein synthesis or cAMP-dependent protein kinase A (PKA) (Fuchs et al. 2009; Arguello et al. 2013). Finally, it should be noted that the amygdala is made up of multiple subregions including the lateral amygdala, basal amygdala, and central amygdala. The vast majority of research has focused on the role of the basolateral amygdala complex in regulating memory reconsolidation. Due to the difficulty in selectively targeting anatomical regions that are so close together, it is hard to know whether one specific part of this complex is more or less responsible for memory reconsolidation. Distinguishing between the basolateral and central amygdala is a bit more feasible, and most studies seem to indicate that the central amygdala is less involved in memory reconsolidation (Thomas et al. 2003; Wang et al. 2008, 2012; Li et al. 2010; Si et al. 2012; Wu et al. 2014a). However, reconsolidation of alcohol-associated cue memories does involve mammalian (aka mechanistic) target of rapamycin complex 1 (mTORC1) signaling in the central amygdala (Barak et al. 2013). Furthermore, the central amygdala may be required to at least orient to or pay attention to conditioned cues during retrieval to initiate reconsolidation processes, particularly for appetitive reinforcers (Olshavsky et al. 2013), but whether or not the central amygdala is responsible for reconsolidation itself is less evident.

3.3 Nucleus Accumbens

In addition to the amygdala, the nucleus accumbens has also been implicated in the reconsolidation of emotional memories and may be particularly important for the reconsolidation of drug-associated memories. The nucleus accumbens receives glutamatergic inputs from the amygdala, hippocampus, and frontal cortex and projects GABAergic outputs to structures responsible for motoric behavior. The nucleus accumbens is thus known as the limbic-motor interface. Therefore, the nucleus accumbens is poised to act on the information provided by memories after retrieval. Interestingly, evidence suggests that the nucleus accumbens is not only responsible for behavioral response output but also participates in the consolidation and reconsolidation of certain types of memory. For example, Hernandez and colleagues (2002) found that consolidation of memory for an instrumental learning task where rats learned to press a lever for food requires protein synthesis in the nucleus accumbens. In this study, the authors did not find an effect of nucleus accumbens protein synthesis inhibition on reconsolidation of the instrumental learning memory; however, the reactivation session used was the same as a regular training session, and there is evidence that reconsolidation processes are not efficiently initiated without the introduction of novelty during the reactivation event (Sevenster et al. 2012, 2013). Miller and Marshall (2005) went on to show that reconsolidation of cocaine conditioned place preference memory did require activation of the extracellular-regulated kinase (ERK) within the nucleus accumbens. A subsequent study of cocaine conditioned reward also found that manipulating the nucleus accumbens, this time via an antisense oligonucleotide targeting the immediate early gene zif268, could inhibit reconsolidation of a place preference memory, but did not affect the conditioned rewarding properties of a cocaine-paired cue when assessed in an instrumental paradigm (Théberge et al. 2010). These data suggest that the nucleus accumbens is either more involved in storing the contextual/spatial aspects of reward than Pavlovian conditioned aspects of reward or that zif268 in the accumbens is not required for Pavlovian cue reconsolidation. Interestingly, using a cocaine self-administration model, which involves an instrumental behavioral component that is lacking in place preference studies, ERK activity in the nucleus accumbens was not found to be necessary for reconsolidation of cocaine-context memory. However, ERK inhibition in the amygdala was able to inhibit this form of reconsolidation (Wells et al. 2013). Therefore, memories associated with instrumental responding may require the nucleus accumbens for execution of behavior, but not for restabilization of the memory. Nevertheless, several other studies have confirmed the importance of the nucleus accumbens for contextual conditioned reward for opioids and stimulants (Milekic et al. 2006; Ren et al. 2012; Wu et al. 2012; Shi et al. 2014). There is less evidence that the nucleus accumbens is important for the reconsolidation of fear-associated memories; however, one study has found an upregulation in zif268 expression in the nucleus accumbens after reactivation of cued and contextual fear memories. Interestingly, in this study contextual memories seemed to engage the nucleus accumbens shell more than the core, while cued memories engaged both subregions of the accumbens (Thomas et al. 2002).

3.4 Frontal Cortex

To date, fewer studies have dissected whether disrupting signaling in various frontal cortical regions, such as the anterior cingulate, prelimbic, infralimbic, or orbitofrontal cortices can interfere with the reconsolidation of memories. However, it is known that once a memory is well established, that is, that the memory has existed for many days, weeks, or years, lesions of the hippocampus can no longer disrupt the memory. Further studies have shown that these "remote" memories have been transferred to the cortex through relatively unknown mechanisms. In addition, there are certain forms of learning that engage the frontal cortex even at the consolidation stage, such as object recognition memory. Indeed, inhibition of protein synthesis or NMDA receptors in the ventromedial prefrontal cortex is able to block both the consolidation and reconsolidation of object recognition memory (Akirav and Maroun 2006). However, this is the sort of "good" memory that we would prefer not to disrupt with pharmacological manipulations. There is somewhat less evidence that the prefrontal cortex regulates the reconsolidation of maladaptive memories associated with psychiatric disorders. Although, the alpha-1-adrenergic receptor antagonist prazosin infused in the prelimbic prefrontal cortex can inhibit reconsolidation of an olfactory-based fear memory (Do Monte et al. 2013). On the other hand, trace fear conditioning, even when the memory is remote, does not undergo protein synthesis-dependent reconsolidation in the medial prefrontal cortex (Blum et al. 2006). Moreover, while retrieval of a cocaine conditioned place preference memory can be persistently disrupted by intra-prelimbic PFC beta-adrenergic receptor blockade, reconsolidation of the memory requires the amygdala and not the PFC (Otis et al. 2013). Thus, while there may be instances where specific remote, fear-associated memories could be disrupted by prefrontal manipulations, the evidence to date suggests that most beneficial effects of reconsolidation blockade on maladaptive memory are mediated by the amygdala. However, the ventromedial PFC is critical for the extinction of maladaptive memories (Peters et al. 2009), and pharmacological manipulations that facilitate extinction could also be useful pharmacotherapies. Mechanisms for facilitating extinction have been discussed extensively in several reviews (Taylor et al. 2009; Myers and Carlezon 2012; Torregrossa and Taylor 2012).

4 Mechanisms for Disrupting Reconsolidation

In this section, we will discuss the current state of knowledge on receptor and signaling systems that can be targeted pharmacologically to disrupt maladaptive memories based on studies using animal models. Depending on whether the study employed a classical fear conditioning, contextual fear conditioning, inhibitory avoidance, drug self-administration, or conditioned place preference paradigm, different brain regions may have been targeted in specific experiments. Because we are particularly interested in manipulations that may be effective clinically, we will highlight when a systemic route of administration was used.

4.1 Methodology

In order to investigate mechanisms for pharmacologically disrupting maladaptive memories, a paradigm that reliably reactivates a memory and returns the memory to a labile state had to be developed. Researchers had already determined many of the molecular mechanisms that are required for initial fear memory *consolidation*, which include protein synthesis and glutamatergic activation of NMDA receptors. In order to study *reconsolidation*, researchers had to first establish a known consolidated memory that could be reactivated by a reminder event and potentially disrupted. Early studies by Judge and Quartermain (1982) and later by Przybyslawski and Sara (1997) demonstrated that this could be accomplished by training animals in an approach-avoidance task or in an eight-arm maze to retrieve palatable food reinforcers. The consolidated avoidance learning could be disrupted by post-retrieval protein synthesis inhibition, while the spatial memory could be disrupted when an NMDA antagonist was given within 2 h of memory retrieval. Subsequently, Nader and colleagues (2000) used fear conditioning, where a tone is paired with foot shock to establish a conditioned fear response to the tone, to determine if fear memories could undergo reconsolidation that could be disrupted by inhibition of protein synthesis. In support of the conclusion that established memories do undergo reconsolidation after reactivation, this study found that when the conditioned tone was presented in the absence of foot shock either one or 14 days after initial conditioning, post-reactivation administration of a protein synthesis inhibitor greatly reduced expression of conditioned fear the following day. The manipulation was not effective if the memory was not reactivated or if the inhibitor was given 6 h after reactivation when the reconsolidation process was presumed to be complete, similar to previous studies. The manipulation also did not affect short-term memory for the tone. Thus, the majority of subsequent studies exploring mechanisms of reconsolidation have used similar experimental designs and include many of the same control conditions as used in these pivotal publications. Follow-up studies have also established some of the parameters required to induce destabilization and reconsolidation of memory, such as the need for new or surprising information to be present that violates expectations, such as experiencing a conditioned cue in a new context and/or without presentation of the US (Pedreira et al. 2004). The studies also pointed to the potential clinical value in disrupting unwanted fear-associated memories. In particular, that reconsolidation disruption can be dissociated from memory extinction by the lack of renewal or reinstatement of memory (cf. Monfils et al. 2009; though also see Lattal and Wood 2013). Figure 13.3 illustrates the general accepted design for experiments testing mechanisms of reconsolidation and/or methods for pharmacologically disrupting maladaptive memories. Many laboratories have used the conditioned fear paradigm to study reconsolidation, and the paradigm has been adapted to study several other forms of memory, including additional types of aversive memory and drug-associated memories. The results from many of these studies will be presented in the following sections.

4.2 NMDA Receptors and Protein Synthesis

First, it is well established that reconsolidation of just about all forms of memory can be disrupted by either intracranial or systemic administration of protein synthesis inhibitors (Nader et al. 2000a; Milekic and Alberini 2002; Pedreira and Maldonado 2003; Valjent et al. 2006) or glutamatergic NMDA receptor antagonists (Torras-Garcia et al. 2005; Przybyslawski and Sara 1997; Akirav and Maroun 2006; Lee et al. 2006; Sadler et al. 2007; Lee and Everitt 2008; Milton et al. 2008a, 2012; Wu et al. 2012; Alaghband and Marshall 2013). These referenced citations have established the effectiveness of NMDAR blockade in attenuating reconsolidation in

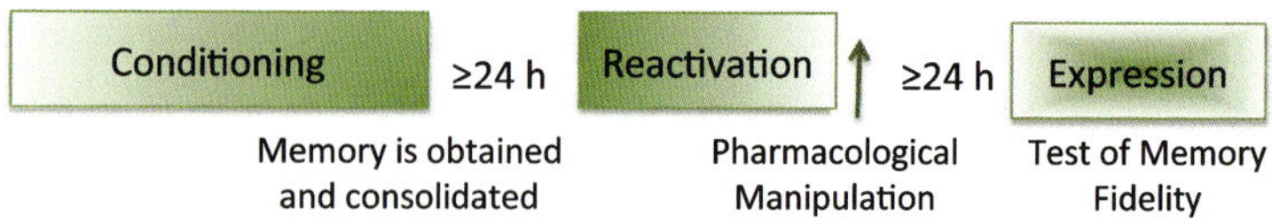

Fig. 13.3 Schematic of typical experimental paradigm utilized to assess whether a pharmacological manipulation can interfere with memory reconsolidation. Solid coloring in the boxes represents a more stabile; consolidated memory, with the lighter color representing relative memory instability or weakness. Positive findings require performance of several control experiments to verify results are dependent upon memory reconsolidation

multiple models of conditioned or contextual fear, odor memories, and taste memories and in multiple models of drug addiction. Manipulations of protein synthesis and NMDAR signaling have generally been used to establish that specific forms of memory actually undergo reconsolidation. Neither manipulation has been tested in a clinical treatment setting, primarily due to the risk of side effects, though one study examining the effects of the NMDAR antagonist ketamine on human conditioned fear reconsolidation will be discussed in the clinical section below.

4.3 Adrenergic/Noradrenergic Signaling

Manipulations of adrenergic or noradrenergic signaling are probably the next most commonly studied mechanism of memory reconsolidation. Research on mechanisms of fear memory consolidation established that stimulation of beta-adrenergic receptors (β-AR) during conditioning can make fear memories stronger and that disruption of adrenergic signaling during conditioning can make fear memories weaker (LaLumiere et al. 2003; Roozendaal et al. 2006a, b; McReynolds et al. 2010). The sympathetic nervous system is activated by stressful situations, such as by foot shock, and the degree of norepinephrine and epinephrine release is associated with the degree of arousal and fright. Therefore, it is thought that adrenergic signaling is responsible for the strength of fearful memories, so that the most arousing and dangerous experiences are remembered best and avoided in the future (McGaugh 2013). However, researchers have found that not only is adrenergic signaling important for initial conditioning but that reconsolidation of memories also required β-AR signaling. Moreover, there are several reports of non-fear-related memories that could be disrupted by β-AR antagonists like propranolol.

Indeed, the first report of adrenergic modulation of memory reconsolidation came from a study by Roullet and Sara (1998), which found that reconsolidation of spatial memory on a radial arm maze was blocked by intracerebroventricular administration of a β-AR antagonist. A follow-up study found that post-retrieval systemic administration of the centrally acting β-AR antagonist was also sufficient to disrupt spatial and passive avoidance memory reconsolidation (Przybyslawski et al. 1999). Interestingly, the time after reactivation required for effective disruption was different for the two routes of administration between the two studies, with the best results after systemic administration found if the β-AR antagonist is given 5 min after memory retrieval. Moreover, the authors found that there was significant "savings" in learning upon reacquisition, suggesting that β-AR blockade did not produce complete amnesia, but rather only produced either a temporary or partial effect. Subsequently, many other studies have examined the ability of systemic β-AR antagonism to prevent the reconsolidation of potentially maladaptive memories.

With regard to fear-related memories, systemic β-AR blockade has been shown to inhibit reconsolidation of auditory fear conditioning (Debiec and Ledoux 2004; Muravieva and Alberini 2010) and inhibitory avoidance as described above (Przybyslawski et al. 1999). Debiec and LeDoux (2004) even found that memory for a remote fear memory (reactivated 60 days after conditioning) could be reduced by post-retrieval propranolol, thus providing evidence for potential therapeutic utility. However, another study was not able to detect an effect of post-retrieval β-AR blockade on reconsolidation of inhibitory avoidance learning (Muravieva and Alberini 2010). The contradictory findings are likely due to methodological differences in the way memory fidelity was assessed, but the studies do agree that propranolol does not produce a complete loss of fear.

A large number of studies have also tested the ability of β-AR antagonists to prevent reconsolidation of appetitive memories, particularly those associated with drugs of abuse. Some of these studies have also found contradictory results. First, however, studies examining whether propranolol could block reconsolidation of sucrose or cocaine-associated cue or contextual memories have been largely consistent and have found an effect of propranolol in reducing memory reconsolidation through systemic or intra-amygdalar manipulation (Bernardi et al. 2006; Diergaarde et al. 2006; Milton et al. 2008b; Otis et al. 2013). On the other hand, one study examining the reconsolidation of food-associated cue memories used to support Pavlovian-to-instrumental transfer (PIT) or Pavlovian approach found no effect of pre-retrieval propranolol administration (Lee and Everitt 2008). This study suggests that the timing of drug administration relative to memory reactivation may affect the efficacy of the manipulation. On the other hand, the ability of propranolol to disrupt the reconsolidation of memories associated with other drugs of abuse has been much more equivocal. Several studies have examined the ability of propranolol to disrupt morphine conditioned place preference (CPP) memories. Robinson and Franklin (2007) first reported that propranolol blocks morphine memory reconsolidation through a central mechanism, but subsequent studies found that this was only true for the first retrieval of a morphine CPP memory and that there was no effect at all if the animals had received prior chronic morphine treatment or if the memory was recently formed and more strongly conditioned (Robinson and Franklin 2007, 2010; Robinson et al. 2011a, b). Moreover, another study also found no effect on morphine CPP, while they did see a blockade of memory for morphine-withdrawal-induced place aversion memory (Wu et al. 2014b). Therefore, the ability of propranolol to disrupt the reconsolidation of memories associated with opioids may be limited. Likewise, attempts to disrupt alcohol-associated memories have also been largely unsuccessful in preclinical models. Propranolol had no effect on the reconsolidation of memories for discrete cues or contextual stimuli associated with alcohol (Milton et al. 2012). However, one study did find that alcohol cue memories could be disrupted, but only with repeated reactivation/propranolol exposure events (Wouda et al. 2010). Thus, alcohol-associated memories may also require a different strategy if disruption of reconsolidation is to be an effective method for preventing relapse. The reason for the different results depending on the type of memory or drug of abuse is unclear. In some cases, the timing of propranolol administration may have prevented an effect from being observed. In addition, it may be that stimulants and passive fear conditioning paradigms are more arousing and induce more adrenergic receptor activity, which makes these memories more dependent on β-AR signaling for future reconsolidation.

4.4 Glucocorticoid Receptors

Glucocorticoid receptors (GR) are activated by the corticosteroid hormones (CORT) released from the adrenal gland in response to stress. Glucocorticoids are released in a circadian pattern throughout the day and night and can also bind to the higher affinity mineralocorticoid receptors (MR); however, due to the higher affinity of MR over GR for CORT, they are generally fully occupied by circulating levels of CORT. Thus, most of the learning and memory-related effects of stress are generally attributed to activation of GR. Like the adrenergic system, stimulation of GR by CORT during or after learning, especially fear-related learning, facilitates memory consolidation. Thus, highly arousing, stressful events are encoded more strongly than less arousing events. Indeed, the combination of GR and adrenergic receptor stimulation creates a strong and enduring memory, and stimulation of these receptors and subsequent plasticity may underlie the development of PTSD. However, while the case for CORT facilitating memory consolidation is strong, there are also reports of acute stress or CORT inhibiting memory, particularly post-retrieval manipulations that presumably affect reconsolidation processes. In contextual fear conditioning paradigms, acute post-retrieval administration of CORT or exposure to a stress, such as cold-water swim, impairs subsequent expression of fear (Cai et al. 2006; Abrari et al. 2008; Yang et al. 2013). A similar result has been found for novel object recognition memory (Maroun and Akirav 2008) and memory for a morphine CPP (Wang et al. 2008). On the other hand, several other studies have found that blockade of GR with the antagonist RU38486, either systemically or in the amygdala, inhibits the reconsolidation of auditory fear memory (Jin et al. 2007; Pitman et al. 2011), inhibitory avoidance memory (Taubenfeld et al. 2009; Nikzad et al. 2011), and morphine CPP memory (Fan et al. 2013). The reason for these apparently opposing results that both activation and blockade of GR are able to inhibit reconsolidation is not clear. One possibility is that normal memory is only retained at optimal levels of GR signaling, such that too little or too much is disruptive. However, the studies that have found that acute stress or CORT can inhibit reconsolidation have found somewhat modest results in general and have reported that these memory disruptions are subject to spontaneous recovery (i.e., the memory returns with time) and are sensitive to reinstatement. These results are more consistent with an interpretation that post-retrieval GR activation facilitates extinction of memory, rather than inhibiting reconsolidation. Extinction involves learning that a cue or context is no longer predictive of shock or reward and this new learning has its own consolidation phase. Generally speaking, most manipulations used to induce retrieval and reconsolidation are not strong enough or long enough to induce sufficient extinction learning. However, in contextual fear conditioning, every exposure to the context without being shocked could induce some degree of extinction learning, and consequently CORT could facilitate the consolidation of this learning. Facilitation of consolidation is an interpretation that is more consistent with the extensive literature on CORT facilitating initial memory consolidation. On the other hand, impairments in novel object recognition memory cannot be explained by facilitated extinction. In this paradigm, animals just need to remember what objects they've seen before, and the retrieval event should strengthen this memory, not weaken it. However, remembering objects that have previously been explored are not as arousing or important for survival, as memories for fearful or rewarding events. Therefore, GR regulation of this type of memory may be opposite to that of arousing memories because the brain may divert resources away from these everyday declarative memories in the face of stress. It is more important for the animal to remember the stressful event than the object. Future studies will have to clarify the role of GR in memory reconsolidation, but to date the preclinical literature is more supportive of the use of GR antagonists to disrupt maladaptive memories.

4.5 GABA

Gamma-amino butyric acid (GABA) is the primary inhibitory neurotransmitter in the brain, so one would predict that facilitating GABA signaling could interfere with the activity of cells necessary for retrieving and reconsolidating maladaptive memories. Therefore, it is somewhat surprising that only a few studies have tested the ability of GABA manipulations to disrupt reconsolidation and they have almost all occurred in the study of contextual fear conditioning. The GABAA receptor agonist at the benzodiazepine binding site, midazolam, given systemically during the post-retrieval component of a contextual fear memory paradigm can disrupt reconsolidation to reduce fear (Bustos et al. 2006; Zhang and Cranney 2008). Subsequent studies found that the effect of midazolam is mediated by GABAA receptors and that a GABAA antagonist can actually slightly facilitate memory reconsolidation (Zhang and Cranney 2008). Moreover, remote memories are more resistant to disruption by midazolam, requiring higher doses and longer retrieval periods (Bustos et al. 2009). Prior stress can also make memories resistant to midazolam but this can be overcome by pre-retrieval administration of the NMDA glycine site partial agonist, D-cycloserine, which presumably facilitated retrieval and induction of the memory destabilization process (Bustos et al. 2010).

Therefore, for GABAA receptor manipulations to be clinically effective, a lot of factors will need to be taken into consideration, including the age of the memory, the patient's stress levels, and the conditions that are required to induce retrieval. In addition, another study found that injections of ethanol given after retrieval of a contextual fear memory actually enhanced reconsolidation, increasing freezing on subsequent testing. Ethanol affects multiple neurotransmitter and signaling systems, but one of its primary mechanisms of action is as a GABAA receptor agonist, and the authors found that the reconsolidation enhancing effect of ethanol could be blocked by the GABAA antagonist picrotoxin (Nomura and Matsuki 2008). Thus, it appears that GABAA receptor positive modulation could potentially facilitate or inhibit reconsolidation. Furthermore, midazolam can block consolidation of extinction, if the retrieval event is long enough to induce extinction (Bustos et al. 2009). Thus, the effect of ethanol could possibly be attributed to extinction blockade rather than inhibition of reconsolidation. Therefore, clinical interventions would also need to ensure that individuals are not extinguishing the memory. The efficacy of midazolam or other GABA manipulations at interfering with reconsolidation of other types of maladaptive memory, such as for addictive drugs, to our knowledge, has not been reported.

4.6 Intracellular Signaling Molecules

In addition to the major neurotransmitter systems, the requirement for many downstream intracellular signaling molecules in memory reconsolidation has been tested extensively. Several kinase and signaling cascades have been implicated in the formation and maintenance of maladaptive memories, but rather than create an exhaustive list, we will focus on those signaling systems that have been reported to regulate multiple forms of memory by multiple laboratories. For the most part these studies have focused on infusing inhibitors or activators in specific brain regions, so it is not clear yet whether any of these manipulations will be translatable to clinical treatment. Nevertheless, we will discuss some of the targets that are most clearly relevant to the reconsolidation of maladaptive memories.

First, the adrenergic receptors discussed above are positively coupled to cAMP, which means that their stimulation leads to increases in cAMP and stimulation of cAMP-dependent kinases like PKA. Tronson and colleagues (2006) found that post-retrieval administration of a PKA inhibitor in the amygdala could inhibit reconsolidation of a conditioned fear memory. Conversely, repeated reactivation followed by administration of a PKA activator could make the memory stronger, consistent with the effects of β-AR blockade and stimulation, respectively (Tronson et al. 2006). PKA inhibition has also been shown to block reconsolidation of memory for a discrete cue (Sanchez et al. 2010) or a context (Arguello et al. 2013) associated with cocaine self-administration.

Several other kinase cascades have been implicated in learning and memory in general, and are required for memory reconsolidation. Extracellular-signal-regulated kinases (ERKs) in various brain regions are required for the reconsolidation of CPP memory for drugs of abuse (Miller and Marshall 2005; Valjent et al. 2006), cocaine-context-response memory (Wells et al. 2013), and auditory fear conditioned memory (Duvarci et al. 2005). Moreover, fear memory can be disrupted with systemic administration of an ERK inhibitor (Cestari et al. 2006), which raises the possibility of clinical utility. However, ERK inhibition has also been reported to impair reconsolidation of object recognition memory (Kelly et al. 2003), which means that non-maladaptive, potentially important memories could unintentionally be disrupted. Manipulations of ERK are also likely to affect many other systems, making it less likely that they will ultimately be used clinically.

Another intracellular signaling cascade that has received a lot of attention recently, as noted above, is the mTOR pathway. mTOR regulates protein translation, and inhibition of mTOR with rapamycin or inhibition of its upstream activators Akt or glycogen synthase kinase 3 (GSK3) can disrupt many types of memory. Rapamycin or a GSK3 inhibitor infused into a specific brain region, usually the amygdala or hippocampus, post-retrieval can disrupt reconsolidation of auditory, contextual, and inhibitory avoidance fear-based memories (Kimura et al. 2008; Gafford et al. 2011; Jobim et al. 2012; Mac Callum et al. 2013) and could block reconsolidation of morphine-, cocaine-, and alcohol-associated memories (Wang et al. 2010; Wu et al. 2011; Barak et al. 2013; Lin et al. 2014; Shi et al. 2014). Systemic inhibition of mTOR has also been reported to disrupt reconsolidation of a contextual fear memory, though it was ineffective for a cued fear memory (Glover et al. 2010). Nevertheless, the mTOR signaling pathway appears to be critically involved in almost all forms of memory reconsolidation, which may not be surprising due to requirement for protein synthesis for reconsolidation of memory, though these studies do suggest that the mTOR pathway is the specific initiator of protein translation that is required to maintain memory. Systemic rapamycin has not been tested clinically for the disruption of maladaptive memory and would likely have unwanted side effects, but it may be possible to find other ways to manipulate this signaling cascade that could be valuable pharmacological targets.

Finally, the downstream target of most kinase signaling cascades is the activation or inhibition of transcription factors that regulate gene expression. The cAMP response element-binding protein (CREB) and nuclear factor kappa B (NFkB) are two transcription factors that have been implicated in learning and memory processes. Activation of CREB is downstream of PKA signaling, so it is not surprising that CREB is important in the regulation of memory. NFkB, on the other hand, is actually most known for its activation by immune system signaling. Therefore, one might not expect NFkB to regulate learning and memory processes. However, the role of immune signaling molecules and NFkB in the brain is becoming increasingly appreciated. Inhibition of NFkB activity in the hippocampus has been shown to interfere with the reconsolidation of contextual fear memory and inhibitory avoidance memory (Boccia et al. 2007; Lubin and Sweatt 2007; de la Fuente et al. 2011) and a morphine CPP memory (Yang et al. 2011). In addition, inhibiting

NFkB in the amygdala can disrupt reconsolidation of auditory fear memory (Si et al. 2012). Lubin and Sweatt (2007) found that reconsolidation disruption could occur through inhibition of binding of the NFkB complex to DNA and through interference with I kappa kinase alpha (IKKalpha)-mediated acetylation of histone H3, which is important for altering chromatin structure to allow gene transcription. In fact, reconsolidation of memory occurred normally if a histone deacetylase inhibitor, which rescued histone acetylation, was combined with an IKK inhibitor. Thus, this study indicates that memory can be regulated by the IKK/NFkB signaling cascade via two distinct mechanisms and points to the importance of chromatin remodeling in memory reconsolidation. Indeed, other studies have shown that inhibiting histone acetylation in general, either through inhibition of DNA methyltransferase (Maddox and Schafe 2011) or histone acetyl transferases (HATs) can disrupt reconsolidation of auditory fear memories (Federman et al. 2012; Maddox et al. 2013a, b). Therefore, manipulations of the IKK/NFkB system and general disruptions in chromatin remodeling provide interesting new avenues for disrupting maladaptive memories. Clinical testing may even be possible as one of the HAT inhibitors was found to be naturally occurring from the rind of the fruit of the kokum tree (Maddox et al. 2013a). However, the clinical utility may be somewhat limited depending on the age of the memory, as one study did find that remote memories may not engage histone acetylation processes as much as recent memories (Gräff et al. 2014).

In summary, a vast amount of preclinical research has been conducted in the last 10+ years that have elucidated many of the neurotransmitter systems and intracellular signaling cascades that regulate the reconsolidation of memory in general and that also apply to maladaptive memories. While the above discussion is certainly not an exhaustive list of the signaling systems that have been reported to modulate reconsolidation, it points to many potential avenues for treatment of clinical disorders involving maladaptive memory. In the next section we will discuss what is known about methods for disrupting the reconsolidation of memories in human subjects.

5 Clinical Efficacy of Reconsolidation Disruption

5.1 Reconsolidation in Healthy Subjects

The vast majority of studies investigating reconsolidation processes in humans have been conducted in healthy subjects. Therefore, most of the experiments we will describe do not necessarily reflect manipulations of maladaptive memory. However, quite a few studies have used human fear conditioning paradigms that are very similar to those used in rodents and many involve an emotionally arousing memory. In addition, many studies in humans have investigated the ability of retrieval or reactivation of memories to induce updating or strengthening mechanisms to, for example, improve memory by repeated studying prior to a test (e.g., Forcato et al. 2011). For the most part we will limit our discussion to studies showing that a pharmacological agent can potentially disrupt memory reconsolidation, though experiments that identify mechanisms of memory strengthening could also be valuable in determining how to avoid unintentionally strengthening maladaptive memories.

One of the clearest demonstrations of memory reconsolidation in humans came from a laboratory study in healthy subjects. Individuals without a history of psychiatric illness underwent a fear conditioning procedure where otherwise neutral stimuli (cues) were paired with an aversive stimulus. After conditioning, subjects returned to the laboratory and were given propranolol or placebo and then either had the fear memory reactivated by presenting the cue or had no memory reactivation. In a later test, the participants were again presented with the conditioned cue, and their expression of anxiety and fear was assessed. Interestingly, the study found that only subjects who received propranolol and experienced memory reactivation reported a reduction in anxiety to the cue. However, while anxiety was reduced, the declarative memory about the relationship between the cue and the aversive stimulus was not changed (Kindt et al. 2009). Thus, while propranolol could potentially inhibit reconsolidation of the emotional component of the memory, explicit knowledge of the association between the cue and aversive stimulus was left intact. The dissociation between effects on emotional and declarative memories and the fact that there is no effect when memory is not reactivated suggests that propranolol + memory reactivation has potential to be a very efficacious treatment strategy with little to no side effects, particularly on the potentially worrisome effects on unreactivated or "important" memories that one would not want to lose. Moreover, the procedure appears to allow the individual to remember that particular traumatic events happened but reduces the negative emotional component of that memory.

The study by Kindt and colleagues has been replicated and expanded in several subsequent publications. The effect of propranolol + memory reactivation has been shown to persist for at least 1 month with a continued dissociation between the emotional and declarative aspects of memory (Soeter and Kindt 2010). The effect can also be observed if propranolol is given after the memory reactivation session (Soeter and Kindt 2012a) and propranolol is effective even if the initial fear learning occurred under conditions of enhanced adrenergic activity (yohimbine administration), which makes memories stronger and resistant to extinction (Soeter and Kindt 2012b). Likewise, the enhanced level of memory that is associated with an emotional episodic memory relative to a neutral memory can be reduced by retrieval after propranolol administration (Schwabe et al. 2013). Furthermore, functional imaging studies of human fear memory have shown activation of the amygdala that is absent after disruption of reconsolidation (Agren et al. 2012). In addition, similar to what has been found in animal studies, only the emotional aspects of the memory trace were associated with amygdala activation, while the declarative aspects were associated with the hippocampus (Soeter and Kindt 2010; Schwabe et al. 2012).

However, there are some reports in the literature that found no effect of propranolol, which point to several boundary conditions that need to be taken into account when designing reconsolidation disruption experiments to treat a disorder. First, propranolol has been reported to interfere with extinction of memory, at least at the cognitive level, making it important to not give propranolol under extinction conditions (Bos et al. 2012). In addition, research has shown that a memory must be successfully reactivated for it to undergo reconsolidation, and that if reactivation does not occur, propranolol is ineffective. What is successful reactivation? Successful reactivation is a reminder event that is sufficient enough to activate the memory trace, destabilize the memory, and initiate the molecular cascades required for restabilization. Research suggests that successful reactivation requires that the memory reminder session involves some sort of novelty or violation of predicted outcomes (Pedreira et al. 2004). This is often referred to as prediction error, and the reason why prediction error is required may be because reconsolidation is fundamentally a memory-updating mechanism, such that if nothing new is being learned, the memory is not reactivated. In animal models, memory reactivation sessions generally involve novelty in terms of either a context shift and/or a presentation of the cue or context in the absence of the shock or reward. Since the cue was conditioned to be predictive of shock or reward, the absence of these outcomes produces a prediction error and allows for memory updating. In experiments in humans, the same principle was found to exist (Sevenster et al. 2012, 2013). If humans were conditioned that a cue was predictive of a shock, which required electrode leads to be hooked up, then when the cue was presented under the same conditions but no shock was given, propranolol could disrupt reconsolidation. However, if the electrode leads were not hooked up during the reminder cue, propranolol was ineffective, presumably because the lack of shock was completely predicted in that situation, so the memory required no updating.

Finally, two studies by Tollenaar and colleagues found no effect of propranolol given prior to re-imagining disturbing memories or prior to retrieving emotional or neutral information on the later integrity of those memories (Tollenaar et al. 2009a, b). In one of the studies, the reactivation session occurred 7 days after the initial learning and testing occurred 7 days after reactivation. Thus, the timing difference between this and other studies may account for the differing results. Alternatively, one could argue that the memory reactivation sessions were not adequately novel, that is, they did not produce a prediction error, so the memory did not destabilize sufficiently. Regardless of the interpretation, more studies in clinical populations suffering from disorders of maladaptive memory are warranted.

In addition to propranolol, a few other studies into manipulations of reconsolidation in healthy human subjects have been conducted, and they have almost all used stress or cortisol as the manipulation. As mentioned above, both cortisol/stress and GR antagonists can potentially inhibit memory reconsolidation, but unfortunately to date only stress or cortisol has been tested in humans, while the effects of GR antagonists are yet to be determined. Nevertheless, stress applied after verbal episodic memory recall or after recall of neutral autobiographical memories impairs later memory recall (Schwabe and Wolf 2010; Strange et al. 2010). Interestingly, no effect was found for autobiographical memories with positive or negative valence (Schwabe and Wolf 2010). Moreover, cortisol administered immediately prior to retrieval of a memory can inhibit initial retrieval, but the reduction in memory persists even after a 1-week washout, suggesting that cortisol can either produce a long-lasting memory retrieval impairment or that cortisol can also disrupt reconsolidation (Tollenaar et al. 2009b). Post-retrieval stress has also been shown to prevent memory updating in the form of incorporating new information into an established memory trace (Dongaonkar et al. 2013). This could also be interpreted as an inhibition of the memory destabilization process. In addition to studies in healthy subjects, the effects of stress and cortisol, as well as propranolol, on maladaptive memories in psychiatric disorders have also been explored and will be discussed in the next sections.

5.2 Reconsolidation Disruption in PTSD and Anxiety Disorders

In contrast to studies of memory reconsolidation in healthy subjects, relatively few studies have examined the ability of pharmacological manipulations to improve symptoms of PTSD or other psychiatric disorders. However, the first randomized, double-blind, placebo-controlled trial of propranolol in subjects with PTSD was very promising. Subjects were asked to describe the traumatic event that led to their PTSD in a script preparation exercise. Immediately following this recall session, subjects received either placebo or propranolol. A week later physiological responses to the script were measured, and a reduction in heart rate, skin conductance, and EMG was found in the propranolol treatment group (Brunet et al. 2008). A follow-up open-label study found that both male and female subjects given 6 reactivations+propranolol exhibited markedly reduced PTSD symptoms, even to the point of not being clinically diagnosable (Brunet et al. 2011). Finally, another open-label trial examined other markers of clinical efficacy of memory reactivation +propranolol to treat PTSD and found an overall improvement in most measures including quality of life and depression symptoms. This study also found that the age of the memory or other comorbidities did not reduce the efficacy of treatment, though women may respond more than men (Poundja et al. 2012). Overall, this series of experiments strongly suggests that reactivation-dependent propranolol administration is a viable treatment strategy.

The effects of cortisol on phobia-based anxiety disorders and PTSD have also been examined. Experiments using cortisol have all given cortisol prior to exposure to the fearful stimulus or reactivation of the traumatic memory. Thus, there have been no experiments that explicitly test the ability of cortisol to interfere with memory reconsolidation. Nevertheless, cortisol has been reported to reduce general feelings of fear, and the reduction in fear is maintained when subjects are exposed to the phobic stimulus (social stress or spider) (Soravia et al. 2006). Fear of heights is also reduced when cortisol is administered prior to exposure therapy sessions (de Quervain et al. 2011). Finally, one month of cortisol treatment in PTSD subjects reduced symptoms overall and reduced the incidence of daily traumatic memories (Aerni et al. 2004). Taken together these studies suggest that cortisol can reduce fear and the intrusion of fear-associated maladaptive memories; however, the mechanism is unclear. The most parsimonious interpretation is that cortisol somehow interferes with memory retrieval rather than inhibiting reconsolidation. In the studies where cortisol was given in conjunction with exposure therapy, it is possible that extinction of the memory was facilitated, which is an interpretation supported by some of the preclinical studies discussed above. Regardless of the mechanism, the promising effect of cortisol in reducing anxiety disorders warrants further investigation.

Finally, the other major pharmaceutical agents that inhibit reconsolidation in animal models have not been tested in clinical populations, primarily due to potential toxic or other side effects. For example, protein synthesis inhibitors and NMDA antagonists are generally not considered viable clinical therapeutics. However, rapamycin (aka sirolimus) is an mTOR inhibitor that is approved for human use as an immunosuppressive agent that helps prevent organ rejection. An immunosuppressive agent would not be good to give to individuals with psychiatric disorders on a long-term basis, but it is plausible that a one-time administration after reactivation of traumatic or fearful memories could be therapeutic with minimal side effects. Thus, one study in combat veterans has been conducted where sirolimus was administered immediately after individuals were asked to remember a traumatic event. Veterans from the Vietnam War era, who presumably had much older traumatic memories, received no benefit from sirolimus treatment. On the other hand, veterans from more recent wars did demonstrate a reduction in symptoms after treatment (Surís et al. 2013). Therefore, the age of a memory—as a boundary condition—may be an important factor in designing treatments. Older traumatic memories may need to be reactivated multiple times or may require disruption with different pharmacological agents than more recent memories.

In summary, PTSD and other anxiety disorders such as social phobia show promise for treatment using a reconsolidation disruption strategy or even by inhibiting memory retrieval. Clinical assessment of mechanisms for reconsolidation disruption is still in an early phase. There is a need for more double-blind, placebo-controlled trials of agents like propranolol and sirolimus and a need to identify new pharmacological targets for reconsolidation disruption that maximize effectiveness and minimize side effects.

5.3 Addictive Disorders

The other major psychiatric disorder that has received some clinical attention with regard to reconsolidation disruption as a treatment strategy is addiction. These studies are still in their infancy, but some experiments have been done to at least establish a proof of principle. Two studies of heroin-dependent subjects have examined the ability of social stress and propranolol to disrupt memory reconsolidation. In these experiments, the individuals are asked to learn word lists that contain either neutral words or words associated with heroin that could have either a positive or negative valence. Post-retrieval social stress (Zhao et al. 2009) and pre-retrieval propranolol (Zhao et al. 2011) were both able to reduce later memory for heroin-associated words, suggesting disruption of reconsolidation. Unfortunately, these studies were not designed to look at whether disrupting heroin-associated memories could reduce later drug use. A more recent study in cocaine-dependent individuals found that post-cocaine memory retrieval administration of propranolol was able to reduce reported cocaine craving a day later. In this experiment the subjects were brought back a week later to assess craving again, and reported drug use was measured. The propranolol-treated group showed no significant reduction in craving after 7 days, and there was no significant reduction in cocaine use; however, the authors report that the study was not powered sufficiently to make any strong claims about effects on drug use. While the results from this study were not dramatic, they do at least point to some short-term effectiveness of post-retrieval propranolol on craving (Saladin et al. 2013). Future studies would need to test the effects of propranolol on a larger sample size, use multiple doses of propranolol, and determine if efficacy can be increased by multiple reactivation sessions as was observed for PTSD symptoms. Finally, drug addiction may be treated more effectively by manipulations other than propranolol, making it imperative that additional pharmaceuticals be identified and tested.

5.4 Other Psychiatric Disorders

In this section, we will briefly note what is known about reconsolidation processes that may be associated with other psychiatric disorders and the potential for using reconsolidation disruption as a therapeutic strategy. First, we are unaware of any studies that have specifically tested the possibility that a pharmacological disruption of reconsolidation could treat disorders like depression, bipolar disorder, or schizophrenia. However, the treatment of depression with electroconvulsive shock therapy (ECT) is actually what gave scientists the first clues that memories are labile and subject to disruption soon after they are acquired, because ECT produced retrograde amnesia for events that occurred right before the therapy session. Thus, ECT appeared able to disrupt memory consolidation (Squire et al. 1976), and some evidence was even reported for clinical benefit from reactivation-dependent amnesia (Rubin 1976). While this has been established for over 50 years, only recently have researchers begun to more specifically test the hypothesis that reconsolidation blockade could be beneficial in depression. In one study, depressed subjects were asked to recall an emotional episodic memory prior to ECT. Control subjects were given ECT but did not reactivate the memory beforehand. The results showed that ECT disrupted reconsolidation of the memory only in the reactivation condition (Kroes et al. 2014). Therefore, emotional memories can be disrupted in depressed subjects using reconsolidation blockade. It will be interesting in future studies to determine if blocking the reconsolidation of memories associated with the negative thoughts and ruminations that characterize depression can improve treatment.

Schizophrenia is a disorder that involves impaired cognition and disordered thoughts that can lead to the formation of bizarre associations and beliefs that impair an individual's ability to function. One could envision a scenario where memory for these bizarre beliefs is disrupted through interfering with reconsolidation of the memory and that this could improve overall functioning. On the other hand, dysfunctional memory consolidation and reconsolidation processes could lead to the creation of these bizarre beliefs, so it may also be possible to improve function by normalizing learning and memory circuitry. While these hypotheses have not been tested directly, one study has examined whether induction of a psychotic-like state can alter memory reconsolidation. Corlett and colleagues gave healthy subjects a subanesthetic, mildly psychotogenic dose of ketamine prior to retrieval of a previously learned fear memory. Ketamine produced dissociative and slight psychosis-like effects that mimicked some of the features of psychosis in schizophrenia. The subjects whose memory was reactivated under ketamine had stronger fear memories the following day in the absence of ketamine than subjects whose reactivated the memory after placebo administration (Corlett et al. 2013). The conclusion from this study is that conditions that produce a psychotic-like state favor enhancement of reconsolidation, which could lead to the abnormal strengthening of delusional thoughts, turning these thoughts into strong beliefs. The learning theory of delusions is intriguing, and future studies will have to validate these results, preferably in subjects with schizophrenia. The mechanism of the effect is also unclear because ketamine is an NMDA antagonist, which, as described above, almost always disrupts memory reconsolidation rather than facilitating it. However, in the clinical study, ketamine was given before the reactivation session, creating a psychotogenic state before the memory was reactivated that may have enhanced destabilization of the memory. It may be that a post-reactivation infusion would disrupt reconsolidation by altering memory restabilization, and indeed, we have unpublished data that suggests this is the case (Honsberger, Corlett, and Taylor, in submission). Ketamine also acts on other neurotransmitter systems, raising the possibility that the effect was not solely mediated by NMDA receptor blockade.

In summary, the use of pharmacological methods to disruptive maladaptive memories in psychiatric disorders is in its infancy, but the few studies described here point to many exciting areas for future research. It will be especially interesting to develop methods for treating disorders like depression and schizophrenia through manipulations of learning and memory, which could provide a whole new avenue for therapeutic development that could overcome the many downsides of the few, only mildly effective treatments that currently exist.

6 Conclusions

Memory reconsolidation, the process by which memories are restabilized after retrieval, is argued to have special relevance to several psychiatric disorders both for the treatment potential and as a mechanism for maladaptive maintenance and strengthening of memory. The study of reconsolidation has been rejuvenated in recent years, and evidence from a variety of memory tasks, and indeed species, has revealed several important implications for biological psychiatry. First, if memories can be disrupted after retrieval, then this has significant potential for treatment of persistent or exaggerated memory in PTSD, addiction, and other psychiatric disorders. Second, given the parallels between the initial storage of memory and an active reconsolidation process after retrieval, it is possible that reconsolidation may not only maintain but also update or strengthen memories after retrieval. Specifically, this latter possibility may be a means by which memories become exaggerated over time, even without additional training trials (or exposures to the reinforcer) and thus may contribute to the etiology of disorders such as PTSD and addiction. Neurobiological studies of reconsolidation have begun to describe the signal transduction and transcription events required for post-retrieval stabilization of memory. In this review, we have discussed these findings and how understanding the behavioral conditions, anatomical substrates, and molecular mechanisms required for reconsolidation will inform our conceptualization of reconsolidation, its relationship with memory consolidation, and its potential role in both treatment and the persistence of pathology in several psychiatric disorders. Specifically, we have argued that the enhancement of memory after retrieval supports the hypothesis that reconsolidation is a real, specific process that maintains, strengthens, and possibly updates memory. Evidence suggests that dysregulation of cellular and molecular mechanisms may lead to ongoing strengthening of memory and potentially cause dysfunctional emotional and behavioral responses, a hallmark of several psychiatric disorders. These findings argue that dysfunctional reconsolidation processes are involved in the etiology of psychiatric disorders.

Several important issues, however, remain. Do stressors or drugs alter destabilization as well as restabilization mechanisms, and how might these processes interact? What are the optimal reactivation conditions for reconsolidation manipulations of newer versus older memories? Do reconsolidation and extinction processes interact? Are there critical age-related constraints or vulnerabilities for maladaptive memory reconsolidation processes and/or therapeutic windows for intervention? Could the development of maladaptive memory processes be prevented pharmacologically? In considering how and why some memories become abnormally strong, and others do not, we have argued that it may be important to distinguish the role of reconsolidation in normal memories versus pathological memories. Under normal conditions, reconsolidation may act to update and maintain memories. In contrast, under altered conditions due to acute or chronic drug use, stress, or genetic predisposition, reconsolidation may act to enhance memories contributing to persistent maladaptive memories. The idea that maladaptive memories are a core feature of most psychiatric disorders may also explain the lack of behavioral and cognitive flexibility, impaired plasticity, generalization, and resistance to extinction of cues and contexts. The rediscovery of memory reconsolidation and its status as a vibrant research area has renewed interest in its potential as a therapeutic target. We, of course, advise that careful attention be given—and additional basic and translational research be targeted—to the nuances and boundaries of reconsolidation processes. Potentials for memory strengthening and hence symptom worsening are critical factors for serious consideration. Nonetheless, the dynamic nature of memory reconsolidation paints a picture that is encouraging for translation into viable treatments and, in our opinion, inspires basic neuroscience research.

Acknowledgments We would like to thank those individuals who worked with us in our studies of memory reconsolidation and thank our funding sources from the National Institutes of Health.

References

Abel T, Lattal KM (2001) Molecular mechanisms of memory acquisition, consolidation and retrieval. Curr Opin Neurobiol 11:180–187

Abrari K, Rashidy-Pour A, Semnanian S, Fathollahi Y (2008) Administration of corticosterone after memory reactivation disrupts subsequent retrieval of a contextual conditioned fear memory: dependence upon training intensity. Neurobiol Learn Mem 89:178–184

Aerni A, Traber R, Hock C, Roozendaal B, Schelling G, Papassotiropoulos A, Nitsch RM, Schnyder U, de Quervain DJ-F (2004) Low-dose cortisol for symptoms of posttraumatic stress disorder. Am J Psychiatry 161:1488–1490

Agren T, Engman J, Frick A, Björkstrand J, Larsson E-M, Furmark T, Fredrikson M (2012) Disruption of reconsolidation erases a fear memory trace in the human amygdala. Science 337:1550–1552

Akirav I, Maroun M (2006) Ventromedial prefrontal cortex is obligatory for consolidation and reconsolidation of object recognition memory. Cereb Cortex 16:1759–1765

Alaghband Y, Marshall JF (2013) Common influences of non-competitive NMDA receptor antagonists on the consolidation and reconsolidation of cocaine-cue memory. Psychopharmacology (Berl) 226:707–719

Arguello AA, Hodges MA, Wells AM, Lara H, Xie X, Fuchs RA (2013) Involvement of amygdalar protein kinase A, but not calcium/calmodulin-dependent protein kinase II, in the reconsolidation of cocaine-related contextual memories in rats. Psychopharmacology (Berl) 231(1):55–65

Barak S, Liu F, Ben Hamida S, Yowell QV, Neasta J, Kharazia V, Janak PH, Ron D (2013) Disruption of alcohol-related memories by mTORC1 inhibition prevents relapse. Nat Neurosci 16:1111–1117

Bernardi RE, Lattal KM, Berger SP (2006) Postretrieval propranolol disrupts a cocaine conditioned place preference. Neuroreport 17:1443–1447

Blum S, Runyan JD, Dash PK (2006) Inhibition of prefrontal protein synthesis following recall does not disrupt memory for trace fear conditioning. BMC Neurosci 7:67

Boccia M, Freudenthal R, Blake M, de la Fuente V, Acosta G, Baratti C, Romano A (2007) Activation of hippocampal nuclear factor-kappa B by retrieval is required for memory reconsolidation. J Neurosci 27:13436–13445

Bos MGN, Beckers T, Kindt M (2012) The effects of noradrenergic blockade on extinction in humans. Biol Psychol 89:598–605

Brown TE, Forquer MR, Cocking DL, Jansen HT, Harding JW, Sorg BA (2007) Role of matrix metalloproteinases in the acquisition and reconsolidation of cocaine-induced conditioned place preference. Learn Mem 14:214–223

Brunet A, Orr SP, Tremblay J, Robertson K, Nader K, Pitman RK (2008) Effect of post-retrieval propranolol on psychophysiologic responding during subsequent script-driven traumatic imagery in post-traumatic stress disorder. J Psychiatr Res 42:503–506

Brunet A, Poundja J, Tremblay J, Bui E, Thomas E, Orr SP, Azzoug A, Birmes P, Pitman RK (2011) Trauma reactivation under the influence of propranolol decreases posttraumatic stress symptoms and disorder: 3 open-label trials. J Clin Psychopharmacol 31:547–550

Bustos SG, Maldonado H, Molina VA (2006) Midazolam disrupts fear memory reconsolidation. Neuroscience 139:831–842

Bustos SG, Maldonado H, Molina VA (2009) Disruptive effect of midazolam on fear memory reconsolidation: decisive influence of reactivation time span and memory age. Neuropsychopharmacology 34:446–457

Bustos SG, Giachero M, Maldonado H, Molina VA (2010) Previous stress attenuates the susceptibility to Midazolam's disruptive effect on fear memory reconsolidation: influence of pre-reactivation D-cycloserine administration. Neuropsychopharmacology 35:1097–1108

Cai W-H, Blundell J, Han J, Greene RW, Powell CM (2006) Postreactivation glucocorticoids impair recall of established fear memory. J Neurosci 26:9560–9566

Cestari V, Costanzi M, Castellano C, Rossi-Arnaud C (2006) A role for ERK2 in reconsolidation of fear memories in mice. Neurobiol Learn Mem 86:133–143

Corlett PR, Fletcher PC (2012) The neurobiology of schizotypy: fronto-striatal prediction error signal correlates with delusion-like beliefs in healthy people. Neuropsychologia 50:3612–3620

Corlett PR, Honey GD, Aitken MRF, Dickinson A, Shanks DR, Absalom AR, Lee M, Pomarol-Clotet E, Murray GK, McKenna PJ, Robbins TW, Bullmore ET, Fletcher PC (2006) Frontal responses during learning predict vulnerability to the psychotogenic effects of ketamine: linking cognition, brain activity, and psychosis. Arch Gen Psychiatry 63:611–621

Corlett PR, Krystal JH, Taylor JR, Fletcher PC (2009) Why do delusions persist? Front Hum Neurosci 3:12

Corlett PR, Taylor JR, Wang X-J, Fletcher PC, Krystal JH (2010) Toward a neurobiology of delusions. Prog Neurobiol 92:345–369

Corlett PR, Cambridge V, Gardner JM, Piggot JS, Turner DC, Everitt JC, Arana FS, Morgan HL, Milton AL, Lee JL, Aitken MRF, Dickinson A, Everitt BJ, Absalom AR, Adapa R, Subramanian N, Taylor JR, Krystal JH, Fletcher PC (2013) Ketamine effects on memory reconsolidation favor a learning model of delusions. PLoS One 8:e65088

De la Fuente V, Freudenthal R, Romano A (2011) Reconsolidation or extinction: transcription factor switch in the determination of memory course after retrieval. J Neurosci 31:5562–5573

De Quervain DJ-F, Bentz D, Michael T, Bolt OC, Wiederhold BK, Margraf J, Wilhelm FH (2011) Glucocorticoids enhance extinction-based psychotherapy. Proc Natl Acad Sci USA 108:6621–6625

Debiec J, Ledoux JE (2004) Disruption of reconsolidation but not consolidation of auditory fear conditioning by noradrenergic blockade in the amygdala. Neuroscience 129:267–272

Diergaarde L, Schoffelmeer ANM, De Vries TJ (2006) Beta-adrenoceptor mediated inhibition of long-term reward-related memory reconsolidation. Behav Brain Res 170:333–336

Do Monte FH, Souza RR, Wong TT, de Padua Carobrez A (2013) Systemic or intra-prelimbic cortex infusion of prazosin impairs fear memory reconsolidation. Behav Brain Res 244:137–141

Dongaonkar B, Hupbach A, Gomez R, Nadel L (2013) Effects of psychosocial stress on episodic memory updating. Psychopharmacology (Berl) 226:769–779

Duvarci S, Nader K (2004) Characterization of fear memory reconsolidation. J Neurosci 24:9269–9275

Duvarci S, Nader K, LeDoux JE (2005) Activation of extracellular signal-regulated kinase-mitogen-activated protein kinase cascade in the amygdala is required for memory reconsolidation of auditory fear conditioning. Eur J Neurosci 21:283–289

Fan Y-D, Niu H-C, Huma T, Li L, Wang G-M, Xu L-Q, Ren H, Ma Y-Y, Yu H-L (2013) Blockage of glucocorticoid receptors during memory acquisition, retrieval and reconsolidation prevents the expression of morphine-induced conditioned place preferences in mice. Dongwuxue Yanjiu 34:E26–34

Federman N, Fustiñana MS, Romano A (2012) Reconsolidation involves histone acetylation depending on the strength of the memory. Neuroscience 219:145–156

Forcato C, Rodríguez MLC, Pedreira ME (2011) Repeated labilization-reconsolidation processes strengthen declarative memory in humans. PLoS One 6:e23305

Fuchs RA, Bell GH, Ramirez DR, Eaddy JL, Su Z (2009) Basolateral amygdala involvement in memory reconsolidation processes that facilitate drug context-induced cocaine seeking. Eur J Neurosci 30:889–900

Gafford GM, Parsons RG, Helmstetter FJ (2011) Consolidation and reconsolidation of contextual fear memory requires mammalian target of rapamycin-dependent translation in the dorsal hippocampus. Neuroscience 182:98–104

Garavan H, Pankiewicz J, Bloom A, Cho JK, Sperry L, Ross TJ, Salmeron BJ, Risinger R, Kelley D, Stein EA (2000) Cue-induced cocaine craving: neuroanatomical specificity for drug users and drug stimuli. Am J Psychiatry 157:1789–1798

Gipson CD, Kupchik YM, Kalivas PW (2013) Rapid, transient synaptic plasticity in addiction. Neuropharmacology 76:276–286

Glover EM, Ressler KJ, Davis M (2010) Differing effects of systemically administered rapamycin on consolidation and reconsolidation of context vs. cued fear memories. Learn Mem 17:577–581

Gräff J, Joseph NF, Horn ME, Samiei A, Meng J, Seo J, Rei D, Bero AW, Phan TX, Wagner F, Holson E, Xu J, Sun J, Neve RL, Mach RH, Haggarty SJ, Tsai L-H (2014) Epigenetic priming of memory updating during reconsolidation to attenuate remote fear memories. Cell 156:261–276

Hall J, Thomas KL, Everitt BJ (2001) Cellular imaging of zif268 expression in the hippocampus and amygdala during contextual and cued fear memory retrieval: selective activation of hippocampal CA1 neurons during the recall of contextual memories. J Neurosci 21:2186–2193

Hernandez PJ, Sadeghian K, Kelley AE (2002) Early consolidation of instrumental learning requires protein synthesis in the nucleus accumbens. Nat Neurosci 5:1327–1331

Jaworska N, Yang X-R, Knott V, Macqueen G (2014) A review of fMRI studies during visual emotive processing in major depressive disorder. J Biol Psychiatry, World

Jin X-C, Lu Y-F, Yang X-F, Ma L, Li B-M (2007) Glucocorticoid receptors in the basolateral nucleus of amygdala are required for postreactivation reconsolidation of auditory fear memory. Eur J Neurosci 25:3702–3712

Jobim PFC, Pedroso TR, Christoff RR, Werenicz A, Maurmann N, Reolon GK, Roesler R (2012) Inhibition of mTOR by rapamycin in the amygdala or hippocampus impairs formation and reconsolidation of inhibitory avoidance memory. Neurobiol Learn Mem 97:105–112

Judge ME, Quartermain D (1982) Characteristics of retrograde amnesia following reactivation of memory in mice. Physiol Behav 28:585–590

Kandel ER, Dudai Y, Mayford MR (2014) The molecular and systems biology of memory. Cell 157:163–186

Kelly A, Laroche S, Davis S (2003) Activation of mitogen-activated protein kinase/extracellular signal-regulated kinase in hippocampal circuitry is required for consolidation and reconsolidation of recognition memory. J Neurosci 23:5354–5360

Kimura T, Yamashita S, Nakao S, Park J-M, Murayama M, Mizoroki T, Yoshiike Y, Sahara N, Takashima A (2008) GSK-3beta is required for memory reconsolidation in adult brain. PLoS One 3:e3540

Kindt M, Soeter M, Vervliet B (2009) Beyond extinction: erasing human fear responses and preventing the return of fear. Nat Neurosci 12:256–258

Kroes MCW, Tendolkar I, van Wingen GA, van Waarde JA, Strange BA, Fernández G (2014) An electroconvulsive therapy procedure impairs reconsolidation of episodic memories in humans. Nat Neurosci 17:204–206

LaLumiere RT, Buen T-V, McGaugh JL (2003) Post-training intra-basolateral amygdala infusions of norepinephrine enhance consolidation of memory for contextual fear conditioning. J Neurosci 23:6754–6758

Lattal KM, Wood MA (2013) Epigenetics and persistent memory: implications for reconsolidation and silent extinction beyond the zero. Nat Neurosci 16:124–129

Lee JLC, Everitt BJ (2008) Appetitive memory reconsolidation depends upon NMDA receptor-mediated neurotransmission. Neurobiol Learn Mem 90:147–154

Lee JLC, Milton AL, Everitt BJ (2006) Reconsolidation and extinction of conditioned fear: inhibition and potentiation. J Neurosci 26:10051–10056

Li F, Xue Y, Wang J, Fang Q, Li Y, Zhu W, He Y, Liu J, Xue L, Shaham Y, Lu L (2010) Basolateral amygdala cdk5 activity mediates consolidation and reconsolidation of memories for cocaine cues. J Neurosci 30:10351–10359

Lin J, Liu L, Wen Q, Zheng C, Gao Y, Peng S, Tan Y, Li Y (2014) Rapamycin prevents drug seeking via disrupting reconsolidation of reward memory in rats. Int J Neuropsychopharmacol 17:127–136

Lubin FD, Sweatt JD (2007) The IkappaB kinase regulates chromatin structure during reconsolidation of conditioned fear memories. Neuron 55:942–957

Mac Callum PE, Hebert M, Adamec RE, Blundell J (2013) Systemic inhibition of mTOR kinase via rapamycin disrupts consolidation and reconsolidation of auditory fear memory. Neurobiol Learn Mem 112:176–185

Maddox SA, Schafe GE (2011) Epigenetic alterations in the lateral amygdala are required for reconsolidation of a Pavlovian fear memory. Learn Mem 18:579–593

Maddox SA, Watts CS, Doyère V, Schafe GE (2013a) A naturally-occurring histone acetyltransferase inhibitor derived from Garcinia indica impairs newly acquired and reactivated fear memories. PLoS One 8:e54463

Maddox SA, Watts CS, Schafe GE (2013b) p300/CBP histone acetyltransferase activity is required for newly acquired and reactivated fear memories in the lateral amygdala. Learn Mem 20:109–119

Mamiya N, Fukushima H, Suzuki A, Matsuyama Z, Homma S, Frankland PW, Kida S (2009) Brain region-specific gene expression activation required for reconsolidation and extinction of contextual fear memory. J Neurosci 29:402–413

Maroun M, Akirav I (2008) Arousal and stress effects on consolidation and reconsolidation of recognition memory. Neuropsychopharmacology 33:394–405

McGaugh JL (2013) Making lasting memories: remembering the significant. Proc Natl Acad Sci USA 110(Suppl 2):10402–10407

McReynolds JR, Donowho K, Abdi A, McGaugh JL, Roozendaal B, McIntyre CK (2010) Memory-enhancing corticosterone treatment increases amygdala norepinephrine and Arc protein expression in hippocampal synaptic fractions. Neurobiol Learn Mem 93:312–321

Milekic MH, Alberini CM (2002) Temporally graded requirement for protein synthesis following memory reactivation. Neuron 36:521–525

Milekic MH, Brown SD, Castellini C, Alberini CM (2006) Persistent disruption of an established morphine conditioned place preference. J Neurosci 26:3010–3020

Miller CA, Marshall JF (2005) Molecular substrates for retrieval and reconsolidation of cocaine-associated contextual memory. Neuron 47:873–884

Milton AL, Everitt BJ (2012) The persistence of maladaptive memory: addiction, drug memories and anti-relapse treatments. Neurosci Biobehav Rev 36:1119–1139

Milton AL, Lee JLC, Butler VJ, Gardner R, Everitt BJ (2008a) Intra-amygdala and systemic antagonism of NMDA receptors prevents the reconsolidation of drug-associated memory and impairs subsequently both novel and previously acquired drug-seeking behaviors. J Neurosci 28:8230–8237

Milton AL, Lee JLC, Everitt BJ (2008b) Reconsolidation of appetitive memories for both natural and drug reinforcement is dependent on {beta}-adrenergic receptors. Learn Mem 15:88–92

Milton AL, Schramm MJW, Wawrzynski JR, Gore F, Oikonomou-Mpegeti F, Wang NQ, Samuel D, Economidou D, Everitt BJ (2012) Antagonism at NMDA receptors, but not β-adrenergic receptors, disrupts the reconsolidation of pavlovian conditioned approach and instrumental transfer for ethanol-associated conditioned stimuli. Psychopharmacology (Berl) 219:751–761

Monfils M-H, Cowansage KK, Klann E, LeDoux JE (2009) Extinction-reconsolidation boundaries: key to persistent attenuation of fear memories. Science 324:951–955

Muravieva EV, Alberini CM (2010) Limited efficacy of propranolol on the reconsolidation of fear memories. Learn Mem 17:306–313

Myers KM, Carlezon WA (2012) D-cycloserine effects on extinction of conditioned responses to drug-related cues. Biol Psychiatry 71:947–955

Nader K, Schafe GE, Le Doux JE (2000a) Fear memories require protein synthesis in the amygdala for reconsolidation after retrieval. Nature 406:722–726

Nader K, Schafe GE, LeDoux JE (2000b) The labile nature of consolidation theory. Nat Rev Neurosci 1:216–219

Nejad AB, Fossati P, Lemogne C (2013) Self-referential processing, rumination, and cortical midline structures in major depression. Front Hum Neurosci 7:666

Nikzad S, Vafaei AA, Rashidy-Pour A, Haghighi S (2011) Systemic and intrahippocampal administrations of the glucocorticoid receptor antagonist RU38486 impairs fear memory reconsolidation in rats. Stress 14:459–464

Nomura H, Matsuki N (2008) Ethanol enhances reactivated fear memories. Neuropsychopharmacology 33:2912–2921

Olshavsky ME, Song BJ, Powell DJ, Jones CE, Monfils M-H, Lee HJ (2013) Updating appetitive memory during reconsolidation window: critical role of cue-directed behavior and amygdala central nucleus. Front Behav Neurosci 7:186

Otis JM, Dashew KB, Mueller D (2013) Neurobiological dissociation of retrieval and reconsolidation of cocaine-associated memory. J Neurosci 33:1271–1281

Parsons RG, Ressler KJ (2013) Implications of memory modulation for post-traumatic stress and fear disorders. Nat Neurosci 16:146–153

Pedreira ME, Maldonado H (2003) Protein synthesis subserves reconsolidation or extinction depending on reminder duration. Neuron 38:863–869

Pedreira ME, Pérez-Cuesta LM, Maldonado H (2004) Mismatch between what is expected and what actually occurs triggers memory reconsolidation or extinction. Learn Mem 11:579–585

Peters J, Kalivas PW, Quirk GJ (2009) Extinction circuits for fear and addiction overlap in prefrontal cortex. Learn Mem 16:279–288

Phillips ML, Swartz HA (2014) A critical appraisal of neuroimaging studies of bipolar disorder: toward a new conceptualization of underlying neural circuitry and a road map for future research. Am J Psychiatry 171(8):829–843

Pitman RK, Milad MR, Igoe SA, Vangel MG, Orr SP, Tsareva A, Gamache K, Nader K (2011) Systemic mifepristone blocks reconsolidation of cue-conditioned fear; propranolol prevents this effect. Behav Neurosci 125:632–638

Pitman RK, Rasmusson AM, Koenen KC, Shin LM, Orr SP, Gilbertson MW, Milad MR, Liberzon I (2012) Biological studies of post-traumatic stress disorder. Nat Rev Neurosci 13:769–787

Poundja J, Sanche S, Tremblay J, Brunet A (2012) Trauma reactivation under the influence of propranolol: an examination of clinical predictors. Eur J Psychotraumatol 3

Przybyslawski J, Sara SJ (1997) Reconsolidation of memory after its reactivation. Behav Brain Res 84:241–246

Przybyslawski J, Roullet P, Sara SJ (1999) Attenuation of emotional and nonemotional memories after their reactivation: role of beta adrenergic receptors. J Neurosci 19:6623–6628

Ramirez DR, Bell GH, Lasseter HC, Xie X, Traina SA, Fuchs RA (2009) Dorsal hippocampal regulation of memory reconsolidation processes that facilitate drug context-induced cocaine-seeking behavior in rats. Eur J Neurosci 30:901–912

Raybuck JD, Lattal KM (2014) Differential effects of dorsal hippocampal inactivation on expression of recent and remote drug and fear memory. Neurosci Lett 569:1–5

Ren Z-Y, Liu M-M, Xue Y-X, Ding Z-B, Xue L-F, Zhai S-D, Lu L (2012) A critical role for protein degradation in the nucleus accumbens core in cocaine reward memory. Neuropsychopharmacology 38:778–790

Robinson MJF, Franklin KBJ (2007) Central but not peripheral beta-adrenergic antagonism blocks reconsolidation for a morphine place preference. Behav Brain Res 182:129–134

Robinson MJF, Franklin KBJ (2010) Reconsolidation of a morphine place preference: impact of the strength and age of memory on disruption by propranolol and midazolam. Behav Brain Res 213:201–207

Robinson MJF, Armson M, Franklin KBJ (2011a) The effect of propranolol and midazolam on the reconsolidation of a morphine place preference in chronically treated rats. Front Behav Neurosci 5:42

Robinson MJF, Ross EC, Franklin KBJ (2011b) The effect of propranolol dose and novelty of the reactivation procedure on the reconsolidation of a morphine place preference. Behav Brain Res 216:281–284

Roozendaal B, Hui GK, Hui IR, Berlau DJ, McGaugh JL, Weinberger NM (2006a) Basolateral amygdala noradrenergic activity mediates corticosterone-induced enhancement of auditory fear conditioning. Neurobiol Learn Mem 86:249–255

Roozendaal B, Okuda S, Van der Zee EA, McGaugh JL (2006b) Glucocorticoid enhancement of memory requires arousal-induced noradrenergic activation in the basolateral amygdala. Proc Natl Acad Sci USA 103:6741–6746

Roullet P, Sara S (1998) Consolidation of memory after its reactivation: involvement of beta noradrenergic receptors in the late phase. Neural Plast 6:63–68

Rubin RD (1976) Clinical use of retrograde amnesia produced by electroconvulsive shock. A conditioning hypothesis. Can Psychiatr Assoc J 21:87–90

Sadler R, Herzig V, Schmidt WJ (2007) Repeated treatment with the NMDA antagonist MK-801 disrupts reconsolidation of memory for amphetamine-conditioned place preference. Behav Pharmacol 18:699–703

Saladin ME, Gray KM, McRae-Clark AL, Larowe SD, Yeatts SD, Baker NL, Hartwell KJ, Brady KT (2013) A double blind, placebo-controlled study of the effects of post-retrieval propranolol on reconsolidation of memory for craving and cue reactivity in cocaine dependent humans. Psychopharmacology (Berl) 226:721–737

Sanchez H, Quinn JJ, Torregrossa MM, Taylor JR (2010) Reconsolidation of a cocaine-associated stimulus requires amygdalar protein kinase A. J Neurosci 30:4401–4407

Sara SJ (2000) Retrieval and reconsolidation: toward a neurobiology of remembering. Learn Mem 7:73–84

Schwabe L, Wolf OT (2010) Stress impairs the reconsolidation of autobiographical memories. Neurobiol Learn Mem 94:153–157

Schwabe L, Nader K, Wolf OT, Beaudry T, Pruessner JC (2012) Neural signature of reconsolidation impairments by propranolol in humans. Biol Psychiatry 71:380–386

Schwabe L, Nader K, Pruessner JC (2013) β-Adrenergic blockade during reactivation reduces the subjective feeling of remembering associated with emotional episodic memories. Biol Psychol 92:227–232

Sevenster D, Beckers T, Kindt M (2012) Retrieval per se is not sufficient to trigger reconsolidation of human fear memory. Neurobiol Learn Mem 97:338–345

Sevenster D, Beckers T, Kindt M (2013) Prediction error governs pharmacologically induced amnesia for learned fear. Science 339:830–833

Shi X, Miller JS, Harper LJ, Poole RL, Gould TJ, Unterwald EM (2014) Reactivation of cocaine reward memory engages the Akt/GSK3/mTOR signaling pathway and can be disrupted by GSK3 inhibition. Psychopharmacology (Berl) 231(16):3109–3118

Si J, Yang J, Xue L, Yang C, Luo Y, Shi H, Lu L (2012) Activation of NF-κB in basolateral amygdala is required for memory reconsolidation in auditory fear conditioning. PLoS One 7: e43973

Soeter M, Kindt M (2010) Dissociating response systems: erasing fear from memory. Neurobiol Learn Mem 94:30–41

Soeter M, Kindt M (2012a) Erasing fear for an imagined threat event. Psychoneuroendocrinology 37:1769–1779

Soeter M, Kindt M (2012b) Stimulation of the noradrenergic system during memory formation impairs extinction learning but not the disruption of reconsolidation. Neuropsychopharmacology 37:1204–1215

Soravia LM, Heinrichs M, Aerni A, Maroni C, Schelling G, Ehlert U, Roozendaal B, de Quervain DJ-F (2006) Glucocorticoids reduce phobic fear in humans. Proc Natl Acad Sci USA 103:5585–5590

Sorg BA (2012) Reconsolidation of drug memories. Neurosci Biobehav Rev 36:1400–1417

Squire LB, Chace PM, Slater PC (1976) Retrograde amnesia following electroconvulsive therapy. Nature 260:775–777

Strange BA, Kroes MCW, Fan JE, Dolan RJ (2010) Emotion causes targeted forgetting of established memories. Front Behav Neurosci 4:175

Surís A, Smith J, Powell C, North CS (2013) Interfering with the reconsolidation of traumatic memory: sirolimus as a novel agent for treating veterans with posttraumatic stress disorder. Ann Clin Psychiatry 25:33–40

Taubenfeld SM, Riceberg JS, New AS, Alberini CM (2009) Preclinical assessment for selectively disrupting a traumatic memory via postretrieval inhibition of glucocorticoid receptors. Biol Psychiatry 65:249–257

Taylor JR, Olausson P, Quinn JJ, Torregrossa MM (2009) Targeting extinction and reconsolidation mechanisms to combat the impact of drug cues on addiction. Neuropharmacology 56(Suppl 1):186–195

Théberge FRM, Milton AL, Belin D, Lee JLC, Everitt BJ (2010) The basolateral amygdala and nucleus accumbens core mediate dissociable aspects of drug memory reconsolidation. Learn Mem 17:444–453

Thomas KL, Hall J, Everitt BJ (2002) Cellular imaging with zif268 expression in the rat nucleus accumbens and frontal cortex further dissociates the neural pathways activated following the retrieval of contextual and cued fear memory. Eur J Neurosci 16:1789–1796

Thomas KL, Arroyo M, Everitt BJ (2003) Induction of the learning and plasticity-associated gene Zif268 following exposure to a discrete cocaine-associated stimulus. Eur J Neurosci 17:1964–1972

Tollenaar MS, Elzinga BM, Spinhoven P, Everaerd W (2009a) Psychophysiological responding to emotional memories in healthy young men after cortisol and propranolol administration. Psychopharmacology (Berl) 203:793–803

Tollenaar MS, Elzinga BM, Spinhoven P, Everaerd W (2009b) Immediate and prolonged effects of cortisol, but not propranolol, on memory retrieval in healthy young men. Neurobiol Learn Mem 91:23–31

Torras-Garcia M, Lelong J, Tronel S, Sara SJ (2005) Reconsolidation after remembering an odor-reward association requires NMDA receptors. Learn Mem 12:18–22

Torregrossa MM, Taylor JR (2012) Learning to forget: manipulating extinction and reconsolidation processes to treat addiction. Psychopharmacology (Berl) 226(4):659–672

Torregrossa MM, Corlett PR, Taylor JR (2011) Aberrant learning and memory in addiction. Neurobiol Learn Mem 96(4):609–623

Tronson NC, Taylor JR (2007) Molecular mechanisms of memory reconsolidation. Nat Rev Neurosci 8:262–275

Tronson NC, Taylor JR (2013) Addiction: a drug-induced disorder of memory reconsolidation. Curr Opin Neurobiol 23:573–580

Tronson NC, Wiseman SL, Olausson P, Taylor JR (2006) Bidirectional behavioral plasticity of memory reconsolidation depends on amygdalar protein kinase A. Nat Neurosci 9:167–169

Valjent E, Corbillé A-G, Bertran-Gonzalez J, Hervé D, Girault J-A (2006) Inhibition of ERK pathway or protein synthesis during reexposure to drugs of abuse erases previously learned place preference. Proc Natl Acad Sci USA 103:2932–2937

Wang H, Hu Y, Tsien JZ (2006) Molecular and systems mechanisms of memory consolidation and storage. Prog Neurobiol 79:123–135

Wang X-Y, Zhao M, Ghitza UE, Li Y-Q, Lu L (2008) Stress impairs reconsolidation of drug memory via glucocorticoid receptors in the basolateral amygdala. J Neurosci 28:5602–5610

Wang X, Luo Y, He Y, Li F, Shi H, Xue L, Xue Y, Lu L (2010) Nucleus accumbens core mammalian target of rapamycin signaling pathway is critical for cue-induced reinstatement of cocaine seeking in rats. J Neurosci 30:12632–12641

Wang Y, Zhang T-Y, Xin J, Li T, Yu H, Li N, Chen Z-Y (2012) Differential involvement of brain-derived neurotrophic factor in reconsolidation and consolidation of conditioned taste aversion memory. PLoS One 7:e49942

Wells AM, Lasseter HC, Xie X, Cowhey KE, Reittinger AM, Fuchs RA (2011) Interaction between the basolateral amygdala and dorsal hippocampus is critical for cocaine memory reconsolidation and subsequent drug context-induced cocaine-seeking behavior in rats. Learn Mem 18:693–702

Wells AM, Arguello AA, Xie X, Blanton MA, Lasseter HC, Reittinger AM, Fuchs RA (2013) Extracellular signal-regulated kinase in the basolateral amygdala, but not the nucleus accumbens core, is critical for context-response-cocaine memory reconsolidation in rats. Neuropsychopharmacology 38:753–762

Wouda JA, Diergaarde L, Riga D, van Mourik Y, Schoffelmeer ANM, De Vries TJ (2010) Disruption of long-term alcohol-related memory reconsolidation: role of β-adrenoceptors and NMDA receptors. Front Behav Neurosci 4:179

Wu P, Xue Y-X, Ding Z-B, Xue L-F, Xu C-M, Lu L (2011) Glycogen synthase kinase 3β in the basolateral amygdala is critical for the reconsolidation of cocaine reward memory. J Neurochem 118:113–125

Wu Y, Li Y, Gao J, Sui N (2012) Differential effect of NMDA receptor antagonist in the nucleus accumbens on reconsolidation of morphine-related positive and aversive memory in rats. Eur J Pharmacol 674:321–326

Wu P, Ding Z-B, Meng S-Q, Shen H-W, Sun S-C, Luo Y-X, Liu J-F, Lu L, Zhu W-L, Shi J (2014a) Differential role of Rac in the basolateral amygdala and cornu ammonis 1 in the reconsolidation of auditory and contextual Pavlovian fear memory in rats. Psychopharmacology (Berl) 231(15):2909–2919

Wu Y, Li Y, Yang X, Sui N (2014b) Differential effect of beta-adrenergic receptor antagonism in basolateral amygdala on reconsolidation of aversive and appetitive memories associated with morphine in rats. Addict Biol 19:5–15

Yang J, Yu J, Jia X, Zhu W, Zhao L, Li S, Xu C, Yang C, Wu P, Lu L (2011) Inhibition of nuclear factor-κB impairs reconsolidation of morphine reward memory in rats. Behav Brain Res 216:592–596

Yang C, Liu J-F, Chai B-S, Fang Q, Chai N, Zhao L-Y, Xue Y-X, Luo Y-X, Jian M, Han Y, Shi H-S, Lu L, Wu P, Wang J-S (2013) Stress within a restricted time window selectively affects the persistence of long-term memory. PLoS One 8:e59075

Zhang S, Cranney J (2008) The role of GABA and anxiety in the reconsolidation of conditioned fear. Behav Neurosci 122:1295–1305

Zhao L-Y, Zhang X-L, Shi J, Epstein DH, Lu L (2009) Psychosocial stress after reactivation of drug-related memory impairs later recall in abstinent heroin addicts. Psychopharmacology (Berl) 203:599–608

Zhao L-Y, Sun L-L, Shi J, Li P, Zhang Y, Lu L (2011) Effects of β-adrenergic receptor blockade on drug-related memory reconsolidation in abstinent heroin addicts. Drug Alcohol Depend 118:224–229

Zlomuzica A, Dere D, Machulska A, Adolph D, Dere E, Margraf J (2014) Episodic memories in anxiety disorders: clinical implications. Front Behav Neurosci 8:131

Non-pharmacological Approaches to Cognitive Enhancement

Áine M. Kelly

Contents

Abstract

Pharmaceuticals and medical devices hold the promise of enhancing brain function, not only of those suffering from neurodevelopmental, neuropsychiatric or neurodegenerative illnesses, but also of healthy individuals. However, a number of lifestyle interventions are proven cognitive enhancers, improving

Á.M. Kelly (✉)
Department of Physiology, School of Medicine, Level 2, Trinity Biomedical Sciences Institute, Trinity College Dublin, 152-160 Pearse Street, Dublin 2, Ireland

Trinity College Institute of Neuroscience, Trinity College Dublin, Dublin 2, Ireland

Trinity Biomedical Sciences Institute, Trinity College Dublin, Dublin 2, Ireland
e-mail: aikelly@tcd.ie

K.M. Kantak, J.G. Wettstein (eds.), *Cognitive Enhancement*, Handbook of Experimental Pharmacology 228, DOI 10.1007/978-3-319-16522-6_14

attention, problem solving, reasoning, learning and memory or even mood. Several of these interventions, such as physical exercise, cognitive, mental and social stimulation, may be described as environmental enrichments of varying types. Use of these non-pharmacological cognitive enhancers circumvents some of the ethical considerations associated with pharmaceutical or technological cognitive enhancement, being low in cost, available to the general population and presenting low risk to health and well-being. In this chapter, there will be particular focus on the effects of exercise and enrichment on learning and memory and the evidence supporting their efficacy in humans and in animal models will be described.

Keywords
Cognitive enhancement • Physical activity • Environmental enrichment • Neurogenesis • BDNF

1 Introduction

1.1 Exercise

Physical inactivity is a risk factor for development of several non-communicable diseases including cardiovascular disease, type 2 diabetes and certain types of cancer and is associated with a decrease in life expectancy (Lee et al. 2012). While regular exercise enhances and preserves general health, it also confers specific benefits on the nervous system that result in measurable cognitive improvement. Such improvements have been seen both in cognitively impaired and in healthy subjects, indicating the potential of exercise to act as a neurotherapeutic (Lautenschlager et al. 2008), a neuroprotectant (Kramer and Erickson 2007a; Rovio et al. 2005) and an enhancer of normal cognitive performance (Griffin et al. 2011). While most forms of exercise promote good health, it appears that aerobic exercise is a more robust enhancer of brain health when compared with static or resistance exercise. The use of rodent models is allowing the cellular and molecular mechanisms underlying these improvements to be characterised (Voss et al. 2013). Increased expression of several growth factors, particularly brain-derived neurotrophic factor (BDNF), is consistently associated with exercise-induced cognitive enhancement (Vaynman and Gomez-Pinilla 2006), while the ability of exercise to remodel brain morphology via angiogenesis, synaptogenesis and neurogenesis may underpin its cognitive-enhancing efficacy (Lista and Sorrentino 2010).

1.2 Enrichment

Environmental enrichment in laboratory rodents is easily defined; it simply means the addition of sources of stimulation to the standard housing environment. Depending on the experiment, this may include sources of social stimulation, such as an increased number of cagemates, physical stimulation such as running wheels or a larger cage environment in which to move around or mental stimulation such as provision of novel objects and toys or participation in learning experiments (van Praag et al. 2000). When reviewing the literature on environmental enrichment, one must be mindful of the profound effects of exercise on brain function; thus care must be taken to distinguish between experimental conditions that include opportunities to engage in increased physical activity, such as provision of running wheels or participation in treadmill running, and those that do not (Bechara and Kelly 2013). What does enrichment mean in the case of humans? Possible sources of enrichment include mental stimulation such as reading, playing chess, solving puzzles or engagement in formal education, social stimulation such as participation in group activities and interaction with family, friends, neighbours and the wider community or, more recently, participation in targeted cognitive training, including computerized training programmes. Again, animal models are proving useful in characterising the biological underpinnings of the cognitive effects of enrichment, at least some of which may be shared with exercise (Pang and Hannan 2013; Brown et al. 2003).

2 Non-pharmacological Cognitive Enhancement in Animals

2.1 Exercise as a Cognitive Enhancer in Rodents

There is a vast and growing literature reporting the ability of exercise to enhance cognitive function, especially learning and memory, in animals and humans (Voss et al. 2013; Gomez-Pinilla and Hillman 2013). The use of rats and mice as experimental subjects has allowed the mechanisms underlying exercise-induced cognitive enhancement to be investigated at cellular and molecular levels. The many published studies have employed different forms of exercise including forced treadmill running and voluntary wheel running, while various tasks such as object recognition memory, spatial learning in mazes and contextual fear conditioning have been used to assess cognitive performance (Voss et al. 2013). At the cellular level, the impact of exercise on forms of synaptic plasticity, particularly long-term potentiation (LTP), has been widely investigated. Exercise appears to have particularly powerful effects on the function of the hippocampus, a region of the medial temporal lobe crucial to spatial navigation and memory formation and one of the few brain regions in which new neurons can develop, a process known as adult hippocampal neurogenesis. The fact that exercise acts as a powerful stimulator of neurogenesis in the dentate gyrus subfield of the hippocampus may explain, at least in part, its profound effects on hippocampus-dependent memory (van Praag et al. 1999b; Creer et al. 2010).

Short-term and long-term exercise protocols, using both forced and voluntary exercise, result in enhanced cognitive performance and synaptic plasticity in rodents at different stages of the life span. Voluntary wheel running enhances LTP and spatial learning in the Morris water maze (van Praag et al. 1999a) and spatial pattern separation (Creer et al. 2010) in young adult mice, while wheel running enhances LTP (Farmer et al. 2004) water maze learning (Vaynman et al. 2004b; Gomez-Pinilla et al. 2008; Ding et al. 2006), fear conditioning (Hopkins and Bucci 2010b) and recognition memory (Hopkins and Bucci 2010a) in young rats. One week of forced treadmill running improves expression of LTP and recognition memory in young rats (Bechara et al. 2014; O'Callaghan et al. 2007; Griffin et al. 2009), while several months of forced exercise enhances water maze learning in the young (Cassilhas et al. 2012) and aged (O'Callaghan et al. 2009; Albeck et al. 2006) rat. It must be noted that in some studies exercise failed to enhance cognitive function (Kennard and Woodruff-Pak 2012), suggesting that the effects of exercise on cognition may depend on variable factors such as the duration of exercise exposure, the modality of the exercise undertaken (forced versus voluntary) and the intensity of the exercise, along with the nature and difficulty of the cognitive task (Berchtold et al. 2010). However, the overall weight of evidence gives powerful support to the hypothesis that exercise is a cognitive enhancer in both young and aged rodents (Vaynman and Gomez-Pinilla 2006).

2.2 Enrichment as a Cognitive Enhancer in Rodents

A complex cage environment that includes provision of running wheels is proven to enhance learning in laboratory rodents (Rosenzweig and Bennett 1996; van Praag et al. 2000, 2002; Pang and Hannan 2013). Such enrichment can also protect against the normal decline in memory associated with ageing and a number of different neurological and psychological pathologies such as depression, Huntington's disease and Alzheimer's disease in both humans and animal models (Mora et al. 2007; Laviola et al. 2008; Brenes et al. 2009; Nithianantharajah and Hannan 2011). Animals housed in enriched environments have improved recognition and spatial memory compared with standard housed controls, while enrichment can rescue cognitive deficits induced by experimentally induced ischaemia or surgical lesions (Gobbo and O'Mara 2004; Mandolesi et al. 2008). It has been suggested that enrichment in early adulthood and throughout one's life might increase the resilience of the brain in old age, resulting in the concept of 'cognitive reserve' (Nithianantharajah and Hannan 2009).

Each different stimulatory factor present in a typical enriched environment may contribute to the resulting improvement in cognitive function; indeed, it has been suggested that different aspects of the environment can induce the same improvements via dissociable pathways (Olson et al. 2006). Several studies have assessed the effects of enrichment in the absence of exercise on various aspects of cognitive function. This type of enrichment is linked with reduced anxiety in rodents (Galani et al. 2007) and has been shown to enhance object recognition and spatial memory in a time-dependent manner in the rat; continuous housing in an enriched environment was necessary for at least 3 weeks before cognitive benefit was detectable (Birch et al. 2013). Other studies have shown that enrichment induces cognitive benefit in middle-aged, but not young, mice, indicating that enrichment may have the particular ability to improve memory where cognitive impairment exists (Harburger et al. 2007a, b). Such observations have also been reported in rats, indicating that enrichment in the absence of exercise may be an efficacious cognitive enhancer in aged animals (Kumar et al. 2012), but a less robust enhancer of cognition in young animals (Gobbo and O'Mara 2004). These observations may be of translational relevance; it is gratifying to note that cognitive stimulation may enhance brain function in older subjects, where participation in regular physical activity may be difficult due to frailty or illness.

2.3 The Potential Mechanisms Underlying Exercise and Enrichment-Induced Cognitive Enhancement

2.3.1 Neurotrophins and Growth Factors

Numerous studies have shown that exercise significantly increases the expression of the neurotrophic factor BDNF in the hippocampus and several neocortical regions (Molteni et al. 2002; Neeper et al. 1996; Vaynman et al. 2004a; Griffin et al. 2009; Ding et al. 2011). BDNF plays a vital role in neurodevelopment, but is now accepted to be a key regulator of synaptic plasticity in the developing and adult brain (Bekinschtein et al. 2014). Exogenous BDNF can induce a form of LTP (Bramham and Panja 2014) and improve hippocampus-dependent learning (Griffin et al. 2009; Bechara et al. 2014), potentially via its ability to stimulate plasticity-related intracellular signalling pathways following binding to its receptor, TrkB (Bekinschtein et al. 2014). BDNF can depolarize neurons, leading to increased neurotransmitter release and therefore rapid modulation of neuronal communication (Lessmann 1998). Some studies have noted significant changes in BDNF mRNA expression within as little as 2 h following exercise (Huang et al. 2006; Soya et al. 2007). Since BDNF expression can be upregulated rapidly and it is released in an activity-dependent manner, it has been proposed that BDNF mediates the rapidly observed aspects of cognitive enhancement induced by exercise (Bechara et al. 2014). It has also been suggested that exercise may increase the brain's resistance to damage and degeneration through the ability of BDNF to regulate neuronal growth and survival (Neeper et al. 1996). Other growth factors may mediate exercise-induced improvements in cognitive function. Some evidence suggests that nerve growth factor (NGF) expression in the hippocampus is increased following exercise (Neeper et al. 1996), and exercise has been shown to ameliorate the age-related decline in expression of both BDNF and NGF, concomitant with improved spatial learning (O'Callaghan et al. 2009). NGF, though first identified as a key regulator of embryonic development of the nervous system, also plays important roles in the adult nervous system, including modulation of hippocampal plasticity (Conner et al. 2009). Enrichment in the absence of exercise also induces a neurotrophic response in the hippocampus, with exposure to an enriched environment increasing expression of NGF (Birch et al. 2013), but not BDNF (Bindu et al. 2007; Kumar et al. 2012). Exogenous NGF has been shown to mimic the cognitive effects of enrichment in young rats (Birch and Kelly 2013). Peripherally produced growth factors such as insulin-like growth factor 1 (IGF-1) may be a link between the systemic and the central changes induced by exercise. IGF-1 receptors are abundantly expressed in the hippocampus and blocking these receptors during exercise has been shown to inhibit exercise-induced enhancements in memory (Ding et al. 2006). Furthermore, spatial memory impairments displayed by serum IGF-1 deficient mice are ameliorated by exogenous IGF-1 administration (Trejo et al. 2008).

2.3.2 Synaptogenesis and Neurogenesis

The formation of new synapses is likely to be critical to storage of new information in the brain and is among the neuroplastic changes induced by physical activity and enrichment. There are several reports of increased expression of the synaptic vesicle proteins synaptophysin and synapsin-I following exercise and some evidence indicates that exercise-induced synaptogenesis may be a BDNF-dependent process (Vaynman et al. 2004a; Ding et al. 2006; Quirie et al. 2012). Voluntary running increases the density of dendritic spines in granule and CA1 pyramidal neurons of the dentate gyrus and layer III pyramidal neurons of the entorhinal cortex (Stranahan et al. 2007), while both forced and voluntary running increase mossy fibre sprouting (Toscano-Silva et al. 2010). Voluntary wheel running increases the expression of the AMPA receptor subunits GluR2/3 and phosphorylation of the NMDA receptor subunits GluN1 and GluN2B (Dietrich et al. 2005), providing further evidence that enhancement of synaptic efficacy may underpin the cognitive enhancements induced by exercise. Enrichment in the absence of physical activity increases expression of synaptic vesicle proteins, indicating that cognitive and social stimulation can also stimulate synaptogenesis (Birch et al. 2013).

Adult hippocampal neurogenesis is defined as the process of generating functional neurons from neuronal precursors in the subgranular zone of the dentate gyrus (Ming and Song 2011). The process involves proliferation of neural precursor cells and their differentiation, migration and integration into the granule cell network of the dentate gyrus (Aimone et al. 2006), a process that takes 3–4 weeks. The majority of proliferating cells are not integrated into the hippocampal circuitry and undergo apoptosis; thus neurogenesis depends on increased cell proliferation in tandem with conditions conducive to cell survival. By the time adult-born neurons are 4–8 weeks old, they are preferentially recruited into circuits supporting spatial memory compared with existing granule cells, consistent with their decreased threshold for plasticity (Kee et al. 2007). Therefore, as adult-generated neurons mature they are increasingly likely to be incorporated into circuits supporting spatial memory (Kee et al. 2007). Although the contribution of neurogenesis to hippocampus-dependent learning and memory has yet to be fully elucidated, it is being increasingly accepted that neurogenesis may be of functional relevance to learning and memory (Gage and Temple 2013). Exercise is by far the most robust neurogenic stimulus yet identified (van Praag 2009). Considering the time frame of the entire process, neurogenesis is a longer-term neuroplastic effect of exercise compared with the rapidly induced effects of exercise on synaptic transmission and plasticity. However, the onset of the neurogenic effect of exercise is rapid with cell genesis reportedly peaking following 3 days of voluntary exercise (Kronenberg et al. 2006) and remaining elevated for up to 32 days before returning to baseline. Numerous other studies have shown that both voluntary and forced exercise can induce an increase in cell proliferation in the dentate gyrus, the survival of neural progenitor cells and their differentiation into neurons rather than glial cells (van Praag et al. 1999a; Fabel et al. 2003; Van der Borght et al. 2009; Wu et al. 2008; Creer et al. 2010). A direct comparison of both exercise modalities indicated that forced exercise is a significantly more robust neurogenic stimulus compared with wheel running, an observation that may explain some of the conflicting results reported in the literature relating to exercise and cognition (Leasure and Jones 2008). The potent effects of exercise on neurogenesis have been observed in the hippocampus of young, middle-aged and aged animals indicating that the brain retains at least some neurogenic ability throughout the life span (van Praag et al. 1999a, 2005; Wu et al. 2008).

Both exercise and cognitive enrichment can bring about similar improvements in learning and memory, at least in aged animals, but it is unclear whether they do so via similar neuroplastic mechanisms, including neurogenesis. An early study in this area demonstrated that enrichment affected cell survival and not cell proliferation whereas exercise increased cell division and net neuronal survival in mice (van Praag et al. 1999a). It has been proposed that enrichment does not stimulate an increase in proliferation per se but promotes increased survival of neuronal progenitor cells and hence increases the number of young neurons available to functionally integrate into neuronal networks (Kempermann and Gage 1999). This would suggest that cell survival and cell proliferation may be regulated by differing mechanisms that can be affected by behavioural and environmental factors (Olson et al. 2006). In contrast, enrichment in the absence of exercise has been shown to increase hippocampus-dependent learning, cell proliferation, but not survival, in the dentate gyrus of the rat (Birch et al. 2013). Another recent study failed to observe any effect of enrichment alone on neurogenesis, but no cognitive measures were assessed in parallel (Gregoire et al. 2014). The relative contribution of the different aspects of an enriched environment to the cognitive enhancement that it stimulates is a complex question that has yet to be answered, but it seems that exercise is the most important element, at least in the case of young animals (Kobilo et al. 2011). Where a cognitive impairment exists, as in the aged or diseased brain, there may be greater capacity for other environmental stimuli such as social and cognitive stimulation to induce a cognitive benefit (Lazarov et al. 2010).

2.3.3 Angiogenesis and Vascular Growth Factors

Physical exercise influences the vasculature, with vasodilation the most obvious and rapid change. Although the brain was originally believed to maintain a constant blood supply in the face of changes in mean arterial pressure, there is now overwhelming evidence to support an exercise-induced increase in cerebral blood flow, possibly due to an increase in brain metabolism (Querido and Sheel 2007). At a cellular level, physical activity has been linked with angiogenesis, the growth of new capillaries from pre-existing blood vessels, in several brain regions including the hippocampus, motor cortex and cerebellum (Isaacs et al. 1992; Swain et al. 2003; Clark et al. 2009; Van der Borght et al. 2009). Vascular endothelial growth factor (VEGF), a hypoxia-inducible secreted protein, plays an important role in the angiogenic effects of exercise (Ferrara 2009; van Praag 2009). VEGF expression is increased in skeletal muscle and hippocampus following a single bout of moderate intensity exercise (Tang et al. 2010), while pharmacological blockade of angiogenesis in the hippocampus impairs spatial learning in the water maze (Kerr et al. 2010). However, other studies report minimal effects of exercise on VEGF expression in the hippocampus suggesting that this angiogenic growth factor must cross the blood–brain barrier (BBB) to induce angiogenesis in the brain (Fabel et al. 2003). Some reports indicate that the BBB permeability may increase in response to exercise, providing a potential route for signaling proteins to enter the brain parenchyma from the circulation (Sharma et al. 1991; Watson et al. 2006). It has also been suggested that increased circulation may permit delivery of more nutrient metabolites, hormones, growth factors and oxygen to the hippocampus while also facilitating metabolic waste disposal, leading to increased cell survival and enhanced neurogenesis (Olson et al. 2006). In humans, cerebral blood volume changes in the dentate gyrus have been correlated with aerobic fitness and cognitive function (Pereira et al. 2007). Collectively, these experiments show that the hippocampus displays remarkable angiogenic plasticity and that the cerebral vasculature responds to physical activity; thus vascular adaptations could be another key mechanism underlying exercise-induced improvement in cognitive function. Unsurprisingly, it appears that angiogenesis may be a more important factor in mediating exercise-induced cognitive enhancement when compared with enrichment in the absence of exercise. However, there is some evidence of modest angiogenic activity in the hippocampus of environmentally enriched rats (Ekstrand et al. 2008). Environmental enrichment that includes access to exercise equipment is widely reported to induce angiogenesis, and it is likely that the major angiogenic stimulus in such complex environments is increased physical activity.

The positive effects of exercise on cognitive function have been attributed to a number of neuroplastic changes in the hippocampus including increased expression of and signaling via growth factors, enhanced synaptic plasticity, synaptogenesis, angiogenesis and neurogenesis. At least some of the cellular mechanisms underlying each of these changes are likely to be shared. For example, administration of BDNF, VEGF and IGF-1 has been shown to increase neurogenesis in the dentate gyrus, while peripheral blockade of VEGF abolishes exercise-induced hippocampal neurogenesis (Fabel et al. 2003). Certainly, it appears that several growth factors whose expression is upregulated by exercise and enrichment have both angiogenic and neurogenic properties, and that both processes may be necessary for at least some of the cognitive-enhancing effects of these non-pharmacological lifestyle factors to be manifested.

3 Non-pharmacological Cognitive Enhancement in the Healthy Human

Regular exercise confers long-term benefits on brain health, but evidence exists to show that those with healthy brain function may also reap short-term cognitive benefits of physical activity. A single bout of exercise can improve cognitive performance in subject groups that are not cognitively impaired (Tomporowski 2003; Lambourne and Tomporowski 2010; Chang et al. 2012). In young healthy adults, high-intensity running is reported to improve vocabulary learning (Winter et al. 2007), while cycling improves performance of the frontal lobe-dependent Stroop colour-word task (Ferris et al. 2007) and the hippocampus-dependent face-name pairs (Griffin et al. 2011) and map recognition tasks (Grego et al. 2005). Aerobic and resistance exercise make different physiological demands on the cardiovascular, musculoskeletal, endocrine and respiratory systems and have been reported to exert different effects on cognitive function; for example aerobic, but not resistance, exercise enhances working memory in young people (Pontifex et al. 2009). This suggests that exercise-induced enhancements in cognitive function are likely to be dependent on exercise modality, intensity and duration as well as the physical fitness of subjects (Grego et al. 2004). During prolonged exercise, fatigue-related factors such as heat stress, dehydration and hypoglycaemia can impair short-term memory (Cian et al. 2001) while fatigue-related increases in circulating cortisol and adrenaline reduced event-related potentials, a measure of cortical activity (Grego et al. 2004). Thus it appears that exercise can enhance or impair cognitive performance in a manner dependent on exercise intensity and duration. Certainly, acute bouts of exercise can confer a cognitive advantage, at least in the short term, since the persistence of these rapid exercise-induced effects is as yet unknown (Griffin et al. 2011; Schmidt-Kassow et al. 2013). Physical activity in children is linked with cognitive development; exercise appears to be of particular benefit to development of executive function (Tomporowski et al. 2011). In addition to physical activity, both cognitive and social stimulation are crucial for normal development in childhood. Additional cognitive stimulation for children from lower socioeconomic backgrounds, at home or in a preschool setting, can significantly improve their academic achievements (Crosnoe et al. 2010).

Evidence from experiments in animals has provided insights into the mechanisms by which exercise may enhance brain function in humans. There are several reports of an increase in circulating BDNF concentration in response to exercise (Gold et al. 2003; Rojas Vega et al. 2006; Goekint et al. 2008; Tang et al. 2008; Rasmussen et al. 2009; Cho et al. 2012); in some studies, a parallel enhancement of cognitive function was observed (Ferris et al. 2007; Griffin et al. 2011). These changes in BDNF concentration appear to be detectable only after aerobic exercise, since resistance training has no such effect (Schiffer et al. 2009; Goekint et al. 2010); hence it may be the case that the inability of resistance exercise to increase BDNF concentration underlies its inability to enhance cognitive function in several studies. Several studies indicate that circulating NGF and IGF-1 do not increase in response to exercise (Gold et al. 2003; Schiffer et al. 2009; Griffin et al. 2011).

The cellular origin of the exercise-induced BDNF response remains to be elucidated. Several reports indicate that the brain itself may be the main contributor of BDNF to the circulation during endurance exercise (Seifert et al. 2010; Rasmussen et al. 2009) although muscle (Matthews et al. 2009), endothelial cells (Nakahashi et al. 2000) and platelets (Fujimura et al. 2002) are also potential sources. The correlation between exercise, cognitive function and BDNF is strong, but the functional relationship between these elements is not yet understood. Based on data from the animal literature, it may be speculated that candidate mechanisms mediating the effects of exercise in humans would include short-term effects on plasticity induced by neurotrophins, angiogenesis, synaptogenesis and neurogenesis. For obvious reasons, direct assessment of these measures in humans is technically difficult; however, it has recently been confirmed that adult hippocampal neurogenesis occurs in humans (Spalding et al. 2013). MRI analysis has revealed that exercise increases blood volume in the dentate gyrus concomitant with improved cognitive function (Pereira et al. 2007); the authors suggest that this may be a correlate of neurogenesis.

4 Cognitive Enhancement in the Cognitively Impaired

Every human will experience some form of physical and mental decline in old age but for many, age-related cognitive decline progresses from mild cognitive impairment to vascular dementia, Alzheimer's disease (AD) and other forms of dementia that impair or even destroy quality of life and the ability to live independently. Others struggle with the cognitive impairment that accompanies Parkinson's Disease (PD), depression, schizophrenia and other mental disorders. The identification and implementation of strategies that promote healthy aging and that improve cognitive performance in specific patient groups, such as physical activity and mental and social stimulation, is thus of broad societal and economic benefit. Worryingly, recent data on lifestyle trends in the United States show that, on average, individuals older than 15 years of age spend almost 55 % of their leisure time watching television, compared with 20 % of free time spent engaging in cognitively stimulating activities such as reading or socializing and only 6 % of their leisure time exercising (American Time Use Study, U.S. Bureau of Labor Statistics 2013); similar trends have been observed globally (Heath et al. 2012; Olafsdottir et al. 2014), with parallel increases in overweight and obesity. Given the impact that a cognitively and physically active lifestyle can have on the health of the mind and body, a lack of stimulation may be causing detrimental effects in the general population that will impact on future mental health.

4.1 Age-Related Cognitive Decline and Dementia

The link between a healthy lifestyle and a healthy old age is indisputable and exercise is a key element of this relationship. The decreased incidence of cognitive impairment and dementia observed in elderly persons who undertake regular physical activity is strong evidence of the neuroprotective effects of exercise (Geda et al. 2010; Laurin et al. 2001; Colcombe et al. 2004). In older adults without dementia the volume of the hippocampus shrinks 1–2 % annually (Raz et al. 2005); such atrophy increases the risk of memory impairment in late adulthood (Jack et al. 2010). The smaller hippocampal volume and poorer memory performance associated with increasing age are paralleled by reduced levels of serum BDNF (Erickson et al. 2010). Exercise in those aged over 65 years reduces the incidence of dementia relative to sedentary controls (Rovio et al. 2005; Larson et al. 2006), and higher levels of aerobic fitness have also been associated with increased hippocampal volume in elderly adults (Erickson et al. 2009), indicating the importance of physical fitness throughout the life span. A 1-year aerobic intervention was shown to increase hippocampal volume by 2 %, effectively offsetting the age-related loss in volume by 1–2 years and improving memory, providing evidence that exercise can act as a neurotherapeutic as well as a neuroprotectant (Erickson et al. 2011). This improvement in function was associated with higher circulating BDNF in these subjects. Meta-analyses of the literature that reveal the rehabilitating effects of exercise in elderly patients suffering from dementia and Alzheimer's disease (Heyn et al. 2004; Farina et al. 2014; Kramer and Erickson 2007b) underline the capacity of exercise to reverse as well as prevent cognitive decline. Cerebral vasoactivity has been correlated with aerobic capacity in older adults, providing a possible physiological mechanism by which exercise impacts cognitive function in old age (Barnes et al. 2013b). The animal literature shows a specific benefit of enrichment and exercise in aged animals (O'Callaghan et al. 2009) and in mouse models of Alzheimer's disease (Valero et al. 2011), delaying or reversing the impairments in neurogenesis and resulting in prevention or reversal of cognitive impairment (Speisman et al. 2013; Lazarov et al. 2010).

Clearly, there may be certain people who are unable to participate in exercise for reasons of frailty, physical infirmity or circumstance and the fact that social and intellectual enrichment can also protect against, delay or reverse age-related cognitive decline is of particular relevance to such individuals. One comprehensive study showed that participation in cognitively stimulating activity by elderly members of Catholic religious orders was associated with a decreased incidence of AD (Wilson et al. 2002c), while participation in leisure activities such as reading and playing board games is correlated with a lower risk of development of dementia in the elderly (Verghese et al. 2003; Wilson et al. 2002b). The number of years spent in formal education is negatively correlated with the risk of dementia (Anstey et al. 2000); conversely, loneliness, social isolation, depression and apathy are increasingly acknowledged risk factors for development of age-related cognitive impairment and dementia (Robert et al. 2008; Shankar et al. 2013; Holwerda et al. 2014; Wilson et al. 2002a). Computerized training programmes are a relatively recently developed source of targeted cognitive stimulation that may enhance brain plasticity in older subjects (Mahncke et al. 2006; Tardif and Simard 2011). These programmes variously target spatial skills, attention, visual skills, working memory and other cognitive functions (Schmiedek et al. 2010; Smith et al. 2009; Nouchi et al. 2012). However, the persistence of these effects and the transfer of the learned skills to real-life situations have not yet been demonstrated.

Taken together, the weight of evidence suggests that in populations with existing cognitive impairment, physical and intellectual activity may help to delay the progression toward more severe dementia. A randomized control trial of the effects of 12 weeks of mental and physical activity in inactive, community-residing older adults with cognitive complaints resulted in significant improvements in global cognitive function in all participants, regardless of whether the exercise intervention was aerobic or anaerobic or whether the mental stimulation consisted of challenging, computer-based activity or watching educational DVDs (Barnes et al. 2013a). These results suggest that regular participation in any type of stimulating physical or mental activity can translate to functional improvement in such groups. It is a reassuring illustration of the capacity of the brain to remain plastic throughout the life span and that exercise or novel intellectual activities begun late in life can result in improved cognitive outcome.

4.2 Parkinson's Disease, Schizophrenia and Depression

The well-known motor symptoms that characterize Parkinson's disease (PD) are often accompanied by cognitive impairment. Despite the motor impairments suffered by PD patients, many retain the ability to participate in exercise activity (Earhart 2013), in some cases resulting in demonstrable cognitive benefit. In a community-dwelling group of PD patients, regular walking resulted in improved motor function, cognition and general quality of life (Uc et al. 2014), while exercise improved cued reaction time, indicative of cognitive improvement, in another group of PD patients (Ebersbach et al. 2014). Another study demonstrated that exercise can increase BDNF in the circulation of PD patients (Frazzitta et al. 2014). A systematic review of the literature has identified a cognitive benefit of exercise training in PD patients, but highlights the need for further research in this promising area (Hindle et al. 2013). With regard to other forms of cognitive enhancement, computer-based cognitive training has benefitted learning and memory in PD patients (Naismith et al. 2013). Exercise and enrichment in animal models of PD can benefit cognitive performance (Faherty et al. 2005; Pothakos et al. 2009; Petzinger et al. 2013) and increase BDNF expression in the brain (Tuon et al. 2012).

Schizophrenia is a major psychiatric disorder whose negative symptoms include cognitive impairments affecting memory, executive functioning and attention.

Pharmacotherapy with antipsychotic medication is of course the most effective treatment for the positive symptoms of this disorder, but there may be cognitive benefit to patients of interventions such as exercise therapy as an adjunct to pharmacotherapy and psychotherapy. Some reports indicate that exercise increases hippocampal volume (Vancampfort et al. 2014; Pajonk et al. 2010) and improves verbal, visual and working memory (Vancampfort et al. 2014; Pajonk et al. 2010; Oertel-Knochel et al. 2014) in people with schizophrenia. In contrast, computer-based brain training improved performance of computer-based tasks, but did not translate to general cognitive benefit in schizophrenic patients (Dickinson et al. 2010). The ability to employ exercise as a cognitive enhancer in those suffering from schizophrenia may be confounded by their lower reported levels of activity (Laursen et al. 2012) and lower cardiorespiratory fitness (Ozbulut et al. 2013). However, participation in sports has been shown to positively affect physical and psychiatric symptoms in schizophrenic patients (Takahashi et al. 2012). Aberrant neurogenesis has been implicated in schizophrenia and thus this process may be a viable clinical target; postmortem analysis demonstrated that cell proliferation was diminished in the dentate gyrus of people who suffered from schizophrenia (Reif et al. 2006), while there is evidence that antipsychotic drugs have neurogenic properties (Keilhoff et al. 2012). Evidence from studies in a mouse model of schizophrenia indicates the ability of exercise to increase neurogenesis and improve behavioural deficits (Wolf et al. 2011).

Depression, and its accompanying cognitive impairment, may present as a primary psychiatric disorder or may be comorbid with conditions such as AD, PD or schizophrenia. Depression is often associated with low levels of physical activity; adults with depression are reported to spend significantly less time in either light or moderate physical activity than non-depressed adults (Song et al. 2012). There is a vast and growing literature on the potential benefits of exercise in the prevention (Mammen and Faulkner 2013) and treatment (Cooney et al. 2013) of depression. Indeed, the UK National Institute for Health and Clinical Excellence recommends structured exercise, three times a week for 10–14 weeks, for the treatment of mild to moderate depression (NICE guidelines [CG90], 2009). A recent Cochrane Review revealed that exercise is associated with a greater reduction in symptoms of depression compared with no treatment, placebo, or active control interventions such as relaxation or meditation (Cooney et al. 2013); though the authors emphasize that the benefits are of small magnitude, any reduction in clinical symptoms is to be welcomed. Reduced expression of BDNF in the brain is associated with depression, a finding that has led to the examination of the efficacy of interventions that may upregulate BDNF as antidepressant strategies (Castren and Rantamaki 2010). While antidepressant medication increases BDNF expression (Russo-Neustadt and Chen 2005), exercise, alone or in combination with antidepressant treatment, is reported to increase BDNF expression and reduce depressive symptoms in animal models of depression (Marais et al. 2009; Sigwalt et al. 2011; Russo-Neustadt et al. 2001; Garza et al. 2004). Antidepressant drugs are also potent stimulators of neurogenesis (Duman et al. 2001; Ota and Duman 2013); thus neurogenesis has been suggested as a key biological mechanism mediating their cognitive effects (Ernst et al. 2006). The links between exercise, BDNF, neurogenesis and cognitive function in the healthy brain are mirrored in depression and major depressive disorder and may explain the ability of exercise to ameliorate at least some of the cognitive dysfunction associated with depression (Ota and Duman 2013).

5 Conclusions

While interventions such as exercise and enrichment are effective cognitive enhancers in their own right, they may also be useful adjuncts to pharmacological treatments of disease, as is the case in depression. Elucidation of the cellular mechanisms underlying cognitive enhancements induced by these non-pharmacological interventions may allow novel molecular drug targets that exploit the same cellular pathways to be developed.

Clinicians routinely recommend or prescribe exercise to those with conditions such as obesity, diabetes and cardiovascular disease. The proven ability of exercise and other forms of environmental stimulation to protect against or treat cognitive impairment associated with normal aging or specific neurodegenerative or neuropsychiatric disorders render these interventions of potential clinical importance. It may be envisaged that current recommendations of physical, mental or social activity for the purpose of maintaining general health may translate to prescription of activity for the specific benefit of brain health. However the optimal type, intensity, frequency and duration of exercise that will benefit specific populations or patient groups have not yet been identified.

While some of these non-pharmacological approaches have shown specific benefits to cognition, especially learning and memory, a generally healthy lifestyle that includes regular exercise, social engagement and mental stimulation has wider impacts on the general health of the individual that also benefits society as a whole. The side effects of the non-pharmacological strategies for cognitive enhancement outlined here are broadly positive.

References

Aimone JB, Wiles J, Gage FH (2006) Potential role for adult neurogenesis in the encoding of time in new memories. Nat Neurosci 9:723–727

Albeck DS, Sano K, Prewitt GE, Dalton L (2006) Mild forced treadmill exercise enhances spatial learning in the aged rat. Behav Brain Res 168:345–348

American Time Use Study, U.S. Bureau of Labor Statistics (2013) http://www.bls.gov/news.release/atus.nr0.htm

Anstey KJ, Matters B, Brown AK, Lord SR (2000) Normative data on neuropsychological tests for very old adults living in retirement villages and hostels. Clin Neuropsychol 14:309–317

Barnes DE, Santos-Modesitt W, Poelke G, Kramer AF, Castro C, Middleton LE, Yaffe K (2013a) The Mental Activity and eXercise (MAX) trial: a randomized controlled trial to enhance cognitive function in older adults. JAMA Intern Med 173:797–804

Barnes JN, Taylor JL, Kluck BN, Johnson CP, Joyner MJ (2013b) Cerebrovascular reactivity is associated with maximal aerobic capacity in healthy older adults. J Appl Physiol (1985) 114:1383–1387

Bechara RG, Kelly AM (2013) Exercise improves object recognition memory and induces BDNF expression and cell proliferation in cognitively enriched rats. Behav Brain Res 245:96–100

Bechara RG, Lyne R, Kelly AM (2014) BDNF-stimulated intracellular signalling mechanisms underlie exercise-induced improvement in spatial memory in the male Wistar rat. Behav Brain Res 275:297–306

Bekinschtein P, Cammarota M, Medina JH (2014) BDNF and memory processing. Neuropharmacology 76:677–683

Berchtold NC, Castello N, Cotman CW (2010) Exercise and time-dependent benefits to learning and memory. Neuroscience 167:588–597

Bindu B, Alladi PA, Mansooralikhan BM, Srikumar BN, Raju TR, Kutty BM (2007) Short-term exposure to an enriched environment enhances dendritic branching but not brain-derived neurotrophic factor expression in the hippocampus of rats with ventral subicular lesions. Neuroscience 144:412–423

Birch AM, Kelly AM (2013) Chronic intracerebroventricular infusion of nerve growth factor improves recognition memory in the rat. Neuropharmacology 75:255–261

Birch AM, Mcgarry NB, Kelly AM (2013) Short-term environmental enrichment, in the absence of exercise, improves memory, and increases NGF concentration, early neuronal survival, and synaptogenesis in the dentate gyrus in a time-dependent manner. Hippocampus 23(6):437–450

Bramham CR, Panja D (2014) BDNF regulation of synaptic structure, function, and plasticity. Neuropharmacology 76 Pt C:601–602

Brenes JC, Padilla M, Fornaguera J (2009) A detailed analysis of open-field habituation and behavioral and neurochemical antidepressant-like effects in postweaning enriched rats. Behav Brain Res 197:125–137

Brown J, Cooper-Kuhn CM, Kempermann G, van Praag H, Winkler J, Gage FH, Kuhn HG (2003) Enriched environment and physical activity stimulate hippocampal but not olfactory bulb neurogenesis. Eur J Neurosci 17:2042–2046

Cassilhas RC, Lee KS, Fernandes J, Oliveira MG, Tufik S, Meeusen R, de Mello MT (2012) Spatial memory is improved by aerobic and resistance exercise through divergent molecular mechanisms. Neuroscience 202:309–317

Castren E, Rantamaki T (2010) The role of BDNF and its receptors in depression and antidepressant drug action: reactivation of developmental plasticity. Dev Neurobiol 70:289–297

Chang YK, Labban JD, Gapin JI, Etnier JL (2012) The effects of acute exercise on cognitive performance: a meta-analysis. Brain Res 1453:87–101

Cho HC, Kim J, Kim S, Son YH, Lee N, Jung SH (2012) The concentrations of serum, plasma and platelet BDNF are all increased by treadmill VO(2)max performance in healthy college men. Neurosci Lett 519:78–83

Cian C, Barraud PA, Melin B, Raphel C (2001) Effects of fluid ingestion on cognitive function after heat stress or exercise-induced dehydration. Int J Psychophysiol 42:243–251

Clark PJ, Brzezinska WJ, Puchalski EK, Krone DA, Rhodes JS (2009) Functional analysis of neurovascular adaptations to exercise in the dentate gyrus of young adult mice associated with cognitive gain. Hippocampus 19:937–950

Colcombe SJ, Kramer AF, Erickson KI, Scalf P, Mcauley E, Cohen NJ, Webb A, Jerome GJ, Marquez DX, Elavsky S (2004) Cardiovascular fitness, cortical plasticity, and aging. Proc Natl Acad Sci U S A 101:3316–3321

Conner JM, Franks KM, Titterness AK, Russell K, Merrill DA, Christie BR, Sejnowski TJ, Tuszynski MH (2009) NGF is essential for hippocampal plasticity and learning. J Neurosci 29:10883–10889

Cooney GM, Dwan K, Greig CA, Lawlor DA, Rimer J, Waugh FR, McMurdo M, Mead GE (2013) Exercise for depression. Cochrane Database Syst Rev 9, CD004366

Creer DJ, Romberg C, Saksida LM, van Praag H, Bussey TJ (2010) Running enhances spatial pattern separation in mice. Proc Natl Acad Sci U S A 107:2367–2372

Crosnoe R, Leventhal T, Wirth RJ, Pierce KM, Pianta RC, NICHD Early Child Care Research Network (2010) Family socioeconomic status and consistent environmental stimulation in early childhood. Child Dev 81:972–987

Dickinson D, Tenhula W, Morris S, Brown C, Peer J, Spencer K, Li L, Gold JM, Bellack AS (2010) A randomized, controlled trial of computer-assisted cognitive remediation for schizophrenia. Am J Psychiatry 167:170–180

Dietrich MO, Mantese CE, Porciuncula LO, Ghisleni G, Vinade L, Souza DO, Portela LV (2005) Exercise affects glutamate receptors in postsynaptic densities from cortical mice brain. Brain Res 1065:20–25

Ding Q, Vaynman S, Akhavan M, Ying Z, Gomez-Pinilla F (2006) Insulin-like growth factor I interfaces with brain-derived neurotrophic factor-mediated synaptic plasticity to modulate aspects of exercise-induced cognitive function. Neuroscience 140:823–833

Ding Q, Ying Z, Gomez-Pinilla F (2011) Exercise influences hippocampal plasticity by modulating brain-derived neurotrophic factor processing. Neuroscience 192:773–780

Duman RS, Nakagawa S, Malberg J (2001) Regulation of adult neurogenesis by antidepressant treatment. Neuropsychopharmacology 25:836–844

Earhart GM (2013) Dynamic control of posture across locomotor tasks. Mov Disord 28:1501–1508

Ebersbach G, Ebersbach A, Gandor F, Wegner B, Wissel J, Kupsch A (2014) Impact of physical exercise on reaction time in patients with Parkinson's disease-data from the Berlin BIG Study. Arch Phys Med Rehabil 95:996–999

Ekstrand J, Hellsten J, Tingstrom A (2008) Environmental enrichment, exercise and corticosterone affect endothelial cell proliferation in adult rat hippocampus and prefrontal cortex. Neurosci Lett 442:203–207

Erickson KI, Prakash RS, Voss MW, Chaddock L, Hu L, Morris KS, White SM, Wojcicki TR, Mcauley E, Kramer AF (2009) Aerobic fitness is associated with hippocampal volume in elderly humans. Hippocampus 19:1030–1039

Erickson KI, Prakash RS, Voss MW, Chaddock L, Heo S, Mclaren M, Pence BD, Martin SA, Vieira VJ, Woods JA, Mcauley E, Kramer AF (2010) Brain-derived neurotrophic factor is associated with age-related decline in hippocampal volume. J Neurosci 30:5368–5375

Erickson KI, Voss MW, Prakash RS, Basak C, Szabo A, Chaddock L, Kim JS, Heo S, Alves H, White SM, Wojcicki TR, Mailey E, Vieira VJ, Martin SA, Pence BD, Woods JA, Mcauley E, Kramer AF (2011) Exercise training increases size of hippocampus and improves memory. Proc Natl Acad Sci U S A 108:3017–3022

Ernst C, Olson AK, Pinel JP, Lam RW, Christie BR (2006) Antidepressant effects of exercise: evidence for an adult-neurogenesis hypothesis? J Psychiatry Neurosci 31:84–92

Fabel K, Fabel K, Tam B, Kaufer D, Baiker A, Simmons N, Kuo CJ, Palmer TD (2003) VEGF is necessary for exercise-induced adult hippocampal neurogenesis. Eur J Neurosci 18:2803–2812

Faherty CJ, Raviie Shepherd K, Herasimtschuk A, Smeyne RJ (2005) Environmental enrichment in adulthood eliminates neuronal death in experimental Parkinsonism. Brain Res Mol Brain Res 134:170–179

Farina N, Rusted J, Tabet N (2014) The effect of exercise interventions on cognitive outcome in Alzheimer's disease: a systematic review. Int Psychogeriatr 26:9–18

Farmer J, Zhao X, van Praag H, Wodtke K, Gage FH, Christie BR (2004) Effects of voluntary exercise on synaptic plasticity and gene expression in the dentate gyrus of adult male Sprague-Dawley rats in vivo. Neuroscience 124:71–79

Ferrara N (2009) VEGF-A: a critical regulator of blood vessel growth. Eur Cytokine Netw 20:158–163

Ferris LT, Williams JS, Shen CL (2007) The effect of acute exercise on serum brain-derived neurotrophic factor levels and cognitive function. Med Sci Sports Exerc 39:728–734

Frazzitta G, Maestri R, Ghilardi MF, Riboldazzi G, Perini M, Bertotti G, Boveri N, Buttini S, Lombino FL, Uccellini D, Turla M, Pezzoli G, Comi C (2014) Intensive rehabilitation increases BDNF serum levels in parkinsonian patients: a randomized study. Neurorehabil Neural Repair 28:163–168

Fujimura H, Altar CA, Chen R, Nakamura T, Nakahashi T, Kambayashi J, Sun B, Tandon NN (2002) Brain-derived neurotrophic factor is stored in human platelets and released by agonist stimulation. Thromb Haemost 87:728–734

Gage FH, Temple S (2013) Neural stem cells: generating and regenerating the brain. Neuron 80:588–601

Galani R, Berthel MC, Lazarus C, Majchrzak M, Barbelivien A, Kelche C, Cassel JC (2007) The behavioral effects of enriched housing are not altered by serotonin depletion but enrichment alters hippocampal neurochemistry. Neurobiol Learn Mem 88:1–10

Garza AA, Ha TG, Garcia C, Chen MJ, Russo-Neustadt AA (2004) Exercise, antidepressant treatment, and BDNF mRNA expression in the aging brain. Pharmacol Biochem Behav 77:209–220

Geda YE, Roberts RO, Knopman DS, Christianson TJ, Pankratz VS, Ivnik RJ, Boeve BF, Tangalos EG, Petersen RC, Rocca WA (2010) Physical exercise, aging, and mild cognitive impairment: a population-based study. Arch Neurol 67:80–86

Gobbo OL, O'Mara SM (2004) Impact of enriched-environment housing on brain-derived neurotrophic factor and on cognitive performance after a transient global ischemia. Behav Brain Res 152:231–241

Goekint M, Heyman E, Roelands B, Njemini R, Bautmans I, Mets T, Meeusen R (2008) No influence of noradrenaline manipulation on acute exercise-induced increase of brain-derived neurotrophic factor. Med Sci Sports Exerc 40:1990–1996

Goekint M, Roelands B, De Pauw K, Knaepen K, Bos I, Meeusen R (2010) Does a period of detraining cause a decrease in serum brain-derived neurotrophic factor? Neurosci Lett 486:146–149

Gold SM, Schulz KH, Hartmann S, Mladek M, Lang UE, Hellweg R, Reer R, Braumann KM, Heesen C (2003) Basal serum levels and reactivity of nerve growth factor and brain-derived neurotrophic factor to standardized acute exercise in multiple sclerosis and controls. J Neuroimmunol 138:99–105

Gomez-Pinilla F, Hillman C (2013) The influence of exercise on cognitive abilities. Compr Physiol 3:403–428

Gomez-Pinilla F, Vaynman S, Ying Z (2008) Brain-derived neurotrophic factor functions as a metabotrophin to mediate the effects of exercise on cognition. Eur J Neurosci 28:2278–2287

Grego F, Vallier JM, Collardeau M, Bermon S, Ferrari P, Candito M, Bayer P, Magnie MN, Brisswalter J (2004) Effects of long duration exercise on cognitive function, blood glucose, and counterregulatory hormones in male cyclists. Neurosci Lett 364:76–80

Grego F, Vallier JM, Collardeau M, Rousseu C, Cremieux J, Brisswalter J (2005) Influence of exercise duration and hydration status on cognitive function during prolonged cycling exercise. Int J Sports Med 26:27–33

Gregoire CA, Bonenfant D, Le Nguyen A, Aumont A, Fernandes KJ (2014) Untangling the influences of voluntary running, environmental complexity, social housing and stress on adult hippocampal neurogenesis. PLoS One 9:e86237

Griffin EW, Bechara RG, Birch AM, Kelly AM (2009) Exercise enhances hippocampal-dependent learning in the rat: evidence for a BDNF-related mechanism. Hippocampus 19:973–980

Griffin EW, Mullally S, Foley C, Warmington SA, O'Mara SM, Kelly AM (2011) Aerobic exercise improves hippocampal function and increases BDNF in the serum of young adult males. Physiol Behav 104:934–941

Harburger LL, Lambert TJ, Frick KM (2007a) Age-dependent effects of environmental enrichment on spatial reference memory in male mice. Behav Brain Res 185:43–48

Harburger LL, Nzerem CK, Frick KM (2007b) Single enrichment variables differentially reduce age-related memory decline in female mice. Behav Neurosci 121:679–688

Heath GW, Parra DC, Sarmiento OL, Andersen LB, Owen N, Goenka S, Montes F, Brownson RC, Lancet Physical Activity Series Working Group (2012) Evidence-based intervention in physical activity: lessons from around the world. Lancet 380:272–281

Heyn P, Abreu BC, Ottenbacher KJ (2004) The effects of exercise training on elderly persons with cognitive impairment and dementia: a meta-analysis. Arch Phys Med Rehabil 85:1694–1704

Hindle JV, Petrelli A, Clare L, Kalbe E (2013) Nonpharmacological enhancement of cognitive function in Parkinson's disease: a systematic review. Mov Disord 28:1034–1049

Holwerda TJ, Deeg DJ, Beekman AT, van Tilburg TG, Stek ML, Jonker C, Schoevers RA (2014) Feelings of loneliness, but not social isolation, predict dementia onset: results from the Amsterdam Study of the Elderly (AMSTEL). J Neurol Neurosurg Psychiatry 85:135–142

Hopkins ME, Bucci DJ (2010a) BDNF expression in perirhinal cortex is associated with exercise-induced improvement in object recognition memory. Neurobiol Learn Mem 94:278–284

Hopkins ME, Bucci DJ (2010b) Interpreting the effects of exercise on fear conditioning: the influence of time of day. Behav Neurosci 124:868–872

Huang AM, Jen CJ, Chen HF, Yu L, Kuo YM, Chen HI (2006) Compulsive exercise acutely upregulates rat hippocampal brain-derived neurotrophic factor. J Neural Transm 113:803–811

Isaacs KR, Anderson BJ, Alcantara AA, Black JE, Greenough WT (1992) Exercise and the brain: angiogenesis in the adult rat cerebellum after vigorous physical activity and motor skill learning. J Cereb Blood Flow Metab 12:110–119

Jack CR Jr, Wiste HJ, Vemuri P, Weigand SD, Senjem ML, Zeng G, Bernstein MA, Gunter JL, Pankratz VS, Aisen PS, Weiner MW, Petersen RC, Shaw LM, Trojanowski JQ, Knopman DS, Alzheimer's Disease Neuroimaging Initiative (2010) Brain beta-amyloid measures and magnetic resonance imaging atrophy both predict time-to-progression from mild cognitive impairment to Alzheimer's disease. Brain 133:3336–3348

Kee N, Teixeira CM, Wang AH, Frankland PW (2007) Preferential incorporation of adult-generated granule cells into spatial memory networks in the dentate gyrus. Nat Neurosci 10:355–362

Keilhoff G, Fusar-Poli P, Becker A (2012) Effects of antipsychotics on dentate gyrus stem cell proliferation and survival in animal models: a critical update. Neural Plast 2012:832757

Kempermann G, Gage FH (1999) Experience-dependent regulation of adult hippocampal neurogenesis: effects of long-term stimulation and stimulus withdrawal. Hippocampus 9:321–332

Kennard JA, Woodruff-Pak DS (2012) A comparison of low- and high-impact forced exercise: effects of training paradigm on learning and memory. Physiol Behav 106:423–427

Kerr AL, Steuer EL, Pochtarev V, Swain RA (2010) Angiogenesis but not neurogenesis is critical for normal learning and memory acquisition. Neuroscience 171:214–226

Kobilo T, Liu QR, Gandhi K, Mughal M, Shaham Y, van Praag H (2011) Running is the neurogenic and neurotrophic stimulus in environmental enrichment. Learn Mem 18:605–609

Kramer AF, Erickson KI (2007a) Capitalizing on cortical plasticity: influence of physical activity on cognition and brain function. Trends Cogn Sci 11:342–348

Kramer AF, Erickson KI (2007b) Effects of physical activity on cognition, well-being, and brain: human interventions. Alzheimers Dement 3:S45–S51

Kronenberg G, Bick-Sander A, Bunk E, Wolf C, Ehninger D, Kempermann G (2006) Physical exercise prevents age-related decline in precursor cell activity in the mouse dentate gyrus. Neurobiol Aging 27:1505–1513

Kumar A, Rani A, Tchigranova O, Lee WH, Foster TC (2012) Influence of late-life exposure to environmental enrichment or exercise on hippocampal function and CA1 senescent physiology. Neurobiol Aging 33(828):e1–e17

Lambourne K, Tomporowski P (2010) The effect of exercise-induced arousal on cognitive task performance: a meta-regression analysis. Brain Res 1341:12–24

Larson EB, Wang L, Bowen JD, Mccormick WC, Teri L, Crane P, Kukull W (2006) Exercise is associated with reduced risk for incident dementia among persons 65 years of age and older. Ann Intern Med 144:73–81

Laurin D, Verreault R, Lindsay J, Macpherson K, Rockwood K (2001) Physical activity and risk of cognitive impairment and dementia in elderly persons. Arch Neurol 58:498–504

Laursen TM, Munk-Olsen T, Vestergaard M (2012) Life expectancy and cardiovascular mortality in persons with schizophrenia. Curr Opin Psychiatry 25:83–88

Lautenschlager NT, Cox KL, Flicker L, Foster JK, van Bockxmeer FM, Xiao J, Greenop KR, Almeida OP (2008) Effect of physical activity on cognitive function in older adults at risk for Alzheimer disease: a randomized trial. JAMA 300:1027–1037

Laviola G, Hannan AJ, Macri S, Solinas M, Jaber M (2008) Effects of enriched environment on animal models of neurodegenerative diseases and psychiatric disorders. Neurobiol Dis 31:159–168

Lazarov O, Mattson MP, Peterson DA, Pimplikar SW, van Praag H (2010) When neurogenesis encounters aging and disease. Trends Neurosci 33:569–579

Leasure JL, Jones M (2008) Forced and voluntary exercise differentially affect brain and behavior. Neuroscience 156:456–465

Lee IM, Shiroma EJ, Lobelo F, Puska P, Blair SN, Katzmarzyk PT, Lancet Physical Activity Series Working Group (2012) Effect of physical inactivity on major non-communicable diseases worldwide: an analysis of burden of disease and life expectancy. Lancet 380:219–229

Lessmann V (1998) Neurotrophin-dependent modulation of glutamatergic synaptic transmission in the mammalian CNS. Gen Pharmacol 31:667–674

Lista I, Sorrentino G (2010) Biological mechanisms of physical activity in preventing cognitive decline. Cell Mol Neurobiol 30:493–503

Mahncke HW, Connor BB, Appelman J, Ahsanuzddin ON, Hardy JL, Wood RA, Joyce NM, Boniske T, Atkins SM, Merzenich MM (2006) Memory enhancement in healthy older adults using a brain plasticity-based training program: a randomized, controlled study. Proc Natl Acad Sci U S A 103:12523–12528

Mammen G, Faulkner G (2013) Physical activity and the prevention of depression: a systematic review of prospective studies. Am J Prev Med 45:649–657

Mandolesi L, De Bartolo P, Foti F, Gelfo F, Federico F, Leggio MG, Petrosini L (2008) Environmental enrichment provides a cognitive reserve to be spent in the case of brain lesion. J Alzheimers Dis 15:11–28

Marais L, Stein DJ, Daniels WM (2009) Exercise increases BDNF levels in the striatum and decreases depressive-like behavior in chronically stressed rats. Metab Brain Dis 24:587–597

Matthews VB, Astrom MB, Chan MH, Bruce CR, Krabbe KS, Prelovsek O, Akerstrom T, Yfanti C, Broholm C, Mortensen OH, Penkowa M, Hojman P, Zankari A, Watt MJ, Bruunsgaard H, Pedersen BK, Febbraio MA (2009) Brain-derived neurotrophic factor is produced by skeletal muscle cells in response to contraction and enhances fat oxidation via activation of AMP-activated protein kinase. Diabetologia 52:1409–1418

Ming GL, Song H (2011) Adult neurogenesis in the mammalian brain: significant answers and significant questions. Neuron 70:687–702

Molteni R, Ying Z, Gomez-Pinilla F (2002) Differential effects of acute and chronic exercise on plasticity-related genes in the rat hippocampus revealed by microarray. Eur J Neurosci 16:1107–1116

Mora F, Segovia G, del Arco A (2007) Aging, plasticity and environmental enrichment: structural changes and neurotransmitter dynamics in several areas of the brain. Brain Res Rev 55:78–88

Naismith SL, Mowszowski L, Diamond K, Lewis SJ (2013) Improving memory in Parkinson's disease: a healthy brain ageing cognitive training program. Mov Disord 28:1097–1103

Nakahashi T, Fujimura H, Altar CA, Li J, Kambayashi J, Tandon NN, Sun B (2000) Vascular endothelial cells synthesize and secrete brain-derived neurotrophic factor. FEBS Lett 470:113–117

Neeper SA, Gomez-Pinilla F, Choi J, Cotman CW (1996) Physical activity increases mRNA for brain-derived neurotrophic factor and nerve growth factor in rat brain. Brain Res 726:49–56

NICE Guidelines (2009) http://www.nice.org.uk/guidance/cg90/evidence

Nithianantharajah J, Hannan AJ (2009) The neurobiology of brain and cognitive reserve: mental and physical activity as modulators of brain disorders. Prog Neurobiol 89:369–382

Nithianantharajah J, Hannan AJ (2011) Mechanisms mediating brain and cognitive reserve: experience-dependent neuroprotection and functional compensation in animal models of neurodegenerative diseases. Prog Neuropsychopharmacol Biol Psychiatry 35:331–339

Nouchi R, Taki Y, Takeuchi H, Hashizume H, Akitsuki Y, Shigemune Y, Sekiguchi A, Kotozaki Y, Tsukiura T, Yomogida Y, Kawashima R (2012) Brain training game improves executive functions and processing speed in the elderly: a randomized controlled trial. PLoS One 7:e29676

O'Callaghan RM, Ohle R, Kelly AM (2007) The effects of forced exercise on hippocampal plasticity in the rat: a comparison of LTP, spatial- and non-spatial learning. Behav Brain Res 176:362–366

O'Callaghan RM, Griffin EW, Kelly AM (2009) Long-term treadmill exposure protects against age-related neurodegenerative change in the rat hippocampus. Hippocampus 19:1019–1029

Oertel-Knochel V, Mehler P, Thiel C, Steinbrecher K, Malchow B, Tesky V, Ademmer K, Prvulovic D, Banzer W, Zopf Y, Schmitt A, Hänsel F (2014) Effects of aerobic exercise on cognitive performance and individual psychopathology in depressive and schizophrenia patients. Eur Arch Psychiatry Clin Neurosci 264(7):589–604

Olafsdottir S, Berg C, Eiben G, Lanfer A, Reisch L, Ahrens W, Kourides Y, Molnar D, Moreno LA, Siani A, Veidebaum T, Lissner L (2014) Young children's screen activities, sweet drink consumption and anthropometry: results from a prospective European study. Eur J Clin Nutr 68:223–228

Olson AK, Eadie BD, Ernst C, Christie BR (2006) Environmental enrichment and voluntary exercise massively increase neurogenesis in the adult hippocampus via dissociable pathways. Hippocampus 16:250–260

Ota KT, Duman RS (2013) Environmental and pharmacological modulations of cellular plasticity: role in the pathophysiology and treatment of depression. Neurobiol Dis 57:28–37

Ozbulut O, Genc A, Bagcioglu E, Coskun KS, Acar T, Alkoc OA, Karabacak H, Sener U, Ucok K (2013) Evaluation of physical fitness parameters in patients with schizophrenia. Psychiatry Res 210:806–811

Pajonk FG, Wobrock T, Gruber O, Scherk H, Berner D, Kaizl I, Kierer A, Muller S, Oest M, Meyer T, Backens M, Schneider-Axmann T, Thornton AE, Honer WG, Falkai P (2010) Hippocampal plasticity in response to exercise in schizophrenia. Arch Gen Psychiatry 67:133–143

Pang TY, Hannan AJ (2013) Enhancement of cognitive function in models of brain disease through environmental enrichment and physical activity. Neuropharmacology 64:515–528

Pereira AC, Huddleston DE, Brickman AM, Sosunov AA, Hen R, McKhann GM, Sloan R, Gage FH, Brown TR, Small SA (2007) An in vivo correlate of exercise-induced neurogenesis in the adult dentate gyrus. Proc Natl Acad Sci U S A 104:5638–5643

Petzinger GM, Fisher BE, Mcewen S, Beeler JA, Walsh JP, Jakowec MW (2013) Exercise-enhanced neuroplasticity targeting motor and cognitive circuitry in Parkinson's disease. Lancet Neurol 12:716–726

Pontifex MB, Hillman CH, Fernhall B, Thompson KM, Valentini TA (2009) The effect of acute aerobic and resistance exercise on working memory. Med Sci Sports Exerc 41:927–934

Pothakos K, Kurz MJ, Lau YS (2009) Restorative effect of endurance exercise on behavioral deficits in the chronic mouse model of Parkinson's disease with severe neurodegeneration. BMC Neurosci 10:6

Querido JS, Sheel AW (2007) Regulation of cerebral blood flow during exercise. Sports Med 37:765–782

Quirie A, Hervieu M, Garnier P, Demougeot C, Mossiat C, Bertrand N, Martin A, Marie C, Prigent-Tessier A (2012) Comparative effect of treadmill exercise on mature BDNF production in control versus stroke rats. PLoS One 7:e44218

Rasmussen P, Brassard P, Adser H, Pedersen MV, Leick L, Hart E, Secher NH, Pedersen BK, Pilegaard H (2009) Evidence for a release of brain-derived neurotrophic factor from the brain during exercise. Exp Physiol 94:1062–1069

Raz N, Lindenberger U, Rodrigue KM, Kennedy KM, Head D, Williamson A, Dahle C, Gerstorf D, Acker JD (2005) Regional brain changes in aging healthy adults: general trends, individual differences and modifiers. Cereb Cortex 15:1676–1689

Reif A, Fritzen S, Finger M, Strobel A, Lauer M, Schmitt A, Lesch KP (2006) Neural stem cell proliferation is decreased in schizophrenia, but not in depression. Mol Psychiatry 11:514–522

Robert PH, Berr C, Volteau M, Bertogliati-Fileau C, Benoit M, Guerin O, Sarazin M, Legrain S, Dubois B, PréAL Study Group (2008) Importance of lack of interest in patients with mild cognitive impairment. Am J Geriatr Psychiatry 16:770–776

Rojas Vega S, Struder HK, Vera Wahrmann B, Schmidt A, Bloch W, Hollmann W (2006) Acute BDNF and cortisol response to low intensity exercise and following ramp incremental exercise to exhaustion in humans. Brain Res 1121:59–65

Rosenzweig MR, Bennett EL (1996) Psychobiology of plasticity: effects of training and experience on brain and behavior. Behav Brain Res 78:57–65

Rovio S, Kareholt I, Helkala EL, Viitanen M, Winblad B, Tuomilehto J, Soininen H, Nissinen A, Kivipelto M (2005) Leisure-time physical activity at midlife and the risk of dementia and Alzheimer's disease. Lancet Neurol 4:705–711

Russo-Neustadt AA, Chen MJ (2005) Brain-derived neurotrophic factor and antidepressant activity. Curr Pharm Des 11:1495–1510

Russo-Neustadt A, Ha T, Ramirez R, Kesslak JP (2001) Physical activity-antidepressant treatment combination: impact on brain-derived neurotrophic factor and behavior in an animal model. Behav Brain Res 120:87–95

Schiffer T, Schulte S, Hollmann W, Bloch W, Struder HK (2009) Effects of strength and endurance training on brain-derived neurotrophic factor and insulin-like growth factor 1 in humans. Horm Metab Res 41:250–254

Schmidt-Kassow M, Deusser M, Thiel C, Otterbein S, Montag C, Reuter M, Banzer W, Kaiser J (2013) Physical exercise during encoding improves vocabulary learning in young female adults: a neuroendocrinological study. PLoS One 8:e64172

Schmiedek F, Lovden M, Lindenberger U (2010) Hundred days of cognitive training enhance broad cognitive abilities in adulthood: findings from the COGITO study. Front Aging Neurosci 2:27

Seifert T, Brassard P, Wissenberg M, Rasmussen P, Nordby P, Stallknecht B, Adser H, Jakobsen AH, Pilegaard H, Nielsen HB, Secher NH (2010) Endurance training enhances BDNF release from the human brain. Am J Physiol Regul Integr Comp Physiol 298:R372–R377

Shankar A, Hamer M, McMunn A, Steptoe A (2013) Social isolation and loneliness: relationships with cognitive function during 4 years of follow-up in the English longitudinal study of ageing. Psychosom Med 75:161–170

Sharma HS, Cervos-Navarro J, Dey PK (1991) Increased blood-brain barrier permeability following acute short-term swimming exercise in conscious normotensive young rats. Neurosci Res 10:211–221

Sigwalt AR, Budde H, Helmich I, Glaser V, Ghisoni K, Lanza S, Cadore EL, Lhullier FL, de Bem AF, Hohl A, de Matos FJ, de Oliveira PA, Prediger RD, Guglielmo LG, Latini A (2011) Molecular aspects involved in swimming exercise training reducing anhedonia in a rat model of depression. Neuroscience 192:661–674

Smith GE, Housen P, Yaffe K, Ruff R, Kennison RF, Mahncke HW, Zelinski EM (2009) A cognitive training program based on principles of brain plasticity: results from the Improvement in Memory with Plasticity-based Adaptive Cognitive Training (IMPACT) study. J Am Geriatr Soc 57:594–603

Song MR, Lee YS, Baek JD, Miller M (2012) Physical activity status in adults with depression in the National Health and Nutrition Examination Survey, 2005–2006. Public Health Nurs 29:208–217

Soya H, Nakamura T, Deocaris CC, Kimpara A, Iimura M, Fujikawa T, Chang H, Mcewen BS, Nishijima T (2007) BDNF induction with mild exercise in the rat hippocampus. Biochem Biophys Res Commun 358:961–967

Spalding KL, Bergmann O, Alkass K, Bernard S, Salehpour M, Huttner HB, Bostrom E, Westerlund I, Vial C, Buchholz BA, Possnert G, Mash DC, Druid H, Frisen J (2013) Dynamics of hippocampal neurogenesis in adult humans. Cell 153:1219–1227

Speisman RB, Kumar A, Rani A, Pastoriza JM, Severance JE, Foster TC, Ormerod BK (2013) Environmental enrichment restores neurogenesis and rapid acquisition in aged rats. Neurobiol Aging 34:263–274

Stranahan AM, Khalil D, Gould E (2007) Running induces widespread structural alterations in the hippocampus and entorhinal cortex. Hippocampus 17:1017–1022

Swain RA, Harris AB, Wiener EC, Dutka MV, Morris HD, Theien BE, Konda S, Engberg K, Lauterbur PC, Greenough WT (2003) Prolonged exercise induces angiogenesis and increases cerebral blood volume in primary motor cortex of the rat. Neuroscience 117:1037–1046

Takahashi H, Sassa T, Shibuya T, Kato M, Koeda M, Murai T, Matsuura M, Asai K, Suhara T, Okubo Y (2012) Effects of sports participation on psychiatric symptoms and brain activations during sports observation in schizophrenia. Transl Psychiatry 2:e96

Tang SW, Chu E, Hui T, Helmeste D, Law C (2008) Influence of exercise on serum brain-derived neurotrophic factor concentrations in healthy human subjects. Neurosci Lett 431:62–65

Tang K, Xia FC, Wagner PD, Breen EC (2010) Exercise-induced VEGF transcriptional activation in brain, lung and skeletal muscle. Respir Physiol Neurobiol 170:16–22

Tardif S, Simard M (2011) Cognitive stimulation programs in healthy elderly: a review. Int J Alzheimers Dis 2011:378934

Tomporowski PD (2003) Effects of acute bouts of exercise on cognition. Acta Psychol (Amst) 112:297–324

Tomporowski PD, Lambourne K, Okumura MS (2011) Physical activity interventions and children's mental function: an introduction and overview. Prev Med 52(Suppl 1):S3–S9

Toscano-Silva M, da Silva Gomes S, Scorza FA, Bonvent JJ, Cavalheiro EA, Arida RM (2010) Hippocampal mossy fiber sprouting induced by forced and voluntary physical exercise. Physiol Behav 101:302–308

Trejo JL, Llorens-Martin MV, Torres-Aleman I (2008) The effects of exercise on spatial learning and anxiety-like behavior are mediated by an IGF-I-dependent mechanism related to hippocampal neurogenesis. Mol Cell Neurosci 37:402–411

Tuon T, Valvassori SS, Lopes-Borges J, Luciano T, Trom CB, Silva LA, Quevedo J, Souza CT, Lira FS, Pinho RA (2012) Physical training exerts neuroprotective effects in the regulation of neurochemical factors in an animal model of Parkinson's disease. Neuroscience 227:305–312

Uc EY, Doerschug KC, Magnotta V, Dawson JD, Thomsen TR, Kline JN, Rizzo M, Newman SR, Mehta S, Grabowski TJ, Bruss J, Blanchette DR, Anderson SW, Voss MW, Kramer AF, Darling WG (2014) Phase I/II randomized trial of aerobic exercise in Parkinson disease in a community setting. Neurology 83(5):413–425

Valero J, Espana J, Parra-Damas A, Martin E, Rodriguez-Alvarez J, Saura CA (2011) Short-term environmental enrichment rescues adult neurogenesis and memory deficits in APP(Sw, Ind) transgenic mice. PLoS One 6:e16832

Van der Borght K, Kobor-Nyakas DE, Klauke K, Eggen BJ, Nyakas C, Van der Zee EA, Meerlo P (2009) Physical exercise leads to rapid adaptations in hippocampal vasculature: temporal dynamics and relationship to cell proliferation and neurogenesis. Hippocampus 19:928–936

van Praag H (2009) Exercise and the brain: something to chew on. Trends Neurosci 32:283–290

van Praag H, Christie BR, Sejnowski TJ, Gage FH (1999a) Running enhances neurogenesis, learning, and long-term potentiation in mice. Proc Natl Acad Sci U S A 96:13427–13431

van Praag H, Kempermann G, Gage FH (1999b) Running increases cell proliferation and neurogenesis in the adult mouse dentate gyrus. Nat Neurosci 2:266–270

van Praag H, Kempermann G, Gage FH (2000) Neural consequences of environmental enrichment. Nat Rev Neurosci 1:191–198 van Praag H, Schinder AF, Christie BR, Toni N, Palmer TD, Gage FH (2002) Functional neurogenesis in the adult hippocampus. Nature 415:1030–1034

van Praag H, Shubert T, Zhao C, Gage FH (2005) Exercise enhances learning and hippocampal neurogenesis in aged mice. J Neurosci 25:8680–8685

Vancampfort D, Probst M, De Hert M, Soundy A, Stubbs B, Stroobants M, De Herdt A (2014) Neurobiological effects of physical exercise in schizophrenia: a systematic review. Disabil Rehabil 36(21):1749–1754

Vaynman S, Gomez-Pinilla F (2006) Revenge of the "sit": how lifestyle impacts neuronal and cognitive health through molecular systems that interface energy metabolism with neuronal plasticity. J Neurosci Res 84:699–715

Vaynman S, Ying Z, Gomez-Pinilla F (2004a) Exercise induces BDNF and synapsin I to specific hippocampal subfields. J Neurosci Res 76:356–362

Vaynman S, Ying Z, Gomez-Pinilla F (2004b) Hippocampal BDNF mediates the efficacy of exercise on synaptic plasticity and cognition. Eur J Neurosci 20:2580–2590

Verghese J, Lipton RB, Katz MJ, Hall CB, Derby CA, Kuslansky G, Ambrose AF, Sliwinski M, Buschke H (2003) Leisure activities and the risk of dementia in the elderly. N Engl J Med 348:2508–2516

Voss MW, Vivar C, Kramer AF, van Praag H (2013) Bridging animal and human models of exercise-induced brain plasticity. Trends Cogn Sci 17:525–544

Watson P, Black KE, Clark SC, Maughan RJ (2006) Exercise in the heat: effect of fluid ingestion on blood-brain barrier permeability. Med Sci Sports Exerc 38:2118–2124

Wilson RS, Barnes LL, de Leon Mendes CF, Aggarwal NT, Schneider JS, Bach J, Pilat J, Beckett LA, Arnold SE, Evans DA, Bennett DA (2002a) Depressive symptoms, cognitive decline, and risk of AD in older persons. Neurology 59:364–370

Wilson RS, Bennett DA, Bienias JL, Aggarwal NT, de Leon Mendes CF, Morris MC, Schneider JA, Evans DA (2002b) Cognitive activity and incident AD in a population-based sample of older persons. Neurology 59:1910–1914

Wilson RS, Mendes De Leon CF, Barnes LL, Schneider JA, Bienias JL, Evans DA, Bennett DA (2002c) Participation in cognitively stimulating activities and risk of incident Alzheimer disease. JAMA 287:742–748

Winter B, Breitenstein C, Mooren FC, Voelker K, Fobker M, Lechtermann A, Krueger K, Fromme A, Korsukewitz C, Floel A, Knecht S (2007) High impact running improves learning. Neurobiol Learn Mem 87:597–609

Wolf SA, Melnik A, Kempermann G (2011) Physical exercise increases adult neurogenesis and telomerase activity, and improves behavioral deficits in a mouse model of schizophrenia. Brain Behav Immun 25:971–980

Wu CW, Chang YT, Yu L, Chen HI, Jen CJ, Wu SY, Lo CP, Kuo YM (2008) Exercise enhances the proliferation of neural stem cells and neurite growth and survival of neuronal progenitor cells in dentate gyrus of middle-aged mice. J Appl Physiol (1985) 105:1585–1594

Optogenetics and Deep Brain Stimulation Neurotechnologies

Krishnakanth Kondabolu, Marek Mateusz Kowalski, Erik Andrew Roberts, and Xue Han

Contents

Abstract

Brain neural network is composed of densely packed, intricately wired neurons whose activity patterns ultimately give rise to every behavior, thought, or emotion that we experience. Over the past decade, a novel neurotechnique, optogenetics that combines light and genetic methods to control or monitor neural activity patterns, has proven to be revolutionary in understanding the functional role of specific neural circuits. We here briefly describe recent advance in optogenetics and compare optogenetics with deep brain stimulation technology that holds the promise for treating many neurological and psychiatric disorders.

Keywords

Optogenetics • Rhodopsin • Channelrhodopsin • Archaerhodopsin • Deep brain stimulation

K. Kondabolu • M.M. Kowalski • E.A. Roberts • X. Han (✉)
Biomedical Engineering Department, Boston University, Boston, MA, USA
e-mail: xuehan@bu.edu

K.M. Kantak, J.G. Wettstein (eds.), *Cognitive Enhancement*, Handbook of Experimental Pharmacology 228, DOI 10.1007/978-3-319-16522-6_15

1 Optogenetics

Optogenetics combines light and genetic methods to control or monitor cellular activities. For rhodopsin based optogenetic control techniques, light-sensitive rhodopsin molecules were genetically introduced into otherwise not-light-sensitive neurons. Upon light illumination, genetically modified neurons that express rhodopsins can then be precisely controlled. Three major classes of rhodopsins, all microbial rhodopsins, have been developed as optogenetic molecular sensors, channelrhodopsins, halorhodopsins, and archaerhodopsins (Fig. 1) (Han 2012a). Because of their small sizes, these rhodopsins can be easily expressed in neurons, and thus, optogenetics has been successfully applied in almost all experiment neural systems, from *Caenorhabditis elegans*, rodents, to nonhuman primates, as well as human retina. With the ability to rapidly and reversibly activate or silence genetically transduced cells, optogenetics has enabled the examination of the causal role of specific cells in neural computation, behavior, and brain disorders. A number of recent reviews and books have summarized various aspects of the current state of this field (Bernstein and Boyden 2011; Miesenbock 2011; Yizhar et al. 2011; Zhang et al. 2011; Chow et al. 2012; Han 2012a, b; Knopfel and Boyden 2012). In parallel, a new generation of genetically encoded calcium/activity optogenetic sensors are being improved, with which neural activity patterns can now be monitored with high spatiotemporal resolution (i.e. Chen et al. 2013). We here will focus our discussion on optogenetic control technologies that are rhodopsin based, and will not further discuss other calcium/activity sensors.

1.1 Rhodopsin Based Optogenetic Sensors

Microbial rhodopsins are photoactive proteins with seven transmembrane domains. They are widely spread in archaea, bacteria, algae, and fungi, where they are critical for light-sensing or photosynthetic functions. Each rhodopsin molecule consists of a protein domain, opsin that binds to the photoactive cofactor all-*trans*-retinal, and thus, rhodopsin refers to the combination of the opsin protein and the bound retinal (Spudich et al. 2000; Spudich 2006). Light-induced photoisomerization of all-*trans*-retinal to 13-*cis*-retinal leads to opsin protein conformational changes that result in direct ion conductance across the membrane. Rhodopsins have been studied since 1970s (Oesterhelt and Stoeckenius 1971; 1973), but they were only recently adopted as optogenetic sensors.

Channelrhodopsin-2 (ChR-2), cloned from green algae *Chlamydomonas reinhardtii*, is the first optogenetic sensor adapted to activate neurons (Boyden et al. 2005). Light-induced photoisomerization of all-*trans*-retinal results in protein conformational changes that lead to a passive conductance to both monovalent and divalent cations such as Na^+, K^+, H^+, and Ca^{2+}. The duration of ion flow is determined by the subsequent ChR2 conformation changes that led to channel closure (Nagel et al. 2003). Because ion flow is independent of photon absorption, engineered ChR2 mutants that alter the process of light-induced protein

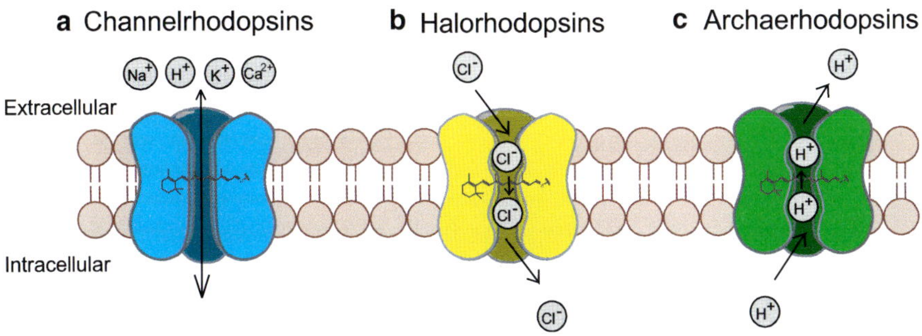

Fig. 1 Optogenetic molecular sensors. Upon light illumination, channelrhodopsins passively transport Na^+, K^+, H^+, Ca^{2+} down their electrochemical gradients to depolarize neurons (**a**); halorhodopsins actively pump Cl^- into the cell to hyperpolarize neurons (**b**); archaerhodopsins actively pump H^+ out of the cell to hyperpolarize neurons (**c**) (Han 2012a)

conformational changes have led to a number of variants that operate on varying timescales from milliseconds to minutes, and present varying permeability to different ions (Bamann et al. 2010; Gunaydin et al. 2010; Berndt et al. 2011, 2014; Wietek et al. 2014).

Two classes of light-activated ion pumps have been used to silence neurons, halorhodopsins, and Archaerhodopsins. Halorhodopsins, such as that from *Natronomonas pharaonis* (Halo, NpHR), are light-activated inward chloride pumps (Han and Boyden 2007; Zhang et al. 2007). Archaerhodopsin, such as that from *Halorubrum sodomense* (Arch), are light-activated outward proton pumps (Chow et al. 2010; Han et al. 2011). These light-activated pumps lead to a net outward current flow in neurons, thereby silencing neural activities. For Halo and Arch, photonic energy is directly coupled to ion transport, and thus photo current depends upon continuous light illumination.

Much effort has been directed to enhance the efficiency, temporal precision, and spectrum properties of these rhodopsin molecules. While it remains to be established whether these optogenetic molecules from archeabacteria or algea produce any side effects in neurons, these molecules are widely used and have so far proven safe and effective in most neural systems.

1.2 Target Rhodopsin Expression Though Genetic Modification

A major advantage of optogenetic technologies over other brain stimulation technologies is the ability to control specific cells with distinct genetic markers. Such specificity is achieved by expressing rhodopsins in desired cell populations through genetic modification. Because of the intrinsic difficulty in transducing neurons, genetic modification of neurons is mainly limited to whole animal transgenic approaches and viral based gene delivery approaches (Han 2012a, b).

Transgenic mice represent a versatile and powerful platform to target a variety of distinct cells of interest, in particular in conjunction with the phage-derived Cre-LoxP recombination technology (Sauer and Henderson 1988; Tsien et al. 1996). Cre recombinase selectively catalyzes the recombination between a pair of LoxP recognition DNA sequences. Through strategic placement of LoxP sequences, a specific gene, such as rhodopsins, can be expressed only in the presence of Cre enzymes. A large number of Cre transgenic mice are available with targeted Cre expression in specific cells. Upon injection of a virus that mediates Cre-dependent expression of rhodopsin molecules, one can selectively express rhodopsins only in cells that also express Cre. Alternatively, Cre transgenic mice can be crossed with transgenic mice with Cre-dependent rhodopsin expressions (Madisen et al. 2012).

In genetically intractable species, viruses remain the most effective methods to transduce brain cells. Over the years, viral based gene delivery methods have been well established and are widely used in basic research and in human gene therapy clinical trials (Waehler et al. 2007; Han 2012a, b). The most commonly used viral vectors, lentivirus and adeno-associated virus (AAV), have been engineered to exhibit little or no toxicity, with excellent transduction efficiency. However, two major limitations remain for viral vectors. First, the packaging ability of a virus is limited, which cannot be easily overcome due to the intrinsic stability of viral particles. Second, different viruses display distinct tropism, likely because specific membrane receptors are required for viral entry into target cells. As a result, it remains difficult to target specific cells with virus, which has presented a major challenge in realizing the full potential of optogenetics in genetically intractable species.

A number of non-viral methods have been developed for gene delivery, i.e. using cationic lipids, cationic polymers, nanoparticles, carbon nanotubes, gene guns, or calcium phosphate (Luo and Saltzman 2000). Although these methods exhibit excellent transduction efficiency in a number of cells, they have largely failed to transduce neurons effectively. A potential advantage of non-viral-based gene delivery method is the ability to introduce large pieces of DNAs into cells, and thus may enable improved targeting to specific cells. However, further optimization of non-viral gene delivery methods is necessary for the use of these methods in the brain.

1.3 Light Illumination of Cells Expressing Rhodopsins

Optogenetic control of neural activities critically relies on the amount of light reaching a neuron, the number of rhodopsin molecules present on the plasma membrane, and the light sensitivity of the rhodopsins. The effectiveness of optogenetic control may also be influenced by the intrinsic neuronal membrane biophysical properties and the surrounding neural network environment—factors that cannot be controlled by experimenters. Having discussed the genetic modification methods that control the expression level of rhodopsin on the plasma membrane, and the molecular properties that dictate a rhodopsin's light sensitivity, we here describe the consideration of light illumination.

Rhodopsins typically operate at visible wavelength light (450–650 nm) that are also highly absorbed by blood hemoglobin. Monte Carlo simulations (Mobley and Vo-Dinh 2003), along with experimental evidence, demonstrated that tissue penetration by visible light drops sharply to less than 10 % within the first few hundred microns (Bernstein et al. 2008; Chow et al. 2010). To circumvent this, rhodopsins with red- and far-red light sensitivity (>650 nm) have been developed to allow for more efficient illumination of brain tissue, thereby improving stimulation volume and reducing possible risk of heat-induced tissue damage associated with high-intensity light illumination (Zhao et al. 2008; Chuong et al. 2014).

A variety of light sources with decent light power are well suited for illuminating neurons expressing rhodopsins. For example, lasers and LEDs that are low cost and easy to handle can provide excellent light illumination for in vivo optogenetic experiments. When coupled with fiberoptics, they allow the delivery of light with a narrow wavelength and high spatiotemporal resolution. The use of thin fibers or fiber arrays is advantageous in reducing mechanical tissue damage (Bernstein et al. 2012). Light-induced tissue heating may alter tissue integrity, cell metabolism, and neuronal excitability (Wells et al. 2005), and thus, the amount of light delivered into brain tissue needs to be properly evaluated and controlled during an optogenetic experiment. Another consideration when using optogenetics in conjunction with metal recording electrode is laser-induced electrical artifact due to photoelectric effects (Han et al. 2009, 2011). Development of novel electrode materials may overcome some of these photoelectric problems (Zorzos et al. 2009).

1.4 Application of Optogenetics

Optogenetics has been used in experimental organisms from *C. elegans* and zebrafish to mice and primates to analyze neural circuits relevant for many behaviors, from motor behavior (Cavanaugh et al. 2012) to learning and memory (Liu et al. 2012). With the proof of principle demonstration that optogenetics can be safely performed in nonhuman primates (Han et al. 2009), optogenetics has been explored for its translational potential in treating blindness (Doroudchi et al. 2011).

2 Deep Brain Stimulation

Deep brain stimulation (DBS) represents a revolutionary brain region-specific neuromodulation therapy that first gained FDA approval in 1997 for treating essential tremor and Parkinson's disease (PD) tremor, and later in 2002 for PD and in 2003 for dystonia. Though invasive, DBS has been proven to be highly effective, and is now actively explored as therapies for a number of brain disorders.

Because of the greater understanding of DBS, we will focus the following discussion on DBS, and compare DBS to optogenetics.

A set of other noninvasive electrical brain stimulation technologies have been historically applied to treat neurological and psychiatric disorders, such as transcranial magnetic stimulation (TMS), transcranial direct current stimulation (tDCS), and transcranial alternating current stimulation (tACS). These noninvasive technologies stimulate a large and often distributed neural network with little spatial resolution, and thus it has been difficult to pinpoint their action mechanisms. However, the noninvasive nature of these technologies attracts much enthusiasm on their therapeutic potentials, and there is much effort on improving the spatial resolution of these tools. Recently, TMS gained FDA approval in the United States for treating migraine in 2013.

2.1 Discovery of DBS Therapy

The use of electrical stimulation can be traced back to Fritsch and Hitzip in the 1870s, and later in the 1940s, Penfield systematically stimulated different parts of the human brain and established the map of human motor and sensory cortices that remain instrumental in our understanding of the functional organization of the brain. With the advance of stereotaxic surgeries and the establishment of Parkinson's disease animal models, much research and clinical effort have finally led to the FDA approval of DBS for treating PD, in 1997 for stimulating thalamus and in 2002 for stimulating STN and GPi.

Current DBS electrode designs consist of four contacts that are 0.5 mm or 1.5 mm apart. Electrical currents are controlled via an integrated pulse generator that can stimulate electrode contacts at certain polarity, amplitude, pulse width, and frequencies. A specific set of stimulation parameters is determined through trial and error for each patient to achieve optimal clinical efficacy with minimal side effects (Volkmann et al. 2006). While DBS has been remarkable in treating several key motor symptoms presented in PD patients, such as bradykinesia, akinesia, and tremor (Anderson et al. 2005; Rodriguez-Oroz et al. 2005), it is often associated with other side effects, such as mood disorders, depression, and impulsivity (Uc and Follett 2007).

2.2 Therapeutic Mechanisms of DBS for PD

The therapeutic mechanisms of DBS for PD remain largely unclear. There are several promising hypothesis. DBS may achieve its therapeutic effects through inhibiting the brain structures being stimulated. This hypothesis is largely based on the understanding that the brain regions targeted by DBS, i.e., STN, thalamus, and GPi, when surgically lesioned, are equally effective in treating PD. This hypothesis is further supported by the observation that the firing rates of STN neurons decreased drastically upon local STN DBS (Dostrovsky et al. 2000; Welter et al. 2004; Foffani et al. 2006). However, electrical stimulation also stimulates the fibers of passage, as well as neurons that project to the site of stimulation antidromically. Thus, while DBS may inhibit local brain structures under stimulation, its effects likely extend to other brain regions that connect to the site of stimulation or have fibers bypassing the stimulation sites.

A second hypothesis is that DBS may reduce pathological oscillations. Implantation of DBS electrodes provides a unique opportunity for recording neural activities from PD brains. Much evidence has suggested the presence of exaggerated oscillations in the cortical–basal ganglion circuit at beta frequencies (oscillations around 20 Hz). Exaggerated beta oscillations closely parallel the key PD motor deficits, bradykinesia, rigidity, akinesia, and tremor (Levy et al. 2000, 2002b; Brown et al. 2001; Boraud et al. 2005; Weinberger et al. 2009a, b), and are largely suppressed by effective dopamine replacement treatment (Brown et al. 2001; Levy et al. 2002a; Williams et al. 2002; Priori et al. 2004; Silberstein et al. 2005) and DBS (Wingeier et al. 2006; Kuhn et al. 2008; Kuhn et al. 2009; Lehmkuhle et al. 2009). It has thus been suggested that DBS therapeutic effect is through reducing beta oscillations. However, it remains unknown whether the exaggerated beta oscillation is a cause or a correlate of motor deficits, and where and how beta oscillations arise in PD.

Finally, it has also been hypothesized that placement of DBS electrodes could recruit glia-related neurotransmission to inhibit neuron activities at the target brain structures (Bekar et al. 2008). While the surgical placement of DBS electrodes presents serious risks intrinsic to any surgery, interestingly electrode implantation within STN may be neuroprotective and could slow down dopamine neuron degeneration in SNpc, as demonstrated in MPTP monkey PD models (Doroudchi et al. 2011). With the development of optogenetics, researchers are now able to start to investigate the specific neural circuit mechanisms underlying DBS therapeutic actions.

3 Optogenetics and DBS

The amazing efficacy of DBS in treating PD has motivated much effort in developing DBS based therapy for many neurological and psychiatric disorders beyond motor deficits, such as major depression, obsessive and compulsive disorders, and Alzheimer's disease. Because of the nonselective nature of electrical stimulation, DBS may not be able to stimulate any specific cell type, or avoid the stimulation of fiber of passages. However, with the simplicity of electrode placement, and the superb resolution of the spatiotemporal specificity, DBS represents a new generation of site-specific neuromodulation therapies, and could revolutionize the treatment of neurological and psychiatric diseases.

Optogenetics however requires both gene therapy to express light activated rhodopsin proteins in neurons and the delivery of light illumination to target neurons. While much progress has been made in improving the efficacy of optogenetics, its clinical translation may be limited by the requirement of gene therapy, and any potential damage from light illumination. In addition, optogenetics controls neuron activities through altering the biophysical properties of a neuron by adding an exogenous light-sensitive ion conductance, and thus the precision may not be as superb as that achieved with DBS. For example, DBS can stimulate at very high frequencies, i.e., >120 Hz, to achieve therapeutic effects for PD, and indeed DBS often needs to be delivered at >120 Hz to be effective. However, it is difficult for optogenetics to stimulate neurons at such high frequency. But as a research tool, optogenetics holds the promise to provide mechanistic understanding of neural circuits underlying behavioral and therapies, and for certain systems, such as the retinal, optogenetics may be proven effective as a therapy.

References

Anderson VC, Burchiel KJ et al (2005) Pallidal vs subthalamic nucleus deep brain stimulation in Parkinson disease. Arch Neurol 62(4):554–560

Bamann C, Gueta R et al (2010) Structural guidance of the photocycle of channelrhodopsin-2 by an interhelical hydrogen bond. Biochemistry 49(2):267–278

Bekar L, Libionka W et al (2008) Adenosine is crucial for deep brain stimulation-mediated attenuation of tremor. Nat Med 14(1):75–80

Berndt A, Schoenenberger P et al (2011) High-efficiency channelrhodopsins for fast neuronal stimulation at low light levels. Proc Natl Acad Sci U S A 108(18):7595–7600

Berndt A, Lee SY et al (2014) Structure-guided transformation of channelrhodopsin into a light-activated chloride channel. Science 344(6182):420–424

Bernstein JG, Boyden ES (2011) Optogenetic tools for analyzing the neural circuits of behavior. Trends Cogn Sci 15(12):592–600

Bernstein JG, Han X et al (2008) Prosthetic systems for therapeutic optical activation and silencing of genetically-targeted neurons. Proc Soc Photo Opt Instrum Eng 6854:68540H

Bernstein JG, Garrity PA et al (2012) Optogenetics and thermogenetics: technologies for controlling the activity of targeted cells within intact neural circuits. Curr Opin Neurobiol 22 (1):61–71

Boraud T, Brown P et al (2005) Oscillations in the basal ganglia: the good, the bad, and the unexpected. In: Bolam JP, Ingham CA, Magill PJ (eds) The basal ganglia VIII, Vol 56. Springer, New York, pp 1–24

Boyden ES, Zhang F et al (2005) Millisecond-timescale, genetically targeted optical control of neural activity. Nat Neurosci 8(9):1263–1268

Brown P, Oliviero A et al (2001) Dopamine dependency of oscillations between subthalamic nucleus and pallidum in Parkinson's disease. J Neurosci 21(3):1033–1038

Cavanaugh J, Monosov IE et al (2012) Optogenetic inactivation modifies monkey visuomotor behavior. Neuron 76(5):901–907

Chen TW, Wardill TJ et al (2013) Ultrasensitive fluorescent proteins for imaging neuronal activity. Nature 499(7458):295–300

Chow BY, Han X et al (2010) High-performance genetically targetable optical neural silencing by light-driven proton pumps. Nature 463(7277):98–102

Chow BY, Han X et al (2012) Genetically encoded molecular tools for light-driven silencing of targeted neurons. Prog Brain Res 196:49–61

Chuong AS, Miri ML et al (2014) Noninvasive optical inhibition with a red-shifted microbial rhodopsin. Nat Neurosci 17(8):1123–1129

Doroudchi MM, Greenberg KP et al (2011) Virally delivered channelrhodopsin-2 safely and effectively restores visual function in multiple mouse models of blindness. Mol Ther 19 (7):1220–1229

Dostrovsky JO, Levy R et al (2000) Microstimulation-induced inhibition of neuronal firing in human globus pallidus. J Neurophysiol 84(1):570–574

Foffani G, Ardolino G et al (2006) Subthalamic oscillatory activities at beta or higher frequency do not change after high-frequency DBS in Parkinson's disease. Brain Res Bull 69(2):123–130

Gunaydin LA, Yizhar O et al (2010) Ultrafast optogenetic control. Nat Neurosci 13(3):387–392

Han X (2012a) In vivo application of optogenetics for neural circuit analysis. ACS Chem Neurosci 3(8):577–584

Han X (2012b) Optogenetics in the nonhuman primate. Prog Brain Res 196:215–233

Han X, Boyden ES (2007) Multiple-color optical activation, silencing, and desynchronization of neural activity, with single-spike temporal resolution. PLoS One 2(3):e299

Han X, Qian X et al (2009) Millisecond-timescale optical control of neural dynamics in the nonhuman primate brain. Neuron 62(2):191–198

Han X, Chow BY et al (2011) A high-light sensitivity optical neural silencer: development and application to optogenetic control of non-human primate cortex. Front Syst Neurosci 5:18

Knopfel T, Boyden ES (eds) (2012) Optogenetics: tools for controlling and monitoring neuronal activity, Progress in brain research. Elsevier, Amsterdam

Kuhn AA, Kempf F et al (2008) High-frequency stimulation of the subthalamic nucleus suppresses oscillatory beta activity in patients with Parkinson's disease in parallel with improvement in motor performance. J Neurosci 28(24):6165–6173

Kuhn AA, Tsui A et al (2009) Pathological synchronisation in the subthalamic nucleus of patients with Parkinson's disease relates to both bradykinesia and rigidity. Exp Neurol 215(2):380–387

Lehmkuhle MJ, Bhangoo SS et al (2009) The electrocorticogram signal can be modulated with deep brain stimulation of the subthalamic nucleus in the hemiparkinsonian rat. J Neurophysiol 102(3):1811–1820

Levy R, Hutchison WD et al (2000) High-frequency synchronization of neuronal activity in the subthalamic nucleus of parkinsonian patients with limb tremor. J Neurosci 20(20):7766–7775

Levy R, Ashby P et al (2002a) Dependence of subthalamic nucleus oscillations on movement and dopamine in Parkinson's disease. Brain 125(Pt 6):1196–1209

Levy R, Hutchison WD et al (2002b) Synchronized neuronal discharge in the basal ganglia of parkinsonian patients is limited to oscillatory activity. J Neurosci 22(7):2855–2861

Liu X, Ramirez S et al (2012) Optogenetic stimulation of a hippocampal engram activates fear memory recall. Nature 484(7394):381–385

Luo D, Saltzman WM (2000) Synthetic DNA delivery systems. Nat Biotechnol 18(1):33–37

Madisen L, Mao T et al (2012) A toolbox of Cre-dependent optogenetic transgenic mice for light-induced activation and silencing. Nat Neurosci 15(5):793–802

Miesenbock G (2011) Optogenetic control of cells and circuits. Annu Rev Cell Dev Biol 27:731–758

Mobley J, Vo-Dinh T (2003) Optical properties of tissue. In: Vo-Dinh T (ed) Biomedical photonics handbook. CRC, Boca Raton, FL, pp 1–72

Nagel G, Szellas T et al (2003) Channelrhodopsin-2, a directly light-gated cation-selective membrane channel. Proc Natl Acad Sci USA 100(24):13940–13945

Oesterhelt D, Stoeckenius W (1971) Rhodopsin-like protein from the purple membrane of Halobacterium halobium. Nat New Biol 233(39):149–152

Oesterhelt D, Stoeckenius W (1973) Functions of a new photoreceptor membrane. Proc Natl Acad Sci USA 70(10):2853–2857

Priori A, Foffani G et al (2004) Rhythm-specific pharmacological modulation of subthalamic activity in Parkinson's disease. Exp Neurol 189(2):369–379

Rodriguez-Oroz MC, Obeso JA et al (2005) Bilateral deep brain stimulation in Parkinson's disease: a multicentre study with 4 years follow-up. Brain 128(Pt 10):2240–2249

Sauer B, Henderson N (1988) Site-specific DNA recombination in mammalian cells by the Cre recombinase of bacteriophage P1. Proc Natl Acad Sci USA 85(14):5166–5170

Silberstein P, Pogosyan A et al (2005) Cortico-cortical coupling in Parkinson's disease and its modulation by therapy. Brain 128(Pt 6):1277–1291

Spudich JL (2006) The multitalented microbial sensory rhodopsins. Trends Microbiol 14 (11):480–487

Spudich JL, Yang CS et al (2000) Retinylidene proteins: structures and functions from archaea to humans. Annu Rev Cell Dev Biol 16:365–392

Tsien JZ, Chen DF et al (1996) Subregion- and cell type-restricted gene knockout in mouse brain. Cell 87(7):1317–1326

Uc EY, Follett KA (2007) Deep brain stimulation in movement disorders. Semin Neurol 27 (2):170–182

Volkmann J, Moro E et al (2006) Basic algorithms for the programming of deep brain stimulation in Parkinson's disease. Mov Disord 21(Suppl 14):S284–S289

Waehler R, Russell SJ et al (2007) Engineering targeted viral vectors for gene therapy. Nat Rev Genet 8(8):573–587

Weinberger M, Hutchison WD et al (2009a) Pathological subthalamic nucleus oscillations in PD: can they be the cause of bradykinesia and akinesia? Exp Neurol 219(1):58–61

Weinberger M, Hutchison WD et al (2009b) Increased gamma oscillatory activity in the subthalamic nucleus during tremor in Parkinson's disease patients. J Neurophysiol 101 (2):789–802

Wells J, Kao C et al (2005) Optical stimulation of neural tissue in vivo. Opt Lett 30(5):504–506

Welter ML, Houeto JL et al (2004) Effects of high-frequency stimulation on subthalamic neuronal activity in parkinsonian patients. Arch Neurol 61(1):89–96

Wietek J, Wiegert JS et al (2014) Conversion of channelrhodopsin into a light-gated chloride channel. Science 344(6182):409–412

Williams D, Tijssen M et al (2002) Dopamine-dependent changes in the functional connectivity between basal ganglia and cerebral cortex in humans. Brain 125(Pt 7):1558–1569

Wingeier B, Tcheng T et al (2006) Intra-operative STN DBS attenuates the prominent beta rhythm in the STN in Parkinson's disease. Exp Neurol 197(1):244–251

Yizhar O, Fenno LE et al (2011) Optogenetics in neural systems. Neuron 71(1):9–34

Zhang F, Wang LP et al (2007) Multimodal fast optical interrogation of neural circuitry. Nature 446(7136):633–639

Zhang F, Vierock J et al (2011) The microbial opsin family of optogenetic tools. Cell 147 (7):1446–1457

Zhao S, Cunha C et al (2008) Improved expression of halorhodopsin for light-induced silencing of neuronal activity. Brain Cell Biol 36(1–4):141–154

Zorzos AN, Dietrich A et al (2009) Light-proof neural recording electrodes. In: Poster presented at 39th society for neuroscience annual meeting, Chicago, 17–21 Oct 2009

Closing Thoughts for Cognitive Enhancement

Kathleen M. Kantak and Joseph G. Wettstein

Contents

Abstract

The wide-ranging field of cognition enhancing research along with its ethics as it stands today is summarized. In the forefront are potentially novel drugs and non-pharmacological treatments for cognitive impairment across many different psychiatric and neurologic indications. Today's research will bring new drugs to patients tomorrow, and tomorrow's research will bring new molecular targets to clinical development that should be cognitive domain-specific. There is the likelihood that special populations may be better treated and that personalized medicine for cognitive impairment could become a reality. It is conceivable that with the current research effort, cognition enhancing drugs will become available to wide-ranging populations of people with neuropsychiatric illness and to those that are healthy. In some cultures, there is a push in society to be more

K.M. Kantak (✉)
Department of Psychological and Brain Sciences, Boston University, 64 Cummington Mall, Boston, MA 02215, USA
e-mail: kkantak@bu.edu

J.G. Wettstein
Roche Pharmaceutical Research and Early Development; Neuroscience, Ophthalmology and Rare Diseases; F. Hoffmann-La Roche Ltd, Grenzacherstrasse 124, 4070 Basel, Switzerland
e-mail: joseph_g.wettstein@roche.com

K.M. Kantak, J.G. Wettstein (eds.), *Cognitive Enhancement*, Handbook of Experimental Pharmacology 228, DOI 10.1007/978-3-319-16522-6_16

intelligent or have more cognitive prowess. Thus, the ethical use of cognitive enhancing drugs should be an area of debate and communication. Neuroethics is a growing field and it intends to bring together key contributors such as physicians, disease experts, regulatory officials, and policy makers to discuss how such medicines can or should be made available. Together with this, one has to consider the possibility that no single medicine or technology will have a great impact on cognition and, therefore, combination therapy of drugs plus other approaches like exercise or transcranial direct-current stimulation may be the path forward. This is another area of scientific inquiry and debate, and the results should be fruitful and helpful to patients. The science of cognition is advancing at a rapid rate, and communication of its progress along with the development of rational and ethical policies for use of cognitive enhancers will be beneficial.

Keywords
Neuroethics • Special populations • Targets for cognitive enhancement

1 Current and Future Status of Cognitive Enhancement

1.1 Targets of Today

Cognitive enhancement, whether achieved through pharmacological or non-pharmacological means, is a reality. Among the drugs approved by the FDA over the past two decades for neuropsychiatric and neurological disorders, 11 are used for the treatment of Attention Deficit Hyperactivity Disorder and 8 for the treatment of Alzheimer's dementia (CenterWatch). Nonetheless, continued research is necessary to achieve functional improvement in all cognitive domains and to ensure specific and lasting changes. It is encouraging that several new clinical trials are underway for improving cognition in Alzheimer's disease, Attention Deficit Hyperactivity Disorder, Autism, Multiple Sclerosis, Parkinson's disease, schizophrenia, and traumatic brain injury (CenterWatch). These trials support the possibility that treatment options for improving attention, executive function, declarative memory, emotional memory, and social cognition may be forthcoming. With a few exceptions (e.g., drugs to reduce beta-amyloid proteins in Alzheimer's disease or oxytocin to enhance social cognition in Autism), most new drugs in current clinical trials focus mainly on altering monoamine and cholinergic systems for improving cognition. The considerable amount of preclinical and clinical laboratory research conducted over the past 2–3 decades supports an approach targeting these mechanisms (reviewed in Callahan and Terry 2015; Talpos and Shoaib 2015; Riedel and Blokland 2015; Sumiyoshi 2015; Nader 2015; Patin and Hurlemann 2015). One striking aspect of the clinical trials now in progress is the evaluation of several non-pharmacological strategies to improve cognition. Strategies include targeted cognitive training, exercise, and transcranial direct-current stimulation. Research in non-pharmacological approaches to improve cognition has escalated over the past decade (reviewed in Kelly 2015). New non-pharmacological technologies, such as deep brain stimulation and optogenetics, are emerging to impact, in a rather precise manner, brain networks that support cognition (reviewed in Kondabolu et al. 2015). Future research may reveal that pharmacological and non-pharmacological approaches may target different cellular and network processes that, when combined, have complementary actions that, in concert, improve cognition more effectively than either approach alone.

1.2 Targets of Tomorrow

Although progress has been made to improve cognition, the development of novel pro-cognitive compounds has not caught up with the advances that have been made in the neurobiology of learning and memory (reviewed in Ménard et al. 2015). Glutamate receptors have been pharmacological targets to improve cognition for decades and drugs that safely and effectively modify NMDAR, AMPAR, and mGluR activity are under development currently. One example is with the class of drugs that potentiate NMDAR activity at the glycine co-agonist binding site, either directly with agonist treatment or indirectly via treatment with a glycine transporter-1 inhibitor. Perhaps the largest explosion of knowledge gained from research in the neurobiology of learning and memory is in the identification of downstream targets that are activated following NMADR and AMPAR stimulation. Targets of interest for cognitive enhancement include: CaMKII, PKC, ERK, and CREB signaling molecules; Arc, Homer 1a, Zif268, and mTOR immediate early genes; and neurotrophins such as BDNF, which all play important roles in neuroplasticity. Neurogenesis is another target of interest; however, adult neurogenesis is maintained in only two discrete regions of the adult mammalian brain: the subventricular zone of the lateral ventricles and the subgranular zone of the dentate gyrus of the hippocampal formation (reviewed in Costa et al. 2015). Thus, drug treatments that increase neurogenesis to improve cognition may only be possible for processes that require the hippocampus (declarative memory, contextual associative memory, and spatial navigation) and therefore be of importance for certain disease states. Development of drugs that interact with these various targets is needed. Once identified, the advances that have been made in elucidating translational animal models of cognitive enhancement (reviewed in Wallace et al. 2015) and in detecting cognitive enhancement in human participants (reviewed in Harvey and Keefe 2015) will help speed the availability of cognitive-enhancing therapeutics for neuropsychiatric and neurological disorders.

1.3 Special Populations

Cognitive improvement in special populations is not always straightforward. Maladaptive memory such as that presenting in individuals with posttraumatic stress disorder and substance use disorder may require treatments and approaches that disrupt memory (reviewed in Taylor and Torregrossa 2015). Mechanistically, this involves the use of drugs with actions that oppose those needed for cognitive enhancement and applying them with cue exposure during the reconsolidation stage of learning. Unclear at this time is whether remote (old) memories can be disrupted as effectively as recent (new) memories when targeting reconsolidation. An alternate approach for lessening the impact of maladaptive memory is to incorporate cognitive enhancing drugs and strategies with cue exposure during the extinction stage of learning. However, if treatment takes place in a clinic setting, then the transfer of facilitated extinction learning to a home environment may be precluded unless the context dependency of the treatment effect also is addressed.

Cognitive enhancement in special populations often involves children, as in the case of Autism Spectrum Disorder (reviewed in Vahabzadeh et al. 2015) and Down syndrome (reviewed in Fernandez and Reeves 2015). Improvement in social cognition is a primary goal for Autism Spectrum Disorder, and many candidate medications are available for testing. Cognitive enhancement for Down syndrome presents a more complicated picture due to diverse brain region impacted and the fact that the effects of the chromosome 21 abnormality are not limited to the central nervous system. Clinical trial designs will be difficult and not without ethical challenges. The medical community's long history of treating Attentional Deficit Hyperactivity Disorder in children with stimulant and non-stimulant cognitive enhancing medications should play an important role in establishing ways to safely and ethically proceed in the testing and treating of other special populations of children with cognition-altering drugs.

2 Neuroethics of Cognitive Enhancement

In some sense, cognition enhancers have been in use for centuries. These were of plant origin and primarily of the psychostimulant type. Three of the better-known examples are caffeine, coca, and nicotine. Much later, drugs of the amphetamine class became available. These substances have minor and relatively nonspecific effects on cognition yet do influence attention and vigilance. Because of their abuse liability, there is an obvious ethical issue with the use of coca extract and amphetamines, and, therefore, these substances are scheduled drugs in most countries, although amphetamines are now widely prescribed for Attention Deficit Hyperactivity Disorder. An important distinction here is that amphetamines for Attention Deficit Hyperactivity Disorder are used at doses much lower than those needed to support abuse.

One of the first focused and successful efforts at finding a drug that was designed to treat memory loss in a disease state was that with relatively selective and brain penetrant acetylcholinesterase inhibitors in the late 1980s and 1990s for patients with Alzheimer's disease. The first of these drugs, tacrine, had what many considered, at best, limited efficacy along with a number of unwanted side effects. Helping it reach market status was public opinion and public demand for drugs to treat Alzheimer's disease. Thus, one could imagine that both the public and the media

Table 1 Ethical issues associated or arising with cognitive enhancers

•Evolving regulatory environment
•Widespread use, overuse, or abuse
•Role of the media and academic literature
•Safety after long-term use
•Safety in healthy individuals
•Use in children
•Parental control and decision making
•Incidental findings
•Access
•Payment and reimbursement
•Need for co-therapy
•Future reality of personalized medicine
•Terminology

had at least some influence, albeit indirect, over regulatory agencies. Fortunately for patients, the next generation of cholinesterase inhibitors was an improvement over tacrine, yet many of the side effects remained and efficacy was limited. What has happened, however, is that one of these compounds, donepezil, is being publicized as a cognitive enhancer for healthy people. This, in light of less than stellar clinical results, could be due in part to the magnification of data in media and academic literature (Wade et al. 2014). As with the development of cholinesterase inhibitors for Alzheimer's, in which the risk:benefit ratio was scrutinized, one wonders if that barrier will be lowered, rather than heightened, for a cognitive enhancer in the general public, thus, exposing more people to risk from a drug. Table 1 summarizes these and other ethical issues surrounding the use of cognitive enhancing drugs and strategies, and each issue is detailed below.

Small molecule drugs and invasive technologies like deep brain stimulation are associated with risk in the form of safety and toxicology. As cognition enhancers become available for individual diseases such as those in the psychiatric area, it is highly likely that these drugs or interventions will be used outside of their regulatory approved indication(s). Although there is reasonably good control of what indications a drug can be used for, neither the respective pharmaceutical company, regulatory organization nor policy maker directly controls the prescribing practices and the lay-media news that may influence public opinion. Thus, in the case of cognitive enhancers perhaps increased scrutiny of safety and toxicology is necessary so that the consumer, the patient, the parents, and the caregivers can be informed and at least have a chance to understand the risks associated with these new drugs and technologies. Physicians will be in the forefront on this ethical issue as they are the chief prescribers. Fortunately, as shown in a survey of physicians in Canada and the USA, physicians, when queried on attitudes towards cognitive enhancers, tended to place the greatest weight on the safety aspects of a given medicine (Banjo et al. 2010).

Associated with risk, of course, is benefit. Specific attention should be given to the real effectiveness of any treatment albeit drug or technology. It is possible that with just a glimmer of positive data in one clinical trial or in a few patients, the media or particular patient advocacy group could blow out of proportion a certain finding when it is in fact not medically justified or truly effective. How this can be controlled is unclear but all sources that relate to public policy, communication, and the contemporary field of neuroethics need to be engaged (Shook et al. 2014). Via advances in pharmacogenomics, going forward, the benefit of cognitive enhancing drugs may be tied closely with an individual's genetic predisposition and, thus, effective and safe treatment may be specifically targeted to that patient (Mohamed and Sahakian 2012); this is the lure of personalized medicine.

Of specific and current interest today is the use of cognitive enhancers in children, particularly those with neurodevelopmental disorders. In disorders such as Autism, Fragile X, or Down syndrome, there is cognitive impairment early on in life. It may seem clear that early, preventive treatment may be the course of action to take so that brain development is not further retarded. However, as a child's central nervous system is still undergoing development, any number of drugs or therapies, although perhaps having a positive effect on one domain of cognition, may have deleterious effects on other domains or on brain maturation, in general. Moreover, one also has to consider the peripheral side effects of such a medication when given to young children. This places parents in rather a precarious role, as they need to be concerned with the promotion of overall health and not be coercive in their approach to cognitive enhancement (Ball and Wolbring 2014).

Over the past 30 years, there has been substantial progress in understanding the dimensions of cognition and the domains that one needs to focus on in individual diseases. The advent of fMRI and the promise of optogenetics will help develop this field at a much quicker pace now. This brings up the likelihood that during the research process, one may identify a drug that is specific for a unique cognitive domain such as executive function. Executive function is markedly impaired in schizophrenia, Autism, and Attention Deficit Hyperactivity Disorder, yet less so in other psychiatric disorders (Millan et al. 2012). This raises practical issues at the clinical development level where study conduct time is lengthy and more ethical issues on the regulatory side during which those organizations must decide on the quality and quantity of data necessary to make a new drug approval across many illnesses. Related to the use of informative technologies like MRI, which allows viewing of the brain in a noninvasive manner, and cognition test batteries is the risk that either brain structural abnormalities or cognitive weaknesses, previously unidentified, may be uncovered. A framework for the handling and communication of such incidental findings is necessary today as there is more research now directed towards generating new knowledge on cognition and cognitive circuitry specifically related to healthy individuals and what might comprise a normal brain and its function. Awareness of this ethical situation is growing and should be manageable (Scott et al. 2012; Di Pietro and Illes 2013).

In disorders of the brain ranging from Parkinson's disease to posttraumatic stress disorder, there are gaps in cognitive processing and function, so it is clear that smart drugs are needed (Sahakian and Morein-Zamir 2011). Both the patients themselves and their caregivers stand the chance to profit by such new drug development. In such an important area, cognition, one wonder how medicines will be distributed so, therein, treatments should be made available to everyone; general access to the appropriate medicines will be important. This opens up questions related to who pays for these treatments and what might be the various country-specific reimbursement structures.

Drugs for cognition enhancement, although possibly marginally effective after a single dose, are likely to affect cognitive domains only after many weeks or months of treatment. Moreover, these domain-specific drugs may not be effective without some kind of co-therapy like cognitive behavioral therapy, biofeedback, or other type of psychotherapy. Not to be discounted in the cognition treatment paradigm is education and the use of exercise as it has a positive effect on neurogenesis and synaptic plasticity (Costa et al. 2015).

As has been identified earlier, the simple use of the phrase "cognition enhancer" may be problematic and is of ethical concern (Wade et al. 2014). What is its definition? The two words alone suggest something positive or beneficial, whereas the drug or treatment under consideration may still be experimental. So, the naming of these drugs is in itself an issue going forward. Nevertheless, with the advances in science today, there is the real possibility that cognitive enhancers will have a significant impact on society, especially in view of the likelihood that such drugs will gain traction as lifestyle-type medicines having broad reach across not only ill but healthy individuals (Forlini et al. 2013; Cabrera 2015). It is possible with correct communication and discussion between policy makers, regulatory authorities, physicians, and disease area experts that many of the ethical issues that are associated with cognition can be addressed and be made open to the general public in an informative and educative manner.

3 Conclusions

The science of cognitive enhancement is likely to soon reach new milestones. Hence ethicists, social scientists, and other policy makers need to join this conversation and work alongside scientists and clinical investigators to develop rational policies for the use of cognitive enhancers in the treatment of neuropsychiatric and neurological disorders as well as in healthy individuals.

References

Ball N, Wolbring G (2014) Cognitive enhancement: perceptions among parents of children with disabilities. Neuroethics 7:345–364

Banjo OC, Nadler R, Reiner PB (2010) Physician attitudes towards pharmacological cognitive enhancement: safety concerns are paramount. PLoS One 5:e14322

Cabrera LY (2015) How does enhancing cognition affect human values? How does this translate into social responsibility? Curr Top Behav Neurosci 19:223–241, Epub ahead of print July 22, 2014

Callahan PM, Terry AV (2015) Attention. In: Kantak KM, Wettstein JG (eds) Cognitive enhancement. Springer, Cham

CenterWatch. A clinical trials and drug information resource website. www.centerwatch.com

Costa V, Lugert S, Jagasia R (2015) Role of adult hippocampal neurogenesis in cognition in physiology and disease: pharmacological targets and biomarkers. In: Kantak KM, Wettstein JG (eds) Cognitive enhancement. Springer, Cham

Di Pietro NC, Illes J (2013) Disclosing incidental findings in brain research: the rights of minors in decision-making. J Magn Reson Imaging 38:1009–1013

Fernandez F, Reeves RH (2015) Assessing cognitive improvement in people with Down syndrome: important considerations for drug efficacy trials. In: Kantak KM, Wettstein JG (eds) Cognitive enhancement. Springer, Cham

Forlini C, Hall W, Maxwell B, Outram SM, Reiner PB, Repantis D, Schermer M, Racine E (2013) Navigating the enhancement landscape. Ethical issues in research on cognitive enhancers for healthy individuals. EMBO Rep 14:123–128

Harvey PD, Keefe RSE (2015) Methods for delivering and evaluating the efficacy of cognitive enhancement. In: Kantak KM, Wettstein JG (eds) Cognitive enhancement. Springer, Cham

Kelly AM (2015) Non-pharmacological approaches to cognitive enhancement. In: Kantak KM, Wettstein JG (eds) Cognitive enhancement. Springer, Cham

Kondabolu K, Kowalski MM, Roberts EA, Han X (2015) Optogenetics and deep brain stimulation neurotechnologies. In: Kantak KM, Wettstein JG (eds) Cognitive enhancement. Springer, Cham

Ménard C, Gaudreau P, Quirion R (2015) Signaling pathways relevant to cognition-enhancing drug targets. In: Kantak KM, Wettstein JG (eds) Cognitive enhancement. Springer, Cham

Millan MJ, Agid Y, Brune M, Bullmore ET, Carter CS, Clayton NS, Connor R, Davis S, Deakin B, DeRubeis RJ, Dubois B, Geyer MA, Goodwin GM, Gorwood P, Jay TM, Joels M, Mansuy IM, Meyer-Lindenberg A, Murphy D, Rolls E, Saletu B, Spedding M, Sweeney J, Whittington M, Young LJ (2012) Cognitive dysfunction in psychiatric disorders: characteristics, causes and the quest for improved therapy. Nat Rev Drug Discov 11:141–168

Mohamed AD, Sahakian BJ (2012) The ethics of elective psychopharmacology. Int J Neuropsychopharmacol 15:559–571

Nader K (2015) Emotional Memory. In: Kantak KM, Wettstein JG (eds) Cognitive enhancement. Springer, Cham

Patin A, Hurlemann R (2015) Social cognition. In: Kantak KM, Wettstein JG (eds) Cognitive enhancement. Springer, Cham

Riedel WJ, Blokland A (2015) Declarative memory. In: Kantak KM, Wettstein JG (eds) Cognitive enhancement. Springer, Cham

Sahakian BJ, Morein-Zamir S (2011) Neuroethical issues in cognitive enhancement. J Psychopharmacol 25:197–204

Scott NA, Murphy TH, Illes J (2012) Incidental findings in neuroimaging research: a framework for anticipating the next frontier. J Empir Res Hum Res Ethics 7:53–57

Shook JR, Galvagni L, Giordano J (2014) Cognitive enhancement kept within contexts: neuroethics and informed public policy. Front Syst Neurosci 8:228

Sumiyoshi T (2015) Verbal memory. In: Kantak KM, Wettstein JG (eds) Cognitive enhancement. Springer, Cham

Talpos J, Shoaib M (2015) Executive function. In: Kantak KM, Wettstein JG (eds) Cognitive enhancement. Springer, Cham

Taylor JR, Torregrossa MM (2015) Pharmacological disruption of maladaptive memory. In: Kantak KM, Wettstein JG (eds) Cognitive enhancement. Springer, Cham

Vahabzadeh A, Landino SM, Finger BC, Carlezon WA Jr, McDougle CJ (2015) Neural targets in the study and treatment of social-cognition in autism spectrum disorder. In: Kantak KM, Wettstein JG (eds) Cognitive enhancement. Springer, Cham

Wade L, Forlini C, Racine E (2014) Generating genius: how an Alzheimer's drug became considered a 'cognitive enhancer' for healthy individuals. BMC Med Ethics 15:37

Wallace TL, Ballard TM, Glavis-Bloom C (2015) Animal paradigms of cognitive enhancement with translation to humans. In: Kantak KM, Wettstein JG (eds) Cognitive enhancement. Springer, Cham

Index

K.M. Kantak, J.G. Wettstein (eds.), *Cognitive Enhancement*, Handbook of Experimental Pharmacology 228, DOI 10.1007/978-3-319-16522-6